T0091681

X86 Assembly Language and C Fundamentals

X86 Assembly Language and C Fundamentals

Joseph Cavanagh

Santa Clara University, Santa Clara, California

CRC Press is an imprint of the
Taylor & Francis Group, an **informa** business

CRC Press
Taylor & Francis Group
6000 Broken Sound Parkway NW, Suite 300
Boca Raton, FL 33487-2742

Printed in the United States of America on acid-free paper
Version Date: 2012928

International Standard Book Number: 978-1-4665-6824-2 (Hardback)

Library of Congress Cataloging-in-Publication Data

Cavanagh, Joseph J. F.
X86 assembly language and C fundamentals / Joseph Cavanagh.
p. cm.
Includes bibliographical references and index.
ISBN 978-1-4665-6824-2 (hardback)
1. X86 assembly language (Computer program language) 2. C (Computer program language) I. Title.

QA76.73.X16C38 2013
005.13'6--dc23 2012036001

Visit the Taylor & Francis Web site at
http://www.taylorandfrancis.com

and the CRC Press Web site at
http://www.crcpress.com

By the same author:

COMPUTER ARITHMETIC and Verilog HDL Fundamentals

DIGITAL DESIGN and Verilog HDL Fundamentals

VERILOG HDL: Digital Design and Modeling

SEQUENTIAL LOGIC: Analysis and Synthesis

DIGITAL COMPUTER ARITHMETIC: Design and Implementation

THE COMPUTER CONSPIRACY
A novel

To Dr. Daniel W. Lewis
Professor, Computer Engineering Department, Santa Clara University,
for his many years of continued encouragement, support, and friendship

CONTENTS

PREFACE

Although assembly language is not as prevalent as a high-level language, such as C or an object-oriented language like C++, it is the predominant language used in embedded microprocessors. A course in a high-level language, such as C, usually precedes a course in assembly language.

Assembly language programming requires a knowledge of number representations, such as fixed-point, decimal, and floating-point; also digital logic, registers, and stacks. In order to thoroughly understand assembly language, it is necessary to be familiar with the architecture of the computer on which the language is being used. For the X86 assembly language, this implies the Intel and Intel-like microprocessors. Programs written in assembly language are usually faster and more compact than programs written in a high-level language and provide greater control over the program application. Assembly language is *machine dependent*; that is, it is used only with a specific type of processor. A high-level language, however, is usually *machine independent*; that is, it can be used with any processor.

Assembly language programs use an *assembler* to convert the assembly language code to the machine language of 1s and 0s. This is in contrast to high-level languages which use compilers to accomplish the transformation.

Assembly languages consist of mnemonic codes, which are similar to English words, making the program easy to read. For example, the MOV instruction moves data from a source location to a destination location; the XCHG instruction exchanges the contents of a source location and a destination location; and the logical AND instruction performs the bitwise AND operation of two operands.

The programs in this book are written using X86 assembly language only, the C programming language only, or by embedding an in-line assembly language module in a C program by using the *_asm* command. The assembly language code immediately follows the *_asm* command and is bracketed by left and right braces, as shown below.

```
#include "stdafx.h"
int main (void)
{
   define variables
   _asm      switch to assembly language
   {
             assembly language code goes here
   }
print results
return 0;
}
```

Assembly languages also have input/output (I/O) instructions to access I/O devices on the computer. Input/output instructions are usually not available for high-level languages. Also, assembly languages can access the stack, general-purpose registers, base pointer registers, segment registers, and execute PUSH and POP operations.

The book presents the binary, octal, decimal, and hexadecimal number systems, as well as the basic X86 processor architecture. The architecture includes the general-purpose registers, the segment registers, the flags register, the instruction pointer, and the floating-point registers. The following topics are also presented: different addressing modes, data transfer instructions, branching and looping operations, stack operations, logic, shift, and rotate instructions. Computer arithmetic topics are presented in detail, including fixed-point, binary-coded decimal, and floating-point instructions. There are additional chapters on procedures, string operations, arrays, macros, and input/output operations. The fundamentals of C programming are covered in a separate chapter.

The book is intended to be tutorial, and as such, is comprehensive and self contained. All program examples are carried through to completion — nothing is left unfinished or partially designed. Also, all programs provide the outputs that result from program execution. Each chapter includes numerous problems of varying complexity to be designed by the reader.

Chapter 1 covers the number systems of different radices, such as binary, octal, binary-coded octal, decimal, binary-coded decimal, hexadecimal, and binary-coded hexadecimal. The chapter also presents the number representations of sign magnitude, diminished-radix complement, and radix complement.

Chapter 2 presents the generic architecture of processors and how the architecture corresponds more appropriately to the X86 architecture execution environment, including the different sets of registers. The chapter also covers the arithmetic and logic unit (ALU), the control unit, and memory, including main memory and cache memory. Error detection and correction is also discussed using the Hamming code developed by Richard W. Hamming. A brief introduction to tape drives and disk drives is also presented. The X86 register set is covered, which includes the general-purpose registers (GPRs), the segment registers, the EFLAGS register containing status flags, system flags, and a control flag. Other registers include the instruction pointer and the floating-point registers. The translation lookaside buffer (TLB) and the assembler are also briefly discussed.

Chapter 3 presents the various addressing modes of the X86 assembly language. The instruction set provides various methods to address operands. The main methods are: register, immediate, direct, register indirect, base, index, and base combined with index. A displacement may also be present. These and other addressing methods are presented in this chapter together with examples. The processor selects the applicable default segment as a function of the instruction: instruction fetching assumes the code segment; accessing data in main memory references the data segment; and instructions that pertain to the stack reference the stack segment. However, a *segment override prefix* can be used to change the default data segment to another segment; that is, to explicitly specify any segment register to be used as the current segment.

Chapter 4 presents a brief introduction to the C programming language, which will be used in most chapters — typically to contain an embedded assembly language module. The main purpose of this chapter is to provide sufficient information regarding C programming in order to demonstrate how a C program can be linked to an assembly language program. The various constants and variables of the C language are presented, plus the input/output functions. The following operators are introduced: arithmetic, relational, **if** and **else** statements, and the logical operators of AND, OR, and NOT. Also included are the conditional operator, the increment and decrement operators, and the bitwise operators. There are two looping statements. The **while** loop executes a statement or block of statements as long as a test expression is true (nonzero). The second looping statement is the **for** loop. The **for** loop repeats a statement or block of statements a specific number of times. Arrays and strings are also covered in this chapter.

Chapter 5 presents the basic data transfer instructions as they apply to the X86 processors. Other data transfer instructions, such as instructions that pertain to stack operations and string operations, are presented in later chapters. This chapter also contains the various data types used in the X86 processors, which include signed binary integers, unsigned binary integers, unpacked binary-coded decimal (BCD) integers, packed BCD integers, and floating-point numbers.

Some of the basic move instructions involving data transfer are also presented. These include register-to-register, immediate-data-to-register, immediate-data-to-memory, memory-to-register, and register-to-memory. This chapter also covers moves with sign extension, moves with zero extension, and conditional moves that move data to a destination depending on the state of a flag. Different types of exchange instructions are discussed, which exchange the contents of a source and destination location. The chapter also presents translate instructions, which change an operand into a different operand in order to translate from one code to another code.

Chapter 6 presents branching and looping instructions as used in the X86 assembly language. These instructions transfer control to a section of the program that does not immediately follow the current instruction. The transfer may be a *backward transfer* to a section of code that was previously executed or a *forward transfer* to a section of code that follows the current instruction. The *unconditional jump* instruction advances the instruction pointer register forward or backward a specific number of instructions. It transfers control to a destination address and provides no return address. The *conditional jump* instruction transfers control to a destination instruction in the same code segment if certain condition codes are met — as determined by a compare instruction. If the condition is not met, then program execution continues with the next instruction that follows the *conditional jump* instruction. Implementing WHILE and FOR loops in assembly is also presented.

Chapter 7 presents stack operations in the X86 processor. The stack is a one-dimensional data structure located in contiguous locations of memory that is used for the temporary storage of data. It is one of the segments in a segmented memory model and is called the *stack segment*, in which the base address of the stack is contained in the *stack segment* register. A data element is placed on top of the stack by a PUSH instruction; a data element is removed from the top of the stack by a POP instruction. A stack builds toward lower addresses. Additional PUSH and POP instructions are also covered in this chapter. The PUSH instruction decrements an explicit number of bytes before the operation is executed, depending on the size of the operand being pushed onto the stack.

Chapter 8 presents the logical operations of AND, OR, exclusive-OR, NOT, and NEG. These instructions execute the Boolean equivalent of the corresponding operations in digital logic circuits. The NOT instruction performs a bitwise 1s complement operation on the destination operand and stores the result in the destination operand. The NEG instruction performs a 2s complement operation on the destination operand and stores the result in the destination operand. There are also bit test instructions that operate on a single bit and are used to scan the bits in an operand and then perform an operation on the selected bit. These instructions include: the *bit test*, *bit test and set*, *bit test and reset*, *bit test and complement*, *bit scan forward*, and *bit scan reverse* operations.

Shift instructions are also presented that perform logical or arithmetic left or right shifts on bytes, words, or doublewords. The number of bits shifted can be specified by an immediate value of 1, an immediate value stipulated in a byte, or a count in general-purpose register CL or CX. The shift instructions are *shift arithmetic left*, *shift logical left*, *shift arithmetic right*, *shift logical right*, *double precision shift left*, and *double precision shift right*. There are also rotate instructions that rotate the operand a number of bits specified by an immediate value of 1, an immediate value stipulated in a byte, or a count in general-purpose register CL or CX. The rotate instructions are *rotate left*, *rotate right*, *rotate through carry left*, and *rotate through carry right*. Also covered is the *set byte on condition* instruction.

There are two bit scan instructions: *bit scan forward* and *bit scan reverse*. These instructions scan the contents of a register or memory location to determine the location of the first 1 bit in the operand.

Chapter 9 covers the four operations of addition, subtraction, multiplication, and division for fixed-point arithmetic. In fixed-point operations, the radix point is in a fixed location in the operand. The operands can be expressed by any of the following number representations: unsigned, sign-magnitude, diminished-radix complement, or radix complement. Addition operations include the *add*, *add with carry*, and the *increment by 1* instructions. Subtraction operations include the *subtract*, *subtract with borrow*, *decrement by 1*, and *twos complement negation* instructions. Multiplication operations include the *unsigned multiply* and *signed multiply* instructions. Division operations include the *unsigned divide* and *signed divide* instructions.

Chapter 10 presents the binary-coded decimal (BCD) operations of addition, subtraction, multiplication, and division. BCD instructions operate on decimal numbers that are encoded as 4-bit binary numbers in the 8421 code. The BCD instructions include the *ASCII adjust after addition* instruction, which adjusts the result of an addition operation of two unpacked BCD operands in which the high-order four bits of a byte contain zeroes; the low-order four bits contain a numerical value; the *decimal adjust AL after addition* instruction, which adjusts the sum of two packed BCD integers to generate a packed BCD result; and the *ASCII adjust AL after subtraction* instruction, which adjusts the result of a subtraction of two unpacked BCD operands.

Other BCD operations include the *decimal adjust AL after subtraction* instruction, which adjusts the result of a subtraction of two packed BCD operands; the *ASCII adjust AX after multiplication* instruction, which adjusts the product in register AX resulting from multiplying two valid unpacked BCD operands; and the *ASCII adjust AX before division* instruction, which converts two unpacked digits in register AX to an equivalent binary value, then divides AX by an unpacked BCD value.

Chapter 11 presents floating-point arithmetic instructions. Floating-point numbers consist of the following three fields: a sign bit, *s*; an exponent, *e*; and a fraction, *f*. These parts represent a number that is obtained by multiplying the fraction, *f*, by a radix, *r*, raised to the power of the exponent, *e*, where *f* and *e* are signed fixed-point numbers, and *r* is the radix (or base). As the exponents are being formed, a *bias constant* is added to the exponents, making all exponents positive, thus allowing exponent comparison to be simplified.

Floating-point operations utilize an 8-register stack in which each register contains 80 bits. There are three main rounding methods used in floating-point operations: truncation rounding, adder-based rounding, and von Neumann rounding. Rounding deletes one or more low-order bits of the fraction and adjusts the retained bits according to a particular rounding technique.

There are several different load instructions that push different types of data onto the register stack. These include pushing a value of +1.0, pushing logarithmic values, pushing the value of π, and pushing the value of +0.0. There are also several different store instructions that pop values off the stack. These include storing operands in the BCD format or as rounded integers, storing an integer then popping the stack, and storing a truncated integer then popping the stack. Operands can also be popped off the stack and stored as floating-point values or stored as floating-point values and then pop the stack.

There are different versions of the add instruction. These include adding a stack register and a memory location then storing the sum in the register stack, or adding two stack registers and storing the sum in the register stack. There are also instructions that add, store, then pop the stack. There are similar versions of the subtract instructions. These include subtracting a memory operand from a stack register and storing the difference in the register stack, or subtracting two stack registers and storing the difference in the register stack. There are also instructions that subtract, store, then pop the stack.

There are different versions of the multiply instruction. These include multiplying a stack register and a memory location then storing the product in the register stack, or multiplying two stack registers and storing the product in the register stack. There are also instructions that multiply, store, then pop the stack. There are similar versions of the divide instructions. These include dividing a stack register by a memory operand and storing the result in the register stack, or dividing two stack registers and storing the result in the register stack. There are also instructions that divide, store, then pop the stack.

There are also several different versions of floating-point instructions that compare different types of data. One version compares an operand in the register stack with an operand in memory and sets condition code flags. Another version compares two operands in the register stack and sets the condition codes. Both versions can also compare operands, set the condition codes, then pop the stack. Another version compares operands, sets the condition codes, then pops the stack twice. There are also different versions that compare integer operands.

This chapter also contains instructions that operate on trigonometric functions, such as sine, cosine, and combined sine and cosine, which calculates both functions. There is also a partial tangent instruction, which calculates the tangent of the source operand in a stack register, then pushes a value of +1.0 onto the stack. The partial arctangent instruction is also included, which is the inverse tangent function.

There are several additional floating-point instructions that perform basic arithmetic operations and have only one syntax. Most of the previous instructions presented above have more than one syntax.

Chapter 12 provides a brief discussion on procedures. A procedure is a set of instructions that perform a specific task. They are invoked from another procedure and provide results to the calling program at the end of execution. Procedures (also called subroutines) are utilized primarily for routines that are called frequently by other procedures. The procedure routine is written only once, but used repeatedly, thereby saving storage space. Procedures permit a program to be coded in modules, thus making the program easier to code and test.

Chapter 13 discusses string instructions. A string is a sequence of bytes, words, or doublewords that are stored in contiguous locations in memory as a one-dimensional array. Strings can be processed from low addresses to high addresses or from high addresses to low addresses, depending on the state of the direction flag. If the direction flag is set, then the direction of processing is from high addresses to low addresses (auto-decrement). If the direction flag is reset, then the direction of processing is from low addresses to high addresses (auto-increment).

There are several *repeat* prefixes, which can be placed before the string instruction, that specify the condition for which the instruction is to be executed. The general-purpose register (E)CX specifies the number of times that the string instruction is to be executed.

The *move string* instructions transfer a string element — byte, word, or doubleword — from one memory location to another memory location. The *load string* instructions transfer a string element from a memory location to general-purpose register AL, AX, or EAX. The *store string* instructions transfer a string element from register AL, AX, or EAX to a destination memory location.

The *compare strings* instructions compare a string element in the first source operand with an equivalent size operand in the second source operand. The status flags reflect the result of the comparison. Both operands are unaltered by the comparison. The *compare strings* instructions are usually followed by a *jump on condition* instruction.

The *scan strings* instructions contain only one operand, which is in a general-purpose register. The instructions compare a string element in general-purpose register AL, AX, or EAX with an equivalent size operand in a memory location. The status flags reflect the result of the comparison. Both the operand in the general-purpose register and the memory location are unaltered by the comparison.

Chapter 14 introduces arrays, which are data structures that contain a list of elements of the same data type (homogeneous) with a common name whose elements can be accessed individually. Array elements are usually stored in contiguous locations in memory, allowing easier access to the array elements. There are two main types of arrays: one-dimensional arrays and multi-dimensional arrays. A one-dimensional array — also called a linear array — is an array that is accessed by a single index. A two-dimensional array — also called a multi-dimensional array — is an array that is accessed by two indexes. One index accesses a row, the other index accesses a column. The two different types of arrays are written in assembly language only, the C programming language only, or an assembly language module embedded in a C program.

Chapter 15 introduces macros, which are segments of code that are written only once, but can be executed many times in the main program. When the macro is invoked, the assembler replaces the *macro call* with the macro code. The macro code is then placed in-line with the main program. Macros generally make the program more readable. Macros and procedures are similar because they both call a sequence of instructions to be executed by the main program; however, there is no **CALL** or **RET** instruction in a macro as there is in a procedure.

Chapter 16 discusses interrupts and input/output operations. When an interrupt occurs, the processor suspends operation of the current program and pushes the contents of specific registers onto the stack. Return from an interrupt is generated by the *interrupt return* instruction, which is similar to the procedure *far return* instruction.

Direct memory access is also covered, which allows an I/O device control unit to transfer data directly to or from main memory without CPU intervention. This is a much faster data transfer operation, allowing both the processor and the I/O device to operate concurrently in most cases.

Memory-mapped I/O is also discussed. For single bus machines, the same bus can be utilized for both memory and I/O devices. Therefore, I/O devices may be assigned a unique address within main memory, which is partitioned into separate areas for memory and I/O devices. Using the memory-mapped technique, I/O devices are accessed in the same way as memory locations, providing significant flexibility in managing I/O operations. Thus, there are no separate I/O instructions and the I/O devices can be accessed utilizing any of the memory read or write instructions and their addressing modes.

There are several instructions used to transfer data between an I/O device and the processor. There are two instructions that transfer data between an I/O port and general-purpose registers: IN and OUT. The IN instruction transfers data from an I/O port to register AL, AX, or EAX. The OUT instruction transfers data from register AL, AX, or EAX to an I/O port. These are referred to as *register I/O* instructions.

There are also two types of instructions that transfer string data between memory and an I/O port: INS and OUTS. The INS instructions transfer bytes, words, or doublewords of string data from an I/O port to memory. The OUTS instructions transfer bytes, words, or doublewords of string data from memory to an I/O port. These are referred to as *string (block) I/O* instructions. The *repeat* prefix may also be used to specify the condition for which the instructions are to be executed.

Chapter 17 presents additional programming examples to provide additional exposure for the reader. The examples include programs written in assembly language only, the C programming language only, and assembly language modules embedded in a C program. The various topics that are covered in the examples include logic instructions, bit test instructions, compare instructions, unconditional and conditional jump instructions, unconditional and conditional loop instructions, fixed-point instructions, floating-point instructions, string instructions, and arrays.

Appendix A lists the American Standard Code for Information Interchange (ASCII) codes for hexadecimal characters 20H through 7FH. These are provided as a reference to be used in certain chapters. **Appendix B** provides solutions to select problems in each chapter.

The outputs obtained from executing the programs in this book are the actual outputs obtained directly from the flat assembler or from the C compiler.

Since there are more than 330 instructions in the X86 Assembly Language, not all instructions are presented in this book — only the most commonly used instructions. For a complete listing of all the X86 assembly language instructions, refer to the following manuals: Intel 64 and IA-32 Architectures Software Developer's Manual, Volumes 2A and 2B.

It is assumed that the reader has an adequate background in C programming, digital logic design, and computer architecture. The book is designed for undergraduate students in electrical engineering, computer engineering, computer science, and software engineering; also for graduate students who require a noncredit course in X86 assembly language to supplement their program of studies.

Although this book does not utilize Verilog HDL, I would like to express my thanks to Dr. Ivan Pesic, CEO of Silvaco International, for allowing use of the SILOS Simulation Environment software for all of my books that use Verilog HDL and for his continued support.

I would like to express my appreciation and thanks to the following people who gave generously of their time and expertise to review the manuscript and submit comments: Professor Daniel W. Lewis, Department of Computer Engineering, Santa Clara University who supported me in all my endeavors; Geri Lamble; and Steve Midford. Thanks also to Nora Konopka and the staff at Taylor & Francis for their support.

X86 Assembly Language code for the figures can be downloaded at:
http://www.crcpress.com/product/isbn/9781466568242

Joseph Cavanagh

1

Number Systems and Number Representations

This chapter discusses positional number systems using various radices (bases), counting in different radices, and conversion from one radix to a different radix. The number systems are presented for both integer and fraction notation. The different number systems covered are binary (radix 2), octal (radix 8), binary-coded octal, decimal (radix 10), binary-coded decimal, hexadecimal (radix 16), and binary-coded hexadecimal. Also, other nontraditional radices are presented to illustrate that the same rules apply to any radix. Various arithmetic operations are provided to demonstrate the four arithmetic operations of addition, subtraction, multiplication, and division.

Number representations are also covered for both positive and negative numbers for the three number representations of sign magnitude, diminished-radix complement, and radix complement. The four arithmetic operations are presented for binary and binary-coded decimal.

1.1 Number Systems

Numerical data are expressed in various positional number systems for each *radix*, or *base*. A *positional number system* encodes a vector of n bits in which each bit is weighted according to its position in the vector. The encoded vector is also associated with a radix r, which is an integer greater than or equal to 2. A number system has exactly r digits in which each bit in the radix has a value in the range of 0 to $r - 1$, thus the highest digit value is one less than the radix. For example, the binary radix has two digits which range from 0 to 1; the octal radix has eight digits which range from 0 to 7. An n-bit integer A is represented in a positional number system as follows:

$$A = (a_{n-1}a_{n-2}a_{n-3} \ldots a_1a_0) \tag{1.1}$$

where $0 \le a_i \le r - 1$. The high-order and low-order digits are a_{n-1} and a_0, respectively. The number in Equation 1.1 (also referred to as a vector or operand) can represent positive integer values in the range 0 to $r^n - 1$. Thus, a positive integer A is written as

$$A = a_{n-1}r^{n-1} + a_{n-2}r^{n-2} + a_{n-3}r^{n-3} + \ldots + a_1r^1 + a_0r^0 \tag{1.2}$$

The value for A can be represented more compactly as

$$A = \sum_{i=0}^{n-1} a_ir^i \tag{1.3}$$

The expression of Equation 1.2 can be extended to include fractions. For example,

$$A = a_{n-1}r^{n-1} + \ldots + a_1r^1 + a_0r^0 + a_{-1}r^{-1} + a_{-2}r^{-2} + \ldots + a_{-m}r^{-m} \tag{1.4}$$

Equation 1.4 can be represented as

$$A = \sum_{i=-m}^{n-1} a_ir^i \tag{1.5}$$

Adding 1 to the highest digit in a radix r number system produces a sum of 0 and a carry of 1 to the next higher-order column. Thus, counting in radix r produces the following sequence of numbers:

$$0, 1, 2, \ldots, (r-1), 10, 11, 12, \ldots, 1(r-1), \ldots.$$

Table 1.1 shows the counting sequence for different radices. The low-order digit will always be 0 in the set of r digits for the given radix. The set of r digits for various radices is given in Table 1.2. In order to maintain one character per digit, the numbers 10, 11, 12, 13, 14, and 15 are represented by the letters A, B, C, D, E, and F, respectively.

Table 1.1 Counting Sequence for Different Radices

Decimal	$r = 2$	$r = 4$	$r = 8$
0	0	0	0
1	1	1	1
2	10	2	2
3	11	3	3
4	100	10	4
5	101	11	5
6	110	12	6
7	111	13	7
8	1000	20	10
9	1001	21	11
10	1010	22	12
11	1011	23	13
12	1100	30	14
13	1101	31	15
14	1110	32	16
15	1111	33	17
16	10000	100	20
17	10001	101	21

Table 1.2 Character Sets for Different Radices

Radix (base)	Character Sets for Different Radices
2	{0, 1}
3	{0, 1, 2}
4	{0, 1, 2, 3}
5	{0, 1, 2, 3, 4}
6	{0, 1, 2, 3, 4, 5}
7	{0, 1, 2, 3, 4, 5, 6}
8	{0, 1, 2, 3, 4, 5, 6, 7}
9	{0, 1, 2, 3, 4, 5, 6, 7, 8}
10	{0, 1, 2, 3, 4, 5, 6, 7, 8, 9}
11	{0, 1, 2, 3, 4, 5, 6, 7, 8, 9, A}
12	{0, 1, 2, 3, 4, 5, 6, 7, 8, 9, A, B}
13	{0, 1, 2, 3, 4, 5, 6, 7, 8, 9, A, B, C}
14	{0, 1, 2, 3, 4, 5, 6, 7, 8, 9, A, B, C, D}
15	{0, 1, 2, 3, 4, 5, 6, 7, 8, 9, A, B, C, D, E}
16	{0, 1, 2, 3, 4, 5, 6, 7, 8, 9, A, B, C, D, E, F}

Example 1.1 Count from decimal 0 to 25 in radix 5. Table 1.2 indicates that radix 5 contains the following set of four digits: {0, 1, 2, 3, 4}. The counting sequence in radix 5 is

$$\begin{aligned}
000, 001, 002, 003, 004 &= (0 \times 5^2) + (0 \times 5^1) + (4 \times 5^0) = 4_{10}\\
010, 011, 012, 013, 014 &= (0 \times 5^2) + (1 \times 5^1) + (4 \times 5^0) = 9_{10}\\
020, 021, 022, 023, 024 &= (0 \times 5^2) + (2 \times 5^1) + (4 \times 5^0) = 14_{10}\\
030, 031, 032, 033, 034 &= (0 \times 5^2) + (3 \times 5^1) + (4 \times 5^0) = 19_{10}\\
040, 041, 042, 043, 044 &= (0 \times 5^2) + (4 \times 5^1) + (4 \times 5^0) = 24_{10}\\
100 &= (1 \times 5^2) + (0 \times 5^1) + (0 \times 5^0) = 25_{10}
\end{aligned}$$

Example 1.2 Count from decimal 0 to 25 in radix 12. Table 1.2 indicates that radix 12 contains the following set of twelve digits: {0, 1, 2, 3, 4, 5, 6, 7, 8, 9, A, B}. The counting sequence in radix 12 is

$$\begin{aligned}
00, 01, 02, 03, 04, 05, 06, 07, 08, 09, 0A, 0B &= (0 \times 12^1) + (11 \times 12^0) = 11_{10}\\
10, 11, 12, 13, 14, 15, 16, 17, 18, 19, 1A, 1B &= (1 \times 12^1) + (11 \times 12^0) = 23_{10}\\
20, 21 &= (2 \times 12^1) + (1 \times 12^0) = 25_{10}
\end{aligned}$$

1.1.1 Binary Number System

The radix is 2 in the *binary number system*; therefore, only two digits are used: 0 and 1. The low-value digit is 0 and the high-value digit is $(r - 1) = 1$. The binary number system is the most conventional and easily implemented system for internal use in a digital computer; therefore, most digital computers use the binary number system. There is a disadvantage when converting to and from the externally used decimal system; however, this is compensated for by the ease of implementation and the speed of execution in binary of the four basic operations: addition, subtraction, multiplication, and division. The radix point is implied within the internal structure of the computer; that is, there is no specific storage element assigned to contain the radix point.

The weight assigned to each position of a binary number is as follows:

$$2^{n-1} 2^{n-2} \ldots 2^3\ 2^2\ 2^1\ 2^0 .\ 2^{-1} 2^{-2} 2^{-3} \ldots 2^{-m}$$

where the integer and fraction are separated by the radix point (binary point). The decimal value of the binary number 1011.101_2 is obtained by using Equation 1.4, where $r = 2$ and $a_i \in \{0,1\}$ for $-m \le i \le n - 1$. Therefore,

$$\begin{array}{cccccccc}
2^3 & 2^2 & 2^1 & 2^0 & . & 2^{-1} & 2^{-2} & 2^{-3}\\
1 & 0 & 1 & 1 & . & 1 & 0 & 1_2
\end{array}
\begin{aligned}
&= (1 \times 2^3) + (0 \times 2^2) + (1 \times 2^1) + (1 \times 2^0) +\\
&\quad (1 \times 2^{-1}) + (0 \times 2^{-2}) + (1 \times 2^{-3})\\
&= 11.625_{10}
\end{aligned}$$

Digital systems are designed using bistable storage devices that are either reset (logic 0) or set (logic 1). Therefore, the binary number system is ideally suited to represent numbers or states in a digital system, since radix 2 consists of the alphabet 0 and 1. These bistable devices can be concatenated to any length n to store binary data. For example, to store 1 byte (8 bits) of data, eight bistable storage devices are required for the value 0110 1011 (107_{10}). Counting in binary is illustrated in Table 1.3, which shows the weight associated with each of the four binary bit positions. Notice the alternating groups of 1s in Table 1.3 for each of the four columns. A binary number is a group of n bits that can assume 2^n different combinations of the n bits. Therefore, the range for n bits is 0 to $2^n - 1$ and the range for four bits is 0 to $2^4 - 1$; that is 0 to 15, as shown in Table 1.3.

Table 1.3 Counting in Binary

Decimal	Binary			
	8	4	2	1
	2^3	2^2	2^1	2^0
0	0	0	0	0
1	0	0	0	1
2	0	0	1	0
3	0	0	1	1
4	0	1	0	0
5	0	1	0	1
6	0	1	1	0
7	0	1	1	1
8	1	0	0	0
9	1	0	0	1
10	1	0	1	0
11	1	0	1	1
12	1	1	0	0
13	1	1	0	1
14	1	1	1	0
15	1	1	1	1

The binary weights for each bit position of an 8-bit integer are shown in Table 1.4 and the binary weights for each bit position of an 8-bit fraction are shown in Table 1.5.

Table 1.4 Binary Weights for an 8-Bit Integer

2^7	2^6	2^5	2^4	2^3	2^2	2^1	2^0
128	64	32	16	8	4	2	1

Table 1.5 Binary Weights for an 8-Bit Fraction

2^{-1}	2^{-2}	2^{-3}	2^{-4}	2^{-5}	2^{-6}	2^{-7}	2^{-8}
1/2	1/4	1/8	1/16	1/32	1/64	1/128	1/256
0.5	0.25	0.125	0.0625	0.03125	0.015625	0.0078125	0.00390625

Each 4-bit binary segment has a weight associated with the segment and is assigned the value represented by the low-order bit of the corresponding segment, as shown in the first row of Table 1.6. The 4-bit binary number in each segment is then multiplied by the value of the segment. Thus, the binary number 0010 1010 0111 1100 0111 is equal to the decimal number $174{,}023_{10}$ as shown below.

$$(2 \times 65536) + (10 \times 4096) + (7 \times 256) + (12 \times 16) + (7 \times 1) = 174{,}023_{10}$$

Table 1.6 Weight Associated with 4-Bit Binary Segments

65536	4096	256	16	1
0001	0001	0001	0001	0001
0010	1010	0111	1100	0111

1.1.2 Octal Number System

The radix is 8 in the *octal number system*; therefore, eight digits are used, 0 through 7. The low-value digit is 0 and the high-value digit is $(r - 1) = 7$. The weight assigned to each position of an octal number is as follows:

$$8^{n-1}\, 8^{n-2} \,\ldots\, 8^3\, 8^2\, 8^1\, 8^0 .\, 8^{-1} 8^{-2} 8^{-3} \,\ldots\, 8^{-m}$$

where the integer and fraction are separated by the radix point (octal point). The decimal value of the octal number 217.6_8 is obtained by using Equation 1.4, where $r = 8$ and $a_i \in \{0,1,2,3,4,5,6,7\}$ for $-m \leq i \leq n - 1$. Therefore,

$$\begin{array}{ccccl} 8^2 & 8^1 & 8^0 & . & 8^{-1} \\ 2 & 1 & 7 & . & 6_8 = (2 \times 8^2) + (1 \times 8^1) + (7 \times 8^0) + (6 \times 8^{-1}) \\ & & & & \quad = 143.75_{10} \end{array}$$

When a count of 1 is added to 7_8, the sum is zero and a carry of 1 is added to the next higher-order column on the left. Counting in octal is shown in Table 1.7, which shows the weight associated with each of the three octal positions.

Table 1.7 Counting in Octal

Decimal	Octal		
	64	8	1
	8^2	8^1	8^0
0	0	0	0
1	0	0	1
2	0	0	2
3	0	0	3
4	0	0	4
5	0	0	5
6	0	0	6
7	0	0	7
8	0	1	0
9	0	1	1
...		...	
14	0	1	6
15	0	1	7
16	0	2	0
17	0	2	1
...		...	
22	0	2	6
23	0	2	7
24	0	3	0
25	0	3	1
...		...	
30	0	3	6
31	0	3	7
...		...	
84	1	2	4
...		...	
242	3	6	2
...		...	
377	5	7	1

Binary-coded octal Each octal digit can be encoded into a corresponding binary number. The highest-valued octal digit is 7; therefore, three binary digits are required to represent each octal digit. This is shown in Table 1.5, which lists the decimal digits and indicates the corresponding octal and binary-coded octal (BCO) digits.

Table 1.8 Binary-Coded Octal Numbers

Decimal	Octal	Binary-Coded Octal		
0	0			000
1	1			001
2	2			010
3	3			011
4	4			100
5	5			101
6	6			110
7	7			111
8	10		001	000
9	11		001	001
10	12		001	010
11	13		001	011
...	...		...	
20	24		010	100
21	25		010	101
...	...		...	
100	144	001	100	100
101	145	001	100	101
...	...		...	
267	413	100	001	011
...	...		...	
385	601	110	000	001

1.1.3 Decimal Number System

The radix is 10 in the *decimal number system*; therefore, ten digits are used, 0 through 9. The low-value digit is 0 and the high-value digit is $(r - 1) = 9$. The weight assigned to each position of a decimal number is as follows:

$$10^{n-1}\,10^{n-2}\,\ldots\;10^3\;10^2\;10^1\;10^0.\;10^{-1}\,10^{-2}\,10^{-3}\;\ldots\;10^{-m}$$

where the integer and fraction are separated by the radix point (decimal point). The value of 6357_{10} is immediately apparent; however, the value is also obtained by using Equation 1.4, where $r = 10$ and $a_i \in \{0,1,2,3,4,5,6,7,8,9\}$ for $-m \leq i \leq n - 1$. That is,

$$\begin{array}{llll} 10^3 & 10^2 & 10^1 & 10^0 \\ 6 & 3 & 5 & 7_{10} = (6 \times 10^3) + (3 \times 10^2) + (5 \times 10^1) + (7 \times 10^0) \end{array}$$

When a count of 1 is added to decimal 9, the sum is zero and a carry of 1 is added to the next higher-order column on the left. The following example contains both an integer and a fraction:

$$\begin{array}{ccccccl} 10^3 & 10^2 & 10^1 & 10^0 & . & 10^{-1} & \\ 5 & 4 & 3 & 6 & . & 5 & = (5 \times 10^3) + (4 \times 10^2) + (3 \times 10^1) + (6 \times 10^0) + (5 \times 10^{-1}) \end{array}$$

Binary-coded decimal Each decimal digit can be encoded into a corresponding binary number; however, only ten decimal digits are valid. The highest-valued decimal digit is 9, which requires four bits in the binary representation. Therefore, four binary digits are required to represent each decimal digit. This is shown in Table 1.9, which lists the ten decimal digits (0 through 9) and indicates the corresponding binary-coded decimal (BCD) digits. Table 1.9 also shows BCD numbers of more than one decimal digit.

Table 1.9 Binary-Coded Decimal Numbers

Decimal	Binary-Coded Decimal
0	0000
1	0001
2	0010
3	0011
4	0100
5	0101
6	0110
7	0111
8	1000
9	1001
10	0001 0000
11	0001 0001
12	0001 0010
...	...
124	0001 0010 0100
...	...
365	0011 0110 0101

1.1.4 Hexadecimal Number System

The radix is 16 in the *hexadecimal number system*; therefore, 16 digits are used, 0 through 9 and A through F, where by convention A, B, C, D, E, and F correspond to decimal 10, 11, 12, 13, 14, and 15, respectively. The low-value digit is 0 and the high-value digit is $(r - 1) = 15$ (F). The weight assigned to each position of a hexadecimal number is as follows:

$$16^{n-1}16^{n-2} \ldots 16^3\ 16^2\ 16^1\ 16^0 .\ 16^{-1}16^{-2}16^{-3} \ldots 16^{-m}$$

where the integer and fraction are separated by the radix point (hexadecimal point). The decimal value of the hexadecimal number $6A8C.D416_{16}$ is obtained by using Equation 1.4, where $r = 16$ and $a_i \in \{0,1,2,3,4,5,6,7,8,9,A,B,C,D,E,F\}$ for $-m \leq i \leq n - 1$. Therefore,

$$\begin{array}{l} 16^3\ 16^2\ 16^1\ 16^0\ .\ 16^{-1}16^{-2}16^{-3}16^{-4} \\ 6\ \ A\ \ 8\ \ C\ .\ D\ \ 4\ \ 1\ \ 6 \end{array} = \begin{array}{l} (6 \times 16^3) + (10 \times 16^2) + (8 \times 16^1) \\ + (12 \times 16^0) + (13 \times 16^{-1}) + (4 \times 16^{-2}) \\ + (1 \times 16^{-3}) + (6 \times 16^{-4}) \end{array}$$

$$= 27{,}276.82846_{10}$$

When a count of 1 is added to hexadecimal F, the sum is zero and a carry of 1 is added to the next higher-order column on the left. Note that when inserting hexadecimal numbers in an assembly language program manually, the first digit must be a number 0 through 9, then the digits A through F, if required, followed by the hexadecimal radix specifier H.

Binary-coded hexadecimal Each hexadecimal digit corresponds to a 4-bit binary number as shown in Table 1.10. All 2^4 values of the four binary bits are used to represent the 16 hexadecimal digits. Table 1.10 also indicates hexadecimal numbers of more than one digit. Counting in hexadecimal is shown in Table 1.11. Table 1.12 summarizes the characters used in the four number systems: binary, octal, decimal, and hexadecimal.

Table 1.10 Binary-Coded Hexadecimal Numbers

Decimal	Hexadecimal	Binary-Coded Hexadecimal
0	0	0000
1	1	0001
2	2	0010
3	3	0011
4	4	0100
5	5	0101
6	6	0110
7	7	0111
8	8	1000
9	9	1001

(Continued on next page)

Table 1.10 Binary-Coded Hexadecimal Numbers

Decimal	Hexadecimal	Binary-Coded Hexadecimal
10	A	1010
11	B	1011
12	C	1100
13	D	1101
14	E	1110
15	F	1111
...	...	...
124	7C	0111 1100
...	...	...
365	16D	0001 0110 1101

Table 1.11 Counting in Hexadecimal

Decimal	Hexadecimal		
	256	16	1
	16^2	16^1	16^0
0	0	0	0
1	0	0	1
2	0	0	2
3	0	0	3
4	0	0	4
5	0	0	5
6	0	0	6
7	0	0	7
8	0	0	8
9	0	0	9
10	0	0	A
11	0	0	B
12	0	0	C
13	0	0	D
14	0	0	E
15	0	0	F
16	0	1	0
17	0	1	1
...		...	
26	0	1	A

(Continued on next page)

Table 1.11 Counting in Hexadecimal

Decimal	Hexadecimal		
	256	16	1
	16^2	16^1	16^0
27	0	1	B
...		...	
30	0	1	E
31	0	1	F
...		...	
256	1	0	0
...		...	
285	1	1	D
...		...	
1214	4	B	E

Table 1.12 Digits Used for Binary, Octal, Decimal, and Hexadecimal Number Systems

Number system	Digits
Binary	0 1
Octal	0 1 2 3 4 5 6 7
Decimal	0 1 2 3 4 5 6 7 8 9
Hexadecimal	0 1 2 3 4 5 6 7 8 9 A B C D E F

1.1.5 Arithmetic Operations

The arithmetic operations of addition, subtraction, multiplication, and division in any radix can be performed using identical procedures to those used for decimal arithmetic. The operands for the four operations are shown in Table 1.13.

Table 1.13 Operands Used for Arithmetic Operations

Addition	Subtraction	Multiplication	Division
Augend	Minuend	Multiplicand	Dividend
+) Addend	–) Subtrahend	×) Multiplier	÷) Divisor
Sum	Difference	Product	Quotient, Remainder

Radix 2 addition Figure 1.1 illustrates binary addition of unsigned operands. The sum of column 1 is 2_{10} (10_2); therefore, the sum is 0 with a carry of 1 to column 2. The sum of column 2 is 4_{10} (100_2); therefore, the sum is 0 with a carry of 0 to column 3 and a carry of 1 to column 4. The sum of column 3 is 3_{10} (11_2); therefore, the sum is 1 with a carry of 1 to column 4. The sum of column 4 is 4_{10} (100_2); therefore, the sum is 0 with a carry of 0 to column 5 and a carry of 1 to column 6. The unsigned values of the binary operands are shown in the rightmost column together with the resulting sum.

Column	6	5	4	3	2	1	Radix 10 values
			1	1	1	0	14
			0	1	1	1	7
			1	0	1	0	10
+)			0_{11}	1_0	0_1	1	5
	1	**0**	**0**	**1**	**0**	**0**	**36**

Figure 1.1 Example of binary addition.

Radix 2 subtraction The rules for subtraction in radix 2 are as follows:

$0 - 0 =$	0
$0 - 1 =$	1 with a borrow from the next higher-order minuend
$1 - 0 =$	1
$1 - 1 =$	0

Figure 1.2 provides an example of binary subtraction using the above rules for unsigned operands. An alternative method for subtraction — used in computers — will be given in Section 1.2 when number representations are presented. In Figure 1.2 column 3, the difference is 1 with a borrow from the minuend in column 4, which changes the minuend in column 4 to 0.

Column	4	3	2	1	Radix 10 values
	1^0	0	1	1	11
–)	0	1	0	1	5
	0	**1**	**1**	**0**	**6**

Figure 1.2 Example of binary subtraction.

Radix 2 multiplication Multiplying in binary is similar to multiplying in decimal. Two n-bit operands produce a $2n$-bit product. Figure 1.3 shows an example of binary multiplication using unsigned operands, where the multiplicand is 7_{10} and the multiplier is 14_{10}. The multiplicand is multiplied by the low-order multiplier bit (0), producing a partial product of all zeroes. Then the multiplicand is multiplied by the next higher-order multiplier bit (1), producing a left-shifted partial product of 0000 111. The process repeats until all bits of the multiplier have been used.

				0	1	1	1	7
			×)	1	1	1	0	14
0	0	0	0	0	0	0	0	
0	0	0	0	1	1	1		
0	0	0	1	1	1			
0	0_1	1_{11}	1_0	1_1				
0	**1**	**1**	**0**	**0**	**0**	**1**	**0**	**98**

Figure 1.3 Example of binary multiplication.

Radix 2 division The division process is shown in Figure 1.4, where the divisor is n bits and the dividend is $2n$ bits. This division procedure uses a sequential shift-subtract-restore technique. Figure 1.4 shows a divisor of 5_{10} (0101_2) and a dividend of 13_{10} ($0000\ 1101_2$), resulting in a quotient of 2_{10} (0010_2) and a remainder of 3_{10} (0011_2).

The divisor is subtracted from the high-order four bits of the dividend. The result is a partial remainder that is negative — the leftmost bit is 1 — indicating that the divisor is greater than the four high-order bits of the dividend. Therefore, a 0 is placed in the high-order bit position of the quotient. The dividend bits are then restored to their previous values with the next lower-order bit (1) of the dividend being appended to the right of the partial product. The divisor is shifted right one bit position and again subtracted from the dividend bits.

This restore-shift-subtract cycle repeats for a total of three cycles until the partial remainder is positive — the leftmost bit is 0, indicating that the divisor is less than the corresponding dividend bits. This results in a no-restore cycle in which the previous partial remainder (0001) is not restored. A 1 bit is placed in the next lower-order quotient bit and the next lower-order dividend bit is appended to the right of the partial remainder. The divisor is again subtracted, resulting in a negative partial remainder, which is again restored by adding the divisor. The 4-bit quotient is 0010 and the 4-bit remainder is 0011.

The results can be verified by multiplying the quotient (0010) by the divisor (0101) and adding the remainder (0011) to obtain the dividend. Thus, $0010 \times 0101 = 1010 + 0011 = 1101$.

								0	**0**	**0**	**1**	**0**	Quotient
	0	1	0	1	0	0	0	0	1	1	0	1	
Subtract					0	1	0	1					
					1	0	1	1					
Restore					0	0	0	0	1				
Shift-subtract						0	1	0	1				
						1	1	0	0				
Restore					0	0	0	0	1	1			
Shift-subtract							0	1	0	1			
							1	1	1	0			
Restore					0	0	0	0	1	1	0		
Shift-subtract								0	1	0	1		
								0	0	0	1		
No restore					0	0	0	0	0	0	1	1	
Shift-subtract									0	1	0	1	
									1	1	1	0	
Restore					0	0	0	0	**0**	**0**	**1**	**1**	Remainder

Figure 1.4 Example of binary division.

Radix 8 addition Figure 1.5 illustrates octal addition. The result of adding column 1 is 17_8, which is a sum of 1 with a carry of 2. The result of adding column 2 is 11_8, which is a sum of 3 with a carry of 1. The remaining columns are added in a similar manner, yielding a result of 21631_8 or 9113_{10}.

		8^3	8^2	8^1	8^0	
Column		4	3	2	1	Radix 10 value
		7	6	5	4	4012
		6	5	4	7	3431
+)		3_1	2_1	0_2	6	1670
	2	**1**	**6**	**3**	**1**	**9113**

Figure 1.5 Example of octal addition.

Radix 8 subtraction Octal subtraction is slightly more complex than octal addition. Figure 1.6 provides an example of octal subtraction. In column 2 (8^1), a 1 is subtracted from minuend 5_8 leaving a value of 4_8; the 1 is then added to the minuend in column 1 (2_8). This results in a difference of 6_8 in column 1, as shown below.

$(1 \times 8^1) + (2 \times 8^0) = 10_{10}$
Therefore, $10 - 4 = 6$

In a similar manner, in column 4 (8^3), a 1 is subtracted from minuend 6_8 leaving a value of 5_8; the 1 is then added to the minuend in column 3 (1_8), leaving a difference of $9 - 5 = 4$, as shown below.

$(1 \times 8^3) + (1 \times 8^2) = 1100_8$
Consider only the 11 of 1100_8, where $(1 \times 8^1) + (1 \times 8^0) = 9_{10}$
Therefore, $9 - 5 = 4$

	8^3	8^2	8^1	8^0	
Column	4	3	2	1	Radix 10 value
	6	1	5	2	3178
–)	5	5	3	4	2908
	0	**4**	**1**	**6**	**0270**

Figure 1.6 Example of octal subtraction.

Radix 8 multiplication An example of octal multiplication is shown in Figure 1.7, where the multiplicand = 7463_8 and the multiplier = 5210_8. The multiplicand is multiplied by each multiplier digit in turn to obtain a partial product. Except for the first partial product, each successive partial product is shifted left one digit. The subscripts in partial products 3 and 4 represent carries obtained from multiplying the multiplicand by the multiplier digits. When all of the partial products are obtained, the partial products are added following the rules for octal addition.

					7	4	6	3	3891_{10}
				×)	5	2	1	0	2696_{10}
Partial product 1	0	0	0	0	0	0	0	0	
Partial product 2	0	0	0	7	4	6	3		
Partial product 3	0	0_1	6_1	0_1	4	6			
Partial product 4	0_4	3_2	4_3	6_1	7				
Carries from addition	1	2	2	2	1				
	5	**0**	**0**	**1**	**0**	**4**	**3**	**0**	**10490136_{10}**

Figure 1.7 Example of octal multiplication.

Radix 8 division An example of octal division is shown in Figure 1.8. The first quotient digit is 3_8 which, when multiplied by the divisor 17_8, yields a result of 55_8. Subtraction of the partial remainder and multiplication of the quotient digit times the divisor are accomplished using the rules stated above for octal arithmetic.

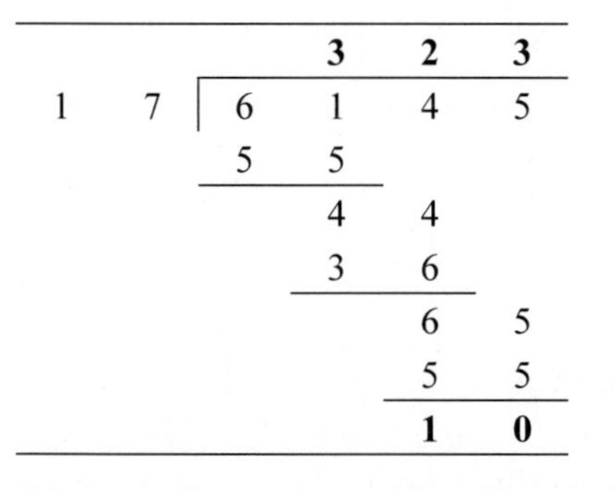

Figure 1.8 Example of octal division.

The results of Figure 1.8 can be verified as follows:

Dividend $= (\text{quotient} \times \text{divisor}) + \text{remainder}$
$= (323_8 \times 17_8) + 10_8$
$= 6145_8$

Radix 16 addition An example of hexadecimal addition is shown in Figure 1.9. The subscripted numbers indicate carries from the previous column. The decimal value of the hexadecimal addition of each column is also shown. To obtain the hexadecimal value of the column, a multiple of 16_{10} is subtracted from decimal value and the difference is the hexadecimal value and the multiple of 16_{10} is the carry. For example, the decimal sum of column 1 is 28. Therefore, 28 – 16 = 12 (C_{16}) with a carry of 1 to column 2. In a similar manner, the decimal sum of column 2 is 40 + 1 (carry) = 41. Therefore, 41 – 32 = 9 (9_{16}) with a carry of 2 to column 3.

Column	4	3	2	1
	A	B	C	D
	9	8	7	6
	E	F	9	4
+)	9_2	A_2	C_1	5
Radix 10 =	44	46	41	28
2	**C**	**E**	**9**	**C**

Figure 1.9 Example of hexadecimal addition.

Radix 16 subtraction Hexadecimal subtraction is similar to subtraction in any other radix. An example of hexadecimal subtraction is shown in Figure 1.10.

Column	4	3	2	1
	C^1	2^1	8	D
−)	8	F	E	9
	3	**2**	**A**	**4**

Figure 1.10 Example of hexadecimal subtraction.

The superscripted numbers indicate borrows from the minuends. For example, the minuend in column 2 borrows a 1 from the minuend in column 3; therefore, column 2 becomes $18_{16} - E_{16} = A_{16}$. This is more readily apparent if the hexadecimal numbers are represented as binary numbers, as shown below.

	1	8	→		0	0	0	1	1	0	0	0
−)		E		−)	0	0	0	0	1	1	1	0
					0	0	0	0	**1**	**0**	**1**	**0**

In a similar manner, column 3 becomes $11_{16} - F_{16} = 2_{16}$ with a borrow from column 4. Column 4 becomes $B_{16} - 8_{16} = 3_{16}$.

Radix 16 multiplication Figure 1.11 shows an example of hexadecimal multiplication. Multiplication in radix 16 is slightly more complex than multiplication in other radices. Each multiplicand is multiplied by each multiplier digit in turn to form a partial product. Except for the first partial product, each partial product is shifted left one digit position. The subscripted digits in Figure 1.11 indicate the carries formed when multiplying the multiplicand by the multiplier digits.

Consider the first row of Figure 1.11 — the row above partial product 1.

$10_{10} \times 4_{10} = 40_{10} = 8_{16}$ with a carry of 2_{16}
$10_{10} \times 13_{10} = 130_{10} = 2_{16}$ with a carry of 8_{16}
$10_{10} \times 9_{10} = 90_{10} = A_{16}$ with a carry of 5_{16}
$10_{10} \times 12_{10} = 120_{10} = 8_{16}$ with a carry of 7_{16}

In a similar manner, the remaining partial products are obtained. Each column of partial products is then added to obtain the product.

					C	9	D	4
				×)	7	8	B	A
				7	8_{51}	A_8	2_2	8
Partial product 1	**0**	**0**	**0**	**7**	**E**	**2**	**4**	**8**
			8	4_6	3_8	F_2	C	
Partial product 2	**0**	**0**	**8**	**A**	**C**	**1**	**C**	
	0	6	4	8_6	8_2	0		
Partial product 3	**0**	**6**	**4**	**E**	**A**	**0**		
	5	4_3	F_5	B_1	C			
Partial product 4	**5**	**8**	**4**	**C**	**C**			
Carries from addition		1	2	3		1		
Product	**5**	**F**	**2**	**E**	**0**	**4**	**0**	**8**

Figure 1.11 Example of hexadecimal multiplication.

Radix 16 division Figure 1.12 (a) and Figure 1.12 (b) show two examples of hexadecimal division. The results of Figure 1.12 can be verified as follows:

$$\text{Dividend} = (\text{quotient} \times \text{divisor}) + \text{remainder}$$

For Figure 1.12 (a): Dividend = $(\text{F0F}_{16} \times 11_{16}) + 0 = \text{FFFF}_{16}$
For Figure 1.12 (b): Dividend = $(787_{16} \times 22_{16}) + 11_{16} = \text{FFFF}_{16}$

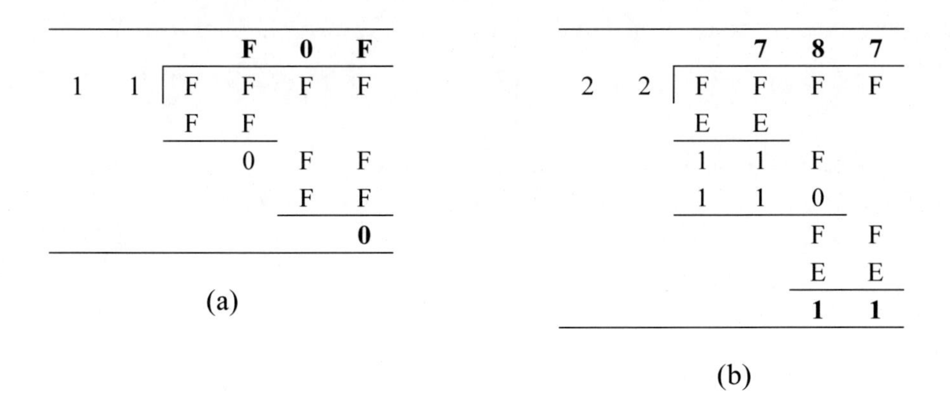

Figure 1.12 Examples of hexadecimal division.

1.1.6 Conversion between Radices

Methods to convert a number in radix r_i to radix r_j will be illustrated in this section. The following conversion methods will be presented:

From		To
Binary	→	Decimal
Octal	→	Decimal
Hexadecimal	→	Decimal
Decimal	→	Binary
Decimal	→	Octal
Decimal	→	Hexadecimal
Binary	→	Octal
Binary	→	Hexadecimal
Octal	→	Binary
Octal	→	Hexadecimal
Hexadecimal	→	Binary
Hexadecimal	→	Octal
Octal	→	Binary-coded octal
Hexadecimal	→	Binary-coded hexadecimal
Decimal	→	Binary-coded decimal

Comparison between the following formats will also be examined:

octal → binary-coded octal and octal → binary
hexadecimal → binary-coded hexadecimal and hexadecimal → binary
decimal → binary-coded decimal and decimal → binary

There will also be examples to illustrate converting between two nonstandard radices and an example to determine the value of an unknown radix for a given radix 10 number.

Binary to decimal Conversion from any radix r to radix 10 is easily accomplished by using Equation 1.2 for integers, and Equation 1.4 for numbers consisting of integers and fractions. Equation 1.2 and Equation 1.4 are reproduced below as Equation 1.6 and Equation 1.7 for convenience.

$$A = a_{n-1}r^{n-1} + a_{n-2}r^{n-2} + a_{n-3}r^{n-3} + \ldots + a_1r^1 + a_0r^0 \tag{1.6}$$

$$A = a_{n-1}r^{n-1} + \ldots + a_1r^1 + a_0r^0 + a_{-1}r^{-1} + a_{-2}r^{-2} + \ldots + a_{-m}r^{-m} \tag{1.7}$$

The binary number 1111000.101_2 will be converted to an equivalent decimal number. The weight by position is as follows:

2^6	2^5	2^4	2^3	2^2	2^1	2^0		2^{-1}	2^{-2}	2^{-3}
1	1	1	1	0	0	0	.	1	0	1

$$\begin{aligned}\text{Therefore, } 1111000.101_2 = \;& (1 \times 2^6) + (1 \times 2^5) + (1 \times 2^4) + (1 \times 2^3) + \\ & (0 \times 2^2) + (0 \times 2^1) + (0 \times 2^0) + \\ & (1 \times 2^{-1}) + (0 \times 2^{-2}) + (1 \times 2^{-3}) \\ =\;& 64 + 32 + 16 + 8 + 0.5 + 0.125 \\ =\;& 120.625_{10}\end{aligned}$$

Octal to decimal The octal number 217.65_8 will be converted to an equivalent decimal number. The weight by position is as follows:

8^2	8^1	8^0		8^{-1}	8^{-2}
2	1	7	.	6	5

$$\begin{aligned}\text{Therefore, } 217.\,65_8 =\;& (2 \times 8^2) + (1 \times 8^1) + (7 \times 8^0) + (6 \times 8^{-1})\,(5 \times 8^{-2}) \\ =\;& 128 + 8 + 7 + 0.75 + 0.078125 \\ =\;& 143.828125_{10}\end{aligned}$$

Hexadecimal to decimal The hexadecimal number $5C2.4D_{16}$ will be converted to an equivalent decimal number. The weight by position is as follows:

16^2	16^1	16^0		16^{-1}	16^{-2}
5	C	2	.	4	D

$$\begin{aligned}\text{Therefore, } 5C2.4D_{16} =\;& (5 \times 16^2) + (12 \times 16^1) + (2 \times 16^0) + (4 \times 16^{-1}) + (13 \times 16^{-2}) \\ =\;& 1280 + 192 + 2 + 0.25 + 0.05078125 \\ =\;& 1474.300781_{10}\end{aligned}$$

Decimal to binary To convert a number in radix 10 to any other radix *r*, repeatedly divide the integer by radix *r*, then repeatedly multiply the fraction by radix *r*. The first remainder obtained when dividing the integer is the low-order digit. The first integer obtained when multiplying the fraction is the high-order digit.

The decimal number 186.625_{10} will be converted to an equivalent binary number. The process is partitioned into two parts: divide the integer 186_{10} repeatedly by 2 until the quotient equals zero; multiply the fraction 0.625 repeatedly by 2 until a zero result is obtained or until a certain precision is reached.

$186 \div 2 =$	quotient = 93,	remainder = 0	(0 is the low-order digit)
$93 \div 2 =$	quotient = 46,	remainder = 1	
$46 \div 2 =$	quotient = 23,	remainder = 0	
$23 \div 2 =$	quotient = 11,	remainder = 1	
$11 \div 2 =$	quotient = 5,	remainder = 1	
$5 \div 2 =$	quotient = 2,	remainder = 1	
$2 \div 2 =$	quotient = 1,	remainder = 0	
$1 \div 2 =$	quotient = 0,	remainder = 1	

$0.625 \times 2 =$	1.25	1	(1 is the high-order digit)
$0.25 \times 2 =$	0.5	0	
$0.5 \times 2 =$	1.0	1	

Therefore, $186.625_{10} = 10111010.101_2$. Converting from decimal to BCD does not yield the same results as converting from decimal to binary, because BCD does not use all 16 combinations of four bits — only ten combinations are used.

Decimal to octal The decimal number 219.62_{10} will be converted to an equivalent octal number. The integer 219_{10} is divided by 8 repeatedly and the fraction 0.62_{10} is multiplied by 8 repeatedly to a precision of three digits.

$219 \div 8 =$	quotient = 27,	remainder = 3	(3 is the low-order digit)
$27 \div 8 =$	quotient = 3,	remainder = 3	
$3 \div 8 =$	quotient = 0,	remainder = 3	

$0.62 \times 8 =$	4.96	4	(4 is the high-order digit)
$0.96 \times 8 =$	7.68	7	
$0.68 \times 8 =$	5.44	5	

Therefore, $219.62_{10} = 333.475_8$.

Decimal to hexadecimal The decimal number 195.828125_{10} will be converted to an equivalent hexadecimal number. The integer is divided by 16 repeatedly and the fraction is multiplied by 16 repeatedly.

$195 \div 16 =$	quotient = 12,	remainder = 3	(3 is the low-order digit)
$12 \div 16 =$	quotient = 0,	remainder = 12 (C)	

$0.828125 \times 16 =$	13.250000	13 (D)	(D is the high-order digit)
$0.250000 \times 16 =$	4.000000	4	

Therefore, $195.828125_{10} = C3.D4_{16}$.

Binary to octal When converting a binary number to octal, the binary number is partitioned into groups of three bits as the number is scanned from right to left for integers and scanned left to right for fractions. If the leftmost group of the integer does not contain three bits, then leading zeroes are added to produce a 3-bit octal digit; if the rightmost group of the fraction does not contain three bits, then trailing zeroes are added to produce a 3-bit octal digit. The binary number 10110100011.11101_2 will be converted to an octal number as shown below.

0 1 0	1 1 0	1 0 0	0 1 1	.	1 1 1	0 1 0
2	6	4	3	.	7	2

Binary to hexadecimal When converting a binary number to hexadecimal, the binary number is partitioned into groups of four bits as the number is scanned from right to left for integers and scanned left to right for fractions. If the leftmost group of the integer does not contain four bits, then leading zeroes are added to produce a 4-bit hexadecimal digit; if the rightmost group of the fraction does not contain four bits, then trailing zeroes are added to produce a 4-bit hexadecimal digit. The binary number 11010101000.1111010111_2 will be converted to a hexadecimal number as shown below.

0 1 1 0	1 0 1 0	1 0 0 0	.	1 1 1 1	0 1 0 1	1 1 0 0
6	A	8	.	F	5	C

Octal to binary When converting an octal number to binary, three binary digits are entered that correspond to each octal digit, as shown below.

2	7	5	4	.	3	6
0 1 0	1 1 1	1 0 1	1 0 0	.	0 1 1	1 1 0

When converting from octal to binary-coded octal (BCO) and from octal to binary, the binary bit configurations are identical. This is because the octal number system uses all eight combinations of three bits.

Octal to hexadecimal To convert from octal to hexadecimal, the octal number is first converted to BCO then partitioned into 4-bit segments to form binary-coded hexadecimal (BCH). The BCH notation is then easily changed to hexadecimal, as shown below.

7			6			3			5			.	4			6				
1	1	1	1	1	0	0	1	1	1	0	1	.	1	0	0	1	1	0	0	0
F				9				D				.	9				8			

Hexadecimal to binary To convert from hexadecimal to binary, substitute the four binary bits for the hexadecimal digits according to Table 1.10 as shown below.

F				A				9				7				.	B				6			
1	1	1	1	1	0	1	0	1	0	0	1	0	1	1	1	.	1	0	1	1	0	1	1	0

When converting from hexadecimal to BCH and from hexadecimal to binary, the binary bit configurations are identical. This is because the hexadecimal number system uses all sixteen combinations of four bits.

Hexadecimal to octal When converting from hexadecimal to octal, the hexadecimal digits are first converted to binary. Then the binary bits are partitioned into 3-bit segments to obtain the octal digits, as shown below.

B				8				E				.	4				D				
1	0	1	1	1	0	0	0	1	1	1	0	.	0	1	0	0	1	1	0	1	0
5			6			1			6			.	2			3			2		

Conversion from radix 5 to radix 10 Equation 1.4 is reproduced below as Equation 1.8 for convenience and will be used to convert the following radix 5 number to an equivalent radix 10 number: 2134.43_5.

$$A = a_{n-1}r^{n-1} + \ldots + a_1r^1 + a_0r^0 + a_{-1}r^{-1} + a_{-2}r^{-2} + \ldots + a_{-m}r^{-m} \tag{1.8}$$

$$\begin{aligned} 2134.43_5 &= (2 \times 5^3) + (1 \times 5^2) + (3 \times 5^1) + (4 \times 5^0) . (4 \times 5^{-1}) + (3 \times 5^{-2}) \\ &= 250 + 25 + 15 + 4 + 0.8 + 0.12 \\ &= 294.92_{10} \end{aligned}$$

Convert from radix *ri* to any other radix *rj* To convert any nondecimal number A_{ri} in radix ri to another nondecimal number A_{rj} in radix rj, first convert the number A_{ri} to decimal using Equation 1.4, then convert the decimal number to radix rj by using repeated division and/or repeated multiplication. The radix 9 number 125_9 will be converted to an equivalent radix 7 number. First 125_9 is converted to radix 10.

$$\begin{aligned} 125_9 &= (1 \times 9^2) + (2 \times 9^1) + (5 \times 9^0) \\ &= 104_{10} \end{aligned}$$

Then, convert 104_{10} to radix 7.

$104 \div 7 =$	quotient = 14,	remainder = 6	(6 is the low-order digit)
$14 \div 7 =$	quotient = 2,	remainder = 0	
$2 \div 7 =$	quotient = 0,	remainder = 2	

Verify the answer.

$$\begin{aligned} 125_9 = 206_7 &= (2 \times 7^2) + (0 \times 7^1) + (6 \times 7^0) \\ &= 104_{10} \end{aligned}$$

Determine the value of an unknown radix The equation shown below has an unknown radix a. This example will determine the value of radix a.

$$\begin{aligned} 44_a^{0.5} &= 6_{10} \\ 44_a &= 36_{10} \\ (4 \times a^1) + (4 \times a^0) &= (3 \times 10^1) + (6 \times 10^0) \\ 4a + 4 &= 30 + 6 \\ 4a &= 32 \\ a &= 8 \end{aligned}$$

Verify the answer.

$$\begin{aligned} 44_8 &= (4 \times 8^1) + (4 \times 8^0) \\ &= 36_{10} \end{aligned}$$

1.2 Number Representations

Computers use both positive and negative numbers, and since a computer cannot recognize a plus (+) or a minus (–) sign, an encoding method must be established to represent the sign of a number in which both positive and negative numbers are distributed as evenly as possible.

There must also be a method to differentiate between positive and negative numbers; that is, there must be an easy way to test the sign of a number. Detection of a number with a zero value must be straightforward. The leftmost (high-order) digit is usually reserved for the sign of the number. Consider the following number A with radix r:

$$A = (a_{n-1}\, a_{n-2}\, a_{n-3}\, \ldots\, a_2\, a_1\, a_0)_r$$

where digit a_{n-1} has the value shown in Equation 1.9.

$$A = \begin{cases} 0 & \text{if } A \geq 0 \\ r-1 & \text{if } A < 0 \end{cases} \tag{1.9}$$

The remaining digits of A indicate either the true magnitude or the magnitude in a complemented form. There are three conventional ways to represent positive and negative numbers in a positional number system: *sign magnitude*, *diminished-radix complement*, and *radix complement*. In all three number representations, the high-order digit is the sign of the number, such that: 0 = positive and $r - 1$ = negative, as shown in Equation 1.9.

1.2.1 Sign Magnitude

In this representation, an integer has the decimal range shown in Equation 1.10.

$$-(r^{n-1} - 1) \text{ to } + (r^{n-1} - 1) \tag{1.10}$$

where the number zero is considered to be positive. Thus, a positive number A is represented as shown in Equation 1.11.

$$A = (0\, a_{n-2} a_{n-3}\, \ldots\, a_1 a_0)_r \tag{1.11}$$

A negative number with the same absolute value as shown in Equation 1.12.

$$A' = [(r-1)\, a_{n-2} a_{n-3}\, \ldots\, a_1 a_0]_r \tag{1.12}$$

In sign-magnitude notation, the positive version, $+A$, differs from the negative version, $-A$, only in the sign digit position. The magnitude portion $a_{n-2}a_{n-3} \ldots a_1a_0$ is identical for both positive and negative numbers of the same absolute value.

There are two problems with sign-magnitude representation. First, there are two representations for the number 0; specifically, +0 and −0. Ideally there should be a unique representation for the number 0. Second, when adding two numbers of opposite signs, the magnitudes of the numbers must be compared to determine the sign of the result. This is not necessary in the other two methods that are presented in subsequent sections. Sign-magnitude notation is used primarily for representing fractions in floating-point notation.

Examples of sign-magnitude notation are shown below using 8-bit binary numbers and decimal numbers that represent both positive and negative values. Notice that the magnitude parts are identical for both positive and negative numbers for the same radix.

Radix 2

```
0   0 0 1   0 1 0 0    +20
1   0 0 1   0 1 0 0    –20
-   -------------
↑         ↑
|         └──────── Magnitude
└────────────────── Sign

0   1 0 0   1 1 0 1 . 1 0 1    +77.625
1   1 0 0   1 1 0 1 . 1 0 1    –77.625

0   1 1 0   0 1 1 1 . 0 1 1    +103.375
1   1 1 0   0 1 1 1 . 0 1 1    –103.375
```

Radix 10

```
0   7 4 5    +745
9   7 4 5    –745
```

where 0 represents a positive number in radix 10, and 9 $(r - 1)$ represents a negative number in radix 10. Again, the magnitudes of both numbers are identical.

```
0   6 7 9 3 8    +67938
9   6 7 9 3 8    –67938
```

1.2.2 Diminished-Radix Complement

This is the *(r – 1) complement* in which the radix is diminished by 1 and an integer has the decimal range shown in Equation 1.13, which is the same as the range for sign-magnitude integers, although the numbers are represented differently. The number zero is again considered to be positive. Thus, a positive number A is represented as shown in Equation 1.14. A negative number is represented as shown in Equation 1.15, in which all the digits are inverted.

$$-(r^{n-1} - 1) \text{ to } +(r^{n-1} - 1) \tag{1.13}$$

$$A = (0\ a_{n-2}a_{n-3}\ \cdots\ a_1a_0)_r \tag{1.14}$$

$$A' = [(r-1)\ a_{n-2}'a_{n-3}'\ \cdots\ a_1'a_0']_r \tag{1.15}$$

For the $r - 1$ complement, individual digits can be determined as shown in Equation 1.16. For example, if $a_i = 1$ in radix 2, then the diminished-radix complement of a_i is $a_i' = [(r-1) - 1] = [(2-1) - 1] = 0$. In a similar manner, if $a_i = 0$ in radix 2, then the diminished-radix complement of a_i is $a_i' = [(r-1) - 0] = [(2-1) - 0] = 1$. In binary notation ($r = 2$), the diminished-radix complement ($r - 1 = 2 - 1 = 1$) is the *1s complement.*

$$a_i' = (r-1) - a_i \tag{1.16}$$

Positive and negative integers have the ranges shown below and are represented as shown in Equation 1.14 and Equation 1.15, respectively.

Positive integers: 0 to $2^{n-1} - 1$
Negative integers: 0 to $-(2^{n-1} - 1)$

To obtain the 1s complement of a binary number, simply complement (invert) all the bits. Thus, $0011\ 1100_2$ ($+60_{10}$) becomes $1100\ 0011_2$ (-60_{10}). To obtain the value of a positive binary number, the 1s are evaluated according to their weights in the positional number system, as shown below.

2^7	2^6	2^5	2^4	2^3	2^2	2^1	2^0	
0	0	1	1	1	1	0	0	$+60_{10}$

To obtain the value of a negative binary number, the 0s are evaluated according to their weights in the positional number system, as shown below.

2^7	2^6	2^5	2^4	2^3	2^2	2^1	2^0	
1	1	0	0	0	0	1	1	-60_{10}

When performing arithmetic operations on two operands, comparing the signs is straightforward, because the leftmost bit is a 0 for positive numbers and a 1 for negative numbers. There is, however, a problem when using the diminished-radix complement. There is a dual representation of the number zero, because a word of all 0s (+0) becomes a word of all 1s (−0) when complemented. This does not allow the requirement of having a unique representation for the number zero to be attained. The examples shown below represent the diminished-radix complement for different radices.

Example 1.3 The binary number 1001 1101_2 will be 1s complemented. The number has a decimal value of −98. To obtain the 1s complement, subtract each digit in turn from 1, the highest number in the radix. Or in the case of binary, simply invert each bit. Therefore, the 1s complement of 1001 1101_2 is 0110 0010_2, which has a decimal value of +98.

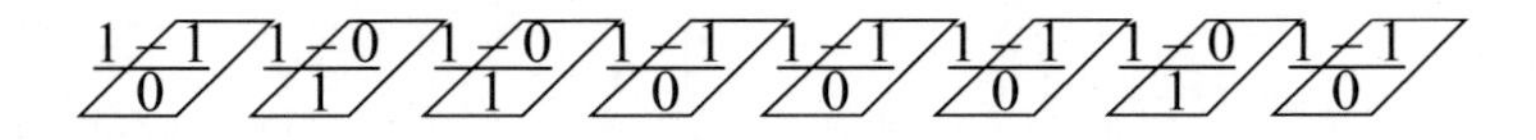

To verify the operation, add the negative and positive numbers to obtain 1111 1111_2, which is zero in 1s complement notation.

	1	0	0	1	1	1	0	1
+)	0	1	1	0	0	0	1	0
	1	1	1	1	1	1	1	1

Example 1.4 The diminished-radix complement (9s complement) of 08752.43_{10}, will be obtained, where 0 is the sign digit indicating a positive number. The 9s complement is obtained by using Equation 1.15 and Equation 1.16. When a number is complemented in any form, the number is negated. Therefore, the sign of the complemented radix 10 number is $(r - 1) = 9$. The remaining digits of the number are obtained by using Equation 1.16, such that each digit in the complemented number is obtained by subtracting the given digit from 9. Therefore, the 9s complement of 08752.43_{10} is

$$\frac{9-0}{9}\ \frac{9-8}{1}\ \frac{9-7}{2}\ \frac{9-5}{4}\ \frac{9-2}{7}\,.\,\frac{9-4}{5}\ \frac{9-3}{6}$$

The sign digit is $(r - 1) = 9$. If the above answer is negated, then the original number will be obtained. Thus, the 9s complement of $91247.56_{10} = 08752.43_{10}$; that is, the 9s complement of -1247.56_{10} is $+8752.43_{10}$, as written in conventional sign magnitude notation for radix 10.

To verify the operation, add the negative and positive numbers to obtain 99999_{10}, which is zero in 9s complement notation.

$$\begin{array}{r} 0\ 8\ 7\ 5\ 2\ .\ 4\ 3 \\ +)\ \underline{9\ 1\ 2\ 4\ 7\ .\ 5\ 6} \\ 9\ 9\ 9\ 9\ 9\ .\ 9\ 9 \end{array}$$

Example 1.5 The diminished-radix complement of the positive radix 8 number 04376_8 will be 7s complemented. To obtain the 7s complement, subtract each digit in turn from 7 (the highest number in the radix), as shown below to obtain the negative number with the same absolute value. The sign of the positive number is 0 and the sign of the negative number is 7.

$$\frac{7-0}{7}\quad \frac{7-4}{3}\quad \frac{7-3}{4}\quad \frac{7-7}{0}\quad \frac{7-6}{1}$$

To verify the operation, add the negative and positive numbers to obtain 77777_8, which is zero in 7s complement notation.

$$\begin{array}{r} 0\ 4\ 3\ 7\ 6 \\ +)\ \underline{7\ 3\ 4\ 0\ 1} \\ 7\ 7\ 7\ 7\ 7 \end{array}$$

Example 1.6 The diminished-radix complement of the positive radix 16 number $0B8E5_{16}$ will be 15s complemented. To obtain the 15s complement, subtract each digit in turn from 15 (the highest number in the radix), as shown below to obtain the negative number with the same absolute value. The sign of the positive number is 0 and the sign of the negative number is F.

$$\frac{F-0}{F}\quad \frac{F-B}{4}\quad \frac{F-8}{7}\quad \frac{F-E}{1}\quad \frac{F-5}{A}$$

To verify the operation, add the negative and positive numbers to obtain $FFFFF_{16}$, which is zero in 15s complement notation.

$$\begin{array}{r} 0\ B\ 8\ E\ 5 \\ +)\ \underline{F\ 4\ 7\ 1\ A} \\ F\ F\ F\ F\ F \end{array}$$

1.2.3 Radix Complement

This is the *r complement*, where an integer has the following decimal range:

$$-(r^{n-1}) \text{ to } +(r^{n-1} - 1) \tag{1.17}$$

where the number zero is positive. A positive number A is represented as

$$A = (0\ a_{n-2}a_{n-3}\ \cdots\ a_1a_0)_r \tag{1.18}$$

and a negative number as

$$(A')_{+1} = \{[(r-1)\ a_{n-2}'a_{n-3}'\ \cdots\ a_1'a_0'] + 1\}_r \tag{1.19}$$

where A' is the diminished-radix complement. Thus, the radix complement is obtained by adding 1 to the diminished-radix complement; that is, $(r-1) + 1 = r$. Note that all three number representations have the same format for positive numbers and differ only in the way that negative numbers are represented, as shown in Table 1.14 for n-bit numbers.

Table 1.14 Number Representations for Positive and Negative Integers of the Same Absolute Value for Radix *r*

Number Representation	Positive Numbers	Negative Numbers
Sign magnitude	$0\ a_{n-2}a_{n-3}\ \cdots\ a_1a_0$	$(r-1)\ a_{n-2}a_{n-3}\ \cdots\ a_1a_0$
Diminished-radix complement	$0\ a_{n-2}a_{n-3}\ \cdots\ a_1a_0$	$(r-1)\ a_{n-2}'a_{n-3}'\ \cdots\ a_1'a_0'$
Radix complement	$0\ a_{n-2}a_{n-3}\ \cdots\ a_1a_0$	$(r-1)\ a_{n-2}'a_{n-3}'\ \cdots\ a_1'a_0' + 1$

Another way to obtain the radix complement of a number is to keep the low-order 0s and the first 1 unchanged and to complement (invert) the remaining high-order bits. Thus, the radix complement (2s complement) of the binary number 0101 1100 (+92) is 1010 0100 (–92). To obtain the value of a negative number in radix 2, the 0s are evaluated according to their weights in the positional number system, then 1 is added to the value obtained. There is a unique zero for binary numbers in radix complement — when the number zero is 2s complemented, the bit configuration does not change; that is, the 2s complement of 0000 0000 is 1111 1111 + 1 = 0000 0000.

1.2.4 Arithmetic Operations

This section will concentrate on fixed-point binary, binary-coded decimal (BCD), and floating point operations, since these are the dominant number representations in computers. Examples of addition, subtraction, multiplication, and division will be presented for these three number representations.

Binary addition Numbers in radix complement representation are designated as signed numbers, specifically as 2s complement numbers in binary. The sign of a binary number can be extended to the left indefinitely without changing the value of the number. For example, the numbers 00001010_2 and 0000000001010_2 both represent a value of $+10_{10}$; the numbers 11110110_2 and 1111111110110_2 both represent a value of -10_{10}.

Thus, when an operand has its sign extended to the left, the expansion is achieved by setting the extended bits equal to the leftmost (sign) bit. This is equivalent to the X86 instructions CBW (convert byte to word), CWDE (convert word to doubleword), and CDQE (convert doubleword to quadword). The maximum positive number consists of a 0 followed by a field of all 1s, depending on the word size of the operand. Similarly, the maximum negative number consists of a 1 followed by a field of all 0s, depending on the word size of the operand.

The radix (or binary) point can be in any fixed position in the number — thus the radix point is referred to as a *fixed-point* radix. For integers, however, the radix point is positioned to the immediate right of the low-order bit position.

Overflow occurs when the result of an arithmetic operation (usually addition) exceeds the word size of the machine; that is, the sum is not within the representable range of numbers provided by the number representation. For numbers in 2s complement representation, the range is from -2^{n-1} to $+2^{n-1} - 1$. For two n-bit numbers

$$A = a_{n-1} a_{n-2} a_{n-3} \ldots a_1 a_0$$

$$B = b_{n-1}\, b_{n-2}\, b_{n-3} \ldots b_1 b_0$$

a_{n-1} and b_{n-1} are the sign bits of operands A and B, respectively. Overflow can be detected by either of the following two equations:

$$\text{Overflow} = (a_{n-1} \bullet b_{n-1} \bullet s_{n-1}') + (a_{n-1}' \bullet b_{n-1}' \bullet s_{n-1})$$

$$\text{Overflow} = c_{n-1} \oplus c_{n-2} \tag{1.20}$$

where the symbol "•" is the logical AND operator, the symbol "+" is the logical OR operator, the symbol "⊕" is the exclusive-OR operator as defined below, and c_{n-1} and c_{n-2} are the carry bits out of positions $n-1$ and $n-2$, respectively.

Thus, overflow produces an erroneous sign reversal and is possible only when both operands have the same sign. An overflow cannot occur when adding two operands of different signs, since adding a positive number to a negative number produces a result that falls within the limit specified by the two numbers.

Binary subtraction The operands used for subtraction are the minuend and subtrahend. The rules for binary subtraction are not easily applicable to computer subtraction. The method used by most processors is to add the radix complement (2s complement) of the subtrahend to the minuend; that is, change the sign of the minuend then add the two operands, as shown below.

					→					
	$A =$	0110	1101	+109			$A =$	0110	1101	+109
−)	$B =$	0011	0111	+55		+)	$B =$	1100	1001	−55
								0011	0110	+54

The diminished-radix complement is rarely used in arithmetic applications because of the dual interpretation of the number zero (0000 ... 0000 and 1111 ... 1111). The example shown below illustrates another disadvantage of the diminished-radix complement. Given the two radix 2 numbers shown below, in 1s complement representation, the difference will be obtained.

$A = 1111\ 1001\ (-6)$
$B = 1110\ 1101\ (-18)$
$A - B = A + B'$ (1s complement of B)

	$A =$	1111 1001	(−6)	
+)	$B' =$	0001 0010	(+18)	1s complement of B
	1 ←	0000 1011	(+11)	Incorrect result
	→	1		End-around carry
	$A - B =$	0000 1100	(+12)	Correct result

When performing a subtract operation using 1s complement operands, an end-around carry will result if at least one operand is negative. As can be seen above, the result will be incorrect (+11) if the carry is not added to the intermediate result. Although 1s complementation may seem easier than 2s complementation, the result that is obtained after an add operation is not always correct. In 1s complement notation, the final carry-out c_{n-1} cannot be ignored. If the carry-out is zero, then the result is correct. Thus, 1s complement subtraction may result in an extra add cycle to obtain the correct result.

Binary multiplication The multiplication of two n-bit operands results in a $2n$-bit product, where $A = a_{n-1}\, a_{n-2}\, a_{n-3} \ldots a_1\, a_0$ and $B = b_{n-1}\, b_{n-2}\, b_{n-3} \ldots b_1\, b_0$ to yield $A \times B = p_{2n-1}\, p_{2n-2}\, p_{2n-3} \cdots p_1\, p_0$. Multiplication of two binary numbers can be accomplished by a variety of methods. These methods include the sequential add-shift technique, the Booth algorithm, bit-pair recoding, a 2-dimension planar array, a table lookup technique, and a method using memory.

The different methods for multiplying two binary numbers are beyond the scope of this book; however, the sequential add-shift technique will be presented in this section as a representative method. Multiplication of two fixed-point binary numbers in 2s complement representation using the add-shift method is done by a process of successive add and shift operations. The process consists of multiplying the multiplicand by each multiplier bit as the multiplier is scanned right to left. If the multiplier bit is a 1, then the multiplicand is copied as a partial product; otherwise, 0s are inserted as a partial product. The partial products inserted into successive lines are shifted left one bit position from the previous partial product. The partial products are then added to form the product.

The sign of the product is determined by the signs of the operands. If the signs are the same, then the sign of the product is plus; if the signs are different, then the sign of the product is minus. In this sequential add-shift technique, however, the multiplier must be positive. When the multiplier is positive, the bits are treated the same as in the sign-magnitude representation. When the multiplier is negative, the low-order 0s and the first 1 are treated the same as a positive multiplier, but the remaining high-order 1s of a negative multiplier are treated as 1s corresponding to their bit position and not as sign bits. Therefore, the algorithm treats the multiplier as unsigned, or positive.

The problem is easily solved by forming the 2s complement of both the multiplier and the multiplicand. An alternative approach is to 2s complement the negative multiplier, leaving the multiplicand unchanged. Depending on the signs of the initial operands, it may be necessary to complement the product.

The example shown below in Figure 1.13 uses the add-shift method to multiply the positive operands of 0111_2 ($+7_{10}$) and 0101_2 ($+5_{10}$). Since both operands are positive, the product will be positive ($+35_{10}$).

				0	1	1	1	(+7)
			×)	0	1	0	1	(+5)
0	0	0	0	0	1	1	1	
0	0	0	0	0	0	0		
0	0	0	1	1	1			
0	0	0	0	0				
0	**0**	**1**	**0**	**0**	**0**	**1**	**1**	(+35)

Figure 1.13 Example of the sequential add-shift technique.

Binary division Division has two operands that produce two results, as shown below. Unlike multiplication, division is not commutative; that is, $A \div B \neq B \div A$, except when $A = B$.

$$\frac{\text{Dividend}}{\text{Divisor}} = \text{Quotient} + \text{Remainder}$$

Like multiplication, there are many ways to perform division on two fixed-point binary operands, all of which are also beyond the scope of this book. These methods include the sequential shift add/subtract technique for restoring and nonrestoring division, SRT division, multiplicative division, and division using a 2-dimension planar array. All operands for a division operation comply with the following equation:

$$\text{Dividend} = (\text{Quotient} \times \text{Divisor}) + \text{Remainder} \tag{1.21}$$

The remainder has a smaller value than the divisor and has the same sign as the dividend. If the divisor B has n bits, then the dividend A has $2n$ bits and the quotient Q and remainder R both have n bits, as shown below.

$$A = a_{2n-1}\, a_{2n-2} \ldots a_n\, a_{n-1} \ldots a_1\, a_0$$

$$B = b_{n-1}\, b_{n-2} \ldots b_1\, b_0$$

$$Q = q_{n-1}\, q_{n-2} \ldots q_1\, q_0$$

$$R = r_{n-1}\, r_{n-2} \ldots r_1\, r_0$$

The sign of the quotient is q_{n-1} and is determined by the rules of algebra; that is,

$$q_{n-1} = a_{2n-1} \oplus b_{n-1}$$

Multiplication is a sequential add-shift operation, whereas division is a shift add/subtract operation. The result of a shift add/subtract operation determines the next operation in the division sequence. If the partial remainder is negative, then the carry-out is 0 and becomes the low-order quotient bit q_0. The partial remainder thus obtained is restored to the value of the previous partial remainder. This technique is referred to as *restoring division*. If the partial remainder is positive, then the carry-out is 1 and becomes the low-order quotient bit q_0. In this case, the partial remainder is not restored.

An example of restoring division using a hardware algorithm is shown in Figure 1.14, where the dividend in register-pair $A\ Q$ is 0000 0111_2 $(+7_{10})$ and the divisor in register B is 0011_2 $(+3_{10})$. The algorithm is implemented with a subtractor and a $2n$-bit shift register.

The first operation in Figure 1.14 is to shift the dividend left 1 bit position, then subtract the divisor, which is accomplished by adding the 2s complement of the divisor. Since the result of the subtraction is negative, the dividend is restored by adding back the divisor, and the low-order quotient bit q_0 is set to 0. This sequence repeats for a total of four cycles — the number of bits in the divisor. If the result of the subtraction was positive, then the partial remainder is placed in register A, the high-order half of the dividend, and q_0 is set to 1.

The division algorithm is slightly more complicated when one or both of the operands are negative. The operands can be preprocessed and/or the results can be postprocessed to achieve the desired results. The negative operands are converted to positive numbers by 2s complementation before the division process begins. The resulting quotient is then 2s complemented, if necessary.

Unlike multiplication, overflow can occur in division. This happens when the high-order half of the dividend is greater than or equal to the divisor. Also, division by zero must be avoided. Both of these problems can be detected before the division process begins by subtracting the divisor from the high-order half of the dividend. If the difference is positive, then an overflow or a divide by zero has been detected.

Divisor *B*		Dividend *A*	*Q*
0011		0000	0111
Shift left 1		0000	111–
Subtract *B*		1101	
	0 ←	1101	
Restore, add *B*		0011	
set $q_0 = 0$		0000	111**0**
Shift left 1		0001	110–
Subtract *B*		1101	
	0 ←	1110	
Restore, add *B*		0011	
set $q_0 = 0$		0001	11**00**
Shift left 1		0011	100–
Subtract *B*		1101	
	1 ←	0000	
No restore			
set $q_0 = 1$		0000	1**001**
Shift left 1		0001	001–
Subtract *B*		1101	
	0 ←	1110	
Restore, add *B*		0011	
set $q_0 = 0$		**0001**	**0010**
		R	*Q*

Figure 1.14 Example of sequential shift add/subtract restoring division.

Binary-coded decimal addition When two binary-coded decimal (BCD) digits are added, the range is 0 to 18. If the carry-in $c_{in} = 1$, then the range is 0 to 19. If the sum digit is ≥ 10 (1010_2), then it must be adjusted by adding 6 (0110_2). This excess-6 technique generates the correct BCD sum and a carry to the next higher-order digit, as shown below in Figure 1.15 (a). A carry-out of a BCD sum will also cause an adjustment to be made to the sum — called the intermediate sum — even though the intermediate sum is a valid BCD digit, as shown in Figure 1.15 (b).

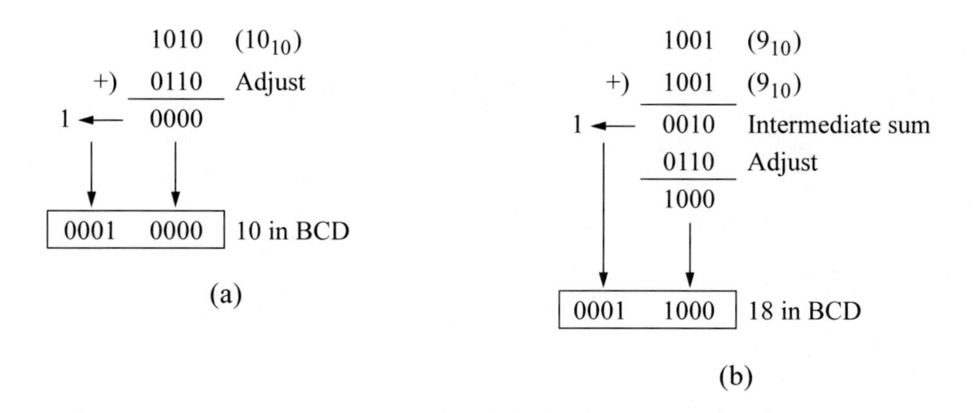

Figure 1.15 Example showing adjustment of a BCD sum.

There are three conditions that indicate when the intermediate sum of a BCD addition should be adjusted by adding six, as shown in Equation 1.22, where c_8 is the carry-out of the high-order bit, $b_8 b_4 = 11$ or $b_8 b_2 = 11$.

$$\text{Carry} = c_8 + b_8 b_4 + b_8 b_2 \tag{1.22}$$

The algorithms used for BCD arithmetic are basically the same as those used for fixed-point arithmetic for radix 2. The main difference is that BCD arithmetic treats each digit as four bits, whereas fixed-point arithmetic treats each digit as a bit. Shifting operations are also different — decimal shifting is performed on 4-bit increments.

Binary-coded decimal subtraction Subtraction in BCD is essentially the same as in fixed-point binary: add the *r*s complement of the subtrahend to the minuend, which for BCD is the 10s complement. Example 1.7 shows a subtract operation using BCD numbers. Negative results can remain in 10s complement notation or be recomplemented to sign-magnitude notation with a negative sign.

Example 1.7 The following decimal numbers will be subtracted using BCD arithmetic: minuend = +30 and subtrahend = +20 to yield a difference of +10, as shown in Figure 1.16. This can be considered as true subtraction, because the result is the difference of the two numbers, ignoring the signs. A carry-out of 1 from the high-order decade indicates a positive number. A carry-out of 0 from the high-order decade indicates a negative number in 10s complement notation. The number can be changed to an absolute value by recomplementing the number and changing the sign to indicate a negative value.

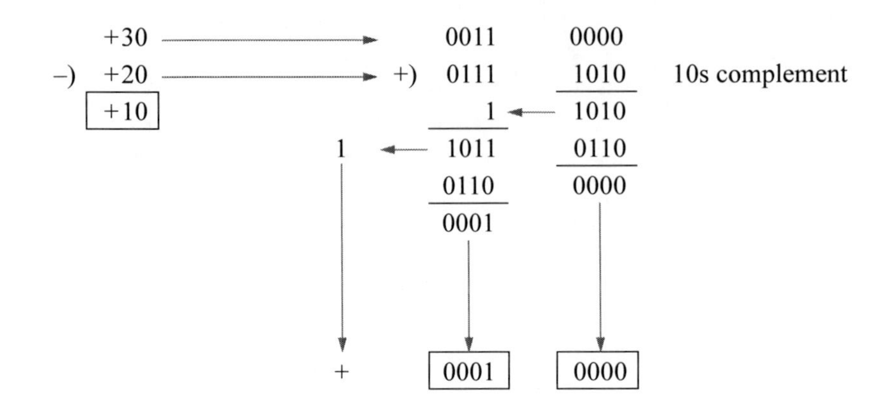

Figure 1.16 Example of BCD subtraction.

Binary-coded decimal multiplication The multiplication algorithms for decimal multiplication are similar to those for fixed-point binary multiplication except in the way that the partial products are formed. In binary multiplication, the multiplicand is added to the previous partial product if the multiplier bit is a 1. In BCD multiplication, the multiplicand is multiplied by each digit of the multiplier and these subpartial products are aligned and added to form a partial product. When adding digits to obtain a partial product, adjustment may be required to form a valid BCD digit. Example 1.8 illustrates BCD multiplication.

Multiplication of two BCD operands can also be accomplished by using high-speed read-only memory (ROM). The concatenated four bits of the multiplicand and the four bits of the multiplier constitute the memory address. The outputs from the memory are valid BCD digits — no correction is required. All corrections (or adjustments) are accomplished by the memory programming.

Decimal multiplication can also be facilitated by using a table lookup method. This is similar to the table lookup method for fixed-point multiplication. The multiplicand table resides in memory.

Example 1.8 The following decimal numbers will be multiplied using BCD arithmetic: multiplicand = +67 and multiplier = +9, resulting in a product of +603, as shown in Figure 1.17.

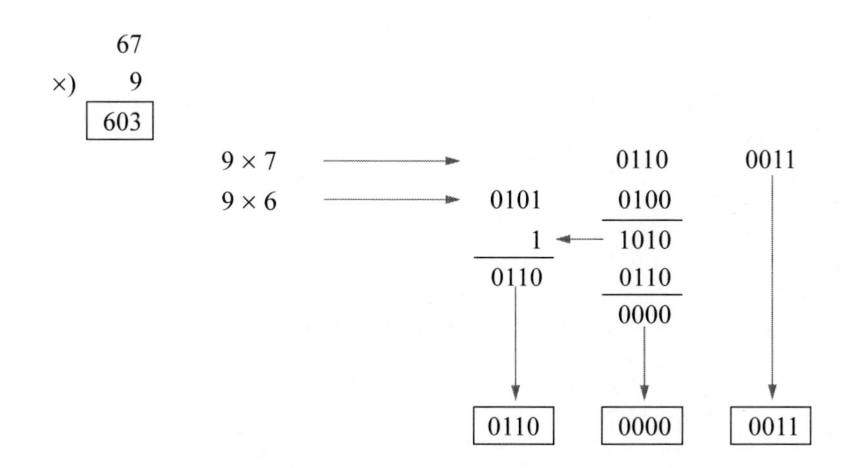

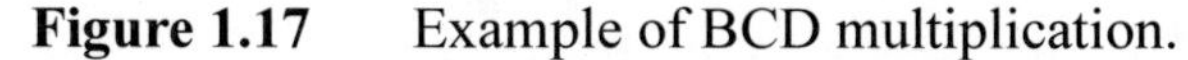
Figure 1.17 Example of BCD multiplication.

Binary-coded decimal division The method of decimal division presented here is analogous to the binary search method used in programming, which is a systematic way of searching an ordered database. The method begins by examining the middle of the database. For division, the method adds or subtracts multiples of the divisor or partial remainder. The arithmetic operation is always in the following order:

$-\,8 \times$ the divisor
$\pm\,4 \times$ the divisor
$\pm\,2 \times$ the divisor
$\pm\,1 \times$ the divisor

This method requires four cycles for each quotient digit. The first operation is $-\,8 \times$ the divisor. If the result is less than zero, then $4 \times$ the divisor is added to the partial remainder; if the result is greater than or equal to zero, then 8 is added to a quotient counter and $4 \times$ the divisor is subtracted from the partial remainder. The process repeats for $\pm\,4 \times$ the divisor, $\pm\,2 \times$ the divisor, and $\pm\,1 \times$ the divisor. Whenever +8, +4, +2, or +1 is added to the quotient counter, the sum of the corresponding additions is the quotient digit. Whenever a partial remainder is negative, the next version of the divisor is added to the partial remainder; whenever a partial remainder is positive, the next version of the divisor is subtracted from the partial remainder. Example 1.9 illustrates this technique.

Example 1.9 Let the dividend = 1945 and the divisor = 20 to yield a quotient of 97 and a remainder of 5. Figure 1.18 shows the decimal version and Figure 1.19 shows the application of the BCD algorithm for the same decimal operands.

$$
\begin{array}{r}
97 \\
20\,\overline{)\,1945} \\
\underline{180} \\
145 \\
\underline{140} \\
5
\end{array}
$$

Figure 1.18 Example of BCD division using decimal operands.

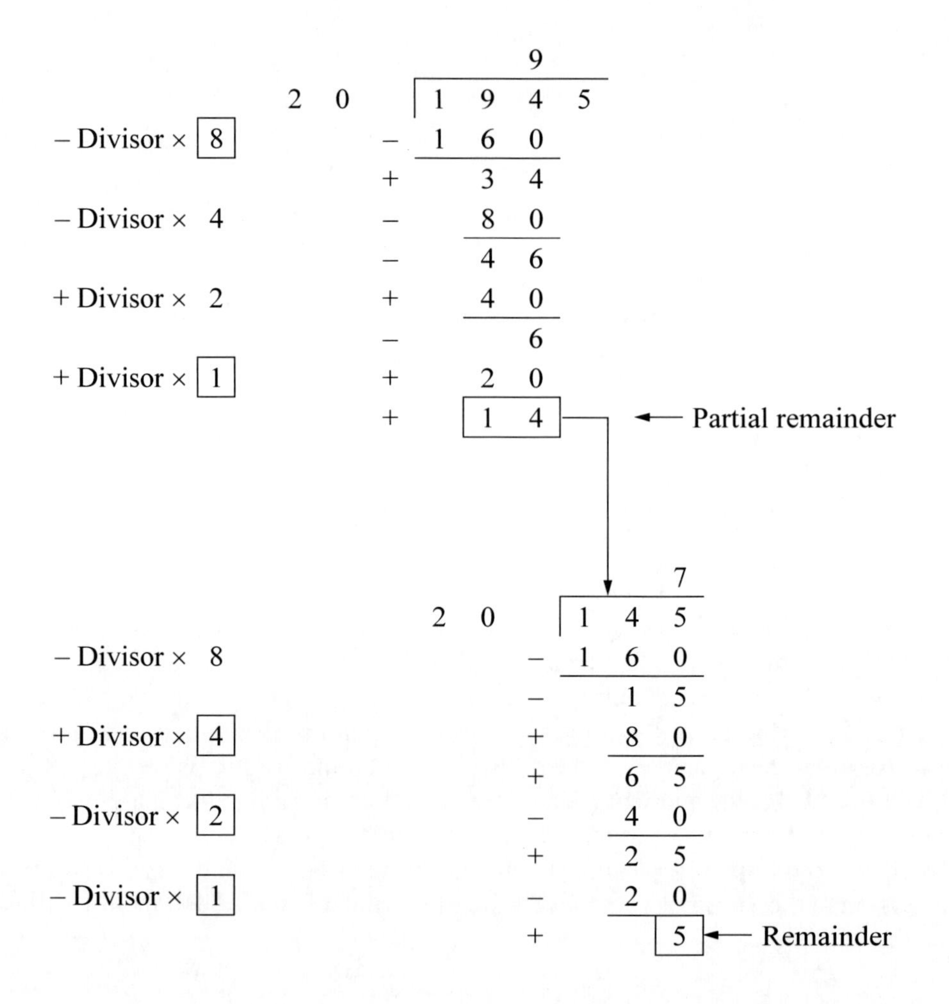

Figure 1.19 Example of BCD division using the BCD algorithm for the decimal operands shown in Figure 1.18.

Floating-point addition Floating-point numbers consist of the following three fields: a sign bit, s; an exponent, e; and a fraction, f. These parts represent a number that is obtained by multiplying the fraction, f, by a radix, r, raised to the power of the exponent, e, as shown in Equation 1.23 for the number A, where f and e are signed fixed-point numbers, and r is the radix (or base).

$$A = f \times r^e \tag{1.23}$$

The exponent is also referred to as the *characteristic*; the fraction is also referred to as the *significand* or *mantissa*. Figure 1.20 shows the format for 32-bit single-precision and 64-bit double-precision floating-point numbers. The single-precision format consists of a sign bit that indicates the sign of the number, an 8-bit signed exponent, and a 24-bit fraction. The double-precision format consists of a sign bit, an 11-bit signed exponent, and a 52-bit fraction.

Although the fraction can be represented in sign-magnitude, diminished-radix complement, or radix complement, the fraction is predominantly expressed in sign-magnitude representation, where the sign bit is bit 31 and the magnitude is bits 22 through 0 for the single-precision format.

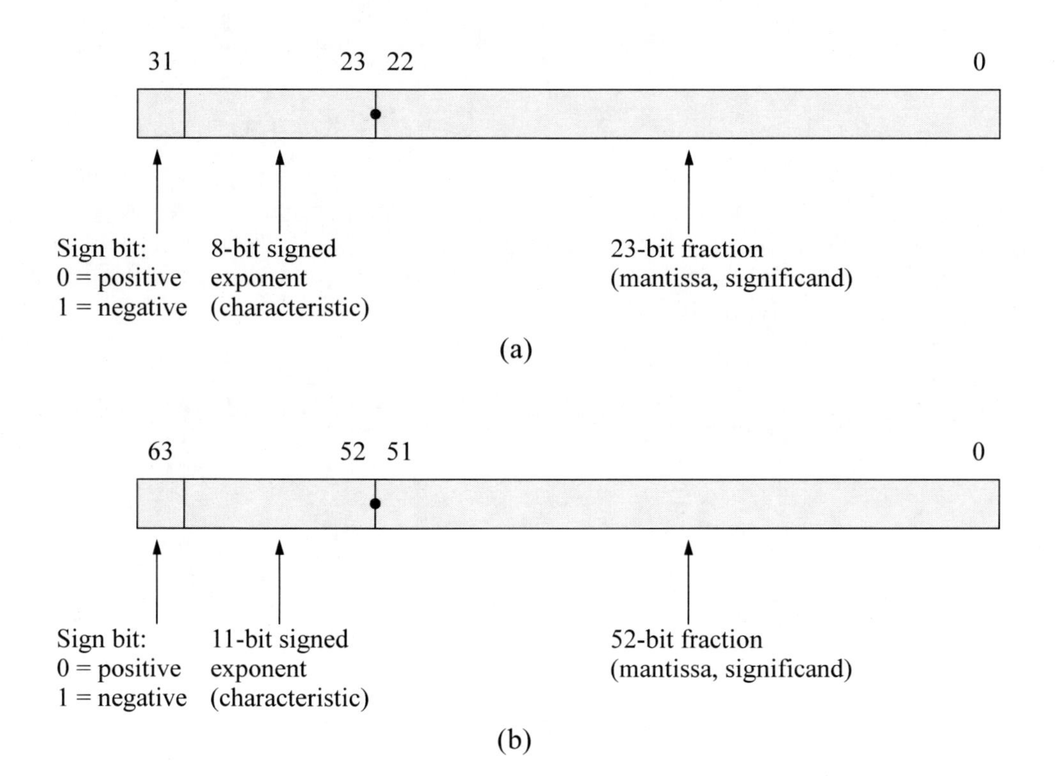

Figure 1.20 Floating-point formats for the IEEE Std 754-1985 (reaffirmed 1990): (a) 32-bit single precision and (b) 64-bit double precision.

When adding or subtracting floating-point numbers, the exponents are compared and made equal. This comparison results in a right shift of the fraction with the smaller exponent. If the exponents are e_A and e_B, then the shift amount is equal to the absolute value of $e_A - e_B$; that is, $|e_A - e_B|$. The comparison is easier if the exponents are unsigned — a simple comparator can be used for the comparison. Therefore, as the exponents are being formed, a *bias* constant is added to the exponents such that all exponents are positive internally.

Fractions in the IEEE format are normalized; that is, the leftmost significant bit is a 1. Since there will always be a 1 to the immediate right of the radix point, the 1 bit is not explicitly shown — it is an *implied 1*.

Floating-point addition is defined as shown in Equation 1.24 for two numbers A and B, where $A = f_A \times r^{e_A}$ and $B = f_B \times r^{e_B}$. The terms shown below are *shifting factors* to shift the fraction with the smaller exponent.

$$r^{-(e_A - e_B)}, \quad r^{-(e_B - e_A)}$$

This is analogous to a divide operation, since $r^{-(e_A - e_B)}$ is equivalent to the term $1/r^{(e_A - e_B)}$, which is a right shift.

$$\begin{aligned} A + B &= (f_A \times r^{e_A}) + (f_B \times r^{e_B}) \\ &= [f_A + (f_B \times r^{-(e_A - e_B)})] \times r^{e_A} \text{ for } e_A > e_B \\ &= [(f_A \times r^{-(e_B - e_A)}) + f_B] \times r^{e_B} \text{ for } e_A \leq e_B \end{aligned} \tag{1.24}$$

Figure 1.21 illustrates an example of floating-point addition. The fractions must be properly aligned before addition can take place; therefore, the fraction with the smaller exponent is shifted right and the exponent is adjusted by increasing the exponent by one for each bit position shifted.

Before alignment

$A =$	0 . 1 1 0 1 0 1 0 0	$\times 2^5$	+26.5
$B =$	0 . 1 0 0 0 1 1 0 0	$\times 2^3$	+4.375

After alignment

$A =$	0 . 1 1 0 1 0 1 0 0	$\times 2^5$	+26.5
$B =$	0 . 0 0 1 0 0 0 1 1	$\times 2^5$	+4.375
$A + B =$	0 . 1 1 1 1 0 1 1 1	$\times 2^5$	+30.875

Figure 1.21 Example of floating-point addition requiring fraction alignment.

Floating-point subtraction The subtraction of two fractions is identical to the subtraction algorithm presented in fixed-point subtraction. If the signs of the operands are the same ($A_{sign} \oplus B_{sign} = 0$) and the operation is subtraction, then this is referred to as *true subtraction* and the fractions are subtracted. If the signs of the operands are different ($A_{sign} \oplus B_{sign} = 1$) and the operation is addition, then this is also specified as *true subtraction*.

All operands will consist of normalized fractions properly aligned with biased exponents. Floating-point subtraction is defined as shown in Equation 1.25 for two numbers A and B, where $A = f_A \times r^{e_A}$ and $B = f_B \times r^{e_B}$. An example of floating-point subtraction is shown in Figure 1.22 for two operands, $A = +36.5$ and $B = +5.75$.

$$
\begin{aligned}
A - B &= (f_A \times r^{e_A}) - (f_B \times r^{e_B}) \\
&= [f_A - (f_B \times r^{-(e_A - e_B)})] \times r^{e_A} \text{ for } e_A > e_B \\
&= [(f_A \times r^{-(e_B - e_A)}) - f_B] \times r^{e_B} \text{ for } e_A \leq e_B
\end{aligned}
\quad (1.25)
$$

Before alignment				
$A =$	0 . 1 0 0 1	0 0 1 0	$\times 2^6$	+36.5
$B =$	0 . 1 0 1 1	1 0 0 0	$\times 2^3$	+5.75
After alignment				
$A =$	0 . 1 0 0 1	0 0 1 0	$\times 2^6$	+36.5
$B =$	0 . 0 0 0 1	0 1 1 1	$\times 2^6$	+5.75
Subtract fractions				
$A =$	0 . 1 0 0 1	0 0 1 0	$\times 2^6$	
+) $B' + 1 =$	0 . 1 1 1 0	1 0 0 1	$\times 2^6$	
1 ←	0 . 0 1 1 1	1 0 1 1	$\times 2^6$	+30.75
Postnormalize (SL1)	0 . 1 1 1 1	0 1 1 0	$\times 2^5$	+30.75

Figure 1.22 Example of floating-point subtraction requiring fraction alignment.

Floating-point multiplication Floating-point multiplication is slightly easier than addition or subtraction, because the exponents do not have to be compared and the fractions do not have to be aligned. For floating-point multiplication, the exponents are added and the fractions are multiplied. Both operations can be done in parallel. Any of the algorithms presented in binary fixed-point multiplication can be used to multiply the floating-point fractions.

Floating-point multiplication is defined as shown in Equation 1.26, which shows fraction multiplication and exponent addition performed simultaneously.

$$\begin{aligned} A \times B &= (f_A \times r^{e_A}) \times (f_B \times r^{e_B}) \\ &= (f_A \times f_B) \times r^{e_A + e_B} \end{aligned} \tag{1.26}$$

Multiplication using the single-precision format generates a double-precision $2n$-bit product; therefore, the resulting fraction is 46 bits. However, the range of a single-precision fraction in conjunction with the exponent is sufficiently accurate so that a single-precision result is usually adequate. Therefore, the low-order half of the fraction can be truncated, which may require a rounding procedure to be performed. The rounding techniques of truncation rounding, adder-based rounding, and von Neumann rounding are covered in Chapter 2.

The sign of the product is determined by the signs of the floating-point numbers. If the signs are the same, then the sign of the product is positive; if the signs are different, then the sign of the product is negative. This can be determined by the exclusive-OR of the two signs, as shown in Equation 1.27.

$$\text{Product sign} = A_{\text{sign}} \oplus B_{\text{sign}} \tag{1.27}$$

The example shown in Figure 1.23 uses the sequential add-shift method with 8-bit operands. A multiplicand fraction *fract_a* $= 0.1010\ 0000 \times 2^3$ (+5) is multiplied by a multiplier *fract_b* $= 0.1100\ 0000 \times 2^2$ (+3) with partial product $D = 0000\ 0000$ to produce a product of *prod* $= 0.1111\ 0000\ 0000\ 0000 \times 2^4$ (+15). Register A contains the normalized multiplicand fraction, *fract_a*; register *prod* contains the high-order n bits of the partial product (initially set to all zeroes); and register B contains the normalized multiplier fraction, *fract_b*.

Since the multiplication involves two n-bit operands, a count-down sequence counter, *count*, is set to a value that represents the number of bits in one of the operands. The counter is decremented by one for each step of the add-shift sequence. When the counter reaches a value of zero, the operation is finished and the product is normalized, if necessary.

If the low-order bit of register *fract_b* is equal to zero, then zeroes are added to the partial product and the sum is loaded into register *prod*. In this case, it is not necessary to perform an add operation — a right shift can accomplish the same result. However, it may require less logic if the same add-shift sequence occurs for each cycle. The sequence counter is then decremented by one. If the low-order bit of register *fract_b* is equal to one, then the multiplicand is added to the partial product. The sum is loaded into register *prod* and the sequence counter is decremented.

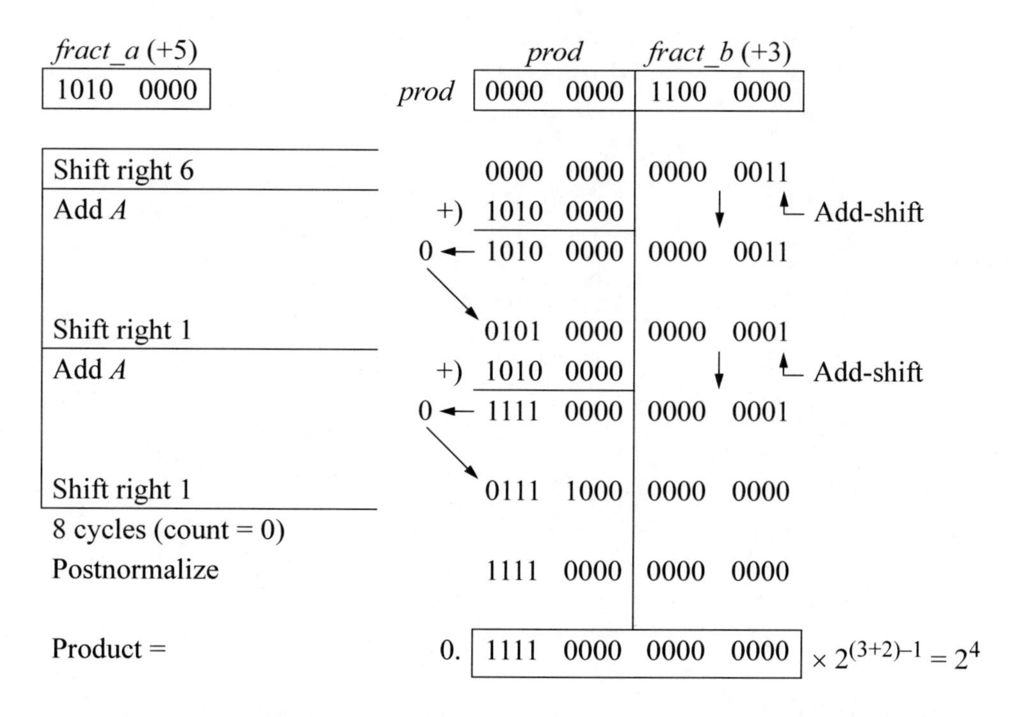

Figure 1.23 Example of floating-point multiplication using the sequential add-shift method for two 8-bit operands.

Floating-point division Floating-point division performs two operations in parallel: fraction division and exponent subtraction. Fraction division can be accomplished using any of the methods for fixed-point division. The dividend is usually $2n$ bits and the divisor is n bits; that is, the dividend conforms to the double-precision format and the divisor conforms to the single-precision format.

Floating-point division is defined as shown in Equation 1.28, which shows fraction division and exponent subtraction performed simultaneously.

$$\begin{aligned} A / B &= (f_A \times r^{e_A}) / (f_B \times r^{e_B}) \\ &= (f_A / f_B) \times r^{e_A - e_B} \end{aligned} \tag{1.28}$$

Figure 1.24 shows floating-point division using the sequential shift-subtract/add restoring division method for a dividend fraction *fract_a* = 0.1000 0110 × 2^7 (+67) and a divisor fraction *fract_b* = 0.1000 × 2^4 (+8) to yield a quotient of 1000 × 2^4 (+8) and a remainder of 0011 × 2^4 (+3).

fract_b (+8) — 1000

fract_a (+67) — 1000 0110

Align — 0100 0011 — $\times 2^{(7+1)} = 2^8$

Shift left 1 — 1000 011–
Subtract *B* — +) 1000
1 ← 0000
No restore — 0000 0111

Shift left 1 — 0000 111–
Subtract *B* — +) 1000
0 ← 1000
Restore — +) 1000
0000 1110

Shift left 1 — 0001 110–
Subtract *B* — +) 1000
0 ← 1001
Restore — +) 1000
0001 1100

Shift left 1 — 0011 100–
Subtract *B* — +) 1000
0 ← 1011
Restore — +) 1000
0011 1000
R *Q* — $\times 2^{(7-4)+1} = 2^4$

Figure 1.24 Example of floating-point division using the sequential shift-subtract/add restoring division method.

Since the high-order half of the dividend is greater than or equal to the divisor, the dividend is shifted right one bit position to prevent overflow. In the first cycle of this 4-cycle example, the dividend is shifted left one bit position and the divisor is subtracted from the high-order four bits of the dividend by adding the 2s complement of the divisor. The difference produces a 0 in the leftmost bit of the partial remainder, indicating that the divisor is less than the corresponding dividend bits. This results in a no-restore cycle in which the partial remainder (0000) is not restored. A 1 bit is placed in the next lower-order quotient bit and appended to the right of the low-order dividend bits.

Then the resulting dividend is shifted left one bit position and the divisor is subtracted from the high-order four bits of the dividend. The result is a partial remainder that is negative — the leftmost bit is 1. Therefore, the high-order four bits of the partial remainder are restored to their previous values by adding the divisor. Then a 0 is placed in the next lower-order bit position of the quotient and appended to the right of the low-order dividend bits.

This restore-shift-subtract cycle repeats for a total of four cycles, resulting in a 4-bit quotient of 1000 and a 4-bit remainder of 0011.

1.3 Problems

1.1 Convert the following unsigned binary numbers to decimal:

(a) $1111\ 0000_2$
(b) $1000\ 0001.111_2$

1.2 Convert the following unsigned binary numbers to decimal and hexadecimal:

(a) $0100\ 1101.1011_2$
(b) $0100\ 1101.1011_2$

1.3 Convert the unsigned binary number $0111\ 1101_2$ to decimal.

1.4 Convert the signed binary number $1100\ 1100_2$ to decimal.

1.5 Convert the octal number 173.25_8 to decimal.

1.6 Convert the binary-coded octal number $110\ 010\ 101_{BCO}$ to decimal.

1.7 Convert the decimal number 4875_{10} to hexadecimal.

1.8 Convert the hexadecimal number $AF6C.B56_{16}$ to decimal.

1.9 Add the following binary numbers to yield a 12-bit sum:

1111 1111 1111
1111 1111 1111
1111 1111 1111
1111 1111 1111

1.10 Obtain the difference of the following binary numbers:

1010 0101 1111
0111 1110 1010

1.11 Convert the hexadecimal number $4A3CB_{16}$ to binary and octal.

1.12 Convert the following octal numbers to hexadecimal: 6536_8 and 63457_8.

1.13 Convert the binary number $0100\ 1101.1011_2$ to decimal and hexadecimal notation.

1.14 Convert 7654_8 to radix 3.

1.15 Represent the decimal numbers +54, –28, +127, and –13 in sign magnitude, diminished-radix complement, and radix complement for radix 2 using eight bits.

1.16 Multiply the unsigned binary numbers 1111_2 and 0011_2.

1.17 Obtain the radix complement of $F8B6_{16}$.

1.18 Obtain the radix complement of 54320_6.

1.19 The numbers shown below are in sign-magnitude representation. Convert the numbers to 2s complement representation for radix 2 with the same numerical value using eight bits.

Sign magnitude	2s complement
1001 1001	
0001 0000	
1011 1111	

1.20 Perform the following binary subtraction using the diminished-radix complement method:

```
    101111
–)  000011
```

1.21 Add the following BCD numbers:

$$\begin{array}{rcc} & 1001 & 1000 \\ +) & 1001 & 0111 \end{array}$$

1.22 Obtain the sum of the following radix 3 numbers:

$$\begin{array}{rcccc} & 1 & 1 & 1 & 1 \\ & 1 & 1 & 1 & 1 \\ & 1 & 1 & 1 & 1 \\ +) & 1 & 1 & 1 & 1 \end{array}$$

1.23 Multiply the two binary numbers shown below, which are in radix complementation.

$$\begin{array}{rccccc} & 1 & 1 & 1 & 1 & 1 \\ \times) & 0 & 1 & 0 & 1 & 1 \end{array}$$

1.24 Let A and B be two binary numbers in 2s complement representation as shown below, where A' and B' are the 1s complement of A and B, respectively. Perform the operation listed below. The answer is to be an 8-bit number in 2s complement representation.

$A = 1011\ 0001$
$B = 1110\ 0100$

$(A' + 1) - (B' + 1)$

1.25 The operands shown below are to be added using decimal (BCD) addition.

$$\begin{array}{ccc} 0111 & 0010 & 0101 \\ 0101 & 0011 & 0110 \end{array}$$

2

X86 Processor Architecture

Microprocessors have evolved considerably over the past four decades, from the Intel 4-bit 4004 to the multi-core processors of today. A multi-core processor consists of two or more separate central processing units (CPUs), called cores. Thus, a quad-core processor consists of four independent CPUs. Multi-core processors can be designed to accommodate very-long-instruction-words (VLIWs), a pipelined reduced-instruction-set computer (RISC), or as a vector processor that operates on sets or vectors. Multi-core processors are designed to operate with multiple threads; that is, to execute multiple tasks simultaneously. Multithreading takes advantage of memory latency in order to improve system performance.

The speed of computers has increased significantly with the advent of multi-core processing and multithreading. IBM recently announced the development of a supercomputer to operate at 20 petaflops; that is, 20×10^{15} floating-point operations per second, but this will undoubtedly be surpassed in the near future. Pipelined RISC processors also increase machine performance by increasing parallelism by fetching, decoding, and executing instructions simultaneously in a multi-stage pipeline.

2.1 General Architecture

All computers have the generic architecture shown in Figure 2.1. This figure does not apply to any specific machine, but is common to all computers. Since there are over twenty different X86 architectures, it would be unreasonable to illustrate all of the architectural variations in this section. Figure 2.1 will be expanded in a subsequent section to more appropriately correspond to the X86 architecture execution environment, including the different sets of registers. A computer has five main functional units: the arithmetic and logic unit (ALU); the control unit (or sequencer), both of which constitute the CPU; the storage unit comprising the main memory and cache; the input devices; and the output devices. The word *memory* will be used throughout the book, although the International Standards Organization (ISO) states that the word *memory* is a deprecated term for main storage, because memory is a human characteristic. However, the word *memory* is used almost exclusively throughout the computer industry.

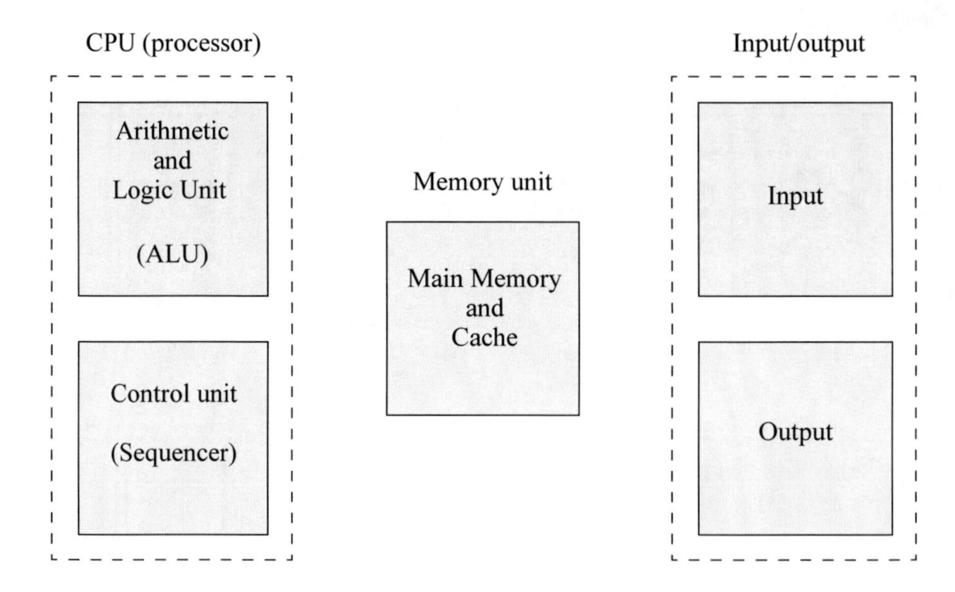

Figure 2.1 General architecture for a computer.

2.2 Arithmetic and Logic Unit

The ALU may consist of a complex-instruction-set computer (CISC), a parallel processor, a pipelined reduced-instruction-set computer (RISC), or any other processor architecture. For a pipelined computer, there are many different stages in a pipeline, with each stage performing one unique operation in the instruction processing. When the pipeline is full, a result is obtained usually every clock cycle.

An example of a simple 4-stage pipeline is shown in Figure 2.2. The *Ifetch* stage fetches the instruction from memory; the *Decode* stage decodes the instruction and fetches the operands; the *Execute* stage performs the operation specified in the instruction; the *Store* stage stores the result in the destination location. The destination is a register file or a reorder queue if instructions are executed out of order. Four instructions are in progress at any given clock cycle. Each stage of the pipeline performs its task in parallel with all other stages.

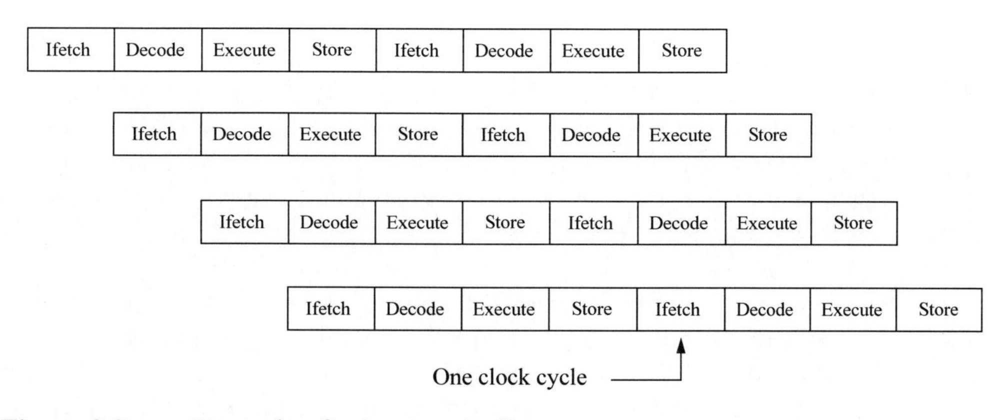

Figure 2.2 Example of a 4-stage pipeline.

If the instruction required is not available in the cache, then a cache miss occurs, necessitating a fetch from main memory. This is referred to as a pipeline *stall* and delays processing the next instruction. Information is passed from one stage to the next by means of a storage buffer, as shown in Figure 2.3. There must be a register in the input of each stage (or between stages) to store information that is transmitted from the preceding stage. This prevents data being processed by one stage from interfering with the following stage during the same clock period.

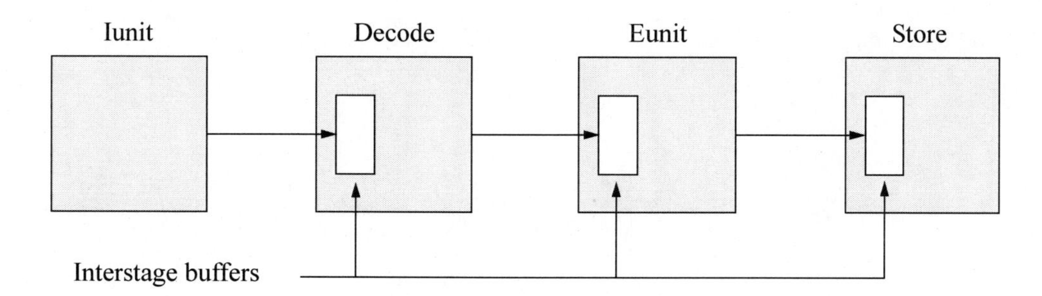

Figure 2.3 Four stages of a pipeline showing the interstage storage buffers.

The ALU performs arithmetic operations, such as add, subtract, multiply, and divide in fixed-point, decimal, and floating-point number representations; increment and decrement; logical operations, such as AND, OR, NOT, exclusive-OR, exclusive-NOR; shifting operations, such as shift right algebraic (SRA), shift right logical (SRL), shift left algebraic (SLA), and shift left logical (SLL); and rotate operations, such as rotate right and rotate left.

2.3 Control Unit

The *control unit*, or control store, is part of the CPU and contains the microprogram, also referred to as microcode. The microprogram is stored in a high-speed memory and accommodates a set of low-level instructions that control the machine's hardware and is machine dependent; that is, it is written for a particular type computer. The instructions in a microprogram represent micro operations that the CPU performs to execute a machine-language instruction. A microprogram is also referred to as *firmware*.

A control word in the microprogram is normally a very-long instruction word (VLIW) — ten bytes or more — to perform the many micro operations required by a CPU instruction, such as effective address generation and load/reset registers. There are two types of control units: hardwired and microprogrammed. A *Hardwired control unit* is too complex for large machines and is inherently inflexible to changes required by design changes or modification of the instruction set. A *microprogrammed control unit* does not have the limitations of a hardwired controller — to change the firmware, simply change the program that resides in a programmable read-only memory (PROM). Figure 2.4 shows the organization of a general microprogrammed control unit.

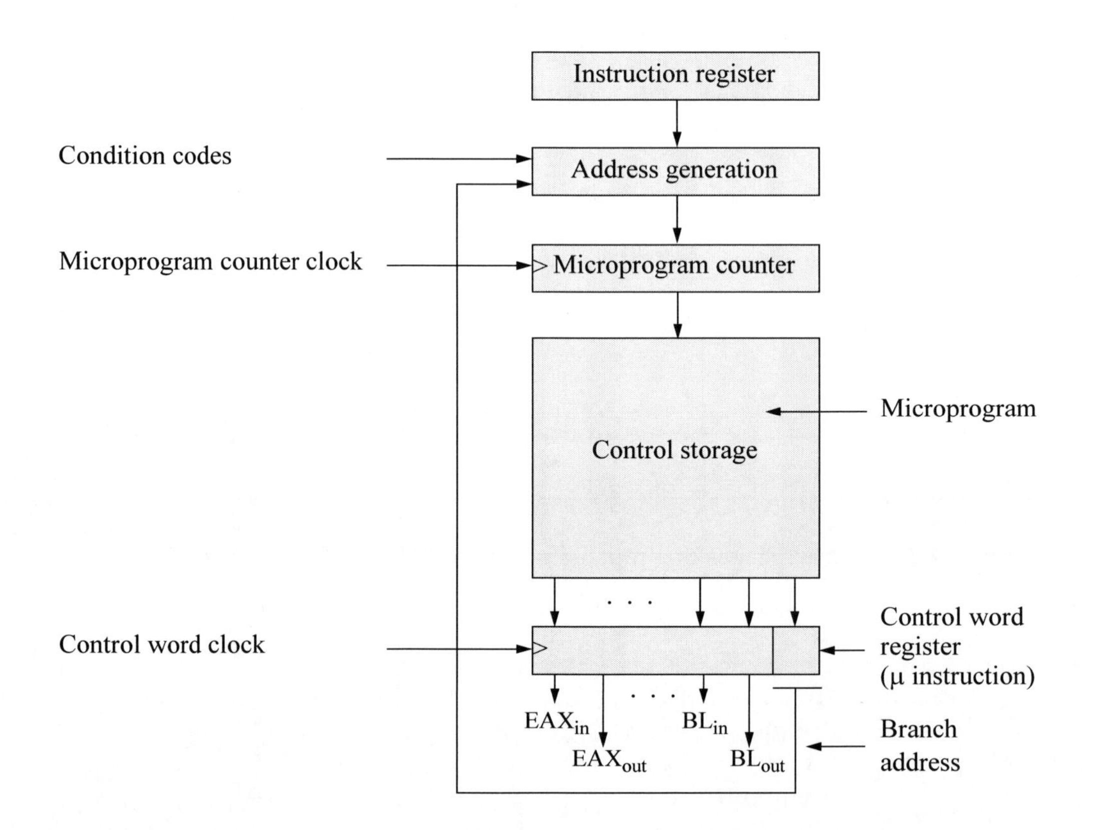

Figure 2.4 Organization of a general microprogrammed control unit.

The instruction register contains the current instruction. The address generation block is set to the starting address of the microprogram for a particular instruction and is a function of the contents of the instruction register, the condition codes, and the branch address of the current control word (microinstruction). The microprogram counter is similar to the program counter (PC) of a computer — it points to the address of the next instruction. The microprogram counter is incremented by 1 or set to the branch address. The control storage block is a PROM that contains the microprogram (or firmware).

The control word register contains the current microinstruction to control the machine's hardware for a specific macroinstruction. Microinstructions are fetched from control storage in a similar manner to instructions fetched from main memory. A microinstruction has two main parts: a *control field* and an *address field*. The control field issues control lines, such as EAX_{in}, which loads the EAX register with data at the next active clock transition; and EAX_{out}, which gates the contents of EAX to the destination bus. The address field (branch address) indicates the address of the next microinstruction in the microprogram if a branch is required.

It is desirable to keep the control word as short as possible to minimize the hardware and yet have as many unique individual bits as possible to obtain high-speed execution of the macro instructions. Microinstructions (control words) are generally classifies as a *horizontal format*, a *vertical format*, or a *combination format*, which is a combination of vertical and horizontal formats.

The horizontal format has no decoding; therefore, it has very long formats. This provides high operating speeds with a high degree of parallelism. An example of the horizontal format is shown in Figure 2.5(a). The vertical format has a large amount of decoding; therefore, it has short formats. It operates at a slower speed due to the inclusion of a decoder and is not highly parallel. An example of the vertical format is shown in Figure 2.5(b). Most microprogrammed computers use a combination of horizontal and vertical formats. An example of the combination format is shown in Figure 2.5(c). Note that decoder output 0 cannot be used, because the field being decoded may be all zeroes.

2.4 Memory Unit

The memory unit consists primarily of main memory and cache. It also contains two registers: memory address register and memory data register. The *memory address register* (MAR) contains the memory address to which data are written of from which data are read. The *memory data register* (MDR) contains the data that is written to memory or read from memory.

2.4.1 Main Memory

The *main memory*, also called *random access memory* (RAM), contains the instructions and data for the computer. There are typically two types RAM: *static RAM* and *dynamic RAM.* Static RAM is designed using flip-flops that store one bit of information. A static RAM does not need refreshing and operates at a higher speed than dynamic RAM, but requires more hardware. Dynamic RAM stores one bit of information in a capacitor and associated hardware. Since the charge in the capacitor leaks and diminishes with time, the charge must be refreshed periodically in order to maintain the state of the data. The density of a dynamic RAM is much greater than the density of a static RAM, but operates at a slower speed.

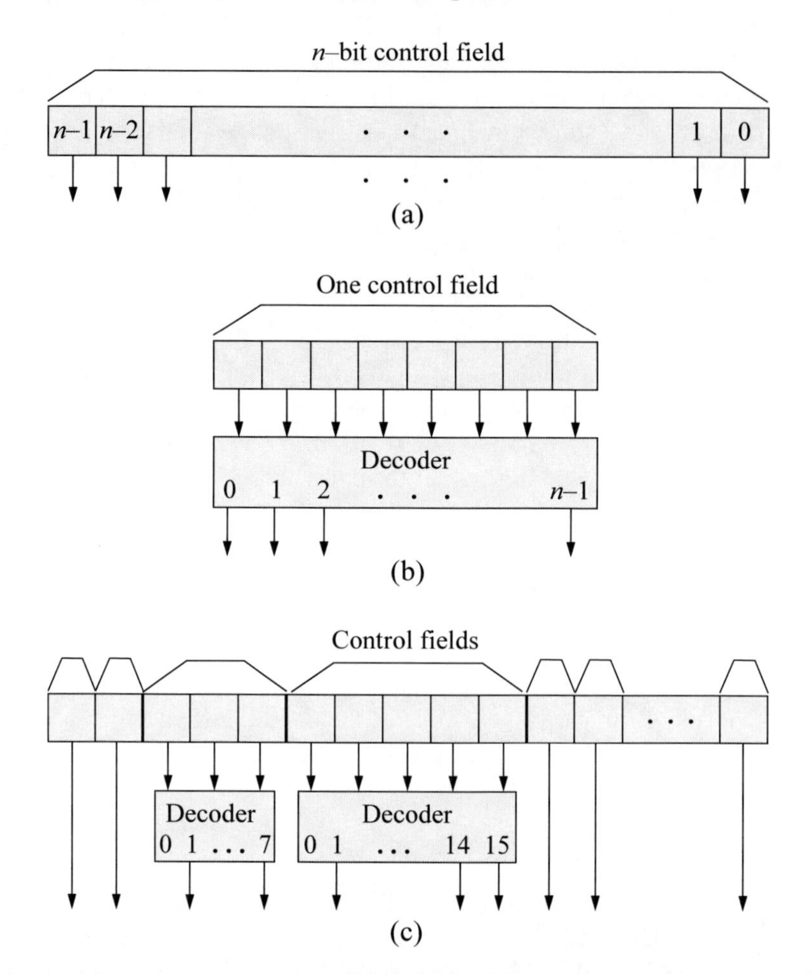

Figure 2.5 Examples of microinstruction formats: (a) horizontal; (b) vertical; and (c) combination.

2.4.2 Hamming Code

Errors can occur in the transmission or storage of information being sent to or from memory. A typical error detection and correction scheme is one developed by Richard W. Hamming. The basic Hamming code can detect single or double errors and can correct single errors. The information sent to memory is coded in the form shown in

Figure 2.6. A *code word* contains n bits consisting of m message bits plus k parity check bits. The m bits represent the information or message part of the code word; the k bits are used for detecting and correcting errors, where $k = n - m$.

Since there can be an error in *any* bit position, including the parity check bits, there must be a sufficient number of k parity check bits to identify any of the $m + k$ bit positions. The parity check bits are normally embedded in the code word and are positioned in columns with column numbers that are a power of two, as shown in Figure 2.7 for a code word containing four message bits (m_3, m_5, m_6, m_7) and three parity bits (p_1, p_2, p_4).

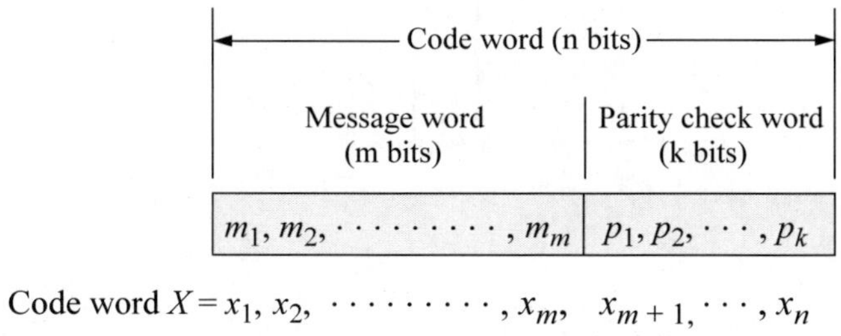

Figure 2.6 Code word to encode message bits using the Hamming code.

Column number	1	2	3	4	5	6	7
Code word =	$\boldsymbol{p_1}$	$\boldsymbol{p_2}$	m_3	$\boldsymbol{p_4}$	m_5	m_6	m_7

Figure 2.7 Code word configuration to encode four bits in Hamming code.

Each parity bit maintains odd parity over a unique group of bits as shown in Figure 2.8 for a code word of four message bits, where E_1, E_2, and E_4 represent the four groups.

Group E_1 =	$\boldsymbol{p_1}$	m_3	m_5	m_7
Group E_2 =	$\boldsymbol{p_2}$	m_3	m_6	m_7
Group E_4 =	$\boldsymbol{p_4}$	m_5	m_6	m_7

Figure 2.8 Parity bit grouping for a code word of seven bits.

The placement of the parity bits in certain columns is not arbitrary. Each of the variables in group E_1 contain a 1 in column 1 (2^0) of the binary representation of the column number as shown in Figure 2.9. Since p_1 has only a single 1 in the binary representation of column 1, p_1 can therefore be used as a parity check bit for a message bit in *any* column in which the binary representation of the column number has a 1 in column 1 (2^0). Thus, group E_1 can be expanded to include other message bits, as shown below.

$$\boldsymbol{p_1}, m_3, m_5, m_7, m_9, m_{11}, m_{13}, m_{15}, m_{17}, \ldots$$

	4	2	1
Group E_1	2^2	2^1	2^0
$\boldsymbol{p_1}$	0	0	1
m_3	0	1	1
m_5	1	0	1
m_7	1	1	1

Figure 2.9 Placement of parity bit $\boldsymbol{p_1}$ for a code word of four bits.

Each of the variables in group E_2 contain a 1 in column 2 (2^1) of the binary representation of the column number as shown in Figure 2.10. Since p_2 has only a single 1 in the binary representation of column 2, p_2 can therefore be used as a parity check bit for a message bit in *any* column in which the binary representation of the column number has a 1 in column 2 (2^1). Thus, group E_2 can be expanded to include other message bits, as shown below.

$$\boldsymbol{p_2}, m_3, m_6, m_7, m_{10}, m_{11}, m_{14}, m_{15}, m_{18}, \ldots$$

	4	2	1
Group E_2	2^2	2^1	2^0
$\boldsymbol{p_2}$	0	1	0
m_3	0	1	1
m_6	1	1	0
m_7	1	1	1

Figure 2.10 Placement of parity bit $\boldsymbol{p_2}$ for a code word of four bits.

Each of the variables in group E_4 contain a 1 in column 4 (2^2) of the binary representation of the column number as shown in Figure 2.11. Since p_4 has only a single 1 in the binary representation of column 4, p_4 can therefore be used as a parity check bit for a message bit in *any* column in which the binary representation of the column number has a 1 in column 4 (2^2). Thus, group E_4 can be expanded to include other message bits, as shown below.

$$\mathbf{p_4}, m_5, m_6, m_7, m_{12}, m_{13}, m_{14}, m_{15}, m_{20}, \ldots$$

	4	2	1
Group E_4	2^2	2^1	2^0
$\mathbf{p_4}$	1	0	0
m_5	1	0	1
m_6	1	1	0
m_7	1	1	1

Figure 2.11 Placement of parity bit $\mathbf{p_4}$ for a code word of four bits.

The format for embedding parity bits in the code word can be extended easily to any size message. For example, the code word for an 8-bit message is encoded as shown below, where $m_3, m_5, m_6, m_7, m_9, m_{10}, m_{11}, m_{12}$ are the message bits and p_1, p_2, p_4, p_8 are the parity check bits for groups E_1, E_2, E_4, E_8, respectively, as shown in Figure 2.12.

$$\mathbf{p_1}, \mathbf{p_2}, m_3, \mathbf{p_4}, m_5, m_6, m_7, \mathbf{p_8}, m_9, m_{10}, m_{11}, m_{12}$$

Group E_1 =	$\mathbf{p_1}$	m_3	m_5	m_7	m_9	m_{11}
Group E_2 =	$\mathbf{p_2}$	m_3	m_6	m_7	m_{10}	m_{11}
Group E_4 =	$\mathbf{p_4}$	m_5	m_6	m_7	m_{12}	
Group E_8 =	$\mathbf{p_8}$	m_9	m_{10}	m_{11}	m_{12}	

Figure 2.12 Parity bit grouping for a code word of twelve bits.

For messages, the bit with the highest numbered subscript is the low-order bit. Thus, the low-order message bit is m_{12} for a byte of data that is encoded using the Hamming code. A 32-bit message requires six parity check bits:

$$p_1, p_2, p_4, p_8, p_{16}, p_{32}$$

There is only one parity bit in each group. The parity bits are independent and no parity bit checks any other parity bit. Consider the following code word for a 4-bit message:

$$p_1, p_2, m_3, p_4, m_5, m_6, m_7$$

The parity bits are generated so that there are an odd number of 1s in the following groups: group E_1, group E_2, and group E_4. For example, the parity bits are generated by the exclusive-NOR function as follows:

$$p_1 = (m_3 \oplus m_5 \oplus m_7)'$$
$$p_2 = (m_3 \oplus m_6 \oplus m_7)'$$
$$p_4 = (m_5 \oplus m_6 \oplus m_7)'$$

Example 2.1 A 4-bit message (0110) will be encoded using the Hamming code then transmitted. The message, transmitted code word, and received code word are shown in Figure 2.13.

	p_1	p_2	m_3	p_4	m_5	m_6	m_7
Message to be sent			0		1	1	0
Code word sent	0	0	0	1	1	1	0
Code word received	0	0	0	1	1	**[0]**	0

Figure 2.13 Detected error in a code word of seven bits.

From the received code word, it is apparent that bit 6 is in error. When the code word is received, the parity of each group is checked using the bits assigned to that group, as shown in Figure 2.14. A parity error is assigned a value of 1; no parity error is assigned a value of 0. The groups are then listed according to their binary weight. The resulting binary number is called the *syndrome word* and indicates the bit in error; in this case, bit 6, as shown in Figure 2.15. The bit in error is then complemented to yield a correct message of 0110.

Group E_1 =	$p_1\, m_3\, m_5\, m_7$ =	0 0 1 0 =	No Error = 0
Group E_2 =	$p_2\, m_3\, m_6\, m_7$ =	0 0 0 0 =	Error = 1
Group E_4 =	$p_4\, m_5\, m_6\, m_7$ =	1 1 0 0 =	Error = 1

Figure 2.14 Generation of the syndrome word for a code word of seven bits.

	2^2	2^1	2^0
Groups	E_4	E_2	E_1
Syndrome word	1	1	0

Figure 2.15 Syndrome word for Example 2.1.

Since there are three groups in this example, there are eight combinations. Each combination indicates the column of the bit in error, including the parity bits. A syndrome word of $E_4E_2E_1 = 000$ indicates no error in the received code word.

Double error detection and single error correction can be achieved by adding a parity bit for the entire code word. The format is shown below, where p_{cw} is the parity bit for the code word.

$$\text{Code word} = \boldsymbol{p_1\, p_2}\, m_3\, \boldsymbol{p_4}\, m_5\, m_6\, m_7\, \boldsymbol{p_{cw}}$$

2.4.3 Cache Memory

The memory block also includes cache memory. Cache memory is used to increase the speed of instruction execution by storing frequently-used information in the form of instructions or data. Each location in cache is referred to as a *cache line*. A cache is a *content-addressable memory* (CAM), because the memory is addressed by contents contained in memory, not by the physical address of the contents. Caches are also referred to as *associative memory.*

There are typically three types of cache memories that are used in current microprocessors: level 1 cache (L1), level 2 cache (L2), and level 3 (L3) cache. The L1 cache, implemented with static RAM, has high-speed access and is the primary cache. It is located on the CPU chip and is used for temporary storage of instructions and data. There are usually two L1 caches: one to store instructions and one to store data.

The L2 cache typically has a larger capacity than the L1 cache but is slower and is logically positioned between the L1 cache and main memory. In many cases, the L2 cache is also located on the CPU chip. The L2 cache is referred to as the secondary cache and is implemented with static RAM. The L2 cache is used to prefetch instructions and data for the processor, thus reducing access time.

The L3 cache is an optional cache and, if used, is usually located on the circuit board logically positioned between the L2 cache and main memory. If the system does have L1, L2, and L3 caches, then the information transfer path is from main memory to L3 to L2 to L1 to CPU.

Figure 2.16 shows the organization of memory priority in relation to the CPU. The highest speed memory is located within the CPU in the form of registers. Next is the cache subsystem (L1, L2, and L3 caches) with slower speeds than the CPU, followed by the relatively low-speed memory but with a much larger capacity. The lowest-speed memory devices are the peripheral I/O units; for example a disk subsystem with the highest capacity.

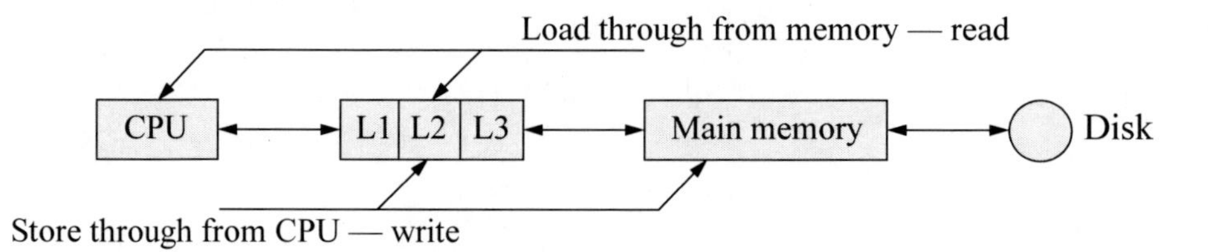

Figure 2.16 Organization of memory priority relative to the CPU.

If the addressed information is not in cache during a read operation from main memory, then a block transfer is performed from main memory to cache. Execution can be increased if the addressed word is passed on to the CPU as soon as it is received from main memory. This is referred to as a *load through*. A *store through* is performed when cache and main memory are updated simultaneously.

When information is transferred from main memory to cache, there are normally three mapping algorithms that can be applied: direct mapping, associative mapping — also referred to as fully associative, and set-associative mapping — also referred to as block-set-associative. When a block is loaded from main memory into cache, a *tag* associated with the main memory location is assigned to the block in cache.

Direct mapping Direct mapping is the most straightforward. Assume that cache has 128 blocks and that there are 4096 blocks in main memory partitioned into 128 blocks per partition. Then block *k* of main memory maps into block *k* modulo-128 of the cache. That is, block 0, block 128, block 256, etc. all map into block 0 of cache. In a similar manner, block 1, block 129, block 257, etc. all map into block 1 of cache.

Associative mapping Associative mapping allows any block in main memory to be loaded into any cache location. This necessitates a large associative cache memory, because any cache location may contain the required information. This mapping technique yields a very high *hit ratio* for an associative search. The hit ratio is the percentage of cache accesses that result in a cache hit. A *cache hit* means that the information is in cache; a *cache miss* means that the information is in main memory or I/O memory.

Set-associative mapping Set-associative mapping combines the advantages of both direct and associative mapping. Blocks — or cache lines — in cache are partitioned into sets. The algorithm assigns a block in main memory to be mapped into any block of a specific set. The cache contains S sets in which each set contains L lines. Let M be the number of lines in cache. Then the set associativity of cache is $L = M/S$ = set size. A 4-way ($L = 4$) set-associative cache with 1024 lines ($M = 1024$) has $S = M/L = 256$ sets. This mapping technique requires an additional field in the memory address: a *set field*. The algorithm first determines which set might contain the desired cache block. Then the tag field in the memory address is compared with the tags in the cache blocks of that set. The technique is direct mapping to set i; however, this is predominantly set-associative mapping, because any memory block to some modulus can map to any line in set i.

Tag field This section describes a general approach for tag comparison to determine if the requested block of data is in cache. Each cache line has a *tag field* associated with the data in the cache line, as shown in Figure 2.17.

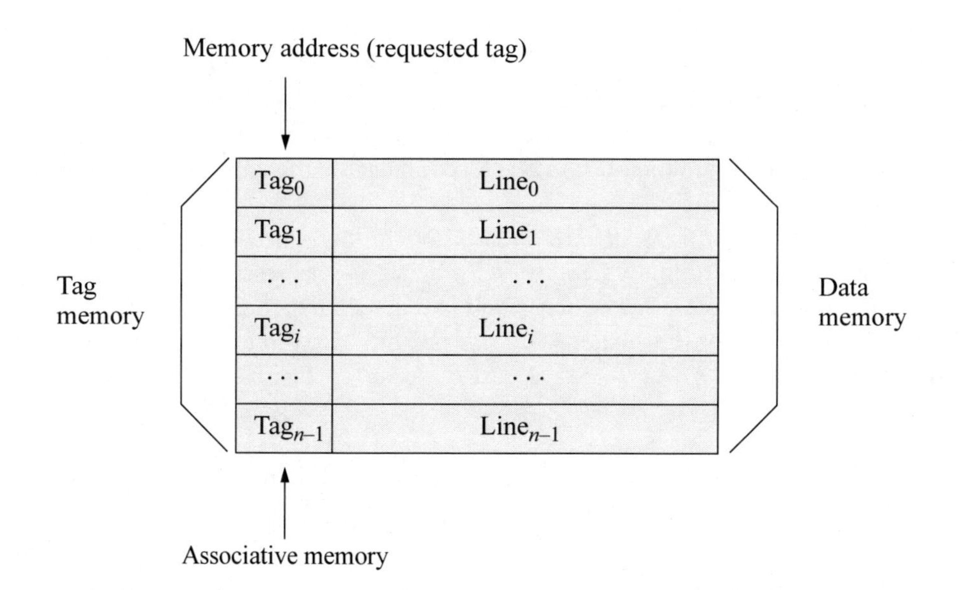

Figure 2.17 The tag associative memory and data cache.

The main memory address consists of three main fields, the tag field, the block field, and the word field which selects the word in the cache. The tag field is usually the high-order bits of the main memory address and connects to the tag memory inputs of a content-addressable memory; thus, data can be identified for retrieval by the contents of the tag field rather than by the address of the data. All tags fields are compared in parallel. If the tag field in cache matches the tag field in the memory address, then a cache hit occurs and the data in that cache line is sent to the CPU. If there is no tag match, then a cache miss occurs and main memory is searched for the required block of data.

Tag fields are used in all three types of mapping: direct mapping, associative mapping, and set-associative mapping. Tag comparison is simply the exclusive-NOR of the tag field bits in the main memory address with the tag field bits in cache. This is the equality function, which generates a logic 1 if the two tags are equal.

There are also two additional bits associated with a cache line: a valid bit and a dirty bit. These bits are usually part of the tag field. The *valid bit* indicates whether a specific cache line contains valid data. When the system is initialized, the valid bit is set to 0. When a cache line is loaded from main memory, the valid bit is set to 1. The *dirty bit* (or *modified bit*) indicates that the cache line has been updated.

Replacement algorithms There are *replacement algorithms* associated with cache. When a new block is to be mapped into a cache that is full, an existing block must be replaced. For direct mapping, there is only one choice: block *k* of main memory maps into block *k* to some modulus of cache and replaces the cache block. For associative and set-associative caches, a different replacement algorithm is used.

One commonly used technique is the *least-recently used* (LRU) algorithm. This method replaces the block in the set which has been in the cache the longest without being referenced. For a two-block set-associative cache, in which there are two blocks per set, each of the two blocks in the set contain a *USE* bit. When a block is referenced, the USE bit for that block is set to 1; the USE bit for the second block is set to 0. When a block is read from main memory to be stored in cache, the cache block whose USE bit equals 0 is replaced.

For a four-block set-associative cache, containing four blocks per set, a 2-bit counter is used to determine which block in the set is least recently used. There are three conditions that determine which block is replaced depending on a cache hit or a cache miss and whether the cache is full or not full. The three conditions are shown below.

1. If there is a cache hit on set_i $block_j$, then the following operations take place:
   If $block_k$ counter < $block_j$ counter, then increment $block_k$ counter.
   If $block_k$ counter ≥ $block_j$ counter, then do not increment $block_k$ counter.
   Set $block_j$ counter = 0.

2. If there is a cache miss and the set is not full, then the following operations take place:
   Memory block is stored in set_i $block_j$ (empty block).
   Set $block_j$ counter = 0.
   Increment all other counters.

3. If there is a cache miss and the set is full, then the following operations take place:
   If $block_j$ counter = 3, then remove $block_j$.
   Memory block is stored in set_i $block_j$ (empty block).
   Set $block_j$ counter = 0.
   Increment all other counters.

Another replacement algorithm is called *first-in, first-out*. This technique replaces the block in the set which has been in the cache the longest and is implemented with a circular buffer technique, as shown below.

	Tag	Data		
		Block 1	Load block 1 first	Remove block 1 first
Set_i		Block 2	Load block 2 second	Remove block 2 second
		Block 3	Load block 3 third	Remove block 3 third
		Block 4	Load block 4 fourth	Remove block 4 fourth

A fourth replacement algorithm is called *random replacement.* A block is randomly chosen to be replaced. This is a very effective method and is only slightly inferior to the other algorithms.

2.5 Input/Output

The input/output (I/0) block of Figure 2.1 incorporates I/O peripheral devices and associated control units. Some examples of I/O device subsystems are tape, disk, keyboard, monitors, plotters, compact disks, and devices attached to a computer by means of a universal serial bus (USB). Many computers also incorporate I/O processors — called channels — to control all data and command transfers between the computer (or memory) and the I/O device. The channels are integrated into the CPU or are standalone units.

2.5.1 Tape Drives

Tape drives A tape drive is sequential access storage device (SASD) and is a low-cost means of obtaining large storage. The disadvantage is that it has long access times, because it reads or writes data sequentially; that is, it cannot go directly to a record on tape, but must progress through previous records to obtain the desired record, depending on the location of the read/write heads. Figure 2.18 shows a typical tape drive head with one write head and one read head per track. Most tape subsystems utilize a *read-after-write* (also called *read-while-write*) procedure to verify that the data were written correctly.

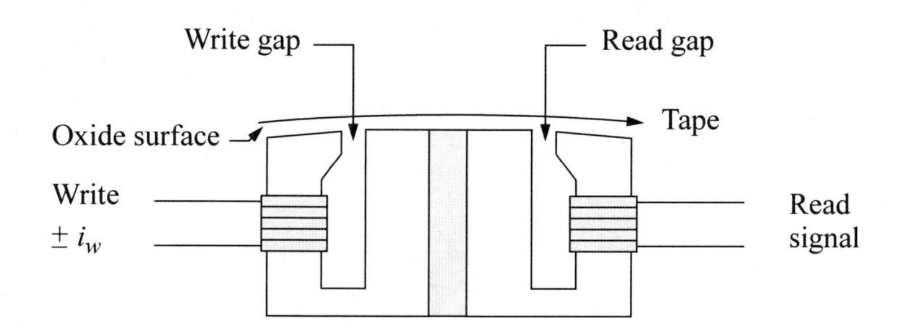

Figure 2.18 Typical read/write head for a tape drive.

The number of tracks per tape varies depending on the format used for the data. There are single-track drives, drives with nine tracks (one byte plus parity) for the Extended Binary-Coded Decimal Interchange Code (EBCDIC) for legacy systems, and drives with seven tracks (including parity) for BCD formats. Most current personal computer (PC) drives use the seven-track American Standard Code for Information Interchange (ASCII) format (0 through 127 decimal), which can be extended to include the remaining codes (128 through 255 decimal).

Data stored on tape can be in a variety of formats, depending on the manufacturer. A typical format is shown in Figure 2.19. The *preamble* is a unique combination of 1s and 0s, which are used to synchronize detection circuits to distinguish 1s from 0s. The *postamble* serves two purposes: It signals the end of data; and it serves as a preamble which permits a read backward operation — the functions of the preamble and postamble are then reversed. The error correcting code (ECC) is used to detect and correct errors in the data stream. The longitudinal redundancy check (LRC) is a form of horizontal redundancy check that is applied to each track. The interrecord gap (IRG) povides a space for starting and stopping the tape. It allows sufficient time to decelerate and accelerate the tape between records.

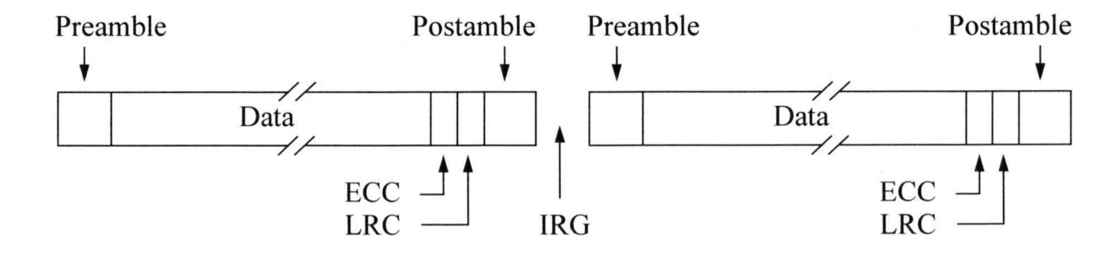

Figure 2.19 Typical tape drive track format.

A tape with many small records requires a large number of IRGs. This decreases the efficiently of tape utilization. This inefficiency can be reduced by grouping the records into blocks with no IRGs between records. The IRGs are placed between the blocks and are called interblock gaps (IBGs). A block and file format is shown in Figure 2.20. There is only one preamble and postamble per block. An *end of file* gap is used to separate files. This is usually blank tape followed by a tape mark (7F hexadecimal).

2.5.2 Disk Drives

Current disk drives have a *mean time between failures* (MTBF) of approximately 40 years. The data rate can be calculated using Equation 2.1. The circumference as a function of the radius is required in order to obtain the data rate in bits per second. The radius changes with each track; however, the density also changes if the method of writing is *fixed data transfer rate* with *varying linear density.* As the radius increases toward the outer tracks. the density decreases. Thus, the data rate remains constant.

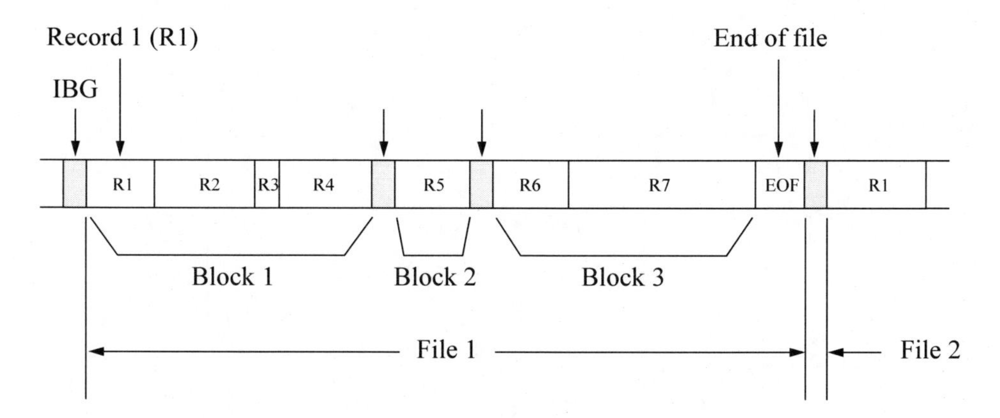

Figure 2.20 Typical block and file format for a tape drive.

$$\text{Data rate (bits/sec)} = \text{Density (bits/in) (rev/min) (min/60 sec)} (2\pi r\text{in/rev}) \qquad (2.1)$$

The read/write head is not in contact with the surface of the disk. The linear density depends on the amount of separation between the head and the disk — higher density requires closer separation. To achieve close separation, the head is *flown* on an *air bearing* and is referred to as the *flying height.* The bearing is simply a cushion of air that is dragged along by the rotating surface. The disk and head in current disk drives are usually contained in a sealed *head-disk assembly* (HDA).

Since the read/write head is expensive, there is usually only one head and one gap per surface. Therefore, it is not possible to read while writing to verify that the data were written correctly, as in tape drives. There are numerous track formats, depending on the manufacturer. One format is shown in Figure 2.21, which contains two fields for each fixed-length sector: the address field and the data field.

The preamble for both the address and the data field consists of two bytes — fifteen 0s followed by a single 1 — that are used for clock synchronization and to differentiate between 1s and 0s. The second two-byte segment of the address field contains the address of the cylinder, head, and sector. This is followed by two bytes containing an *error correcting code* (ECC). A common error correcting code is a *cyclic redundancy check* (CRC) code. Cyclic redundancy check codes can detect both single-bit errors and multiple-bit errors and are especially useful for large strings of serial binary data found on single-track storage devices, such as disk drives. They are also used in serial data transmission networks and in 9-track magnetic tape systems, where each track is treated as a serial bit stream.

The postamble consists of sixteen 0s and is used to separate the address and data fields. It is possible to verify a sector address then read or write the data in that sector by switching from read mode to write mode between the address and data fields.

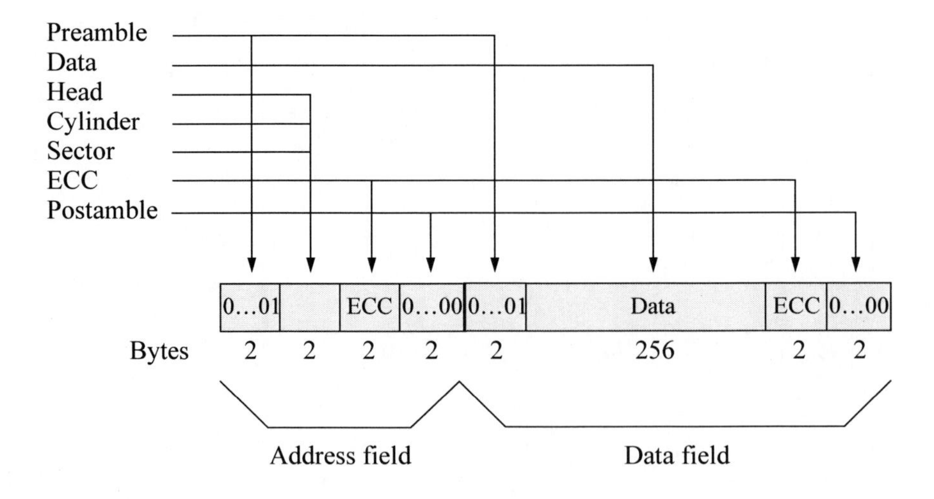

Figure 2.21 An example of a sector format for a disk drive.

The areal density (bits/in^2) can be increased by using thin film heads which are manufactured using semiconductor technology. This technology can double the areal density and has a flying height of less than one-half the wavelength of red light — approximately 0.000075 centimeters.

2.6 Register Set

Many of the X86 microprocessors use a *superscalar* architecture that executes more than one instruction per clock cycle in a parallel implementation. A superscalar processor has multiple parallel pipelines in which each pipeline processes instructions from a separate instruction thread. Superscalar processors are different from multicore processors which process multiple threads from multiple cores, one thread per core. Superscalar processors are also different than pipelined processors, which operate on multiple instructions in various stages of execution.

There are a variety of X86 processors from Intel Corporation, Advanced Micro Devices (AMD), and many other microprocessor companies. Intel processors range from the Pentium family to the dual-core and quad-core processors that support 32- and 64-bit architectures. Figure 2.22 shows a typical X86 processor register set used for most common applications in the IA-32 processor basic execution environment. There are other registers that are not shown, such as those used for memory management; single-instruction multiple-data (SIMD); packed floating-point operations; control registers; debug registers; machine check registers; and the 64-bit registers with R prefix, among others.

General-Purpose Registers

32-bit mode	31 ... 16	15 ... 8	7 ... 0	16-bit mode
EAX		AH	AL	AX
EBX		BH	BL	BX
ECX		CH	CL	CX
EDX		DH	DL	DX
ESP		SP		
EBP		BP		
ESI		SI		
EDI		DI		

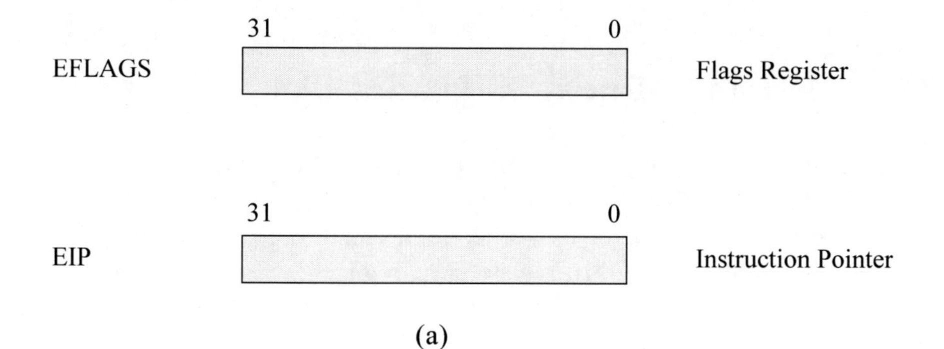

(a)

Figure 2.22 Typical X86 register set for most common applications in the IA-32 basic execution environment: (a) general-purpose registers and (b) floating-point registers.

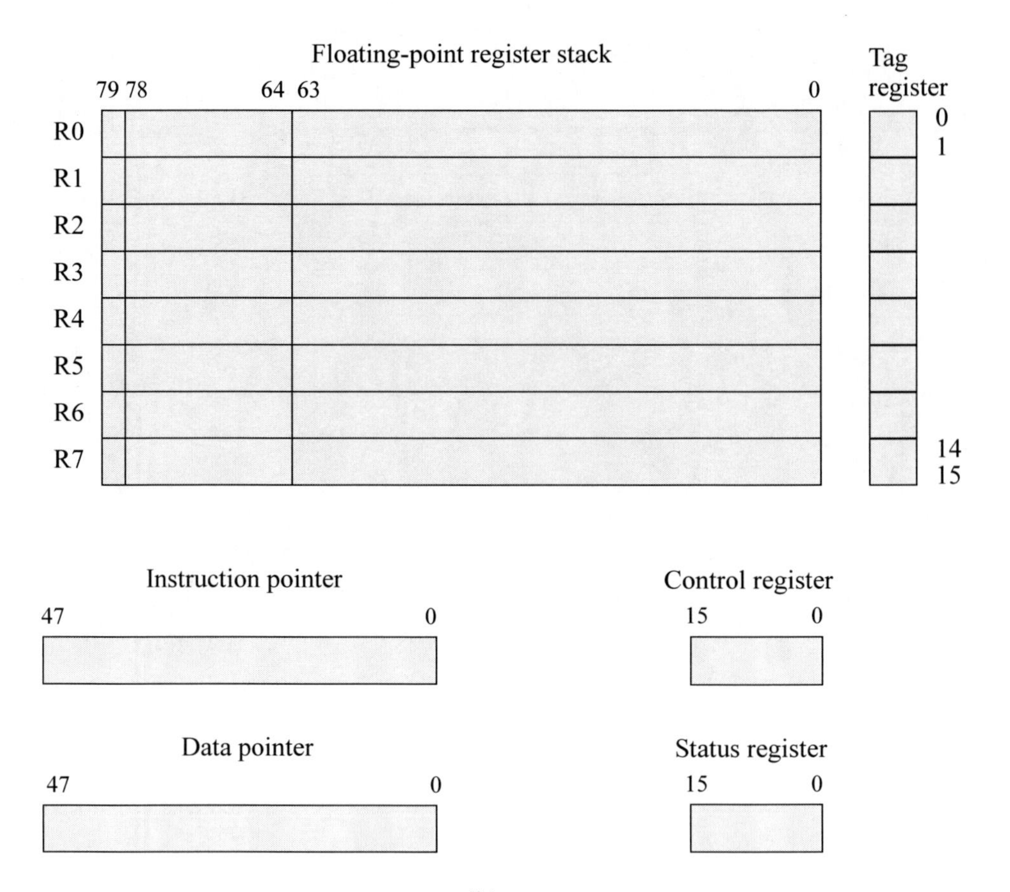

Figure 2.22 (Continued)

2.6.1 General-Purpose Registers

The eight *general-purpose registers* (GPRs) can be addressed separately as 8-bit registers, as 16-bit registers, or as 32-bit registers, such as AL, AX, or EAX, for example. Register AL specifies the low-order eight bits of register AX; register AH specifies the high-order eight bits of register AX; register AX specifies the low-order 16 bits of register EAX. The general-purpose registers are used for arithmetic operations, logical operations, and for addressing memory.

Although they are referred to as general-purpose registers, each register has a specific application. Register EAX and its constituent registers are used as the accumulator; specifically, as the dividend for unsigned and signed divide operations, as the multiplicand for unsigned and signed multiply operations, and for input/output operations. Register AL is also used for the translate (XLAT) instruction.

Register EBX and its constituent registers are used as index registers for base addressing and as pointers to data in the data segment (DS). The effective address (EA) is the sum of a displacement value and the contents of one of the EBX registers. When these registers are used, the default location of the operand resides in the current data segment.

The ECX register and its constituent registers are the implied counters for certain instructions; for example, to count the number of iterations in a loop. The CL register is also used to indicate the shift amount when shifting or rotating bits left or right. For string operations, the ECX register set denotes the number of operands that are moved from the source string to the destination string.

The EDX register and its associated registers — generally referred to as the data registers — are used for input/output operations and for multiply and divide operations for both unsigned and signed operands. The multiplicand (destination operand — AL, AX, or EAX) is multiplied by the multiplier (source operand in a GPR or memory) and the product is stored in the destination operand. The product is stored in register AX, concatenated register pair DX:AX, or concatenated register pair EDX:EAX, depending on the size of the $2n$-bit result. The destination operand in 64-bit mode is RDX:RAX. For a divide operation, the $2n$-bit dividend in registers AX, DX:AX, EDX:EAX, or RDX:RAX is divided by the n-bit divisor (source operand in a GPR or memory). The resulting quotient is stored in AL, AX, EAX, or RAX. The remainder is stored in register AH, DX, EDX, or RDX.

Register SP (for 16-bit stack words) or register ESP (for 32-bit stack doublewords) are stack pointers that point to the current top of stack by providing an offset from the stack segment register address to point to the last valid entry in the stack. Figure 2.23 shows a stack of 16-bit words using SP as the stack pointer. When a word is pushed onto the stack in this figure, SP is first decremented by two, then the word is stored on the stack. SP points to the new top of stack (TOS) containing the last valid entry. When a word is popped off the stack, the word is first stored in the destination, then SP is incremented by two.

The BP and EBP registers are base pointers that are used to point to data on the stack; for example, parameters pushed onto the stack by one program to be accessed by another program via the stack. The base pointer registers provide an offset from the stack segment register address to point to the required data.

The SI and ESI registers are source index registers that are used for string operations. They provide an offset in the data segment (DS) that points to the beginning of a source string that is to be transferred to a destination string in the extra segment (ES).

The DI and EDI registers are destination index registers that are used for string operations. They provide an offset in the extra segment that points to the beginning address of a destination string that is to be transferred from a source string in the data segment.

2.6.2 Segment Registers

When using a *segmented memory model* in the IA-32 processors, memory is partitioned into independent segments for code, data, and stack. The code segment (CS) register — segment selector — contains the starting address of the computer's code segment. The offset in the instruction pointer (IP or EIP) points to a particular instruction in the code segment. The combination of the segment selector and the offset is the logical address of the next instruction to be executed.

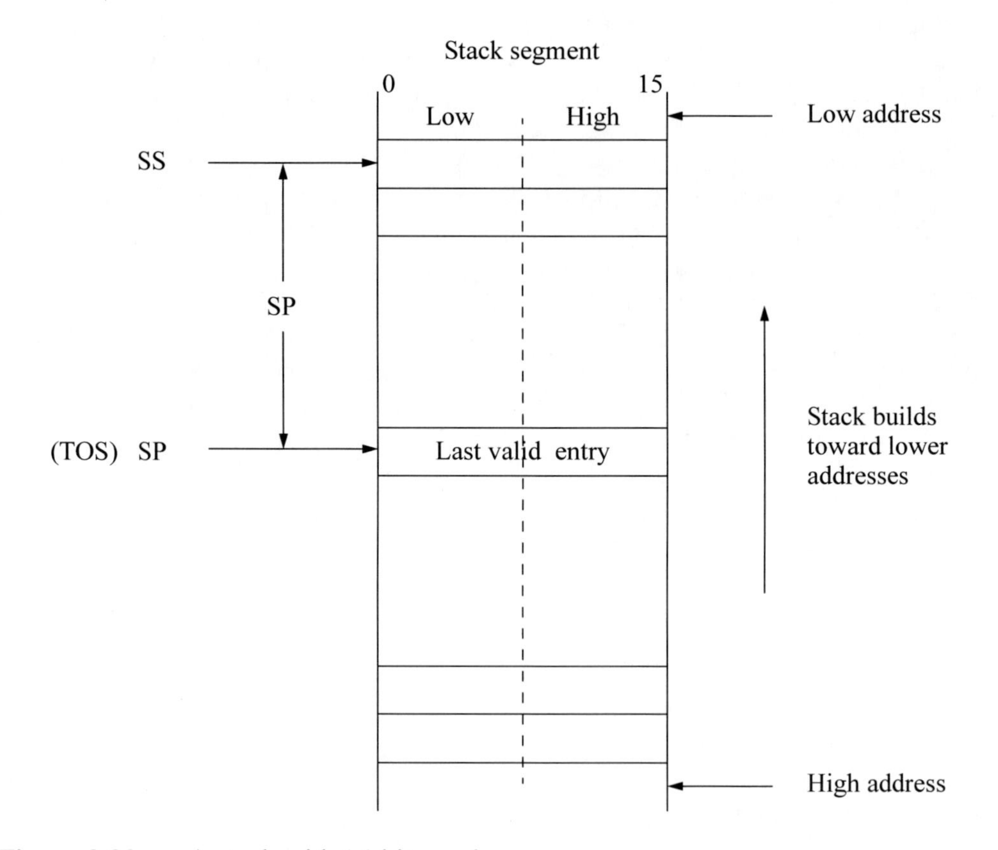

Figure 2.23 A stack with 16-bit words.

There are four data segments: DS, ES, FS, and GS. The data segment (DS) register contains the starting address of the program's data segment. This address plus the offset value indicates the logical address of the required data. The contents of the extra segment (ES) register plus an offset in the destination index (DI) register points to a destination address in the extra segment for string operations. The FS and GS registers are available for different data structures.

2.6.3 EFLAGS Register

The 32-bit EFLAGS register is partitioned into three groups of flags: status flags, control flags, and system flags, as shown in Figure 2.24.

31	30	29	28	27	26	25	24
0	0	0	0	0	0	0	0

23	22	21	20	19	18	17	16
0	0	ID	VIP	VIF	AC	VM	RF

15	14	13	12	11	10	9	8
0	NT	IOPL		OF	DF	IF	TF

7	6	5	4	3	2	1	0
SF	ZF	0	AF	0	PF	1	CF

Figure 2.24 EFLAGS register.

Status flags The carry flag (CF) is set if there is a carry out of or a borrow into the high-order bit position of an arithmetic operation; otherwise it is reset. It is also used with shift and rotate instructions.

The parity flag (PF) is set if the low-order byte of the result has an even number of 1s; otherwise it is reset. The adjust flag (AF or auxiliary carry) is set if there is a carry out of or a borrow into the low-order four bits — used primarily for a binary-coded decimal arithmetic operation; otherwise it is reset. The zero flag (ZF) is set if the result is zero; otherwise it is reset. The sign flag (SF) is set to the value of the high-order bit position, which is the sign bit for signed operands. If the sign flag is 0, then the result is positive; if the sign flag is 1, then the result is negative.

The overflow flag (OF) is set if the result of an operation is too large or too small to be contained in the destination operand; that is, the number is out of range for the size of the result. An *interrupt-on-overflow* (INTO) instruction generates an interrupt if the overflow flag is set. The equations to detect overflow for two n-bit operands A and B are shown in Equation 2.2, where a_{n-1} and b_{n-1} are the sign bits of A and B, respectively and s_{n-1} is the sign of the result. The carry bits out of positions $n-1$ and $n-2$ are c_{n-1} and c_{n-2}, respectively.

$$\text{Overflow} = a_{n-1}\, b_{n-1}\, s_{n-1}' + a_{n-1}'\, b_{n-1}'\, s_{n-1}$$

$$\text{Overflow} = c_{n-1} \oplus c_{n-2} \tag{2.2}$$

Control flag The direction flag (DF) indicates the direction for string operations and is used for the following string instructions: move string (MOVS), compare strings (CMPS), scan string (SCAS), load string (LODS), and store string (STOS). If the direction flag is reset, then this causes the string instruction to auto-increment the index registers (SI and DI); that is, to process strings from left (low address) to right (high address). If the direction flag is set, then this causes the string instruction to auto-decrement the index registers (SI and DI); that is, to process strings from right (high address) to left (low address).

System flags The system flags are used to control the operating system and CPU operations. For example, if the interrupt enable flag (IF) is set and an interrupt occurs, then this causes the CPU to transfer control to a memory location specified by an interrupt vector. The CPU can temporarily ignore maskable interrupts, but must respond immediately to non-maskasble interrupts, such as non-recoverable hardware errors, time-critical interrupts, or power failure. If the interrupt enable flag is reset, then interrupts are disabled.

The trap flag (TF) allows single-step mode for debugging; that is, single instruction execution. The CPU generates an interrupt after each instruction so that the program and results can be examined.

The nested task (NT) flag is set if the present task is nested within another task. The two bits of the input/output privilege level (IOPL) flag contain the I/O privilege level of the current task and supports multitasking. An IOPL of 00 is the highest level; an IOPL of 11 is the lowest level. The privilege levels are used with the privileged instructions, such as input/output instructions and segment accessibility.

The resume flag (RF), when set, indicates to the CPU to resume debugging. The debug software sets the flag prior to returning to the interrupted program. When the virtual-8086 mode (VM) flag is set, the CPU emulates the program environment of the 8086 processor; when reset, the processor returns to protected mode. Protected mode permits the CPU to use virtual memory, paging, and multi-tasking features.

The alignment check (AC) flag is used to check the alignment of memory references. It is set if a word (16 bits) or doubleword (32 bits) is not on a word or doubleword boundary. The virtual interrupt flag (VIF) is a virtual image of the interrupt flag and is used in combination with the virtual interrupt pending (VIP) flag. The VIP flag is set if an interrupt is pending and reset if there is no interrupt pending.

If a program has the capability to set or reset the identification (ID) flag, then the processor supports the CPU identification (CPUID) instruction. The CPUID instruction returns the processor identification in certain general registers.

2.6.4 Instruction Pointer

The instruction pointer (EIP) contains the offset for the current code segment that points to the next instruction to be executed. The EIP is updated as shown below for sequential execution, where (EIP) specifies the '*contents of*' EIP. The EIP register is used for 32-bit mode; the RIP register is used for 64-bit mode.

$$\text{EIP} \leftarrow \text{(EIP) + instruction length}$$

The EIP progresses from one instruction to the next in sequential order, or forwards/backwards a specified number of instructions for jump (JMP), jump on condition (J_{CC}), return from a procedure/subroutine (RET), or return from interrupt (IRET) instructions.

2.6.5 Floating-Point Registers

The floating-point unit (math coprocessor) provides a high-performance floating-point processing component for use with the IEEE Standard 754. It provides support for floating-point, integer, and binary-coded decimal operands for use in engineering and other applications. The floating-point registers were shown in Figure 2.22(b).

Data registers There are eight 80-bit data registers, R0 through R7, whose functions are similar to that a stack. Any R_i can be assigned as top of stack (T0S). Data are stored in these registers in the double extended-precision floating-point format consisting of a sign bit (79), a 15-bit exponent field (78 through 64), and a 64-bit significand/fraction (63 through 0). The implied 1 is to the immediate left of bit position 63, but is not shown. Data that are loaded into the floating-point data registers are converted to the double extended-precision floating-point format.

There are no push and pop instructions the floating-point register stack. A load (push) operation is accomplished decreasing the stack top by 1, then loading the operand into the new top of stack (T0S). This is similar to a regular stack which builds toward lower addresses. A store (pop) operation is accomplished by storing the operand from the current stack top to the destination, then increasing stack top by 1. If a load is to be executed when the stack top is R0, then the registers wrap around making the new stack top R7. This may generate an overflow condition if the contents of R7 were not previously saved.

Tag register The 16-bit tag register specifies the condition of the individual data registers. There are two bits per register that are defined as shown in Figure 2.25. A tag of 10 indicates a special floating-point number, such as not-a-number (NaN), a value of infinity, a denormal number, or unsupported format.

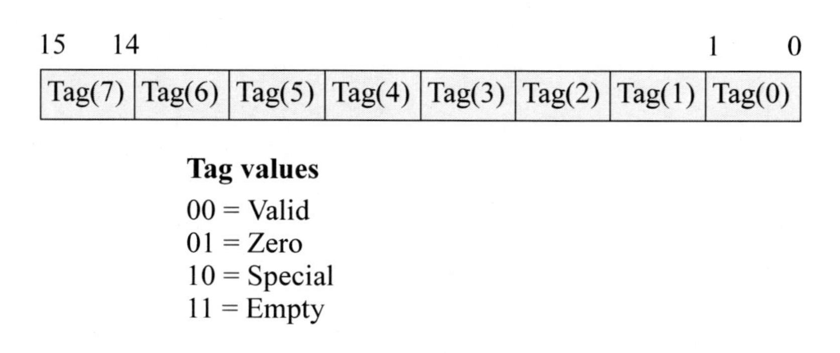

Figure 2.25 Floating-point unit tag register organization.

A denormal number has a biased exponent of zero. NaN is a value that does not depict a numeric or infinite quantity. It is a value that is generally produced as the result of an arithmetic operation using invalid operands. For example, calculating the square root of a negative number. NaNs can also be used to represent missing values in arithmetic calculations.

Status register The format for the 16-bit status word register is shown in Figure 2.26. The status word indicates the condition of the floating-point unit. It can be inspected by first storing the status word into a memory location and then transferring it to a general register. If a bit is set, then this indicates an active condition.

15	14	13	12	11	10	9	8
B	C3		TOS		C2	C1	C0

7	6	5	4	3	2	1	0
ES	SF	PE	UE	OE	ZE	DE	IE

Figure 2.26 Floating-point unit status word format.

Bit 15 indicates that the floating-point unit is busy (B). The condition code bits C3 through C0 are specified by bits 14, 10, 9, and 8, respectively. They represent the result of certain floating-point arithmetic operations. The condition code is used primarily for conditional branching. Bits 13 through 11 indicate the top of stack (TOS) pointer, as shown below.

Bits				
13	12	11		Stack top
0	0	0	=	Register 0
0	0	1	=	Register 1
0	1	0	=	Register 2
0	1	1	=	Register 3
1	0	0	=	Register 4
1	0	1	=	Register 5
1	1	0	=	Register 6
1	1	1	=	Register 7

Bit 7 is the exception summary status (ES) bit, which is set if any of the unmasked exception flags — bits 5 through 0 — in the status word are set. The exception flags in the status word can be prevented from being set by setting the corresponding exception mask bit in the control word register.

Bit 6 is the stack fault (SF) bit, which is set if a stack overflow or stack underflow has been detected. Bit 5 is the precision exception (PE) bit, which is set if the result of an operation cannot be precisely represented in the destination format. Bit 4 is the underflow exception (UE) bit, which is set whenever the rounded normalized result of an arithmetic operation is too small to be contained in the destination operand. Bit 3 is the overflow exception (OE) bit, which is set whenever the rounded normalized result of an arithmetic operation is too large to be contained in the destination operand. Bit 2 is the divide-by-zero exception (ZE) bit, which is set whenever an instruction attempts to divide an operand by zero, where the dividend is a nonzero operand.

Bit 1 is the denormalized operand exception (DE) bit. If the biased exponent of a floating-point number is zero, then extremely small numbers can be represented by setting the high-order bits of the fraction (significand) to zero, thus allowing very small numbers to be realized. Numbers in this range are referred to as *denormalized numbers*. A number that is denormalized may result in a loss of precision. If the exponent is zero and the significand is also zero, then this is usually specified as a value of zero. Therefore, a number is denormalized if the exponent is zero and the fraction is not all zeroes. Denormalized numbers are located around the value of zero.

Bit 0 is the invalid operation exception (IE), which is set whenever there is a stack overflow, a stack underflow, or when an invalid arithmetic operand has been detected. When the stack fault bit is set in the status word register, the invalid operation was caused by a stack overflow or underflow. When the stack fault bit is reset in the status word register, the invalid operation was caused by an invalid operand.

Control register The format for the control word register is shown in Figure 2.27. The control word provides several processing options by loading a word from memory into the 16-bit control word register. Blank entries in the register are reserved. The control word register manages the precision, rounding methods, and masking. The control word register is initialized to the following states: round to nearest, 64-bit double extended precision, and set the six exception mask flags to one.

15	14	13	12	11	10	9	8
			IC	RC		PC	

7	6	5	4	3	2	1	0
		PM	UM	OM	ZM	DM	IM

Figure 2.27 Floating-point unit control word format.

Bit 12 is the infinity control (IC) bit, which is used to allow compatibility with older versions of floating-point units. Bit 11 and bit 10 are the rounding control field bits that determine the rounding method used in arithmetic operations. The four different types of rounding techniques are shown in Table 2.1 and explained in the paragraphs that follow.

Table 2.1 Rounding Control Field

Rounding Method	11	10
Round to nearest (default mode)	0	0
Round down toward minus infinity	0	1
Round up toward positive infinity	1	0
Round toward zero (truncate)	1	1

The *round to nearest* mode is the default rounding method, which provides the nearest approximation to the result. The result of a floating-point arithmetic operation can be rounded to the nearest number that contains n bits. This method is also called *adder-based rounding* and rounds the result to the nearest approximation that contains n bits. The operation is as follows: The bits to be deleted are truncated and a 1 is added to the retained bits if the high-order bit of the deleted bits is a 1. When a 1 is added to the retained bits, the carry is propagated to the higher-order bits. If the addition results in a carry out of the high-order bit position, then the fraction is shifted right one bit position and the exponent is incremented.

Consider the fraction $0.b_{-1}\, b_{-2}\, b_{-3}\, b_{-4}\, 1\, x\, x\, x$ — where the xs are 0s or 1s — which is to be rounded to four bits. Using the adder-based rounding technique, this rounds to $0.b_{-1}\, b_{-2}\, b_{-3}\, b_{-4} + 0.0001$. The retained bits of fraction $0.b_{-1}\, b_{-2}\, b_{-3}\, b_{-4}\, 0\, x\, x\, x$ round to $0.b_{-1}\, b_{-2}\, b_{-3}\, b_{-4}$. The first fraction approaches the true value from above; the second fraction approaches the true value from below. Examples of adder-based rounding are shown in Figure 2.28 and illustrate approaching the true value from above and approaching the true value from below.

```
                                   Delete
                  0. 0 1 0 1 | [1] 0 0 0 |  × 2^8 = +88    True value
              +)  0. 0 0 0 1
Rounded result    0. 0 1 1 0                × 2^8 = +96    Approach from above
                                (a)

                                   Delete
                  0. 0 1 1 1 | [0] 1 1 1 |  × 2^8 = +119   True value
                        ↓
Rounded result    0. 0 1 1 1                × 2^8 = +112   Approach from below
                                (b)
```

Figure 2.28 Adder-based rounding: (a) true value approached from above and (b) true value approached from below.

The rounding mode of *round down toward minus infinity* (also referred to as *directed rounding*) produces a solution that is nearest to but no greater than the result. This technique rounds towards negative infinity. It stipulates that the result of an arithmetic operation should be the value closest to negative infinity; that is, a value that is algebraically less than the precise result.

The rounding mode of *round up toward positive infinity* (also referred to as *directed rounding*) produces a solution that is nearest to but no less than the result. This technique rounds towards positive infinity. It stipulates that the result of an arithmetic operation should be the value closest to positive infinity; that is, a value that is algebraically greater than the precise result.

The rounding mode of *round toward zero* is also referred to as *truncation* or *chopping*. Truncation deletes extra bits and makes no changes to the retained bits. This makes the truncated value less than or equal to the original value. Aligning fractions during addition or subtraction could result is losing several low-order bits, so there is obviously an error associated with truncation. Assume that the following fraction is to be truncated to four bits:

$$0.b_{-1}\ b_{-2}\ b_{-3}\ b_{-4}\ b_{-5}\ b_{-6}\ b_{-7}\ b_{-8}$$

Then all fractions in the range $0.b_{-1}\ b_{-2}\ b_{-3}\ b_{-4}\ 0000$ to $0.b_{-1}\ b_{-2}\ b_{-3}\ b_{-4}\ 1111$ will be truncated to $0.b_{-1}\ b_{-2}\ b_{-3}\ b_{-4}$. The error ranges from 0 to .00001111. In general, the error ranges from 0 to approximately 1 in the low-order position of the retained bits. Truncation is a fast and easy method for deleting bits resulting from a fraction underflow and requires no additional hardware. Fraction underflow can occur when aligning fractions during addition or subtraction when one of the fractions is shifted to the right. There is one disadvantage in that a significant error may result.

There is another method of rounding called *von Neumann rounding*. The von Neumann rounding method is also referred to as *jamming* and is similar to truncation. If the bits to be deleted are all zeroes, then the bits are truncated and there is no change to the retained bits. However, if the bits to be deleted are not all zeroes, then the bits are deleted and the low-order bit of the retained bits is set to 1.

Thus, when 8-bit fractions are rounded to four bits, fractions in the range

$$0.b_{-1}\ b_{-2}\ b_{-3}\ b_{-4}\ 0001 \text{ to } 0.b_{-1}\ b_{-2}\ b_{-3}\ b_{-4}\ 1111$$

will all be rounded to $0.b_{-1}\ b_{-2}\ b_{-3}\ 1$. Therefore, the error ranges from approximately -1 to $+1$ in the low-order bit of the retained bits when

$$0.b_{-1}\ b_{-2}\ b_{-3}\ b_{-4}\ 0001 \text{ is rounded to } 0.b_{-1}\ b_{-2}\ b_{-3}\ 1$$

and when

$$0.b_{-1}\ b_{-2}\ b_{-3}\ b_{-4}\ 1111 \text{ is rounded to } 0.b_{-1}\ b_{-2}\ b_{-3}\ 1$$

Although the error range is larger in von Neumann rounding than with truncation rounding, the error range is symmetrical about the ideal rounding line and is an unbiased approximation. Assuming that individual errors are evenly distributed over the error range, then positive errors will be inclined to offset negative errors for long sequences of floating-point calculations involving rounding. The von Neumann rounding method has the same total bias as adder-based rounding; however, it requires no more time than truncation.

Bit 9 and bit 8 represent the precision control field and set the precision at 64 bits, 53 bits, or 24 bits, as defined in Table 2.2. The default precision is double extended precision using a 64-bit significand — bit 63 through bit 0 — thereby providing a high degree of precision.

Table 2.2 Precision Control Field

Precision	9	8
Single precision (24 bits)	0	0
Reserved	0	1
Double precision (53 bits)	1	0
Double extended precision (64 bits)	1	1

Bit 5 through bit 0 are designated as exception mask bits for certain exceptions; if a bit is set, then the exception is masked. The exception mask bits in the control word register correspond directly to the exception flag bits in the same position in the status word register. When a mask bit is set, the corresponding exception is blocked from being produced.

Instruction pointer and data pointer The instruction and data pointer registers contain pointers to the instruction and data for the last non-control instruction executed; if a control instruction is executed, then the register contents remain unchanged for both the instruction pointer and the data pointer. The data register contains a pointer for a memory operand. These pointers are stored in 48-bit registers — a segment selector is stored in bit 47 through bit 32; a memory offset is stored in bit 31 through bit 0.

2.7 Translation Lookaside Buffer

Another common component that is inherent in the architecture of current microprocessors is a translation lookaside buffer (TLB). A TLB is a cache used by the memory management unit to increase the speed of virtual-to-physical address translation. A TLB is generally a content-addressable memory in which each virtual (or logical) address is a tag in cache that is associated with a physical address. Since the TLB is an associative memory, comparison of the virtual address with the corresponding tags is accomplished in parallel. If the desired virtual address is in cache, then this is a TLB hit and the physical address thus obtained is used to address main memory.

TLBs contain instruction addresses and data addresses of the most recently used pages. A page is a fixed-length area of memory that consists of blocks of contiguous memory locations. The virtual address is formed by a virtual page number and an offset — the high-order and low-order bits, respectively. The virtual address is applied to the TLB and to the system cache concurrently. If the address is not in main memory, then the contents are retrieved from a direct-access storage device and stored in main memory; this is referred to as *page-in* technique. When transferring a page in main memory to an external storage device, this is referred to as a *page-out* technique. At any given time, main memory contains only a portion of the total contents of virtual memory.

There are advantages in utilizing a virtual memory technique: More efficient use of main memory is achieved; only the amount of main memory that is needed at the time is used; that is, programs are not present in memory if they are not being used; an application program can be designed that exceeds the main memory size, thus a program's address space is not bound by the amount of main (real) memory. When a program does not fit completely into memory, the parts that are not currently being executed are stored in secondary storage, such as a disk subsystem.

The relative (virtual) address of an instruction or operand is translated into a real (physical) address only when the virtual address is referenced. This type of translation is called *dynamic relocation* and is executed by a hardware component called *dynamic address translation*. Address translation occurs whenever an instruction or operand is referenced (addressed) during program execution.

2.8 The Assembler

There are various levels of programming in a computer. The lowest level is *machine level programming* in which the program is entered using a binary, octal, or hexadecimal number system. This is a tedious method and is prone to errors. The next higher level is *assembly language programming* in which the program is entered using symbolic instructions, such as MOV, ADD, SAR, etc., which represent a move instruction, an add instruction, and a shift arithmetic right instruction, respectively. Assembly language programming is machine dependent and requires an *assembler*, which translates the instructions into the required bit pattern of 1s and 0s.

The next popular higher level language currently in use is *C programming*, which is machine independent; that is, unlike an assembly language, it can be run on any computer with little or no alteration. C programs require a compiler to generate machine code. Other languages in this category are Fortran, PL1, Basic, and Pascal, among others. The next higher level language is the popular *C++ programming* language, which is classified as an *object-oriented programming* language and maintains the integrity and support for C. Another object-oriented programming language is *Java*, which is used for general-purpose programming and World Wide Web programming.

An *interpreter* translates a single instruction, written in a high-level language, to machine code before executing the instruction. A *compiler* translates all instructions in the source code program to machine (object) code before executing the program. A compiler translates source code from a high-level language to a low-level language, such as assembly language and may generate many lines of machine code. An *assembler* translates one line of source code to one line of machine code.

2.8.1 The Assembly Process

There are many different versions of an X86 assembler, such as Microsoft Assembler (MASM), Turbo Assembler (TASM), Flat Assembler (FASM), and Netwide Assembler (NASM), among others. The X86 assembler translates an assembly language program into a relocatable object file that can be linked with other object files to generate an executable file.

An editor is used to create an assembler source program which is saved as a *.asm* file. Then the assembler translates the source program to machine code and generates an object program *.obj*. For example, if the source code was to move an immediate operand of 0123H to register AX, then the source code would be MOV AX, 23 01. This translates to the following hexadecimal machine code: B8 23 01.

The next task is to link the *.obj* program and create an executable program *.exe*. Because a program can be loaded anywhere in memory, the assembler may not have generated all the addresses. Also, there may be other programs to link. Therefore, the link program

(1) Completes address generation.
(2) Combines more than one assembled module into an executable program.
(3) Initializes the *.exe* module for loading the program for execution.

For a simple two-pass assembler, the following steps take place during each pass:

Pass 1:
(1) The assembler reads the entire symbolic program.
(2) The assembler makes a symbol table of names and labels; for example, data field labels.
(3) The assembler determines the amount of code to be generated.

Pass 2:
(1) The assembler now knows the length and relative position of each data field and instruction.
(2) The assembler can now generate the object code with relative addresses.

2.9 Problems

2.1 Three code words, each containing a message of four bits which are encoded using the Hamming code, are received as shown below. Determine the correct 4-bit messages that were transmitted using odd parity.

	Received Code Words						
Bit Position	1	2	3	4	5	6	7
(a)	0	1	0	1	0	1	0
(b)	1	1	0	0	1	1	0
(c)	0	0	1	0	1	1	1

2.2 An 11-bit message is to be encoded using the Hamming code with odd parity. Write the equations for all of the groups that are required for the encoding process.

2.3 The 7-bit code words shown below are received using Hamming code with odd parity. Determine the syndrome word for each received code word.

(a)								
	Bit Position =	1	2	3	4	5	6	7
	Received Code Word =	1	1	1	1	1	1	1

(b)								
	Bit Position =	1	2	3	4	5	6	7
	Received Code Word =	0	0	0	0	0	0	0

2.4 A code word containing one 8-bit message, which is encoded using the Hamming code with odd parity, is received as shown below. Determine the 8-bit message that was transmitted.

Bit Position	1	2	3	4	5	6	7	8	9	10	11	12
Received Code Word	0	1	0	1	1	0	0	1	0	0	1	1

2.5 A code word containing one 8-bit message, which is encoded using the Hamming code with odd parity, is received as shown below. Determine the 8-bit message that was transmitted.

Bit Position	1	2	3	4	5	6	7	8	9	10	11	12
Received Code Word	1	1	0	0	1	0	1	1	0	0	0	1

2.6 A code word containing one 12-bit message, which is encoded using the Hamming code with odd parity, is received as shown below. Determine the syndrome word.

1	2	3	4	5	6	7	8	9	10	11	12	13	14	15	16	17
1	0	1	1	1	1	1	0	0	0	1	0	1	1	1	0	1

2.7 Obtain the code word using the Hamming code with odd parity for the following message word: 1 1 0 1 0 1 0 1 1 1 1.

2.8 Determine the relative merits of horizontal and vertical microinstruction formats.

2.9 Discuss the advantages and disadvantages of hardwired and microprogrammed control.

2.10 A block-set-associative cache consists of a total of 64 blocks (lines) divided into four-block sets. Determine the number of sets in cache.

2.11 Perform the arithmetic operations shown below with fixed-point binary numbers in 2s complement representation. In each case, indicate if there is an overflow.

(a)
```
    0100 0000
+)  0100 0000
```

(b)
```
    0011 0110
+)  1110 0011
```

(c) $+64_{10}$
$+63_{10}$

2.12 Let A and B be two binary integers in 2s complement representation as shown below, where A' and B' are the diminished radix complement of A and B, respectively. Determine the result of the operation and indicate if an overflow exists.

$A = 1011\ 0001$ $\quad$ $A' + 1 + B' + 1$
$B = 1110\ 0100$

2.13 A fraction and the bits to be deleted (low-order three bits) are shown below. Determine the rounded results using truncation (round toward zero), adder-based rounding (round to nearest), and von Neumann rounding.

Fraction											
.	0	0	1	0	1	0	0	1	1	0	1

2.14 Let the augend be $A = 0.1100\ 1101 \times 2^6$
Let the addend be $B = 0.1011\ 0001 \times 2^4$

Perform the addition operation and round the result using all three rounding methods.

2.15 Let the augend be $A = 0.1110\ 1001 \times 2^6$
Let the addend be $B = 0.1001\ 0111 \times 2^4$

Perform the addition operation and round the result using all three rounding methods.

3

Addressing Modes

X86 instructions can have zero or more operands; for example, a return (RET) instruction from a procedure (subroutine) can have no operands or an immediate operand; a negate (NEG) instruction forms the 2s complement of a single operand, NEG AX; a move instruction has two operands MOV AX, BX, which moves the contents of register BX to register AX. In the X86 instruction set, the first operand listed is the destination operand; the second operand listed is the source operand.

The instruction set provides various methods to address operands. The main methods are: register, immediate, direct, register indirect, base, index, and base combined with index. A displacement may also be present. These and other addressing methods are presented in this chapter together with examples. All arithmetic instructions with the exception of unsigned division (DIV) and signed division (IDIV) permit the source operand to be an immediate value.

The source operand can be obtained from the instruction as immediate data, from a general-purpose register, from a memory location, or from an input/output (I/O) port. The source operand is normally unchanged by the instruction execution. The result of an instruction execution can be sent to a destination in a register, a location in memory, or to an I/O port. Memory-to-memory operations are not permitted with single instructions — this type of operation is reserved for string instructions, which require that the source and destination locations be set up before the transfer takes place.

An instruction has the format shown below, where a *label* is an identifier for a line of code or a block of code, followed by a colon (:). The *mnemonic* is the name of the instruction, for example, ADD, SUB, MOV, etc. This is followed by zero to three *operands* — depending on the operation — separated by commas.

LABEL: MNEMONIC OPERAND1, OPERAND2, OPERAND3

Some instructions require an operand to be a quadword (eight bytes), such as a dividend for a divide operation, or generate a quadword result, such as the product for a multiply operation. The dividend or product are contained in a concatenated register pair, such as EDX:EAX, where the colon specifies concatenation.

Addressing modes provide different ways to access operands. A *displacement* is an 8-bit, 16-bit, or 32-bit immediate value in the instruction and is used to address memory. A *base* is the contents of a general-purpose register (GPR). The registers normally used for this purpose are (E)BX and (E)BP. An *index* is the contents of a GPR. The registers that are normally used for this purpose are (E)SI and (E)DI.

b

3.1 Register Addressing

Using register addressing, the instruction selects one or more registers which represent the operand or operands. This is the most compact addressing method, because the register addresses are encoded in the instruction. It achieves fast execution since the operation is performed entirely within the central processing unit (CPU); that is, there are no bus transfers to or from memory. Examples of register addressing are shown below, where the semicolon indicates a comment.

```
INC  BH          ;increment register BH
MOV  AX, BX      ;move contents of register BX to register AX
SUB  EBX, EBX    ;subtract EBX from EBX (EBX = 0)
XOR  EAX, EAX    ;exclusive-OR EAX with EAX (EAX = 0)
```

The source and destination registers can be any of the general-purpose registers, either 8-bit, 16-bit, 32-bit, or 64-bit GPRs. The segment registers CS, DS, SS, ES, FS, or GS can also be used in register addressing as well as the flag register and the floating-point registers, among others. The length of the operand is determined by the name of the GPR.

3.2 Immediate Addressing

Some instructions have a data value encoded in the instruction so that it is available immediately as the source operand. The immediate data can be 8 bits, 16 bits, or 32 bits, which are sign-extended to fit the size of the destination operand. Examples of immediate addressing are shown below for both positive and negative values.

```
ADD   EAX, 3              ;add 3 to register EAX
MOV   AH, 00              ;reset AH; AH = 0000 0000
MOV   EAX, 0ABCDFFFFH     ;32-bit operand ABCD FFFF(Hex) = EAX
MOV   AX, 302             ;AX = 0000 0001 0010 1110
MOV   AX, -40             ;AX = 1111 1111 1101 1000
```

3.3 Direct Memory Addressing

The memory operand can be either the source or the destination operand and is addressed by a segment selector and a 32-bit or 64-bit offset that is part of the instruction. The combination of the selector and the offset generates an effective address that points *directly* to the memory operand. The segment selector can be any of the following segment registers: code segment (CS), data segment (DS), stack segment (SS), or the extra segment (ES).

A segment selector is loaded with the appropriate value, then the CPU addresses memory using the current (implicit) selector and the offset. Explicit segment selectors are discussed in later section using a segment override prefix. Examples of direct memory addressing using an implicit segment selector are shown below.

```
MOV  AL, [1A33D4H]       ;AL = contents of memory at 1A33D4H
INC  DWORD PTR [17H]     ;incr 32-bit operand at offset 17H
```

3.4 Base (Register Indirect) Memory Addressing

Base, or register indirect, addressing is used when a register contains the address of the data, rather than the actual data to be accessed. This method of addressing uses the general-purpose registers within brackets; for example, [BX], [BP], [EAX], [EBX], etc., where the brackets specify that the indicated register contains an offset that points to the data within a specific segment. Registers (E)BX, (E)SI, and (E)DI contain offsets for processing operands in the data segment; for example, DS:BX, DS:ESI, and DS:DI. Register (E)BP is used to reference data in the stack segment, such as SS:EBP.

Examples of indirect memory addressing are shown below. In the first example, the notation [[EBX]] in the comments reads "the contents of the contents of EBX"; that is, the contents of memory specified by the contents of EBX are loaded into register ECX. In the second example, assume that BX = 0002H and that memory locations 0002H and 0003H contain 00H and FFH, respectively. Then register AX = FF 00H after completion of the move instruction, where 00H is the low-order byte in AL. This is illustrated in Figure 3.1.

```
MOV     ECX, [EBX]      ;ECX = [[EBX]]; If [EBX] = 1000H, then
                        ;ECX = bytes [1000H through 1003H]
MOV     AX, [BX]        ;AX = [0002H] , [0003H]
```

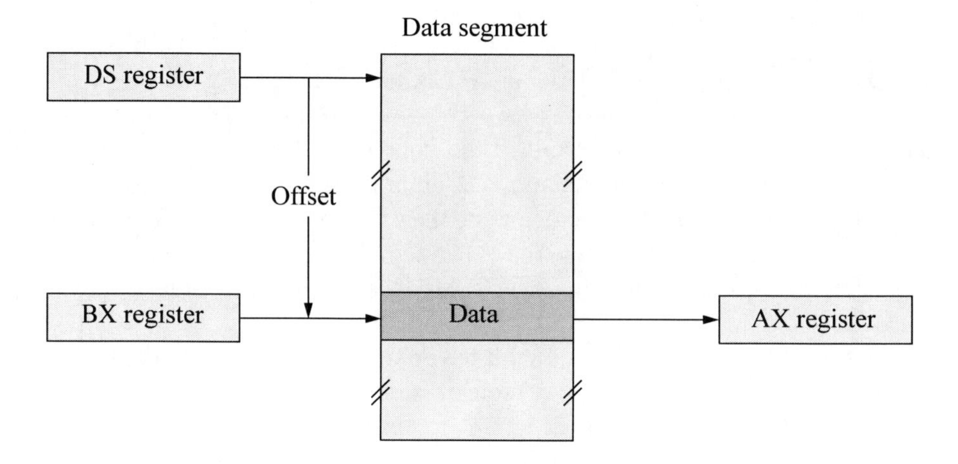

Figure 3.1 Figure to illustrate indirect memory addressing.

3.5 Base or Index Plus Displacement Addressing

This is a variation of base addressing and uses the general-purpose registers (E)AX, (E)BX, (E)CX, (E)DX, (E)SP, (E)BP, (E)SI, or (E)DI as the base address or (E)AX, (E)BX, (E)CX, (E)DX, (E)BP, (E)SI, or (E)DI as an index address, plus a displacement to access memory. A scaling factor may also be used in conjunction with the index register. The displacement consists of a signed number in 2s complement representation that is added to the value in the base register or index register to produce an effective address. When (E)BP is utilized as the base register, the stack segment is the default segment.

An expression using a displacement is usually written as [BX + 4] using BX as the base register with a displacement of 4, or as [ESI + 8] using ESI as the index register with a displacement of 8. Both methods indicate a displacement of four or eight bytes from the base or index address stored in register BX or register ESI. This addressing method can be used as an index into an array, as shown in the code segment shown below, where the DB directive defines the five values as bytes. The compare instruction compares the contents of AL with the byte at location TABLE + 4, which is the *fifth* byte (4) of the list. The first element (15) of the list is at location TABLE, which has a displacement of zero. Changing the value of the displacement accesses different fields in the same record. Changing the base register accesses the same field in different records.

```
TABLE:     DB     15, 7, 6, 10, 4
           ...
           LEA    BX, TABLE          ;loads the effective
                                     ;addr of TABLE into BX
           CMP    AL, [BX + 4]
```

Base or index addressing provides a way to address tables, arrays, or lists which may be located in different areas of memory. A base register can be set to point to the base of a table. The elements of the table can then be addressed by their displacement from the base. Different tables can be addressed by simply changing the value in the base register. A graphical representation of this concept is illustrated in Figure 3.2, which shows relevant information regarding an employee. The contents of register BX can be changed to access a different employee, where the same displacement addresses the rate.

Alternatively, this addressing method can be used to access elements in an array. The displacement value locates the beginning of the array and the value stored in the index register selects an element in the array, as shown in Figure 3.3.

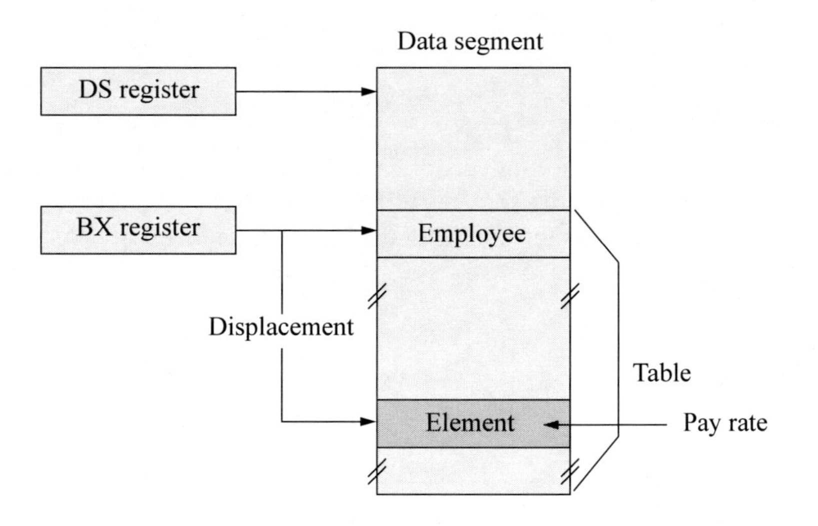

Figure 3.2 Illustration of base addressing with displacement.

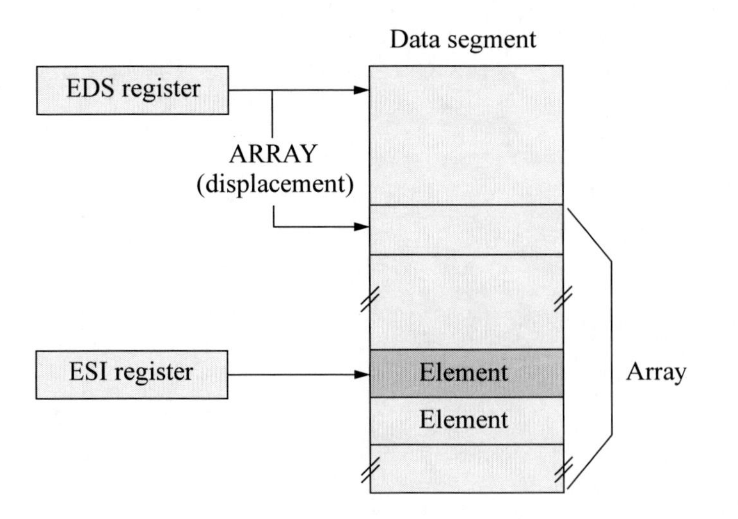

Figure 3.3 Illustration of index addressing with displacement.

The effective address in Figure 3.3 is obtained from the sum of the displacement and the contents of index register ESI. Indexed addressing is used to access elements in an array at execution time. The value of the index register selects an element in the array; the first element is selected if the index register contains a value of zero. The elements are the same length; therefore, simple arithmetic on the index register will select any element. An instruction to move the contents of an array element to register EAX is shown below, where ARRAY is the address (displacement) of the array (or table) and the element is accessed whose index is in ESI.

```
MOV     EAX, ARRAY[ESI]
```

ARRAY[ESI] references the contents of the element in the location addressed by the displacement ARRAY plus the contents of ESI — ARRAY + 0 references the first element in the array. To access the next element, simply increment register ESI.

3.6 Base and Index Plus Displacement Addressing

The effective address of the operand is obtained by adding the contents of the base register, the contents of the index register, and the displacement (which may be zero). Base indexed addressing without displacement is frequently used to access the elements of a dynamic array. A dynamic array is an array whose base address can change during program execution. This addressing method can also be used to access a 2-dimensional array in which the beginning of the array is determined by the displacement and the base register; the index register addresses the array elements.

Another application is to access elements in a 1-dimensional array on a stack, as shown in Figure 3.4. The base register can point to the stack top; the contents of the base register plus the displacement points to the beginning of the array. The index register then selects elements in the array. The effective address (EA) is calculated as shown below.

```
EA = EBP + displacement + ESI
```

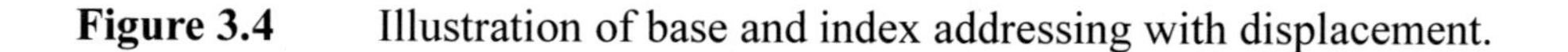

Figure 3.4 Illustration of base and index addressing with displacement.

Example 3.1 A 2-dimensional array is shown in Figure 3.5 in which the elements are stored row by row. The elements can be accessed by base-index with displacement addressing. The second element in the first row (56) can be accessed by

```
[EBX/EBP][ESI/EDI]
```

by storing the base address of the first element in BX/EBX (or BP/EBP) and storing a 1 in SI/ESI (or DI/EDI) with no displacement — the index register selects the *i*th element. A 1 rather than a 2 is stored in the index register, because the element whose address is in the base register is the first element (12). To access the second element in the second row (89), a value of 5 is added to the base register— the number of elements in a row — as shown below assuming a value of 1 in the index register.

`5[EBX/EBP][ESI/EDI]` or `[EBX/EBP + ESI/EDI + 5]`

12	**56**	78	5	44
54	**89**	65	92	6
37	66	15	77	19

Figure 3.5 An illustration of a 2-dimensional array.

3.7 Scale Factor

A scale factor can be used to access elements in an array in which the size of the elements is two, four, or eight bytes. The address calculation can be considered as

(Index register $\times 2^{scale}$) + Displacement or
Base register + (index register $\times 2^{scale}$) + Displacement

where the *scale* is 0, 1, 2, or 3. This addressing mode is useful for processing arrays and provides an index into the array when the array elements are two, four, or eight bytes. If the array elements are 8-byte operands, then the index register is multiplied by eight prior to calculating the effective address. Since the scale factor is a power of two, multiplication can be accomplished by shifting left rather than by a multiply operation.

(Index times scale) plus displacement The displacement points to the beginning of the array, while the index contains the address of the appropriate array element; that is, the index is the subscript of the array element. The effective address of the operand is obtained by multiplying the index by a scale factor of one, two, four, or eight and then adding the displacement.

Base plus (index times scale) plus displacement The effective address of the operand is obtained by multiplying the index register by a scale factor of one, two, four, or eight, then adding the base register, and then adding the displacement. This addressing mode is useful for accessing array elements in a 2-dimensional array when the operands are one, two, four, or eight bytes. An example is shown in Figure 3.6 using EBX as the base register, ESI as the index register with a scale factor of 4, and a displacement for the following move instruction that moves the addressed operand into register EAX:

```
MOV  EAX, [EBX + ESI * 4 + DISPLACEMENT]
```

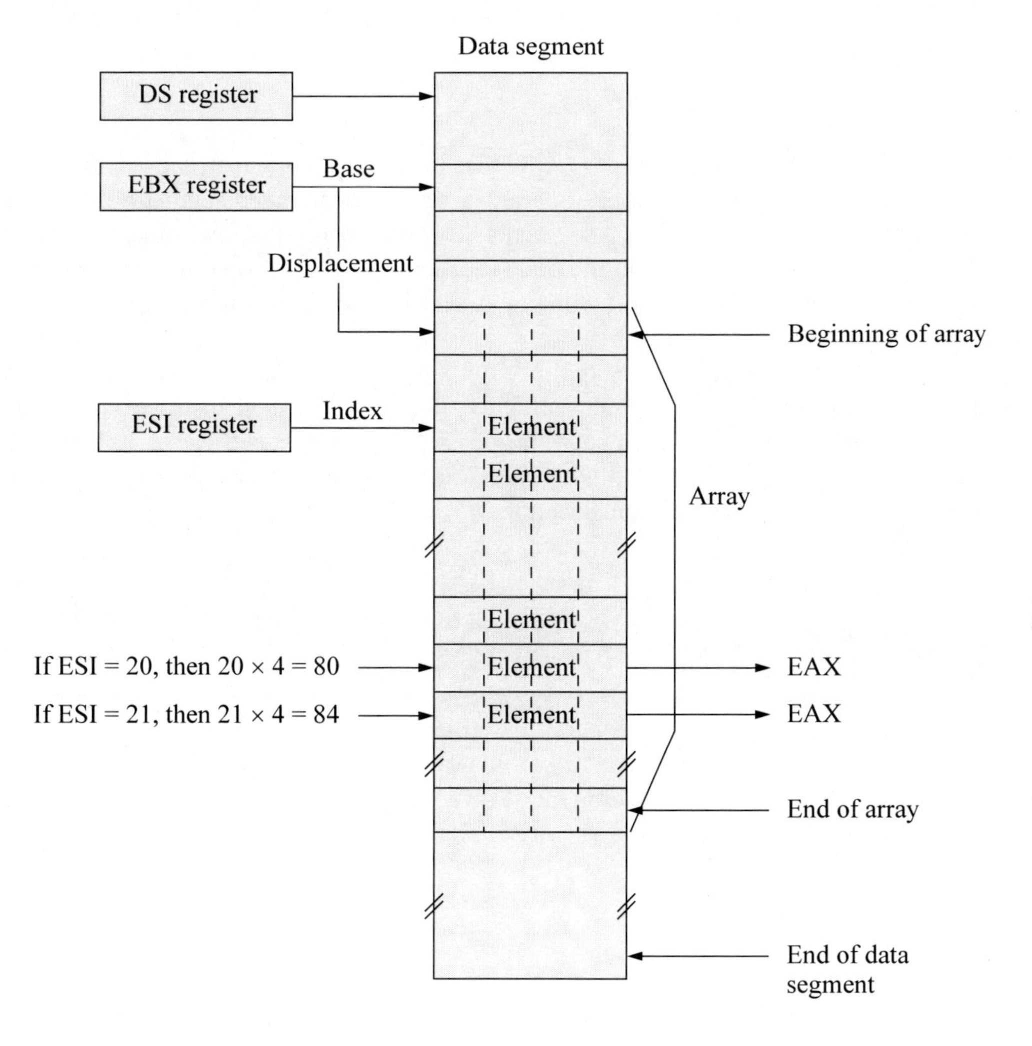

Figure 3.6 Illustration of base plus (index times scale) plus displacement.

3.8 Segment Override Prefix

The processor selects the applicable default segment as a function of the instruction: instruction fetching assumes the code segment; accessing data in main memory references the data segment; and instructions that pertain to the stack reference the stack segment. When transferring data to or from memory, the processor usually references the current data segment as the default segment. However, a *segment override prefix* can be used to change the default data segment to another segment; that is, to explicitly specify any segment register to be used as the current segment.

This is accomplished by listing the new segment to be used for source or destination operands followed by a colon and is placed in front of the offset variable. For example, to override the data segment with the extra segment, the coding shown below applies to a move instruction, which moves the 32-bit operand addressed by register EBX in the extra segment to register EAX. When the move instruction is executed, the offset in register EBX pertains to the extra segment rather than the data segment.

```
MOV   EAX, ES:[EBX]
```

The segment override prefix is a byte that is inserted in the instruction and causes the processor to fetch the memory operand from the designated segment rather than from the default segment. Some segments, however, cannot be overridden. Instruction fetches must be made from the code segment; push and pop operations must be made with reference to the stack segment; and string operations must provide a destination address in the extra segment.

The segment override prefix applies *only* to the instruction in which it is inserted; that is, it is restricted to a single instruction — all subsequent instructions use the default segment. Thus, the segment override prefix must be specified for each instruction that is to be overridden. Utilizing a segment override prefix increases the execution time of the corresponding instruction.

3.9 X86 Operation Modes

This section presents two of the operating modes for the X86 architecture for the basic execution environment: *protected mode* that allows for multiple tasks and multiple users and *real mode*. The operating modes are set by software and specify the instructions and architectural attributes that are available.

3.9.1 Protected Mode

Protected mode allows the system software to use features, such as virtual memory and multitasking, to increase system performance. The architecture provides four levels of protection: level 0 through level 3. Level 0 has the highest priority and is reserved for the operating system; levels 1 and 2 are for less critical operating system functions; level 3 is reserved for application programs.

Multitasking can execute several programs simultaneously. These programs (or tasks) may have separate segments or may share segments. The programs are written as if on a dedicated microprocessor. The multitasking software then simulates a dedicated microprocessor by allocating a *virtual processor*. The substitution of these virtual processors creates the appearance of a dedicated processor, where each task has an allotment of CPU time. If a task must wait for an I/O operation, it suspends its operation. The computer must be able to switch rapidly between tasks, save and load the entire machine state, prevent interference of tasks, and prioritize tasks.

Protected mode allows reliable multitasking and prevents tasks from overwriting code or data of another task. Thus, if a program fails, the effects are confined to a limited area and the operating system will not be affected. Protection is applied to segments and pages.

3.9.2 Real Mode

Real mode (or real-address mode), is an operating mode that implements the 8086 architecture and is initialized when power is applied or the system is reset. Thus, real mode allows compatibility with programs that were written for the 8086 processor. Real mode allows access to the 32-bit register set and protected mode when processor extensions are in effect. Real mode does not support multitasking.

The real-address mode implements a segmented memory consisting of 64 kilobytes in each segment. Each segment register is associated with either code, data, or stack, which are contained in separate segments. Real mode segments begin on 16-byte boundaries. Memory is accessed using a 20-bit address that is calculated by adding a right-aligned 16-bit offset to a segment register that is multiplied by 16; that is, the segment register is shifted left four bits. This provides an address space of 1 megabyte.

3.10 Problems

3.1 Specify the addressing mode for each operand in the instructions shown below.

(a) ADD AX, [BX]

(b) ADD CX, [BP + 8]

(c) ADD EBX, ES:[ESI – 4]

(d) ADD [EBP +EDI + 6], 10

3.2 Let DS = 1000H, SS = 2000H, BP = 1000H, and DI = 0100H. Using the real addressing mode with a 20-bit address, determine the physical address memory address for the instruction shown below.

```
MOV   AL, [BP + DI]
```

3.3 Let DS = 1100H, DISPL = –126, and SI = 0500H. Using the real addressing mode with a 20-bit address, determine the physical address memory address for the instruction shown below.

```
MOV   DISPL[SI], DX
```

3.4 Let the directive shown below be located in the data segment at offset 04F8H, where DW 10 DUP(?) specifies that a word is defined with an unknown value and duplicated ten times.

Let BX = 04F8H, SI = 04FAH, and DI = 0006H. Let the 10 reserved words beginning at location VALUE be labelled

WORD1, WORD2, . . . , WORD10

Indicate the word that is referenced by the memory operand in each of the move instructions shown below.

```
(a)   MOV   AX, VALUE + 2
(b)   MOV   AX, [BX]
(c)   MOV   AX, [SI]
(d)   MOV   AX, [BX + DI – 2]
```

3.5 Differentiate between the operation of the following two move instructions:

```
(a)   MOV   EAX, 1234ABCDH
(b)   MOV   AX, [1234H]
```

3.6 Differentiate between the operation of the following two move instructions:

```
(a)   MOV   EBX, 1234ABCDH
(b)   MOV   [EBX], 1234H
```

3.7 Use base and index plus displacement addressing to obtain the physical address of the memory operand for the following conditions:

BX = 6F30H, SI = 1000H, DS = 85H, Displacement = 2106H

3.8 Obtain the real (physical) address that corresponds to each segment:offset pair shown below.

	Segment: Offset
(a)	2B8C: 8D21
(b)	F000: FFFF
(c)	3BAC: 90DF

4

C Programming Fundamentals

The C programming language was developed by Bell Laboratories in 1972 and has become the predominant high-level programming language. As a high-level language, C programs are machine independent; that is, they can be run on different computer systems. Assembly language uses an assembler to convert the symbolic program to machine code; whereas, the C programming language uses a compiler to convert the source code into machine code.

The C language is case sensitive; therefore, add ( ) and ADD ( ) are different function names. It is common practice to use lowercase for both the code statements and the comments. The main purpose of this chapter is to provide sufficient information regarding C programming in order to demonstrate how a C program can be linked to an assembly language program. This concept will be used in subsequent chapters.

An *editor* is used to create a disk file for the *source* code, which is saved with a *.c* or a *.cpp* extension. The source code is then compiled to create an *object* file, which is linked with *library* files to assign addresses and to create an *executable* file that can be run on a computer. The flowchart of Figure 4.1 illustrates this development cycle.

4.1 Structure of a C Program

A simple C program is shown in Figure 4.2 displaying both the program code and the outputs obtained from the program. The program contains some basic statements and symbols that are inherent to C programs. Line 1 is a comment line indicated by double forward slashes //. This comment syntax is used for single-line comments only. A second syntax to specify a multiple-line comment uses the symbol /* before the first comment character, followed by the comment, and terminated by the symbol */. This syntax can also be used for single-line comments. Examples of different types of comments are shown below.

```
//This is a single-line comment.
/*This is a single-line comment.*/

printf ("Hello World\n");  //comment for this line of code

/*
This is a multiple-line comment
used to accommodate information
that spans several lines.
*/
```

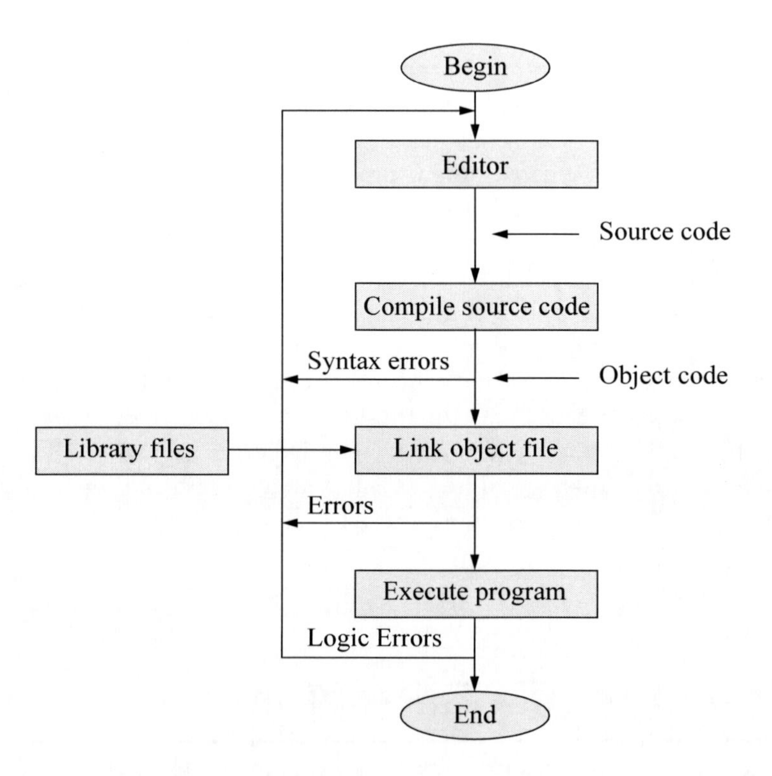

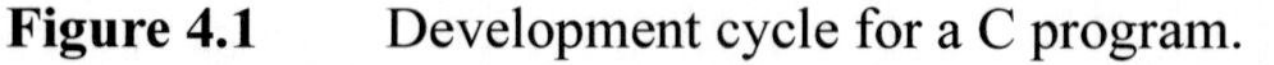
Figure 4.1 Development cycle for a C program.

```
1  //hello world2, a simple C program
2  #include <stdio.h>
3
4  main ()
5  {
6     printf ("Hello World\n");
7     return;
8  }
```

(a)

```
Hello World
Press any key to continue_
```

(b)

Figure 4.2 A simple C program to illustrate some basic characteristics: (a) the C program and (b) the resulting output generated by the program.

Line 2 is a preprocessor directive used to read in another file and *include* it in the program. Preprocessor directives begin with symbol #. It informs the compiler to include at this location in the program information that is contained in the **stdio.h** file. The angle brackets < > indicate that the **stdio.h** file is located in a specific machine-dependent location. The **stdio.h** files are specified by the American National Standards Institute (ANSI) and include the **printf ( )** function used in this program. Line 3 is a separation line.

Line 4 consists of a function called **main ( )**, where the parentheses indicate to the compiler that it is a function. The keyword **void** can also be included within the parentheses to indicate that no arguments are being passed to the function. Line 5 contains a left brace {, which marks the beginning of the body of the function. Left and right braces are also used as delimiters to group statements together.

Line 6 invokes the print function **printf ( )**, which displays information on the monitor screen. This function is part of the standard library of functions contained in the C programming environment. Information pertaining to the **printf ( )** function is contained in the **stdio.h** header file. The data within the parentheses is a series of characters called a string argument contained within double quotation marks. The string is called an argument and is displayed on the screen, except for the symbol (\n), which is a newline character. This character is not displayed, but simply places the cursor at the beginning of the next line. Line 6 is terminated with a semicolon — all declarations and statements are terminated with a semicolon.

Line 7 indicates that the function returns to the calling function **main ( )**. This is part of the ANSI requirement for a C program. An expression can be passed to the calling function, in which case the expression is usually enclosed in parentheses. The expression may be the result of a calculation requested by the calling function. A value of zero means that the program was executed correctly — a nonzero integer indicates that the function **main ( )** was unsuccessful. Line 8 is the closing right brace and indicates the end of the program specified by the function **main ( )**.

It is important to make the program format easy to read. This makes the program easier to understand and to correct or modify. One technique to make the program easier to understand is to make the variable names meaningful. Comments also help to clarify certain tasks in the program and indicate more clearly how the program functions. Comments should be placed before a code segment to illustrate the function of the segment and also on individual lines of code, where applicable.

Blank lines help to delineate different segments of the program by separating them. Although C is classified as a *free-form* language — allowing more than one statement per line — the code is easier to read if there is only one statement per line. The braces should be placed on the extreme left of the page with the body of the segment right-indented a few spaces. These techniques make the program easier to read and understand at a later date for both the programmer who wrote the code and others.

4.2 Variables and Constants

This section describes how the C programming language defines and stores variables and constants. Variables store values in memory and the *type* of each variable must be specified; that is, the variable must be declared as type character, integer, and so forth. The focus will be on numeric variables and numeric constants.

4.2.1 Variables

A variable is a named memory location and must be declared before it is used in the program. The declaration of a variable with an assigned type permits the compiler to assign an appropriate amount of storage for the variable. A variable is defined by indicating the type of variable followed by the name of the variable, which can range from a single first letter to 31 characters, including underscore characters. Some typical variables are shown in Table 4.1. The typical ranges for the variables listed in Table 4.1 are shown in Table 4.2.

Table 4.1 Some Typical C Variables

Type	Keyword	Meaning
Character	**char**	Signed character (one byte)
Unsigned character	**unsigned char**	Unsigned character (one byte)
Integer	**int**	Signed integer (two bytes)
Unsigned integer	**unsigned int**	Unsigned integer (two bytes)
Float	**float**	Single-precision floating-point numbers
Double	**double**	Double-precision floating-point numbers

Table 4.2 Typical Ranges for Some Typical C Variables

Type	Range	Bits required
char	–128 to +127	8 bits
unsigned char	0 to 255	8 bits
int	–32,768 to +32,767	16 bits
unsigned int	0 to 65,535	16 bits
float	10^{-38} to 10^{+38} (approximate range)	32 bits
double	10^{-308} to 10^{+308} (approximate range)	64 bits

Variables that are declared outside a function — defined before the function brace — are global variables and can be accessed by any function in the remainder of the program. It is preferable to have global variables declared at the beginning of the program prior to **main ( )**, because this makes them easier to notice.

Variables that are declared inside a function — defined after the brace at the start of the function — are local variables and can be accessed only by the function in which they are defined. They are unique from other variables of the same name that are declared at other locations in the program; that is, they are distinct and separate variables with the same name. Variables are declared as follows: *type var_name*; for example: **int** *counter*;. Examples of global and local variable declarations are shown in Figure 4.3.

```
main ()
{
                    int a       //local
                                //(declared after the brace)
                    ...
}
```

(a)

```
#include <stdio.h>
int b                           //global
                                //(declared before a function)
main ()                         //main () is a function
{
                    ...
}
```

(b)

Figure 4.3 Examples of declared variables: (a) local variable and (b) global variable.

More than one variable can be declared by using commas to separate the variable names; for example, the floating-point numbers x, y, and z: **float** x, y, z;. Variables can be assigned values by using the assignment operator (=) in an assignment statement, as follows: *var_name* = *value* (*or expression*); for example: *num* = *100*;. The variable name is any variable that has been previously defined. Commas are not allowed in numerical values; thus, the number 36,000 is invalid.

Variables can be initialized when they are declared; for example, **int** *num* = *100*;. Since an integer has a range of –32,768 to +32,767, a value of 50,000 would be beyond the range of an integer. Errors of this type are undetected by the compiler. An integer value has no fractional part; however, a floating-point value must include a decimal point. Floating-point numbers are segmented into the sign, the exponent (characteristic), and the fraction (mantissa or significand), as shown in Figure 4.4 for single-precision and double-precision formats.

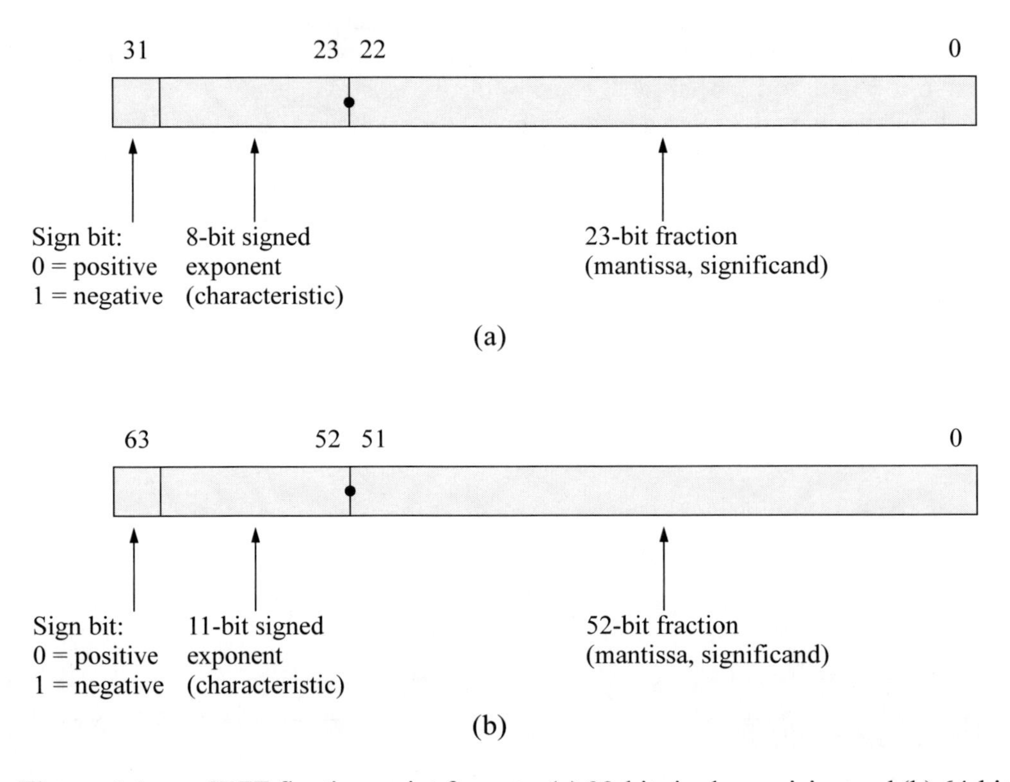

Figure 4.4 IEEE floating-point formats: (a) 32-bit single precision and (b) 64-bit double precision.

The result of a floating-point operation can be displayed on the monitor by using the print function **printf ( )**, as shown in the short program of Figure 4.5, which multiplies two floating-point numbers a and b, then displays the product. The symbol *%f* is a conversion specifier that causes the result of the multiply operation to be displayed as a floating-point number. Conversion specifiers are covered in more detail in a later section. The symbol (*) is the multiplication operator. Operators are also covered in a later section.

```
//float_mul
//multiplies two floating-point numbers
#include <stdio.h>

main ( )
{
   float a, b;

   a = 3.5;
   b = 4.5;

   //product of 3.5 x 4.5 = 15.750000
   printf ("Product = %f\n", a * b);

   return;
}
```

(a)

```
Product = 15.750000
Press any key to continue_
```

(b)

Figure 4.5 Example of floating-point multiplication: (a) the C program and (b) the resulting output generated by the program.

4.2.2 Constants

Unlike variables, constants are fixed values that cannot be altered by the program. There are primarily four types of constants in the C programming language: integer constants, floating-point constants, character constants, and string constants.

Integer constants An integer is a whole number without a decimal point, such as the natural numbers and their negatives, including the number zero. Examples of integer constants are 10 and –100. Integer constants can be represented in three different radices: decimal (radix 10, with digits 0 through 9), octal (radix 8, with digits 0 through 7), and hexadecimal (radix 16, with digits 0 through 9 and letters A through F that represent digits 10 through 15 as one character).

A decimal number in C is entered as shown above for 10 and –100 with no prefix. If there is no leading plus or minus sign, then the compiler assumes that the number is positive. An octal number is preceded by a zero (0) prefix; for example, $024_8 = 20_{10}$. Octal numbers can be preceded by a plus or minus sign. A hexadecimal number is preceded by a 0x prefix; for example, $0xA5B_{16} = 2651_{10}$ and can have a plus or minus sign.

An integer constant can also be written as a *literal constant* in which the integer constant is entered directly into the statement that specifies the integer; for example: **int** *num = 35*;.

Floating-point constants A floating-point number consists of a sign, either plus or minus — if there is no sign indicated, then the number is assumed to be positive; an exponent character — either *E* or *e*; an integer exponent; and a significand containing a decimal point.

Floating-point constants are decimal digits that are of type **float**, **double**, **long**, or **long double**. Unless otherwise specified, floating-point numbers are stored as **double** as the default type; that is, they are stored in the double-precision format. A suffix can be added to the constant, however, to uniquely specify the type. A suffix of *f* or *F* specifies a single-precision number; a suffix of *l* or *L* specifies a double-precision number. The C compiler maps **long double** to type **double**.

Floating-point numbers can also be written in scientific notation. Thus, 2.45E3 = 2.45×10^3 = 2,450; 3.7256e4 = 3.7256×10^4 = 37,256; and 0.92e–3 = 0.92×10^{-3} = 0.00093. When converting a floating-point number to an integer, the fraction is truncated, as shown in Figure 4.6, where the conversion specifier *%f* is a floating-point number and *%d* is a signed decimal integer.

```
//convert_flp_to_int
//convert a floating-point number to an integer

#include <stdio.h>
main ( )

{
    int i;
    float f;

    f = 1234.009766;
    i = f;

    printf (" float = %f\n integer = %d \n", f, i);
    return;
}                                   //continued on next page
```

(a)

Figure 4.6 Program to illustrate converting a floating-point number to an integer: (a) the C program and (b) the outputs obtained from the program.

```
float = 1234.009766
integer = 1234
Press any key to continue_
```

(b)

Figure 4.6 (Continued)

When using the Microsoft Visual C++ Express Edition, the **#include** directive is slightly different and **main** is preceded by the **int** data type, as shown in Figure 4.7 for the floating-point-to-decimal conversion of Figure 4.6. The **stdafx.h** directive is a precompiled header directive, which is an abbreviation for Standard Application Framework Extensions. This directive contains standard and project-specific files that are to be included in the program.

The construct **int main (void)** is a function prototype that has two components. The **int** is the return type, which is the *type* of variable that is returned to the operating system at the completion of the program and represents the status of the program execution. The **main (void)** indicates to the compiler where program execution begins and that no arguments are being passed to the invoked function. The left and right braces indicate the beginning and end of the function.

```
// convert_flp_to_int.cpp

#include "stdafx.h"
int main (void)

{
   int i;
   float f;

   f = 1234.009766;
   i = f;

   printf (" float = %f\n integer = %d \n ", f, i);
   return 0;
}
```

(a)

```
float = 1234.009766
integer = 1234
Press any key to continue . . . _
```

(b)

Figure 4.7 Program to illustrate converting a floating-point number to an integer using C++ Express Edition: (a) the C program and (b) the outputs obtained from the program.

Character constants Unlike variables, the value of a constant cannot be changed. The keyword **char** indicates the type used for letters and other characters, such as !, &, $, %, and *, among others. When characters are stored, they are actually stored as integers using the American Standard Code for Information Interchange (ASCII). A different code is used for IBM mainframes, called the Extended Binary Coded Decimal Interchange Code (EBCDIC).

For example, to store an uppercase A represented in ASCII, 65_{10} (41_{16}) is stored; whereas, to store a lowercase a, 97_{10} (61_{16}) is stored. The standard ASCII code has a range of decimal 0 to 127 (hexadecimal 00 to 7F). The extended ASCII code has a range of decimal 128 to 255 (hexadecimal 80 to FF) for special characters, such as $\geq$ or $\leq$. Both standard ASCII and extended ASCII characters are stored as one-byte integers.

Character constants are written within single quotation marks, such as '#' and 'R', which act as delimiters for the character. The single quotation marks indicate to the compiler that a single character is specified. Note that the number 7 and the character constant '7' are different — the number 7 has an integer value of seven; the character constant '7' has an integer value of 55_{10} (37_{16}).

The program shown in Figure 4.8 illustrates printing a standard ASCII character and an extended ASCII character. The conversion specifier *%c* specifies a character value. An extended ASCII character must be unsigned.

```
//ascii char.cpp
//print standard ascii and extended ascii

#include "stdafx.h"

int main (void)
{
   char x1;
   unsigned char x2;

   x1 = 'R';
   x2 = 193;    //extended ascii character is ⊥

   printf ("x1 = %c, x1= %d, x2 = %c \n", x1, x1, x2);
   return 0;
}
```

(a)

```
x1 = R, x1 = 82, x2 =⊥
Press any key to continue . . . _
```

(b)

Figure 4.8 Program to illustrate a program to print standard ASCII and extended ASCII characters: (a) the C program and (b) the outputs obtained from the program.

If an attempt is made to place a 16-bit character into an 8-bit integer, only the low-order eight bits will be retained — the high-order eight bits will be removed. For example, $1000_{10} = 03E8_{16}$. This value is represented in eight bits as $E8_{16} = 1110\ 1000_2 = -24$ in 2s complement representation. Figure 4.9 illustrates this concept.

```
//truncate.cpp
//truncate 16 bits to 8 bits

#include "stdafx.h"

int main (void)
{
    char ch;        //char is 8 bits
    int i;          //int is 16 bits

    i = 1000;       //03E8H is > 8 bits
    ch = i;         //cannot put 16 bits into 8 bits

    printf ("%d\n", ch);
    return 0;
}
```

(a)

```
-24
Press any key to continue . . . _
```

(b)

Figure 4.9 Program to illustrate high-order truncation when attempting to place a 16-bit variable into an 8-bit variable: (a) the C program and (b) the outputs.

The program shown in Figure 4.10 illustrates declaring three different types: **char**, **float**, and **double**, then assigns a value to each type and then prints the output to the screen using one print statement.

```
//char_flp_dbl.cpp
//create variables of type char, float, and double
//assign each a value and print to the screen
#include "stdafx.h"                    //continued on next page
```

(a)

Figure 4.10 Program to illustrate assigning values to types **char**, **float**, and **double**: (a) the C program and (b) the outputs.

```
int main (void)

{
    char ch;
    float flp;
    double dbl;

    ch = 'X';
    flp = 100.123;
    dbl = 123.009;

    printf (" ch = %c\n flp = %f\n dbl = %f\n ",
             ch, flp, dbl);

    return 0;
}
```

```
 ch = X
 flp = 100.123001
 dbl = 123.009000
Press any key to continue . . . _
```

(b)

Figure 4.10 (Continued)

String constants String constants are a set of characters enclosed by double quotation marks. The string characters are terminated by a null zero (\0). This is a binary zero (0000 0000) and is inserted by the compiler to mark the end of the string — it is not the number zero (0011 0000). The null zero is also called the *string delimiter.* Since the quotation marks are not stored as part of the string, the only indication that the compiler has that the string has ended is the null zero.

Strings produce a one-dimensional array in memory, including the terminating null zero. To use double quotation marks in the string, precede the quotation mark with a backslash. Examples of strings are shown in Table 4.3 together with the string lengths.

Table 4.3 String Examples and their Lengths

String	Length in Bytes
"012"	3
"Hello World"	11
"Temperature is 74 degrees Fahrenheit."	36

4.3 Input and Output

This section presents additional detail on the **printf ( )** function and introduces the **scanf ( )** function. These are the only two input/output (I/O) functions that will be used in later chapters when linking C to an assembly language program. Other I/O functions that will not be covered in this section include the **getchar ( )** function, which reads the next character typed from the keyboard as an integer; the **putchar ( )** function, which writes a single character to the screen; the **getche ( )** function, which reads the next character and then displays the character on the screen; and the **getch ( )** function, which is similar to the **getche ( )** function, but does not display the character on the screen, among others.

4.3.1 Printf ( ) Function

The **printf ( )** function, together with the **scanf ( )** function, are two of the most versatile I/O functions in the C repertoire. The **printf ( )** function has the general format shown below. It prints the characters that are contained within the double quotation marks. Some of the more common conversion specifiers are shown in Table 4.4.

```
printf ("format_string and conversion specifier(s)\n",
        one or more arguments);
```

Table 4.4 Some Common Conversion Specifiers

Conversion Specifiers	Resulting Output
%c	Single character
%d	Signed decimal integer
%u	Unsigned decimal integer
%f	Decimal floating-point number
%s	Character string
%o	Unsigned octal integer
%x	Unsigned hexadecimal integer (a–f)
%X	Unsigned hexadecimal integer (A–F)
%%	Percent sign

The escape character for a newline (\n) has already been discussed. This character causes a carriage return and a line feed. Another escape character is the tab character (\t), which advances the output horizontally to the next tab stop. Use of these two escape characters is shown in the program of Figure 4.11, which displays decimal and floating-point numbers in various configurations using the newline and tab characters.

Three integers are defined: *int1* = *4*, *int2* = *20*, and *int3* = *40*. These are displayed on separate lines with and without tabs. The double newline characters cause the next printed line to be two lines below the current line. The floating-point numbers are displayed on three separate lines and indented by placing a newline character and a tab character before each floating-point conversion specifier. The sum of the two floating-point numbers *flp1* = *1.15* and *flp2* = *15.50* is obtained by the mathematical operator (+) to yield a sum of *flp1* + *flp2* = *16.650000*. The subtraction of the two floating-point numbers positive *flp1* = *1.15* and negative *flp3* = *–1.5* is obtained by the mathematical operator (–) to yield a result of *flp1* – *flp3* = *2.650000*. The mathematical operators are discussed in detail in the next section.

```
//displ_val_tabs.cpp
//display decimal and floating-point values
//using newline and tabs

#include "stdafx.h"

   int int1 = 4, int2 = 20, int3 = 40;
   float flp1 = 1.15, flp2 = 15.50, flp3 = -1.5;

int main (void)
{
   printf ("Decimal values without tabs:
            %d, %d, %d\n\n", int1, int2, int3);

   printf ("Decimal values with tabs:
            \t%d, \t%d, \t%d\n\n", int1, int2, int3);

   printf ("Floating-point values on three lines:
            \n\t%f, \n\t%f, \n\t%f\n\n", flp1, flp2, flp3);

   printf ("Floating-point addition of %f + %f = %f\n\n",
            flp1, flp2, flp1+flp2);

   printf ("Floating-point subtraction of %f - %f
            = %f\n\n", flp1, flp3, flp1-flp3);

   return 0;
}
                                 //continued on next page
```

(a)

Figure 4.11 Program to illustrate the use of the newline character and the tab character: (a) the C program and (b) the outputs.

```
Decimal values without tabs: 4, 20, 40

Decimal values with tabs:      4,     20,    40

Floating-point values on three lines:
      1.150000,
      15.500000,
      -1.500000

Floating-point addition  of  1.150000 + 15.500000 = 16.650000

Floating-point subtraction of 1.150000 - -1.500000 = 2.650000

Press any key to continue . . . _
```
(b)

Figure 4.11 (Continued)

Figure 4.12 illustrates a method to align columns using the newline and tab characters. The columns display decimal numbers in both the uppercase and lowercase hexadecimal number representation. The conversion specifiers *%X* and *%x* specify uppercase and lowercase hexadecimal, respectively.

```
//create_col.cpp
//creates three columns using the newline and tab characters

#include "stdafx.h"

int main (void)
{
  printf ("hex uppercase \t\thex lowercase \t\tdecimal\n");
  printf ("%X \t\t\t%x \t\t\t%d\n", 8, 8, 8);
  printf ("%X \t\t\t%x \t\t\t%d\n", 11, 11, 11);
  printf ("%X \t\t\t%x \t\t\t%d\n", 14, 14, 14);

  return 0;
}
                                   //continued on next page
```
(a)

Figure 4.12 Program to illustrate the use of the newline character and the tab character to create columns: (a) the C program and (b) the outputs.

```
hex uppercase          hex lowercase          decimal
8                      8                      8
B                      b                      11
E                      e                      14
Press any key to continue . . . _
```

(b)

Figure 4.12 (Continued)

One final example to illustrate alignment is shown in Figure 4.13, which adds a minimum field width integer between the percent sign (%) and the letter (*d*) of a conversion specifier. The integer is referred to as a *minimum field width specifier* and provides a minimum field width for the output. For example, a minimum right-aligned field width of six spaces is specified by *%6d*, as shown in the program of Figure 4.13.

If no minimum field width is specified, the output is left-aligned. A minimum field width of *%06d* will right-align the output with high-order zeroes, if necessary, to fill in the minimum field width. If the minimum field width is less than the width of the output, then the entire output is printed since the output is greater than the specified minimum width.

```
//min_field_width.cpp
//add an integer between the % sign and the letter of the
//conversion specifier to indicate a minimum field width

#include "stdafx.h"
   int num1, num2;            //global variables

int main (void)
{
   num1 = 92;
   num2 = 56789;

   printf ("%d\n", num1);   //prints 92 width 6 left-aligned
   printf ("%d\n", num2);   //prints 56789 width 6
                            //left-aligned
   printf ("%6d\n", num1); //prints 92 width 6 right-aligned

                                 //continued on next page
```

(a)

Figure 4.13 Program to illustrate the use of the minimum field width specifier: (a) the C program and (b) the outputs.

```
    printf ("%06d\n", num1);//prints 92 width 6 with
                            //0s right-aligned
    printf ("%3d\n", num2); //prints entire number
                            //left-aligned even though
                            //width of number is greater than
                            //the minimum width specified

    return 0;
}
```

```
92
56789
    92
000092
56789
Press any key to continue . . . _
```

(b)

Figure 4.13 (Continued)

4.3.2 Scanf ( ) Function

The **scanf ( )** function is the most commonly used input function of the several input functions in the C programming language. The customary use for the **scanf ( )** function is to input data from the keyboard. It converts inputs to the following formats: decimal integers, floating-point numbers, characters, and strings. The **scanf ( )** function can contain spaces and tabs — which are ignored — to make the string more readable. The **scanf ( )** function is similar to the **printf ( )** function — it contains a control string of format specifiers (also called conversion specifiers) *%d*, *%c*, *%f*, and so forth, which describe the format of each input variable. The general format for the **scanf ( )** function is shown below, together with a specific format to read a decimal integer.

```
scanf ( ) ("conversion specifiers", argument names);

                scanf ("%d", &num);
```

The ampersand (&) is the *address-of* operator, which indicates that the decimal input number is to be stored in the location assigned to the variable *num*. The conversion specifiers for the **scanf ( )** function are the same as for the **print ( )** function and are reproduced in Table 4.5 for convenience. When the **scanf ( )** stores input data, it converts the data to the corresponding conversion specifier in the format string.

The **scanf ( )** function is normally preceded by the **printf ( )** function, which requests the user to enter information. The **scanf ( )** function then assigns the input data to a specific location.

Table 4.5 Some Common Conversion Specifiers

Conversion Specifiers	Resulting Output
%c	Single character
%d	Signed decimal integer
%u	Unsigned decimal integer
%f	Decimal floating-point number
%s	Character string
%o	Unsigned octal integer
%x	Unsigned hexadecimal integer (a–f)
%X	Unsigned hexadecimal integer (A–F)
%%	Percent sign

The example shown in Figure 4.14 requests the user to enter an integer number and a floating-point number using the **printf ( )** function, then stores the data in specific locations using the **scanf ( )** function, then prints the input data to the monitor screen.

```
//input_int_float
//input an integer and a floating-point
//number and display the values

#include "stdio.h"

int main (void)
{
    int int_num;
    float float_num;

    printf ("Enter an integer: ");
    scanf (" %d", &int_num);

    printf ("Enter a floating-point number: ");
    scanf(" %f", &float_num);

    printf ("%d\n", int_num);
    printf ("%f\n", float_num);

    return 0;
}
```

//continued on next page

(a)

Figure 4.14 Program to illustrate the use of the printf ( ) and scanf ( ) functions to enter and print an integer and a floating-point number: (a) the C program and (b) the outputs.

```
Enter an integer: 7
Enter a floating-point number: 4.5
7
4.500000
Press any key to continue_
```

(b)

Figure 4.14 (Continued)

The example shown in Figure 4.15 uses the **printf ( )** function and the **scanf ( )** function to request the user to enter two integer numbers, perform an addition and subtraction on the numbers, then print the sum and difference. This program uses the arithmetic operators of addition (+) and subtraction (–), which are covered in more detail in the next section.

```
//add_sub_int.cpp
//enter two integers, then add and subtract them

#include "stdafx.h"

int main (void)
{
	int num1, num2;
	printf ("Enter first integer:");
	scanf ("%d", &num1);

	printf ("Enter second integer:");
	scanf ("%d", &num2);

	printf ("Sum is %d\n", num1 + num2);
	printf ("Difference is %d\n", num1 - num2);

	return 0;
}
```

//continued on next page

(a)

Figure 4.15 Program to illustrate the use of the **printf( )** and **scanf ( )** functions to enter two integers, perform addition and subtraction on the two numbers, then print the sum and difference: (a) the C program and (b) the outputs.

```
Enter first integer: 9
Enter second integer: 3
Sum is 12
Difference is 6
Press any key to continue . . . _
----------------------------------------------------------
Enter first integer: 45
Enter second integer: 63
Sum is 108
Difference is -18
Press any key to continue . . . _
```

(b)

Figure 4.15 (Continued)

4.4 Operators

Operators are symbols that direct C to execute specific operations. Some of the more common operators are used to perform the following operations: an arithmetic operation, a relational operation to determine the relationship between two expressions, a logical operation of AND, OR, or NOT, a conditional operation to replace the **if-else** statement, an increment or decrement operation, and bitwise operations that perform logical operations on a bit-by-bit basis. Table 4.6 lists the more common operators.

Table 4.6 Common Operators Used in C Programming

Operator	Function
=	Assignment
+	Addition
–	Subtraction
*	Multiplication
/	Division
%	Modulus (remainder)
==	Equal to
>	Greater than
<	Less than
>=	Greater than or equal to
<=	Less than or equal to
!=	Not equal to //continued on next page

Table 4.6 Common Operators Used in C Programming

<table>
<tr><th>Operator</th><th>Function</th></tr>
<tr><td>&&</td><td>Logical AND</td></tr>
<tr><td>||</td><td>Logical OR</td></tr>
<tr><td>!</td><td>Logical NOT</td></tr>
<tr><td>? :</td><td>Conditional</td></tr>
<tr><td>++</td><td>Increment the operand by 1</td></tr>
<tr><td>--</td><td>Decrement the operand by 1</td></tr>
<tr><td>&</td><td>Bitwise AND</td></tr>
<tr><td>|</td><td>Bitwise OR</td></tr>
<tr><td>^</td><td>Bitwise exclusive-OR</td></tr>
<tr><td>~</td><td>Bitwise 1s complement</td></tr>
</table>

4.4.1 Arithmetic Operators

An expression is a conjunction of operators and operands — or expressions — and may also contain variables and constants. Expressions can contain multiple operators that are evaluated in a specific sequence, depending on the precedence of the operators; for example, multiplication, division, and modulus have a higher precedence than addition. However, parentheses can be utilized to alter the precedence, as shown in the following example:

```
result = (6 + 7) * 8;
```

Normally, the multiplication operator (*) would have a higher precedence than the addition operator (+). The use of parentheses, however, places a higher precedence on the addition operation than on the multiply operation. This provides a value of 104 for the right-hand expression, which is assigned to the variable *result*.

Assignment operator The assignment operator (=) does not signify *equality* — it means to copy the right-hand side expression to the left-hand side tar expression. For example, the statement $x = y$ means to copy the value of y into x, that is, y is *assigned* to x, where the variable x refers to a memory location. However, a value cannot be assigned to a constant.

There can be multiple assignments in a single statement, where the order of assignments is in a right-to-left sequence. For example, the statement shown below first assigns 75 to z, then assigns 75 to y, and finally assigns 75 to x. The program shown in Figure 4.16 illustrates this concept. Multiple assignments is a simple way to initialize variables to a known value.

```
x = y = z = 75;
```

```
//mult_assign.cpp
//multiple assignments are made to initialize
//variables to a known value

#include "stdafx.h"

int    main (void)
{
       int x, y , z;
       x = y = z = (75 + 25);

//the \n prints each variable on a separate line
       printf ("x = %d, \ny = %d, \nz = %d\n", x, y, z);

       return 0;
}
```

(a)

```
x = 100,
y = 100,
z = 100
Press any  key to continue . . . _
```

(b)

Figure 4.16 Program to illustrate using multiple assignments: (a) the C program and (b) the outputs.

Addition and subtraction operators The operands used for addition are the *augend* and the *addend.* The addend is added to the augend to form the sum, which replaces the augend in most computers — the addend is unchanged. The addition operator (+) performs the addition operation on the two operands, which can be either constants or variables. The rules for binary addition are shown in Table 4.7. A carry of 1 indicates a carry-out to the next higher-order column.

Table 4.7 Rules for Binary Addition

Augend		Addend	Carry	Sum
0	+	0	0	0
0	+	1	0	1
1	+	0	0	1
1	+	1	1	0

An example of binary addition is shown in Figure 4.17. The sum of column 1 is 2_{10} (10_2); therefore, the sum is 0 with a carry of 1 to column 2. The sum of column 2 is 4_{10} (100_2); therefore, the sum is 0 with a carry of 0 to column 3 and a carry of 1 to column 4. The sum of column 3 is 4_{10} (100_2); therefore, the sum is 0 with a carry of 0 to column 4 and a carry of 1 to column 5. The sum of column 4 is 2_{10} (10_2); therefore, the sum is 0 with a carry of 1 to column 5. The sum of column 5 is 2 (10_2); therefore, the sum is 0 with a carry of 1 to column 6. The unsigned radix 10 values of the binary operands are shown in the rightmost column together with the resulting sum.

Column	6	5	4	3	2	1	Radix 10 values
			0	1	1	0	6
			0	1	0	1	5
			1	1	1	1	15
+)		$_{11}$	0_{10}	1_0	1_1	0	6
	1	0	0	0	0	0	32

Figure 4.17 Example of binary addition.

The operands used for subtraction are the *minuend* and the *subtrahend.* The subtrahend is subtracted from the minuend to obtain the difference. The subtraction operator (–) performs the subtraction operation on the two operands, which can be either constants or variables. The rules for binary subtraction are shown in Table 4.8. A borrow of 1 indicates a borrow from the minuend in the next higher-order column.

Table 4.8 Rules for Binary Subtraction

Minuend		Subtrahend	Borrow	Difference
0	–	0	0	0
0	–	1	1	1
1	–	0	0	1
1	–	1	0	0

Two 8-bit operands are shown in Figure 4.18 to illustrate the rules for radix 2 subtraction in which all four combinations of two bits are provided. The borrow from the minuend in column 2^1 to the minuend in column 2^0 changes column 2^1 from 10 to 00; that is, the operation of column 2^1 then becomes $0 - 0 = 0$. The program shown in Figure 4.19 illustrates the addition and subtraction operations utilizing three integers.

		2^7	2^6	2^5	2^4	2^3	2^2	2^1	2^0
	A (Minuend) = +54	0	0	1	1	0	1	1	0
–)	*B* (Subtrahend) = +37	0	0	1	0	0	1	0	1
	D (Difference) = +17	0	0	0	1	0	0	0	1

Figure 4.18 Example of binary subtraction.

```
//add_sub_int2.cpp
//addition and subtraction of integers

#include "stdafx.h"

int main (void)
{
        int int1, int2, int3;

        printf ("Enter three integers: ");
        scanf ("%d %d %d", &int1, &int2, &int3);

        printf ("\nInteger sum = %d", int1 + int2 + int3);
        printf ("\nInteger difference = %d\n",
                    (int1 + int2) - int3);

        return 0;
}
```

(a)

```
Enter three integers: 125 50 200

Integer sum = 375
Integer difference = -25
Press any key to continue . . . _
```

(b)

Figure 4.19 Program to illustrate the addition and subtraction of integers: (a) the C program and (b) the outputs.

The program shown in Figure 4.20 illustrates addition and subtraction operations utilizing integer numbers and floating-point numbers. The keyword **int** establishes *int1* and *int2* as integers; in the same way, *flp1* and *flp2* are declared as type **float**. The

scanf ( ) function stores the two integers in the locations specified by *&int1* and *&int2*, where the ampersand indicates the *address-of*. The arguments (or parameters) of the second print function specify an addition of *int1*and *int2*, in the same way that the arguments of the third print function specify a subtraction of *int1*and *int2*. In both print functions, the newline character (\ *n*) at the beginning of the format strings places the sum and difference on separate lines. The double newline characters insert a blank line between the integer difference result and the print function to enter two floating-point numbers. The conversion specifier (*%d*) in both cases indicate a decimal integer value.

The second **scanf ( )** function stores the two floating-point numbers in the locations specified by *&flp1* and *&flp2*. The conversion specifier (*%f*) in both cases indicates a decimal floating-point value. The arguments of the print function specify an addition of *flp1* and *flp2*, in the same way that the arguments in the next print function specify a subtraction of *flp1* and *flp2*.

```
//add_sub_int_flp3.cpp
//add and subtract integer numbers
//and floating-point numbers

#include "stdafx.h"

int main (void)
{
      int int1, int2;
      float flp1, flp2;

      printf ("Enter two integer numbers: ");
      scanf ("%d %d", &int1, &int2);

      printf ("\nInteger sum = %d", int1 + int2);
      printf ("\nInteger difference = %d\n\n", int1 - int2);

      printf ("Enter two floating-point numbers: ");
      scanf ("%f %f", &flp1, &flp2);

      printf ("\nFloating-point sum = %f", flp1 + flp2);
      printf ("\nFloating-point difference = %f\n\n",
                  flp1 - flp2);

      return 0;
}                                     //continued on next page
```

(a)

Figure 4.20 Program to illustrate the addition and subtraction of integer numbers and floating-point numbers: (a) the C program and (b) the outputs.

```
Enter two integer numbers: 45 70

Integer sum = 115
Integer difference = -25

Enter two floating-point numbers: 4.5 5.75

Floating-point sum = 10.250000
Floating-point difference = -1.250000

Press any key to continue . . ._
------------------------------------------------------------
Enter two integer numbers: -125 25

Integer sum = -100
Integer difference = -150

Enter two floating-point numbers: 6.25 -8.50

Floating-point sum = -2.250000
Floating-point difference = 14.750000

Press any key to continue . . ._
```

(b)

Figure 4.20 (Continued)

Multiplication and division operators The multiplication operator (*) multiplies the multiplicand by the multiplier to produce a product. In a hardware multiplication unit, the multiplicand and multiplier are both n-bit operands that produce a $2n$-bit result. If the operands are in 2s complement notation, then the sign bit is treated in a manner identical to the other bits; however, the sign bit of the multiplicand is extended left in the partial product to accommodate the $2n$-bits of the product.

The only requirement is that the multiplier must be positive — the multiplicand can be either positive or negative. This can be resolved by either 2s complementing both operands or by 2s complementing the multiplier, performing the multiplication, then 2s complementing the result.

A simple example of multiplying two 4-bit operands is shown in Figure 4.21. Let the multiplicand and multiplier be *a[3:0]* = 0111 (+7) and *b[3:0]* = 0101 (+5), respectively to produce a product *p[7:0]* = 0010 0011 (+35). A multiplier bit of 1 copies the multiplicand to the partial product; a multiplier bit of 0 enters 0s in the partial product. This is the sequential add-shift multiplication algorithm. There are numerous methods to perform multiplication; however, since the topic of this book is programming, the multiplication algorithms are not important.

Multiplicand *A*					0	1	1	1	+7
Multiplier *B*				×)	0	1	0	1	+5
Partial products	0	0	0	0	0	1	1	1	
	0	0	0	0	0	0	0		
	0	0	0	1	1	1			
	0	0	0	0	0				
Product *P*	0	0	1	0	0	0	1	1	+35

Figure 4.21 Example of the sequential add-shift multiply algorithm.

An example of integer and floating-point multiplication is shown in Figure 4.22, which prompts the user to enter an integer length and width, then prints the integer area. Next a floating-point length and width are entered, then the program prints the floating-point area.

```
//area.cpp
//calculates the area of a flat surface
//using integers and floating-point numbers

#include "stdafx.h"

int main (void)
{
	int int_length, int_width;
	float flp_length, flp_width;

	printf ("Enter integer length: ");
	scanf ("%d", &int_length);

	printf ("Enter integer width: ");
	scanf ("%d", &int_width);

	printf ("Integer area = %d\n", int_length * int_width);

                              //continued on next page
```

(a)

Figure 4.22 Program to illustrate the multiplication of integer numbers and floating-point numbers: (a) the C program and (b) the outputs.

```
        printf ("\nEnter floating-point length: ");
        scanf ("%f", &flp_length);

        printf ("Enter floating-point width: ");
        scanf ("%f", &flp_width);

        printf ("Floating-point area = %f\n\n",
                    flp_length * flp_width);

        return 0;
}
```

```
Enter integer length: 40
Enter integer width: 40
Integer area = 1600

Enter floating-point length: 10.5
Enter floating-point width: 10.5
Floating-point area = 110.250000

Press any key to continue . . . _
```

(b)

Figure 4.22 (Continued)

The division operator (/) divides the dividend by the divisor to produce a quotient when used in integer division. The integer division operation in C programming produces a quotient result only — the remainder is discarded; that is, any fraction is truncated, because integer division produces an integer result. If a remainder is desired, then this can be obtained by using the modulus operator (%) for integer division. Division can take place using either integer operands or floating-point operands. In a hardware division unit, a $2n$-bit dividend is divided by an n-bit divisor to produce an n-bit quotient and an n-bit remainder.

The example shown in Figure 4.23 is a program to illustrate integer division, the modulus operator, and floating-point division. The integer quotient is obtained by the operands that were entered by the user, as shown below.

```
Integer quotient = int_dvdnd / int_dvsr
```

The integer remainder is obtained using the modulus operator (%) as shown below.

```
Integer remainder = int_dvdnd % int_dvsr
```

The floating-point quotient is obtained by the division operator (/). Note that the conversion specifier for the result is *%.2f*, which means that the result is to be rounded up to two digits to the right of the decimal point.

```
//div_int_float.cpp
//program to illustrate integer division
//and floating-point division

#include "stdio.h"

int main (void)
{
   int int_dvdnd, int_dvsr;
   float flp_dvdnd, flp_dvsr;

   printf ("Enter an integer dividend: ");
   scanf ("%d", &int_dvdnd);

   printf ("Enter an integer divisor: ");
   scanf ("%d", &int_dvsr);

   printf ("Integer quotient = %d\n", int_dvdnd / int_dvsr);
   printf ("Integer remainder = %d\n\n",
            int_dvdnd % int_dvsr);

//--------------------------------------------------------

   printf ("Enter a floating-point dividend: ");
   scanf ("%f", &flp_dvdnd);

   printf ("Enter a floating-point divisor: ");
   scanf ("%f", &flp_dvsr);

   printf ("Floating-point quotient = %.2f\n\n",
         flp_dvdnd / flp_dvsr);

   return 0;
}
```

//continued on next page

(a)

Figure 4.23 Program to illustrate integer division, the modulus operator, and floating-point division: (a) the C program and (b) the outputs.

```
Enter an integer dividend: 128
Enter an integer divisor: 6
Integer quotient = 21
Integer remainder = 2

Enter a floating-point dividend: 50000.0
Enter a floating-point divisor: 75.0
Floating-point quotient = 666.67

Press any key to continue . . . _
------------------------------------------------------------
Enter an integer dividend: 125
Enter an integer divisor: 100
Integer quotient = 1
Integer remainder = 25

Enter a floating-point dividend: 65000.0
Enter a floating-point divisor: 85.0
Floating-point quotient = 764.71

Press any key to continue . . . _
------------------------------------------------------------
Enter an integer dividend: 625
Enter an integer divisor: 35
Integer quotient = 17
Integer remainder = 30

Enter a floating-point dividend: 44444.0
Enter a floating-point divisor: 65.0
Floating-point quotient = 683.75

Press any key to continue . . . _
------------------------------------------------------------
Enter an integer dividend: 625
Enter an integer divisor: 25
Integer quotient = 25
Integer remainder = 0

Enter a floating-point dividend: 75.0
Enter a floating-point divisor: 6.0
Floating-point quotient = 12.50

Press any key to continue . . . _
```

(b)

Figure 4.23 (Continued)

The program shown in Figure 4.24 illustrates the order of precedence of the arithmetic operators. The order of precedence is reproduced below for convenience. The value of the *flp1* variable is straightforward, it is simply the sum of 3.0 + 3.0 = 6.000000. The value of the *flp2* uses the operator precedence in the calculation, where the value of the variable is obtained by first multiplying 3.0 × 10.5 to yield 31.5; this result is then added to a value of 2.0 to yield a result of 33.500000.

In the calculation of *flp3*, both the division operator and the multiplication operator have the same precedence. Therefore, the operation proceeds in a left-to-right sequence. The division of 3/5 yields a result of zero; therefore, 0 × 22.0 = 0.00000. In the equation for *flp4*, both operators again have the same precedence and the operation proceeds in a left-to-right sequence. Thus, 22.0 × 3 = 66.0, which is then divided by 5.0 to yield a result of 13.200000. To obtain the value of *flp5*, the modulus operator takes precedence over the addition operator. The remainder of 3/5 is 3, which is added to 2.0 to yield a value of 5.000000.

Precedence Order	Arithmetic Operator
Highest order	Multiplication (*), division (/), modulus remainder (%)
Next highest order	Addition (+), subtraction (–)

```
//op_precedence.cpp
//shows the precedence of the arithmetic operators

#include "stdafx.h"
int main (void)
{
      float flp1, flp2, flp3, flp4, flp5;

      flp1 = 3.0 + 3.0;
      flp2 = 2.0 + 3.0 * 10.5;        //* is higher precedence
      flp3 = 3/5 * 22.0;              //same precedence
      flp4 = 22.0 * 3.0/5.0;          //same precedence
      flp5 = 2.0 + 3 % 5;             //% is higher precedence

      printf ("%f\n", flp1);
      printf ("%f\n", flp2);
      printf ("%f\n", flp3);
      printf ("%f\n", flp4);
      printf ("%f\n\n", flp5);
      return 0;                       //continued on next page
}
```

(a)

Figure 4.24 Program to illustrate operator precedence: (a) the C program and (b) the outputs.

```
6.000000
33.500000
0.000000
13.200000
5.000000

Press any key to continue . . . _
```

(b)

Figure 4.24 (Continued)

4.4.2 Relational Operators

Relational operators compare data to determine the relationship between the data. There are six relational operators that are used for comparing data, where the data can be constants, variables, or expressions, or a combination of these. When determining the relationship between data, if the relationship is true, then a value of 1 (or nonzero) is returned; if the relationship is false, then a value of 0 is returned. The six relational operators are defined in Table 4.9.

Table 4.9 Relational Operators

Operator Symbol	Relationship	Example
= =	Equal	x = = y
! =	Not equal	x ! = y
>	Greater than	x > y
> =	Greater than or equal	x > = y
<	Less than	x < y
< =	Less than or equal	x < = y

Relational operators can be used for both integers and floating-point numbers. Relational operators have a lower precedence than the arithmetic operators. The relational operators (>), (> =), (<), and (< =) have a higher precedence than (= =) or (! =). Note that (= =) and (=) have distinct meanings:

```
x = 5;        Assigns a value of 5 to x; whereas
x == 5;       Checks to determine if x = 5
```

An expression generated by relational operators is also referred to as a *relational expression*. Relational operators are placed between the expressions that are being compared. The program shown in Figure 4.25 provides an example of utilizing the six relational operators. The user is prompted to enter two integers, then the program applies the relational operators to determine the relationship between the integers.

```
//relational_ops.cpp
//determines the relationship between two integers

#include "stdio.h"

int main (void)
{
   int int1, int2;

   printf ("Enter first integer: ");
   scanf ("%d", &int1);

   printf ("Enter second integer: ");
   scanf ("%d", &int2);

//determine relationship: true = 1, false = 0
   printf ("\nint1 == int2 %d\n", int1 == int2);
   printf ("int1 != int2 %d\n", int1 != int2);
   printf ("int1 > int2 %d\n", int1 > int2);
   printf ("int1 >= int2 %d\n", int1 >= int2);
   printf ("int1 < int2 %d\n", int1 < int2);
   printf ("int1 <= int2 %d\n\n", int1 <= int2);

   return 0;
}
```

(a)

```
Enter first integer: 4
Enter second integer: 9

int1 == int2 0
int1 != int2 1
int1 > int2 0
int1 >= int2 0
int1 < int2 1
int1 <= int2 1

Press any key to continue . . . _
```

//continued on next page

(b)

Figure 4.25 Program to illustrate using the relational operators: (a) the C program and (b) the outputs.

```
Enter first integer: 92
Enter second integer: 75

int1 == int2 0
int1 != int2 1
int1 > int2 1
int1 >= int2 1
int1 < int2 0
int1 <= int2 0

Press any key to continue . . . _
------------------------------------------------------------
Enter first integer: 25
Enter second integer: 25

int1 == int2 1
int1 != int2 0
int1 > int2 0
int1 >= int2 1
int1 < int2 0
int1 <= int2 1

Press any key to continue . . . _
```

Figure 4.25 (Continued)

As a final example on relational operators, the program in Figure 4.26 illustrates that the arithmetic operators have a higher precedence than relational operators.

```
//relational_ops2.cpp
//program to illustrate that the arithmetic operators
//have a higher precedence than the relational operators

#include "stdafx.h"

int main (void)
{
	int int1, int2;
	double flp1;
										//continued on next page
```

(a)

Figure 4.26 Program to illustrate using the relational operators: (a) the C program and (b) the outputs.

```
        int1 = 10;
        int2 = 40;
        flp1 = 37.65;

        printf ("Let int1 = %d, int2 = %d, flp1 = %.2f\n\n",
                    int1, int2, flp1);

        printf ("int1 >= int2 produces: %d\n", int1 >= int2);
        printf ("int1 == int2 produces: %d\n", int1 == int2);
        printf ("int1 < flp1 produces: %d\n", int1 < flp1);
        printf ("int2 > flp1 produces: %d\n", int2 > flp1);

//arithmetic operations execute before relational operations
        printf ("int1 != int2 - 30 produces: %d\n",
                    int1 != int2 - 30);
        printf ("int1 + int2 != flp1 produces: %d\n\n",
                    int1 + int2 != flp1);

        return 0;
}
```

```
Let int1 = 10, int2 = 40, flp1 = 37.65

int1 >= int2 produces: 0
int1 == int2 produces: 0
int1 < flp1 produces: 1
int2 > flp1 produces: 1
int1 != int2 - 30 produces: 0    //arithmetic ops execute 1st
int1 + int2 != flp1 produces: 1  //arithmetic ops execute 1st

Press any key to continue . . . _
```

(b)

Figure 4.26 (Continued)

4.4.3 The If Statement

Conditional statements, such as the **if** statement, alter the flow within a program based on certain conditions. The **if** statement is also referred to as a *program control statement*, a *selection statement*, or a *decision statement*. The **if** statement makes a decision based on the result of a test. The choice among alternative statements depends on the

Boolean value of an expression. The alternative statements can be a single statement or a block of statements. The syntax for the **if** statement is shown below, where a true value is 1 or any nonzero value; a false value is 0. If the expression is true, then the statement, or block of statements, is executed; if the expression is false, then the statement, or block of statements, is not executed, then the next statement in the program flow is executed.

```
if (expression/condition)
   {statement or block of statements}
```

Figure 4.27 illustrates an example of a C program to determine if an integer entered by the user is greater than, equal to, or less than 100. This program requires three **if** statements, one for each comparison.

```
//chk_value.cpp
//determine if a value is greater than,
//equal to, or less than 100

#include "stdafx.h"

int main (void)
{
        int num;

        printf ("Enter an integer: ");
        scanf ("%d", &num);

        if (num > 100)
              printf ("Number is greater than 100\n\n");

        if (num == 100)
              printf ("Number is equal to 100\n\n");

        if (num < 100)
              printf ("Number is less than 100\n\n");

        return 0;
}
                                        //continued on next page
```

(a)

Figure 4.27 Program to illustrate using the **if** statement: (a) the C program and (b) the outputs.

```
Enter an integer: 101
Number is greater than 100

Press any key to continue . . . _
-----------------------------------------------------------
Enter an integer: 100
Number is equal to 100

Press any key to continue . . . _
-----------------------------------------------------------
Enter an integer: 99
Number is less than 100

Press any key to continue . . . _
```

(b)

Figure 4.27 (Continued)

A final example to illustrate the **if** statement is shown in Figure 4.28, which converts Fahrenheit to centigrade or centigrade to Fahrenheit, depending on the user's request. The conversion equations are shown.

$$\text{Fahrenheit} = (\text{Centigrade} \times 1.8) + 32$$

$$\text{Centigrade} = (\text{Fahrenheit} - 32) \times 5/9$$

```
//fahr_cent_conv.cpp
//converts fahrenheit to centigrade or
//centigrade to fahrenheit

#include "stdafx.h"

int main (void)
{
      int choice;
      float tempf;
      float temp_fahr;
      float tempc;
      float temp_cent;             //continued on next page
```

(a)

Figure 4.28 Program to illustrate using the **if** statement: (a) the C program and (b) the outputs.

```
        printf ("Enter fahrenheit temperature: ");
        scanf ("%f", &tempf);
        temp_cent = ((tempf - 32) * 5)/9;


        printf ("Enter centigrade temperature: ");
        scanf ("%f", &tempc);
        temp_fahr = (tempc * 1.8) + 32;

        printf ("\n1: fahr to cent, 2: cent to fahr \n");
        printf ("Enter choice: ");
        scanf ("%d", &choice);

        if (choice == 1)
              printf ("Centigrade = %f\n", temp_cent);

        if (choice == 2)
              printf ("Fahrenheit = %f\n", temp_fahr);

        return 0;
}
```

```
Enter fahrenheit temperature: 32
Enter centigrade temperature: 25

1: fahr to cent, 2: cent to fahr
Enter choice: 1
Centigrade = 0.000000

Press any key to continue . . . _
------------------------------------------------------------
Enter fahrenheit temperature: 68
Enter centigrade temperature: 45

1: fahr to cent, 2: cent to fahr
Enter choice: 1
Centigrade = 20.000000

Press any key to continue . . . _
                                        //continued on next page
```

(b)

Figure 4.28 (Continued)

```
Enter fahrenheit temperature: 45
Enter centigrade temperature: 22

1: fahr to cent, 2: cent to fahr
Enter choice: 2
Fahrenheit = 71.599998

Press any key to continue . . . _
-----------------------------------------------------------
Enter fahrenheit temperature: 72
Enter centigrade temperature: 41

1: fahr to cent, 2: cent to fahr
Enter choice: 2
Fahrenheit = 105.800003

Press any key to continue . . . _
-----------------------------------------------------------
Enter fahrenheit temperature: 0
Enter centigrade temperature: 10

1: fahr to cent, 2: cent to fahr
Enter choice: 1
Fahrenheit = -17.777779

Press any key to continue . . . _
```

Figure 4.28 (Continued)

4.4.4 The Else Statement

The **else** statement is used in conjunction with the **if** statement to provide alternative paths through the program. The syntax for the conditional branching **if-else** construct is shown below. When the **if** expression is true, then the statement(s) following the **if** statement will be executed. When the **if** expression is false, then the statement(s) contained in the **else** block are executed.

```
if (expression/condition)
      {statement or block of statements}

else
      {statement or block of statements}
```

The statement(s) in the **if** block will be executed only when the condition is true (1). However, when the condition is false (0), the **else** statement(s) will be executed. When the **if** expression/condition returns a true value, then the statements in the **else** block are not executed. This technique provides a two-way decision path.

A variation of the **if-else** construct is the nested **if** statements in which only one block of statements is executed. The syntax is shown below. If expression_1 is true, then statement_1 (or block of statements) is executed and the program exits the nested **if** statements. If expression_1 is false and expression_2 is true, then statement_2 (or block of statements) is executed and the program exits the nested **if** statements. If expression_2 is false, then the program executes statement_3 (or block of statements).

```
if (expression_1/condition_1)
      {statement_1 or block of statements}

else if (expression_2/condition_2)
      {statement2 or block of statements}

else {statement_3 or block of statements}
```

The program of Figure 4.27 will be redesigned using the nested **if** technique to determine if a number that is entered from the keyboard is greater than, equal to, or less than 100. The program is listed in Figure 4.29.

```
//chk_value_if_else.cpp
//use nested if-else to determine if a number
//is greater than, equal to, of less than 100

#include "stdafx.h"

int main (void)
{
      int num;

      printf ("Enter an integer: ");
      scanf ("%d", &num);

      if (num > 100)
      printf ("Number is greater than 100\n\n");
                                    //continued on next page
```

(a)

Figure 4.29 Program to illustrate using the nested **if** statement: (a) the C program and (b) the outputs.

```
        else if (num == 100)
                printf ("Number is equal to 100\n\n");

        else
                printf ("Number is less than 100\n\n");

        return 0;
}
```

```
Enter an integer: 222
Number is greater than 100

Press any key to continue . . . _
------------------------------------------------------------------
Enter an integer: 100
Number is equal to 100

Press any key to continue . . . _
------------------------------------------------------------------
Enter an integer: 45
Number is less than 100

Press any key to continue . . . _
```

(b)

Figure 4.29 (Continued)

The program shown in Figure 4.30 depicts an example of integer division. If the divisor is zero, then the division operation is invalid. The user is requested to enter a dividend and a divisor. If the divisor is not zero, then the divide operation is performed; otherwise, a message is displayed stating that division by zero is invalid.

```
//div_integer.cpp
//perform division of two integers using if-else.
//detect divide by zero

#include "stdafx.h"

int main (void)                         //continued on next page
```

(a)

Figure 4.30 Program to illustrate using the **if-else** construct for a divide operation: (a) the C program and (b) the outputs.

```
{
      int dvdnd, dvsr;

      printf ("Enter dividend: ");
      scanf ("%d", &dvdnd);

      printf ("Enter divisor: ");
      scanf ("%d", &dvsr);

      if (dvsr == 0)
            printf ("\nCannot divide by zero\n");

      else
            printf ("\nQuotient = %d\n", dvdnd/dvsr);
            printf ("Remainder = %d\n\n", dvdnd % dvsr);

      return 0;
}
```

```
Enter dividend: 45
Enter divisor: 6

Quotient = 7
Remainder = 3

Press any key to continue . . . _
------------------------------------------------------------
Enter dividend: 64
Enter divisor: 8

Quotient = 8
Remainder = 0

Press any key to continue . . . _
------------------------------------------------------------
Enter dividend: 115
Enter divisor: 4

Quotient = 28
Remainder = 3

Press any key to continue . . . _
                                  //continued on next page
```

(b)

Figure 4.30 (Continued)

```
Enter dividend: 75
Enter divisor: 0

Cannot divide by zero
Press any key to continue . . . _
```

Figure 4.30 (Continued)

As a final example in this section, Figure 4.31 provides an example illustrating using the **if-else** construct to either add two operands or subtract two operands. For addition, the addend is added to the augend; for subtraction, the subtrahend is subtracted from the minuend. The variable named *opnd1* is the augend/minuend; the variable named *opnd2* is the addend/subtrahend. If the user enters a choice of 1, then the operation is addition; otherwise, the operation is subtraction. The blocks of code for the **if** statement and the **else** statement are delimited by beginning and ending braces.

```
//add_sub_if_else.cpp
//addition and subtraction using if-else

#include "stdafx.h"

int main (void)
{
   int choice, opnd1, opnd2;

   printf ("Enter: 1 for addition, 2 for subtraction: ");
   scanf ("%d", &choice);

   if (choice == 1)
      {
         printf ("Enter augend: ");
         scanf ("%d", &opnd1);

         printf ("Enter addend: ");
         scanf ("%d", &opnd2);

         printf ("Sum = %d\n", opnd1 + opnd2);
      }                                  //continued on next page
```

(a)

Figure 4.31 Program to illustrate using the **if-else** construct for addition and subtraction operations: (a) the C program and (b) the outputs.

```
   else
      {
         printf ("Enter minuend: ");
         scanf ("%d", &opnd1);

         printf ("Enter subtrahend: ");
         scanf ("%d", &opnd2);

         printf ("Difference = %d\n", opnd1 - opnd2);
      }

   return 0;
}
```

```
Enter: 1 for addition, 2 for subtraction: 1
Enter augend: 45
Enter addend: 57
Sum = 102

Press any key to continue . . . _
------------------------------------------------------------
Enter: 1 for addition, 2 for subtraction: 1
Enter augend: -258
Enter addend: 200
Sum = -58

Press any key to continue . . . _
------------------------------------------------------------
Enter: 1 for addition, 2 for subtraction: 2
Enter minuend: 79
Enter subtrahend: 33
Difference = 46

Press any key to continue . . . _
------------------------------------------------------------
Enter: 1 for addition, 2 for subtraction: 2
Enter minuend: -340
Enter subtrahend: -200
Difference = -140

Press any key to continue . . . _
```

(b)

Figure 4.31 (Continued)

4.4.5 Logical Operators

There are three logical operators, as shown in Table 4.10. The AND operation evaluates as true (1) only if both expression 1 AND expression 2 are true; otherwise, it evaluates as false (0). The OR operation evaluates as true (1) if either expression 1 OR expression 2 is true; it evaluates as false (0) if both expressions are false. The OR operator is also referred to as the *inclusive* OR; the *exclusive* OR is discussed later. The NOT operation evaluates as false (0) if expression 1 is true; otherwise, it evaluates as true (1) if expression 1 is false. Table 4.11 is a truth table that illustrates the three logical operators using binary values.

Table 4.10 Logical Operators

Operator Symbol	Operation	Example
&&	AND	expression 1 && expression 2
||	OR	expression 1 || expression 2
!	NOT	! expression 1

Table 4.11 Truth Table for the Logical Operators

x_1	x_2	x_1 && x_2	x_1 || x_2	! x_1
0	0	0	0	1
0	1	0	1	1
1	0	0	1	0
1	1	1	1	0

Logical operators can be combined with relational operators to form a single expression that evaluates to either true (1) or false (0). When used in this context, they are sometimes referred to as *compound relational* operators. When relational and logical operators are combined, relational operators have a higher precedence. Logical operators are used with expressions, not with individual bits — individual bits are evaluated using *bitwise* operators. The arithmetic operators have a higher precedence than the relational operators, which have a higher precedence than the logical operators. The precedence of relational and logical operators is shown in Table 4.12.

Relational and logical operators can be used in **if** statements. For example, in the statement shown below, since the first expression is true, C will not evaluate the second expression. This is because a logic 1 ORed with anything will generate a logic 1 regardless of the value of the second expression.

```
if (10 > 9) || (2 > 1)
```

In a similar manner, in the statement shown below, since the first expression is false, C will not evaluate the second expression. This is because a logic 0 ANDed with anything will generate a logic 0 regardless of the value of the second expression.

```
if (10 < 9) && (1< 2)
```

Table 4.12 Precedence of Relational and Logical Operators

Operator Symbol	Precedence
!	Highest
>, >=, <, <=	
= =, ! =	
&&	
||	Lowest

The exclusive-OR function is defined for two variables x_1 and x_2 as shown below; that is, only when the variables are different will a logic 1 be generated.

$$x_1x_2' + x_1'x_2$$

The C programming language does not provide an exclusive-OR operator; however, the function can be easily obtained by utilizing the logical operands && and | |, as follows:

$$(x_1 \ \&\& \ !x_2) \ || \ (!x_1 \ \&\& \ x_2)$$

Examples of logical operators are shown below together with the resulting evaluation.

Expression	Evaluates as
(14 = = 14) && (20 != 10)	True (1), because both expressions are true
(20 > 10) | | (15 < 10)	True (1), because one expression is true
(10 = = 10) && (30 < 20)	False (0), because one expression is false
!(10 = = 15)	True (1), because the expression is false
(30 > 10) && (20 = = 20)	True (1), because both expressions are true
!(10 = = 10)	False (0), because the expression is true
!(20 > 10) && (30 = = 30)	False (0), because both expressions are true

Figure 4.32 shows a program to illustrate the logical operators, including the exclusive-OR function. The user enters two binary digits and the program performs the appropriate logical operations. All combinations of two variables are entered.

```
//logical_ops.cpp
//program to illustrate the use of
//the logical operators, AND, OR, and NOT

#include "stdafx.h"

int main (void)
{
        int x1, x2;

        printf ("Enter binary digit for x1: ");
        scanf ("%d", &x1);

        printf ("Enter binary digit for x2: ");
        scanf ("%d", &x2);

        printf ("\nx1 AND x2 = %d\n", x1 && x2);
        printf ("x1 OR x2 = %d\n", x1 || x2);
        printf ("x1 XOR x2 = %d\n", (x1 && !x2) || (!x1 && x2));
        printf ("NOT x1 = %d\n\n", !x1);

        return 0;
}
```

(a)

```
Enter binary digit for x1: 0
Enter binary digit for x2: 0

x1 AND x2 = 0
x1 OR x2 = 0
x1 XOR x2 = 0
NOT x1 = 1

Press any key to continue . . . _
```

//continued on next page

(b)

Figure 4.32 Program to illustrate using the logical operators: (a) the C program and (b) the outputs.

```
Enter binary digit for x1: 0
Enter binary digit for x2: 1

x1 AND x2 = 0
x1 OR x2 = 1
x1 XOR x2 = 1
NOT x1 = 1

Press any key to continue . . . _
------------------------------------------------------------
Enter binary digit for x1: 1
Enter binary digit for x2: 0

x1 AND x2 = 0
x1 OR x2 = 1
x1 XOR x2 = 1
NOT x1 = 0

Press any key to continue . . . _
------------------------------------------------------------
Enter binary digit for x1: 1
Enter binary digit for x2: 1

x1 AND x2 = 1
x1 OR x2 = 1
x1 XOR x2 = 0
NOT x1 = 0

Press any key to continue . . . _
```

Figure 4.32 (Continued)

4.4.6 Conditional Operator

The conditional operator (**? :**) has three operands, as shown in the syntax below. The *conditional_expression* is evaluated. If the result is true (1), then the *true_expression* is evaluated; if the result is false (0), then the *false_expression* is evaluated.

```
conditional_expression ? true_expression : false_expression;
```

The conditional operator can be used when one of two expressions is to be selected. For example, in the statement below, if x_1 is greater than or equal to x_2, then z_1 is assigned the value of x_3; if x_1 is less than x_2, then z_1 is assigned the value of x_4.

```
z1 = (x1 >= x2) ? x3 : x4;
```

Since the conditional operator selects one of two values, depending on the result of the conditional_expression evaluation, the operator can be used in place of the **if-else** construct. Conditional operators can be nested; that is, each true_expression and false_expression can be a conditional operation, as shown below.

```
conditional_expression ? (cond_expr1 ? true_expr1 : false_expr1)
            : (cond_expr2 ? true_expr2 : false_expr2);
```

The program shown in Figure 4.33 illustrates a method to determine whether one integer is less than, equal to, or greater than the other integer using the conditional operator.

```
//cond_op.cpp
//determine the relationship between two variables
//using the conditional operator.
//print the largest variable.
//if the variables are equal, then set result to 0

#include "stdafx.h"

int main (void)
{
      int x1, x2, result, rslt;

      printf ("Enter first integer: ");
      scanf ("%d", &x1);

      printf ("Enter second integer: ");
      scanf ("%d", &x2);

      result = (x1 > x2) ? (rslt = x1)
                  : ((x1 < x2) ? rslt = x2 : rslt = 0);
      printf ("\nLargest = %d\n\n", result);
      return 0;
}
```

(a)

```
Enter first integer: 3
Enter second integer: 4

Largest = 4
Press any key to continue . . . _
```

//continued on next page

(b)

Figure 4.33 Program to illustrate using the conditional operator: (a) the C program and (b) the outputs.

```
Enter first integer: 425
Enter second integer: 54

Largest = 425
Press any key to continue . . . _
-------------------------------------------------------------
Enter first integer: 78
Enter second integer: 78

Largest = 0
Press any key to continue . . . _
```

Figure 4.33 (Continued)

As a final example in this section, consider the program listed in Figure 4.34, which detects a division by zero. Division by zero generates a result of infinity or *Not a Number* (NaN). Most calculators will display an error message when an attempt is made to divide by zero, because division of any number by zero is not defined. A NaN is a value that is unrepresentable in floating-point calculations and is defined in the IEEE 754-1985 (Reaffirmed 1990) floating-point standard.

```
//cond_ops2.cpp
//detect a division of zero (infinity or NaN)
//by using the conditional operator
#include "stdafx.h"
int main (void)
{
      float flp1, flp2, result;

      printf ("Enter floating-point dividend: ");
      scanf ("%f", &flp1);

      printf ("Enter floating-point divisor: ");
      scanf ("%f", &flp2);

      result = flp2 ? flp1/flp2 : 0;
      printf ("\nQuotient = %f\n\n", result);
      return 0;
}                                    (a)   //continued on next page
```

Figure 4.34 Program to illustrate using the conditional operator to detect a division by zero: (a) the C program and (b) the outputs.

```
Enter floating-point dividend: 10.0
Enter floating-point divisor: 5.0

Quotient = 2.000000

Press any key to continue . . . _
------------------------------------------------------------
Enter floating-point dividend: 156.25
Enter floating-point divisor: 4.5

Quotient = 34.722221

Press any key to continue . . . _
------------------------------------------------------------
Enter floating-point dividend: 76.34
Enter floating-point divisor: 85.29

Quotient = 0.895064

Press any key to continue . . . _
------------------------------------------------------------
Enter floating-point dividend: 87.45
Enter floating-point divisor: 0.0

Quotient = 0.000000

Press any key to continue . . . _
```

(b)

Figure 4.34 (Continued)

4.4.7 Increment and Decrement Operators

There are four versions of the increment and decrement operators, as shown below. The increment/decrement operators are *unary* operators and can be placed before or after the variable — they cannot be used with constants.

Prefix Mode		Postfix Mode	
$++x$	Increment operand x *before* it is used	$x++$	Increment operand x *after* it is used
$--x$	Decrement operand x *before* it is used	$x--$	Decrement operand x *after* it is used

The increment operator An example of a preincrement operator is shown below. After execution, $x = 11$ and $y = 11$; that is, x was incremented, *then* its value was assigned to y.

$$x = 10$$
$$y = ++x$$

An example of a postincrement operator is shown below. After execution, $x = 11$ and $y = 10$; that is, the value of x was assigned to y, *then* x was incremented. The program shown in Figure 4.35 illustrates using the preincrement and postincrement operators.

$$x = 10$$
$$y = x++$$

```
//incr_ops.cpp
//program to illustrate the preincrement
//and postincrement operators

#include "stdafx.h"

int main (void)

{
      int int1, int2, int3, int4, result;

      printf ("Enter an integer for int1: ");
      scanf ("%d", &int1);
      int2 = ++int1; //the current value of int1 is incr,
                     //then the new value is assigned to int2
      printf ("int1 and int2 = %d %d\n\n", int1, int2);

      printf ("Enter an integer for int3: ");
      scanf ("%d", &int3);
      int4 = int3++; //the current value of int3 is assigned
                     //to int4, then int3 is incr
      printf ("int3 and int4 = %d %d\n\n", int3, int4);

      return 0;
}                                       //continued on next page
```

(a)

Figure 4.35 Program to illustrate using the preincrement and postincrement operators: (a) the C program and (b) the outputs.

```
Enter an integer for int1: 100
int1 and int2 = 101 101

Enter an integer for int3: 100
int3 and int4 = 101 100

Press any key to continue . . . _
-----------------------------------------------------------
Enter an integer for int1: 75
int1 and int2 = 76 76

Enter an integer for int3: 75
int3 and int4 = 76 75

Press any key to continue . . . _
```

(b)

Figure 4.35 (Continued)

The decrement operator An example of a predecrement operator is shown below. The current value of x is decremented, then the new value is assigned to y. After execution, $x = 9$ and $y = 9$; that is, x was decremented, *then* its value was assigned to y.

$$x = 10$$
$$y = --x$$

An example of a postdecrement operator is shown below. After execution, $x = 9$ and $y = 10$; that is, the value of x was assigned to y, *then* x was decremented. The program shown in Figure 4.36 illustrates using the predecrement and postdecrement operators.

$$x = 10$$
$$y = x--$$

As can be seen in both the programs of Figure 4.35 and Figure 4.36, there is only one operand utilized in the unary arithmetic operators. An expression cannot be incremented or decremented, thus the statement shown below is invalid.

```
z = ++(x * y);
```

```
//decr_ops.cpp
//program to illustrate the predecrement
//and postdecrement operators

#include "stdafx.h"

int main (void)

{
      int int1, int2, int3, int4;

      printf ("Enter an integer for int1: ");
      scanf ("%d", &int1);
      int2 = --int1; //the current value of int1 is decr,
                     //then the new value is assigned to int2
      printf ("int1 and int2 = %d %d\n\n", int1, int2);

      printf ("Enter an integer for int3: ");
      scanf ("%d", &int3);
      int4 = int3--; //the current value of int3 is assigned
                     //to int4, then int3 is decr
      printf ("int3 and int4 = %d %d\n\n", int3, int4);

      return 0;
}
```

(a)

```
Enter an integer for int1: 10
int1 and int2 = 9 9

Enter an integer for int3: 10
int3 and int4 = 9 10

Press any key to continue . . . _
--------------------------------------------------------------
Enter an integer for int1: 50
int1 and int2 = 49 49

Enter an integer for int3: 50
int3 and int4 = 49 50

Press any key to continue . . . _
```

(b)

Figure 4.36 Program to illustrate using the predecrement and postdecrement operators: (a) the C program and (b) the outputs.

As a final example in this section, the program shown in Figure 4.37 illustrates using the unary increment/decrement operators with the binary multiplication operator. Note that parentheses are not required for the prefix unary operators when used with the binary operator, because the prefix operators have a higher precedence than the multiplication operator in a left-to-right sequence.

```
//pre_post_incr_decr.cpp
//using pre_post_incr_decr with multiplication
#include "stdafx.h"
int main (void)
{
    int int1, int2;

    printf ("Enter an integer for int1 for pre-incr: ");
    scanf ("%d", &int1);

    int2 = 2*++int1;    //incr int1 by 1, then mul by 2
                        //and assign to int2
    printf ("int2 = %d", int2);
//-----------------------------------------------------------
    printf ("\n\nEnter an integer for int1 for post-incr: ");
    scanf ("%d", &int1);

    int2 = 2*int1++;    //mul by 2, assign to int2
                        //then incr int1 by 1
    printf ("int2 = %d\n\n", int2);
//===========================================================
    printf ("Enter an integer for int1 for pre-decr: ");
    scanf ("%d", &int1);

    int2 = 2*--int1;    //decr int1 by 1, then mul by 2
                        //and assign to int2
    printf ("int2 = %d", int2);
//-----------------------------------------------------------
    printf ("\n\nEnter an integer for int1 for post-decr: ");
    scanf ("%d", &int1);

    int2 = 2*int1--;    //mul by 2, assign to int2
                        //then decr int1 by 1
    printf ("int2 = %d\n\n", int2);
    return 0;
}                               (a)      //continued on next page
```

Figure 4.37 Program to illustrate using the pre/postincrement and pre/postdecrement operators with the multiply operator: (a) the C program and (b) the outputs.

```
Enter an integer for int1 for pre-incr: 1
int2 = 4

Enter an integer for int1 for post-incr: 1
int2 = 2

Enter an integer for int1 for pre-decr: 1
int2 = 0

Enter an integer for int1 for post-decr: 1
int2 = 2

Press any key to continue . . . _
------------------------------------------------------------
Enter an integer for int1 for pre-incr: 50
int2 = 102

Enter an integer for int1 for post-incr: 50
int2 = 100

Enter an integer for int1 for pre-decr: 50
int2 = 98

Enter an integer for int1 for post-decr: 50
int2 = 100

Press any key to continue . . . _
```

(b)

Figure 4.37 (Continued)

4.4.8 Bitwise Operators

There are three bitwise operators: AND, OR, and the exclusive-OR, that operate on the individual bits of two operands; the NOT operator performs the 1s complement on one operand. The operators (or symbols) used for the bitwise operations and the corresponding function definitions are listed in Table 4.13. Table 4.14 illustrates the truth tables for the bitwise operators, where z_1 is the result of the operation.

The AND operator corresponds to the Boolean product and generates a logic 1 output if both bits are a logic 1; otherwise a logic 0 is generated. The OR operator corresponds to the Boolean sum and generates a logic 1 output if either or both bits are a logic 1 — if both bits are a logic 0, then a logic 0 is generated. The exclusive-OR operator generates a logic 1 if both bits are different — if both bits are the same, then the output is a logic 0.

Table 4.13 Boolean Operators for Variables x_1 and x_2

Operator	Function	Definition
&	AND	x_1 & x_2
|	OR	$x_1 \mid x_2$
^	Exclusive-OR	$x_1 \wedge x_2 = (x_1x_2') + (x_1'x_2)$
~	NOT (1s complement)	$\sim x_1$

Table 4.14 Truth Table for AND, OR, Exclusive-OR, and NOT

AND		OR		Exclusive-OR		NOT	
x_1x_2	z_1	x_1x_2	z_1	x_1x_2	z_1	x_1	z_1
0 0	0	0 0	0	0 0	0	0	1
0 1	0	0 1	1	0 1	1	1	0
1 0	0	1 0	1	1 0	1		
1 1	1	1 1	1	1 1	0		

The program shown in Figure 4.38 illustrates using the three bitwise operators of AND, OR, and exclusive-OR for various user-generated inputs. Integers are entered by the user and the values are displayed in both decimal and hexadecimal number representations. The bitwise operator is then executed and the result is displayed in both decimal and hexadecimal.

For the first AND operation, a value or 100_{10} (64_{16}) was entered. This value was ANDed with 255_{10} (FF_{16}), as shown below, to yield a result of 100_{10} (64_{16}). Thus, the initial value was unchanged, since 1 & 1 = 1 and 1 & 0 = 0.

```
      0110    0100
&)    1111    1111
      ------------
      0110    0100
```

For the second OR operation, a value or 58_{10} $(3A_{16})$ was entered. This value was ORed with 97_{10} (61_{16}), as shown below, to yield a result of 123_{10} $(7B_{16})$. Thus, the initial value was changed, since 1 | 1 = 1 and 1 | 0 = 1.

```
      0011    1010
|)    0110    0001
      ------------
      0111    1011
```

For the second exclusive-OR operation, a value or 125_{10} ($7D_{16}$) was entered. This value was exclusive-ORed with 255_{10} (FF_{16}), as shown below, to yield a result of 130_{10} (82_{16}). Thus, the initial value was changed, because the result is a logic 1 only when the bits being exclusive-ORed are different. The NOT operator produces the 1s complement of the operand. This will be left as a problem in which a program is written to illustrate the operation of the NOT operator.

```
      0111    1101
^)    1111    1111
      ------------
      1000    0010
```

```
//bitwise_and_or_xor.cpp
//use the bitwise operators

#include "stdafx.h"

int main (void)
{
   int int1, int2;

   printf ("Enter an integer: ");
   scanf ("%d", &int1);
   printf ("Initial value of int1 = %d decimal, %X hex\n",
            int1, int1);

   int2 = int1 & 255;     //FFH
   printf ("After AND 255 (FFH) int2 = %d decimal,
            %X hex\n\n", int2, int2);
//----------------------------------------------------------
   printf ("Enter an integer: ");
   scanf ("%d", &int1);
   printf ("Initial value of int1 = %d decimal, %X hex\n",
            int1, int1);

   int2 = int1 & 15;//0FH
   printf ("After AND 15 (0FH) int2 = %d decimal,
            %X hex\n\n", int2, int2);
//==========================================================
                              //continued on next page
```

(a)

Figure 4.38 Program to illustrate using the bitwise operators of AND, OR, and exclusive-OR: (a) the C program and (b) the outputs.

```
//===========================================================
   printf ("Enter an integer: ");
   scanf ("%d", &int1);
   printf ("Initial value of int1 = %d decimal, %X hex\n",
            int1, int1);

   int2 = int1 | 188;//BCH
   printf ("After OR 188 (BCH) int2 = %d decimal,
            %X hex\n\n", int2, int2);
//-----------------------------------------------------------
   printf ("Enter an integer: ");
   scanf ("%d", &int1);
   printf ("Initial value of int1 = %d decimal, %X hex\n",
            int1, int1);

   int2 = int1 | 97;//61H
   printf ("After OR 97 (61H) int2 = %d decimal,
            %X hex\n\n", int2, int2);
//===========================================================
   printf ("Enter an integer: ");
   scanf ("%d", &int1);
   printf ("Initial value of int1 = %d decimal, %X hex\n",
            int1, int1);

   int2 = int1 ^ 85;//55H
   printf ("After XOR 85 (55H) int2 = %d decimal,
            %X hex\n\n", int2, int2);
//-----------------------------------------------------------
   printf ("Enter an integer: ");
   scanf ("%d", &int1);
   printf ("Initial value of int1 = %d decimal, %X hex\n",
            int1, int1);

   int2 = int1 ^ 255;//FFH
   printf ("After XOR 255 (FFH) int2 = %d decimal,
            %X hex\n\n", int2, int2);

   return 0;
}
                                       //continued on next page
```

Figure 4.38 (Continued)

```
Enter an integer: 100
Initial value of int1 = 100 decimal, 64 hex
After AND 255 (FFH) int2 = 100 decimal, 64 hex

Press any key to continue . . . _
-----------------------------------------------------------
Enter an integer: 240
Initial value of int1 = 240 decimal, F0 hex
After AND 15 (0FH) int2 = 0 decimal, 0 hex

Press any key to continue . . . _
=========================================================++
Enter an integer: 75
Initial value of int1 = 75 decimal, 4B hex
After OR 188 (BCH) int2 = 255 decimal, FF hex

Press any key to continue . . . _
-----------------------------------------------------------
Enter an integer: 58
Initial value of int1 = 58 decimal, 3A hex
After OR 97 (61H) int2 = 123 decimal, 7B hex

Press any key to continue . . . _
=========================================================++
Enter an integer: 170
Initial value of int1 = 170 decimal, AA hex
After XOR 85 (55H) int2 = 255 decimal, FF hex

Press any key to continue . . . _
-----------------------------------------------------------
Enter an integer: 125
Initial value of int1 = 125 decimal, 7D hex
After XOR 255 (FFH) int2 = 130 decimal, 82 hex

Press any key to continue . . . _
```

(b)

Figure 4.38 (Continued)

4.5 While Loop

The **while** loop (or **while** statement) executes a statement or block of statements as long as a test expression is true (nonzero). The statements that are controlled by the **while** statement loop repeatedly until the expression becomes false (0). If a block of statements is executed, then the block is delimited by beginning and ending braces; for a single statement, braces are not required. The syntax for the **while** loop is shown below. The expression is checked at the beginning of the loop and may contain relational and logical operators.

```
while (expression)
    {block of statements}
```

The statements in the block must change the value of the expression, since this determines the duration of the loop. If the expression does not change (always true), then the loop would never stop, resulting in an infinite loop. When the expression becomes false (0), the program exits the loop and transfers control to the first statement following the block of statements. If the expression is initially false, then the block is not executed.

Figure 4.39 shows a simple program that uses the **while** statement in conjunction with the preincrement and postincrement operators. The **while** loop adds 3 to the integer *int1* during each iteration of the loop. The integer *int1* is initially assigned a value of 1. The **printf ( )** argument adds 1 to *int1* in the preincrement mode, increasing *int1* to 2. Then the two postincrement operators add 2 to *int1*, increasing *int1* to 4. At the next iteration of the loop, the **printf ( )** function increases *int1* to a value of 5.

```
//while.cpp
//use the while loop in conjunction with the
//preincrement and postincrement operators

#include "stdafx.h"
int main (void)
{
        int int1;
        int1 = 1;

        while (int1 <= 10)
        {
                printf ("%d \n", ++int1);
                int1++;
                int1++;
        }
        return 0;
}
```

(a)

```
2 5 8 11 Press any key to continue . . . _
```

(b)

Figure 4.39 Program to illustrate using the **while** loop with the preincrement and postincrement operators: (a) the C program and (b) the outputs.

The program shown in Figure 4.40 illustrates the **while** loop to multiply an integer from the keyboard by 2 until the number becomes equal to or greater than 2048. The integer *int1* is defined as a local variable, is printed, then multiplied by 2. This process repeats until the value reaches or exceeds 2048.

```
//powers_of_2.cpp
//multiplies by 2 up to a limit of 2048

#include "stdafx.h"
int main (void)
{
	int int1;

	printf ("Enter an integer: ");
	scanf ("%d", &int1);

	while (int1 < 2048)
	{
		printf ("%d ", int1);
		int1 = int1 * 2;
	}
	printf ("\n\n");
	return 0;
}
```
(a)

```
Enter an integer: 1
1 2 4 8 16 32 64 128 256 512 1024

Press any key to continue . . . _
--------------------------------------------------------------
Enter an integer: 64
64 128 256 512 1024

Press any key to continue . . . _
--------------------------------------------------------------
Enter an integer: 23
23 46 92 184 368 736 1472

Press any key to continue . . . _
--------------------------------------------------------------
Enter an integer: 2048

Press any key to continue . . . _
```
(b)

Figure 4.40 Program to illustrate using the **while** loop to multiply by 2 up to a limit of 2048: (a) the C program and (b) the outputs.

4.6 For Loop

The **for** loop repeats a statement or block of statements a specific number of times. This is different than the **while** loop, which repeats the loop as long as a certain condition is met. When the **for** loop has completed the final loop, the program exits the loop and transfers control to the first statement following the block of statements. The syntax for the **for** loop is shown below.

```
for (initialization; conditional test; increment)
   {one or more statements}
```

The initialization step sets a variable to a specific value; this is the loop control variable and is executed only once. The conditional test is performed at the start of the loop to determine if the loop should be entered. This is usually a relational expression that tests the loop control variable against a target value. If the test is true (nonzero), then the loop is entered; otherwise, the loop is exited and the next instruction following the loop is executed. The increment step is performed at the end of the loop; this can also be a decrement depending on the initialization and conditional test.

Figure 4.41 and Figure 4.42 show two examples of using a **for** loop to increment an integer. Figure 4.41 has no user input. The integer *int1* is initialized to a value of 0; the loop continues until the value is equal to or less than 30; the integer is incremented by 3 at the beginning of the loop. Figure 4.42 requires a user input. A space for the initialization step of the **for** loop indicates that a user input is required. An integer is entered for the initialization step and is stored in location *int1*. The conditional test specifies an end result of less than or equal to 50; the integer is incremented by 5 at the beginning of the loop.

```
//for_loop2.cpp
//use a for loop to increment a variable
//by 3 until it is less than or equal to 30

#include "stdafx.h"
int main (void)
{
      int int1;

      for (int1 = 0  ; int1 <= 30; int1 = int1 + 3)
            printf ("Number = %d\n", int1);

      return 0;
}                                  //continued on next page
```

(a)

Figure 4.41 Program to illustrate using the **for** loop to increment a variable by 3 up to a limit of less than or equal to 30: (a) the C program and (b) the outputs.

```
Number = 0
Number = 3
Number = 6
Number = 9
Number = 12
Number = 15
Number = 18
Number = 21
Number = 24
Number = 27
Number = 30
Press any key to continue . . . _
```

(b)

Figure 4.41 (Continued)

```
//for_loop.cpp
//use a for loop to increment a variable
//by 5 until it is less than or equal to 50


#include "stdafx.h"

int main (void)
{
	int int1;

	printf ("Enter an integer less than 50: ");
	scanf ("%d", &int1);


//a blank initialization indicates that
//the user enters a number
	for (  ; int1 <= 50; int1 = int1 + 5)
		printf ("Number = %d\n", int1);

	return 0;
}
```

//continued on next page

(a)

Figure 4.42 Program to illustrate using the **for** loop to increment a variable by 5 up to a limit of less than or equal to 50: (a) the C program and (b) the outputs.

```
Enter an integer less than 50: 25
Number =  25
Number =  30
Number =  35
Number =  40
Number =  45
Number =  50
Press any key to continue . . . _
------------------------------------------------------------
Enter an integer less than 50: 27
Number =  27
Number =  32
Number =  37
Number =  42
Number =  47
Press any key to continue . . . _
------------------------------------------------------------
Enter an integer less than 50: 49
Number =  49
Press any key to continue . . . _
```

(b)

Figure 4.42 (Continued)

As a final example, the program shown in Figure 4.43 illustrates a **for** loop consisting of a user-entered integer and the postdecrement operator, which decrements the integer by 1. As long as the integer *int1* is nonzero, as determined by the conditional test, the looping continues. When the value of *int1* reaches a value of zero, the program exits the loop. The **printf ("\n");** simply places the cursor at the beginning of a new line. Since there is only one statement in the **for** loop, braces are not required.

```
//for_loop3.cpp
//use a for loop to decrement a user-entered
//integer by 1 using the post-decrement operator
#include "stdafx.h"
int main (void)
{
      int int1;
      printf ("Enter an integer: ");
      scanf ("%d", &int1);              //continued on next page
```

(a)

Figure 4.43 Program to illustrate using the **for** loop to postdecrement a variable by 1 until its value is 0: (a) the C program and (b) the outputs.

```
//a blank initialization indicates that
//the user enters a number
      for ( ; int1; int1--)
            printf ("%d ", int1);

            printf ("\n");

      return 0;
}
```

```
Enter an integer: 10
10 9 8 7 6 5 4 3 2 1
Press any key to continue . . . _
-------------------------------------------------------------
Enter an integer: 15
15 14 13 12 11 10 9 8 7 6 5 4 3 2 1
Press any key to continue . . . _
```

(b)

Figure 4.43 (Continued)

4.7 Additional C Constructs

This section gives a short introduction to arrays, strings, pointers, and functions. Arrays are one-dimensional or multidimensional structures that contain a series of elements of a specific data type; that is, an array is a list of variables with the same name. The declaration and initialization of arrays is discussed in which arrays are used to declare a set of variables of the same data type.

Strings are usually one-dimensional character arrays that are terminated with a null character (\0) — the null character is automatically appended to the end of the string by the compiler to mark the end of the string. Strings are part of the **printf ( )** function and are contained in the *format_string and conversion* section of the print function.

Pointers are variables that address (or *point* to) a memory location where a data type is stored. The memory location can be selected by the pointer to access or modify the data. Pointers provide an effective method to access arrays.

A function is an independent procedure (or subroutine) which is invoked by a calling function. The invoked function receives arguments from the calling program, performs specific operations on the arguments, then returns the results to the calling program. A function is identified by a pair of parentheses. Functions are useful in avoiding repetitive programming — a function can be defined once then used by several invoking functions.

4.7.1 Arrays

An array can be initialized by simply listing the array elements, as shown below together with examples. The *type* can be **char**, **int**, **float**, **double**, or any valid C type. The *value list* is a sequence of constants that are the same type as the *array_name* type and enclosed in braces.

```
type array_name [size] = {value list};
int array[5] = {1, 4, 9, 16, 25};
char array[3] = {'A', 'B', 'C'};
```

The size of the array does not have to be specified, as shown below for an array of eight elements — the compiler creates the array based upon the number of elements. The length of the array is one element longer than the *value list*; the compiler supplies the null terminator.

```
int powers[ ] = {1, 2, 4, 8, 16, 32, 64, 128};
```

An array is an ordered structure, in which the location of the individual elements is known. An array is *homogeneous*; that is, every element is of the same type. The array can be an array of integers, an array of floating-point numbers, or an array of any other data types — array types cannot be mixed. An *automatic* array is an array that is defined inside the **main ( )** function. An *external* array is defined outside a function, usually before **main ( )**. A *static* array is defined inside a function with the keyword **static**, but retains its values between function calls.

One-dimensional array A *one-dimensional array* is declared as shown below, where the *type* is a valid C type, *var_name* is the name of the array, and *size* is the number of elements in the array. An example of a one-dimensional array of type **int**, called *array* with 20 elements, is also shown below.

```
type var_name [size]
int array [20]
```

Arrays start at element 0; therefore, the second element is accessed as follows: *array [1]*. The first element of the array can be set to a value of 200 by the following statement: *array [0] = 200*. One-dimensional arrays are stored in contiguous memory locations, with the first array element at the lowest address.

Figure 4.44 illustrates a method to initialize a one-dimensional array with ten elements to values 1 through 10. The integer *i* is set to a value of 1 in the **for** loop, then the first element (*array[i–1]*), which is *array[0]*, is set to a value of 1 by the following statement: *array[i–1] = i*; that is, *array[0] = 1*. The array is then printed, where the first element (*i–1*) is assigned the value of *array[i–1]*, which is the contents of the first element — a value of 1. A similar structure is shown in the program of Figure 4.45, which generates the cubes of integers 1 through 10.

```
//array_init.cpp
//initialize a one-dimensional array
//with values 1 through 10

#include "stdafx.h"

int main (void)
{
	int array[10];		//declare an array of 10 elements
	int i;			//declare an integer to count

//initialize and print the array
	for (i=1; i<11; i++)
	{
		array[i-1] = i;
		printf ("array[%d] = %d\n", i-1, array[i-1]);
	}
	return 0;
}
```
(a)

```
array[0] = 1
array[1] = 2
array[2] = 3
array[3] = 4
array[4] = 5
array[5] = 6
array[6] = 7
array[7] = 8
array[8] = 9
array[9] = 10
Press any key to continue . . . _
```
(b)

Figure 4.44 Program to illustrate initializing and printing a one-dimensional array: (a) the C program and (b) the outputs.

```
//array_cubes.cpp
//obtain the cubes of numbers 1 through 10
#include "stdafx.h"
int main (void)				//continued on next page
```
(a)

Figure 4.45 Program to illustrate an array to obtain the cubes of integers 1 through 10: (a) the C program and (b) the outputs.

```
{
        int cubes[10];      //declare an array of 10 elements
        int i;              //declare an integer to count

//initialize and print the array
        for (i=1; i<11; i++)
        {
                cubes[i-1] = i*i*i;
                printf ("cubes[%d] = %d\n", i-1, cubes[i-1]);
        }
        return 0;
}
```

```
cubes[0] = 1
cubes[1] = 8
cubes[2] = 27
cubes[3] = 64
cubes[4] = 125
cubes[5] = 216
cubes[6] = 343
cubes[7] = 512
cubes[8] = 729
cubes[9] = 1000
Press any key to continue . . . _
```

(b)

Figure 4.45 (Continued)

Two-dimensional array A *two-dimensional array* has the syntax shown below for a 10×12 array, where the type is **int**, the array name is *count*, the number of rows is 10, and the number of columns is 12.

```
int count [10][12];
```

A two-dimensional array is an array of one-dimensional arrays that is accessed one row at a time from left to right. The rightmost index (columns) will change faster than the leftmost index (rows). Figure 4.46 shows a 3×3 array consisting of nine elements with addresses for select elements. Figure 4.47 shows a program to initialize and print the two-dimensional array of Figure 4.46 using nested **for** loops. The array elements can also be listed in a single line. The row index does not have to be specified — C will index the array properly. This permits the construction of arrays of varying lengths; the compiler allocates storage automatically.

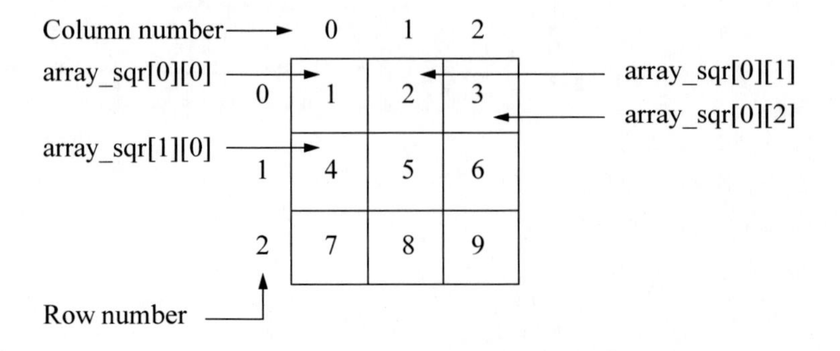

Figure 4.46 A multidimensional array consisting of three rows and three columns.

```
//array_sqr.cpp
//initialize and print a two-dimensional array

#include "stdafx.h"

int main (void)
{
        int array_sqr[3][3] = {1,2,3,4,5,6,7,8,9};
        int i, j;

        for (i=0; i<3; i++)                 //i is row index
        {
                for (j=0; j<3; j++)         //j is column index

//"%4d" in the printf ( ) is 4-digit spacing right aligned
                        printf ("%4d", array_sqr[i][j]);
                        printf ("\n");
        }
        return 0;
}
```

(a)

```
   1   2   3
   4   5   6
   7   8   9
Press any key to continue . . . _
```

(b)

Figure 4.47 Program to illustrate initializing and printing a two-dimensional array: (a) the C program and (b) the outputs.

Figure 4.48 illustrates a program that loads a 3×4 array with the products of the indices, then displays the array in a row–column format. This program also uses nested **for** loops. Note that a **for** loop block that has only one statement does not require delimiting braces. Integers i and j represent the row and column indices, respectively.

```
//array_mul_indices.cpp
//load a 3 x 4 array with the products
//of the indices, then display the
//array in a row-column format

#include "stdafx.h"
int main (void)
{
	int array_mul[3][4];
	int i, j;

	for (i=0;i<3; i++)			//i is row index
		for (j=0; j<4; j++)		//j is column index
			array_mul[i][j] = i*j;

	for (i=0; i<3; i++)
	{
		for (j=0; j<4; j++)
			printf ("%4d", array_mul[i][j]);
			printf ("\n");
	}
			printf ("\n");
	return 0;
}
```

(a)

```
   0   0   0   0
   0   1   2   3
   0   2   4   6

Press any key to continue . . . _
```

(b)

Figure 4.48 Program to illustrate loading a 4×5 array with the product of the indices then printing the array: (a) the C program and (b) the outputs.

4.7.2 Strings

A string is a one-dimensional character array that is terminated by a null character (\0). The string can be defined to be a certain length and must include the null character as part of the overall length. There are many library functions that apply to strings, including input/output functions. The following string functions are described in this section: **gets ( )**, **puts ( )**, **strcpy ( )**, **strcat ( )**, **strcmp ( )**, and **strlen ( )**.

Gets ( ) The **gets** (string) function stores input data — primarily from the keyboard — into the string named *string*. The **gets ( )** function inputs the string data until a newline (\n) character is detected, which is generated by pressing the *Enter* key. The **gets ( )** function discards the newline character and adds the null character (\0) or (00_{16}), which indicates the end of the string. The length of the string array must be large enough to hold the entire string, because the bounds of the string are not checked. Figure 4.49 shows a simple program to illustrate the use of the **gets ( )** function.

```
//string_input.cpp
//use the gets ( ) function to input a string

#include "stdafx.h"

int main (void)
{
    char string [20];      //declare a string for 20 elements
    int i;                 //declare an integer to count

    printf ("Enter the day and month: \n");
    gets (string);
    for (i=0; string[i]; i++)
        printf ("%c", string[i]);

    printf ("\n");

    return 0;
}
```

(a)

```
Enter the day and month:
monday september
monday september
Press any key to continue . . . _
```

(b)

Figure 4.49 Program to illustrate using the **gets ( )** function: (a) the C program and (b) the outputs.

Puts ( ) The **puts ( )** function displays text on the monitor screen and automatically adds a newline character (\n) at the end of the string; therefore, a newline character does not have to be inserted by the programmer. When a string is to be output to the screen, the **puts ( )** function is faster than the **printf ( )**. The program shown in Figure 4.50 illustrates the use of the **puts ( )** function to tell the user to enter an integer number followed by a floating-point number. The numbers are read by the **scanf ( )** function and displayed on the screen by the **printf ( )** function.

```
//string_puts.cpp
//use the puts ( ) function to output data
//to the monitor screen

#include "stdafx.h"

int main (void)
{
        int int1;
        float flp1;

        puts ("Enter an integer: ");
        scanf ("%d", &int1);

        puts ("\nEnter a float: ");
        scanf ("%f", &flp1);

        printf ("\ninteger entered: %d,
                    \nfloat entered: %f\n\n", int1, flp1);
        return 0;
}
```

(a)

```
Enter an integer:
759

Enter a  float:
372.125

integer entered: 759
float entered: 372.125000

Press any key to continue . . . _
```

(b)

Figure 4.50 Program to illustrate using the **puts ( )** function: (a) the C program and (b) the outputs.

Strcpy ( ) The **strcpy ( )** function copies the source string to the destination string, including the null character (\0). Both the source and destination strings are arrays. Since the **strcpy ( )** function does not perform bounds checking on the destination string, the destination array must be of sufficient length to accommodate the source string plus the terminating null character. The syntax for the **strcpy ( )** function is shown below.

```
strcpy (destination_string, source-string);
```

Figure 4.51 shows a program that illustrates a user-entered source string being copied to a destination string. The **"string.h"** header is required in order to make some string operations valid. The **gets ( )** function stores the keyboard input string into the source string *src_str*, which is initialized to 25 characters. This is then transferred to the destination string *dst_str*, also initialized to 25 characters.

```
//string_copy2.cpp
//copy a user-entered source string to a
//destination string using the strcpy ( ) function
#include "stdafx.h"
#include "string.h"
int main (void)
{
	char src_str[25];
	char dst_str[25];

	puts ("Enter a short string: ");
	gets (src_str);

	strcpy(dst_str, src_str);

	printf("\nSource string: %s\n", src_str);
	printf ("Destination string: %s\n\n", dst_str);
	return 0;
}
```
(a)

```
Enter a short string:
This is a short string.

Source string: This is a short string.
Destination string: This is a short string.

Press any key to continue . . . _
```
(b)

Figure 4.51 Program to illustrate using the **strcpy ( )** function: (a) the C program and (b) the outputs.

Strcat ( ) String concatenation is performed by the **strcat ( )** function, shown in Figure 4.52. The syntax is shown below, containing two arguments: *string1* and *string2*, where *string2* is concatenated to the right of *string1* to generate a new string. The terminating null character is moved to the right of the new string. The **strcat ( )** function performs no bounds checking; therefore, the array of *string1* must be of sufficient length to accommodate both strings plus the null character. Figure 4.52 shows a program illustrating the use of the **strcat ( )** function.

```
strcat (string1, string2);
```

```
//string_concat.cpp;
//concatenate strings using the strcat ( ) function
#include "stdafx.h"
#include "string.h"

int main (void)
{
      char str1 [30];
      char str2 [12];

      printf ("Enter 10 letters for the first string: ");
      gets (str1);        //gets ( ) reads characters
                          //until enter is pressed
                          //the carriage return is replaced
                          //by the null character
      printf ("Enter 10 numbers for the second string: ");
      gets (str2);

      strcat (str1, str2);

      printf ("\nThe concatenated strings are:
                  %s\n\n", str1);
      return 0;
}
```

(a)

```
Enter 10 letters for the first string: abcdefghij
Enter 10 numbers for the second string: 0123456789

The concatenated strings are: abcdefghij0123456789

Press any key to continue . . . _
```

(b)

Figure 4.52 Program to illustrate using the **strcat ( )** function: (a) the C program and (b) the outputs.

Strcmp ( ) String comparison is performed by the **strcmp ( )** function. The syntax is shown below, which contains two arguments: *string1* and *string2*, where *string1* is compared to *string2*. The string arrays do not have to be the same size, because comparison is not based on the length of the strings. The **strcmp ( )** function returns a value of 0 if the strings are the same; it returns a value of < 0 if *string1* is less than *string2*; it returns a value of > 0 if *string1* is greater than *string2*.

```
strcmp (string1, string2);
```

The strings are compared character by character; that is, lexicographically, as in a dictionary. The comparison is case sensitive and in the same order as shown in the ASCII character code chart in Appendix A for hexadecimal 20 through 7F. Lowercase letters are greater than uppercase letters, because they have a higher decimal value.

Figure 4.53 shows a program illustrating the use of the **strcmp ( )** function to compare two strings from a set of three strings. The result of the string comparison is contained in the *comp_rslt* variable. The value of *comp_rslt* will be either 0, less than 0, or greater than 0. This is determined by the conditional **if**, **else-if**, and **else** statements. If *comp_rslt* is equal to zero, then this indicates that *string1* is equal to *string2*; if *comp_rslt* is less than zero, then this indicates that *string1* is less than *string2*; if *comp_rslt* is greater than zero, then this indicates that *string1* is greater than *string2*.

```
//string_comp.cpp
//compare strings using the strcmp ( ) function
#include "stdafx.h"
#include "string.h"

int main (void)
{
	char str1 [10];
	char str2 [10];
	char str3 [10];
	int comp_rslt;		//the string comparison result

	printf ("Enter a 6-digit string for string str1: ");
	gets (str1);

	printf ("Enter a 6-digit string for string str2: ");
	gets (str2);

	printf ("Enter a 6-digit string for string str3: ");
	gets (str3);			//continued on next page
```

(a)

Figure 4.53 Program to illustrate using the **strcmp ( )** function: (a) the C program and (b) the outputs.

```
//---------------------------------------------------------------
      comp_rslt = strcmp (str1, str2);
      if (comp_rslt == 0)
            printf ("String1 = string2.\n");
      else if (comp_rslt < 0)
            printf ("String1 < string2\n");
      else
            printf ("String1 > string2\n");
//---------------------------------------------------------------
      comp_rslt = strcmp (str1, str3);
      if (comp_rslt == 0)
            printf ("String1 = string3.\n");
      else if (comp_rslt < 0)
            printf ("String1 < string3\n");
      else
            printf ("String1 > string3\n");
//---------------------------------------------------------------
      comp_rslt = strcmp (str2, str3);
      if (comp_rslt == 0)
            printf ("String2 = string3.\n");
      else if (comp_rslt < 0)
            printf ("String2 < string3\n");
      else
            printf ("String2 > string3\n");
            return 0;
}
```

```
Enter a 6-digit string for string str1: 121212
Enter a 6-digit string for string str2: 121212
Enter a 6-digit string for string str3: 121212
String1 = string2
String1 = string3
String2 = string3
Press any key to continue . . . _
-----------------------------------------------------------------
Enter a 6-digit string for string str1: abcDef
Enter a 6-digit string for string str2: abcdEf
Enter a 6-digit string for string str3: abcdeF
String1 < string2
String1 < string3
String2 < string3
Press any key to continue . . . _
                                    //continued on next page
```

(b)

Figure 4.53 (Continued)

```
Enter a 6-digit string for string str1: 345678
Enter a 6-digit string for string str2: 234567
Enter a 6-digit string for string str3: 123456
String1 > string2
String1 > string3
String2 > string3
Press any key to continue . . . _
-----------------------------------------------------------
Enter a 6-digit string for string str1: 123456
Enter a 6-digit string for string str2: ABCDEF
Enter a 6-digit string for string str3: abcdef
String1 < string2
String1 < string3
String2 < string3
Press any key to continue . . . _
```

Figure 4.53 (Continued)

Strlen ( ) The **strlen ( )** function returns the length of a string, but does not include the null character. The number of characters defined for the string is irrespective of the length of the string. The syntax for the **strlen ( )** function is shown below. Figure 4.54 shows a program illustrating the use of the **strlen ( )** function to determine the length of two strings.

```
strlen (string);
```

```
//string_length.cpp
//determine the length of two strings using strlen ( )
#include "stdafx.h"
#include "string.h"

int main (void)
{
   char str1 [20];
   char str2 [20];

   printf ("Enter 10 or less characters for string str1: ");
   gets (str1);

   printf ("Enter 10 or less characters for string str2: ");
   gets (str2);                          //continued on next page
```

(a)

Figure 4.54 Program to illustrate using the **strlen ( )** function: (a) the C program and (b) the outputs.

```
//determine the length of the two strings
//and print the result
   printf ("\n%s is %d characters in length\n",
               str1, strlen(str1));
   printf ("%s is %d characters in length\n\n",
               str2, strlen(str2));

   return 0;
}
```

```
Enter 10 or less characters for string str1: 88888888
Enter 10 or less characters for string str2: 4444

88888888 is 8 characters in length
4444 is 4 characters in length

Press any key to continue . . . _
```

(b)

Figure 4.54 (Continued)

4.7.3 Pointers

A pointer is a variable that contains the address of (*points* to) another variable. The syntax for a pointer declaration is shown below, where *type* is the type of variable addressed by the pointer, such as **char**, **int**, and **float**, for example. The asterisk is a pointer operator that returns the contents of the variable to which it points — the asterisk, as used here, has a different meaning than the asterisk used for multiplication.

```
type *pointer_name;
```

Another pointer operator that was presented previously is the *address-of* operator (&), which returns the address of the variable. Examples of pointers that return the contents of the variable to which they point are as follows:

```
char *character_pointer;
int *integer_pointer;
float *flp_pointer;
```

The asterisk is also referred to as the *indirection operator*. A simple program to illustrate one use of a pointer is shown in Figure 4.55. The program declares two floating-point variables: a pointer **ptr* and a floating-point number *flp*. The floating-point number, *flp*, is assigned a value of 75.25 and *ptr* is assigned the address of *flp*. The floating-point value is then printed using the **ptr* pointer that points to the address of *flp*, which contains 75.25. Pointers are also an effective way to access arrays, in which the array name is the address of the first element [0] of the array. Pointers used in this manner, however, are beyond the scope of this book.

```
//pointer_ex.cpp
//program to illustrate the use of a pointer

#include "stdafx.h"

int main (void)
{
   float *ptr;
   float flp;

   flp = 75.25;
   ptr = &flp;

//print the floating-point value of the variable flp
   printf ("Floating-point value of flp: %f\n\n", *ptr);

   return 0;
}
```

(a)

```
Floating-point value of flp: 75.250000

Press any key to continue . . . _
```

(b)

Program to illustrate one use of a pointer: (a) the C program and (b) the outputs.

4.7.4 Functions

A function is a procedure, or subroutine, that is written once and can be executed several times by calling routines. Functions perform specific tasks and may return results to the calling program. Each function is assigned a specific name that is different from the names of other functions. The use of functions is referred to as *structured programming* or *modular programming*. The general syntax for a function is shown below, where *type* is the return type of the function, followed by the unique function name, followed by a parameter list, or arguments, which can be **int**, **char**, **float**, and so forth.

```
type function_name (parameter list)
{
   Statements
}
```

The return type is specified as a global variable — also called an external variable — and indicates the type of value that the function will return to the calling program after execution of the function. In the program listing of Figure 4.56, *a*, *b*, *sum*, and *prod* are declared as type **int** prior to the **main ( )** function. Integers *a* and *b* are the arguments that are passed to the functions; the return values from the functions are assigned to *sum* and *prod*. The function statements are delimited by a beginning and an ending brace.

The *addition* function is called by the following statement, where *a* and *b* are the arguments that are passed to the function — the *a* and *b* values were entered by the user:

```
sum = addition (a, b);
```

The *addition* function is defined as shown below, where the type is **int** and the arguments, *int x* and *int y*, are declared as local variables to perform the addition operation. The **return** statement returns the sum of the integers that were entered by the user.

```
int addition (int x, int y)
{
   return (x + y);
}
```

```
//arith_ops.cpp
//use functions to perform addition and multiplication
#include "stdafx.h"

      int a, b, sum, prod;

//declare the add function
      int addition (int x, int y);

//declare the multiply function
      int multiplication (int x, int y);
//--------------------------------------------------------------
                              (a)    //continued on next page
```

Figure 4.55 Program to illustrate one use of functions to perform addition and multiplication: (a) the C program and (b) the outputs.

```
int main (void)
{
//input the first number
      printf ("Enter the first number: ");
      scanf ("%d", &a);

//input the second number
      printf ("Enter the second number: ");
      scanf ("%d", &b);

//calculate and display the sum
      sum = addition (a, b);        //call the add fctn
      printf ("\n%d plus %d = %d\n", a, b, sum);

//calculate and display the product
      prod = multiplication (a, b); //call the multiply fctn
      printf ("\n%d times %d = %d\n\n", a, b, prod);
}

//------------------------------------------------------------
//define the add function
      int addition (int x, int y)
{
      return (x + y);               //return the sum
}

//------------------------------------------------------------
//define the multiply function
      int multiplication (int x, int y)
{
      return (x * y);               //return the product
}
```

```
Enter the first number: 12
Enter the second number: 12

12 plus 12 = 24

12 times 12 = 144

Press any key to continue . . . _
-------------------------------------------------------------
```

(b) //continued on next page

Figure 4.56 (Continued)

```
Enter the first number: 75
Enter the second number: 136

75 plus 136 = 211

75 times 136 = 10200

Press any key to continue . . . _
```

(b)

Figure 4.56 (Continued)

This chapter has presented only a minimal introduction to the C programming language, but is sufficient for the purpose of this book, which focuses on X86 assembly language programming. As stated previously, this book will link C programs with assembly language programs — this will occur in the remaining chapters of the book.

There are over 330 instructions in the X86 assembly language; therefore, in order to keep the number of pages to a reasonable number, only the most commonly used instructions will be presented. The remaining chapters in the book will discuss both the 16-bit general-purpose register (GPR) set and the 32-bit extended GPR register set to give variety to the presentation. Both sets of GPRs will incorporate linkage of C programs with assembly language.

4.8 Problems

4.1 Enter three floating-point numbers from the keyboard, add them, then print the sum.

4.2 Write a program to create three columns that show the binary numbers 1 through 256 as powers of 2.

4.3 Write a program to change the following Fahrenheit temperatures to centigrade temperatures: 32.0, 100.0, and 85.0. Then print the results.

4.4 Write a program that displays on three lines:
the number 10 as five digits right-aligned;
the number 99.95 as a floating-point number preceded by a $;
the first 10 characters of a 20-character string.

4.5 Write a program to calculate the volume of a parallelepiped. A parallelepiped is a six-faced polyhedron all of whose faces are parallelograms lying in pairs of parallel planes. Prompt for length, width, and height as integers, calculate the volume, then display the volume.

4.6 Write a program to determine the quotient and remainder of the following operations: $216 \div 5$, $76 \div 6$, and $744 \div 9$.

4.7 Write a program to illustrate the difference between integer division and floating-point division regarding the remainder, using the following dividends and divisors: $742 \div 16$ and $1756 \div 24$. For the floating-point result, let the integer be two digits and the fraction be four digits.

4.8 Write a program to prompt for five integers, then display their average. Then display the five numbers in five-space increments.

4.9 Indicate which of the following expressions return a value of 1 (true) or 0 (false).

(a) (5 == 5) && (6 != 2)
(b) (5 > 1) || (6 < 1)
(c) (2 == 1) && (5 == 5)
(d) !(5 == 4)

4.10 Indicate the value to which each of the following expressions evaluate.

(a) (1 + * 3)
(b) 10 % 3 * 3 – (1 + 2)
(c) ((1 + 2) * 3)
(d) (5 == 5)
(e) (5 = 5)

4.11 Given x = 7, y = 25, and z = 24.46, write a program that generates either a 1 (true) or a 0 (false) for the following relational operators:

x >= y,   x == y,   x < z,   y > z,   x != y – 18,   x + y != z

4.12 Write a program to convert from feet to meters or from meters to feet. Prompt for a value to be entered for feet or meters, then prompt for a choice: 1 means feet to meters; 2 means meters to feet. Check for both conversions.

4.13 Write a program to compare two integers that are entered from the keyboard. Indicate the relationship between the two integers — less than, equal, or greater than.

4.14 Write a program that prompts for two integers, then prompts whether the arithmetic operation is to be addition, subtraction, multiplication, or division. Display the resulting sum, difference, product, or quotient and remainder. Enter numbers for all four operations.

4.15 After execution of the following program, indicate the value of the expression for x:

(a) Without parentheses.
(b) With parentheses

```
//  log_op_precedence.cpp

/*  Determine to what the expression evaluates
without and with parentheses.
*/

#include <stdio.h>

    int a = 5, b = 6, c = 5, d = 1;
    int x;

main (void)
{
    x = a < b || a < c && c < d;
    printf ("\nWithout parentheses, the expression
                evaluates as %d", x);

    x = (a < b || a < c) && c < d;
    printf ("\nWith parentheses, the expression
                evaluates as %d\n\n", x);
    return 0;
}
```

4.16 Write a program to illustrate the relational, arithmetic, and logical operators. Prompt for three integers separated by spaces, then use the three operators on combinations of the inputs. Print the resulting outputs.

4.17 Given the values $a = 3$ and $b = 4$, write a program to evaluate the following two conditional statements:

```
(5 + 2 * (a > b)) ? a : b;
(6>(a > b)) ? a : b;
```

4.18 To what does the expression 5 + 3 * 8 / 2 + 2 evaluate?
Rewrite the expression, adding parentheses, so that it evaluates to 16.

4.19 Given the program shown below, obtain the output for *a*, *b*, and *c*.

```
//pre_post_incr.cpp
//example of preincrement and postincrement

#include "stdafx.h"

int main (void)
{
      int a, b;
      int c = 2;

      a = ++c;
      b = c++;

      printf ("%d %d %d\n\n", a, b, ++c);

      return 0;
}
```

4.20 Given the program shown below, obtain the output for *a* and *x*.

```
//pre_post_incr2.cpp
//evaluate preincrement and postincrement expressions

#include "stdafx.h"

int main (void)
{
      int a;
      int x = 10;

      a = x++;
      printf ("a = %d\nx = %d\n\n", a, x);

      a = ++x;
      printf ("a = %d\nx = %d\n\n", a, x);

      return 0;
}
```

4.21 Display the outputs for variables *int1* and *int2* after the program shown below has executed.

```
//pre_post_incr3.cpp
//illustrates pre- and postincrement operators
//utilizing integer multiplication

#include "stdafx.h"

int main (void)

{
        int int1, int2;

        int1 = int2 = 10;

        printf ("%d   %d", int1++, ++int2);
        printf ("\n%d   %d", int1++, ++int2);
        printf ("\n%d   %d\n\n", 5*int1++, 5*++int2);

        return 0;
}
```

4.22 Given the program shown below, obtain the outputs for variables *a* and *b* after the program has executed.

```
//pre_post_incr4.cpp
//example to illustrate pre- and postincrement

#include "stdafx.h"

int main (void)

{
        int a, b;
        int c = 0;

        a = ++c;
        b = c++;

        printf ("%d   %d   %d\n\n", a, b, ++c);

        return 0;
}
```

4.23 This problem uses the bitwise operators AND, OR, and XOR. Prompt for two 2-digit lowercase hexadecimal characters, convert them to uppercase hexadecimal characters, and display them. Then perform the operations of AND, OR, and XOR on the two characters. Display the results of the three bitwise operations.

4.24 Determine the returned values for the six **printf ( )** statements in the program shown below by entering one even and one odd integer.

```
//logical_and_or_xor.cpp
//using the logical and, or, and
//exclusive-or operators

#include "stdafx.h"

int main (void)
{
   int int1, int2;

   printf ("Enter one even and one odd integer: ");
   scanf ("%d %d", &int1, &int2);

   printf ("\nThe AND operator returns a value of:
            %d\n", (int1%2 == 0) && (int1%4 == 0));

   printf ("The AND operator returns a value of:
            %d\n", (int2%2 == 0) && (int2%4 == 0));

//------------------------------------------------------
   printf ("\nThe OR operator returns a value of:
            %d\n", (int1%2 == 0) || (int1%4 == 0));

   printf ("The OR operator returns a value of:
            %d\n", (int2%2 == 0) || (int2%3 == 0));
//------------------------------------------------------
   printf ("\nThe XOR operator returns a value of:
            %d\n", (int1%2 == 0) ^ (int1%4 == 0));

   printf ("The XOR operator returns a value of:
            %d\n\n", (int2%2 == 0) ^ (int2%3 == 0));

   return 0;
}
```

4.25 Determine the returned values (True or False) for the **printf ( )** functions in the program shown below.

```
//if_else_statements.cpp
//if-else statements returning a value of true or false

#include "stdafx.h"

int main (void)
{
   int a = 10, b = 12, c = 15;
//-------------------------------------------------------
   if (c == 15)
      printf ("True");
   else
      printf ("\nFalse");
//-------------------------------------------------------
   if (a != c + b)
      printf ("\nTrue");
   else
      printf ("\nFalse");
//-------------------------------------------------------
   if (c == 10)
      printf ("\nTrue");
   else
      printf ("\nFalse");
//-------------------------------------------------------
   if (b == a * c - 2)
      printf ("\nTrue\n\n");
   else
      printf ("\nFalse\n\n");
//-------------------------------------------------------
   return 0;
}
```

4.26 Write a program to illustrate the bitwise NOT operator using the conversion specifier %c for characters entered from the keyboard.

4.27 Determine the number of times that the program shown below will execute the **while ( )** loop.

```
#include "stdafx.h"
int main (void)
{
    int n = 0;
    while (n < 3)
        printf ("n = %d\n", n);
        n++;
    printf ("Have a good day.\n");
    return 0;
}
```

4.28 Prompt for a single character from the keyboard, then display the character. Continue looping with a **while ( )** loop until a lowercase *x* is entered. Give the user instructions on using the program.

4.29 Write a program to display the characters A through G as numeric values using a **while ( )** loop. Use the decimal value of G as the limit for the **while ( )** loop.

4.30 Use a **for ( )** loop to count from 10 to 100 in increments of 10. Display the numbers on a single line.

4.31 Use a **for ( )** loop to display the numbers that are evenly divisible by 4 and by 6 up to a value of 60.

4.32 Write a program to generate i^1, i^2, and i^3, where i represents the integers 1, 2, 3, 4, and 5. Display the results in three columns that specify i^1, i^2, and i^3.

4.33 Determine the number of *A*s that are printed by the program shown below.

```
//for_loop_nested.cpp
#include "stdafx.h"
int main (void)
{
    int x, y;
    for (x = 0; x < 5; x++)
        for (y = 5; y > 0; y--)
            printf ("A");
    printf ("\n\n");
    return 0;
}
```

4.34 Initialize an array with the squares of numbers 1 through 10, then print the array. Copy the array to a second array, then print the second array.

4.35 Write a program that defines a 9×2 array with powers of 2. Prompt the user to enter a number from 0 to 8, then print the number raised to the power of 2. Use the **break** statement to exit from any loops that are used. The **break** statement permits the program to exit a loop immediately from any location within the body of the loop.

4.36 Write a program that defines a string, then prints the string. Then reverse the order of the original string and print the new string.

4.37 Write a program that prompts for uppercase and lowercase letters to be inserted into a string. Then print the letters that were entered, plus their decimal equivalent and their hexadecimal equivalent.

4.38 Write a program to illustrate pointer postincrement. Assign an integer *int1* a value of 50 and assign a pointer variable the address of *int1*.

4.39 Write a program to illustrate pointer addition and subtraction from two integers that are entered from the keyboard.

4.40 Write a program containing a function, which adds two integers that are entered from the keyboard.

4.41 Write a program containing two functions: one to add two integers and one to add two floating-point numbers. The integers and floating-point numbers are entered from the keyboard.

4.42 Write a program to create five functions to perform arithmetic operations on two integers. Call the functions *add*, *sub*, *mul*, *divq*, and *divr*, where *divq* and *divr* represent divide (quotient) and divide (remainder), respectively. Prompt for two operands, then execute all five functions in sequence and display the results.

5

Data Transfer Instructions

This chapter presents the basic data transfer instructions as they apply to the X86 processors. Other data transfer instructions, such as instructions that pertain to stack operations and string operations, are presented in later chapters. This chapter also includes the various data types used in the X86 processors.

5.1 Data Types

The data types that are covered in this section are signed binary integers, unsigned binary integers, unpacked binary-coded decimal (BCD) integers, packed BCD integers, and floating-point numbers.

5.1.1 Signed Binary Integers

A signed integer is a binary number that is interpreted as a number in 2s complement representation, where the high-order (leftmost) bit is the sign bit. Signed integers can occupy a byte (8 bits: 7 through 0), a word (16 bits: 15 through 0), a doubleword (32 bits: 31 through 0), or a quadword (64 bits: 63 through 0), where the sign bits are bits 7, 15, 31, and 63, respectively. A sign bit of 0 indicates a positive number; a sign bit of 1 indicates a negative number. An integer has the following range:

$$-(2^{n-1}) \text{ to } +(2^{n-1} - 1)$$

Therefore, a signed integer byte has a range from –128 to +127, where $n = 8$; a signed integer word has a range from –32,768 to + 32,767, where $n = 16$; a signed doubleword has a range from –2,147,483,648 to +2,147,483,647, where $n = 32$; and a signed quadword has a range from $-9.223372037 \times 10^{18}$ to $+9.223372037 \times 10^{18} - 1$, where $n = 64$.

Bytes of a multibyte number are stored in the *little endian* format; that is, the low-order byte is stored first at the lower address and subsequent bytes are stored in successively higher addresses in memory. Little endian can also refer to the way that the bits are ordered in a byte; for example, bits 7 through 0 in a left-to-right sequence.

The *big endian* format stores the high-order byte first at the lower address and subsequent bytes are stored in successively higher addresses in memory. Big endian can also refer to the way that the bits are ordered in a byte; for example, bits 0 through 7 in a left-to-right sequence. Examples of signed integers in 2s complement representation are shown in Table 5.1.

Table 5.1 Examples of Numbers in 2s Complement Representation

Positive Numbers	Decimal Value	Negative Numbers	Decimal Value
0111	+7	**1**001	–7
0101 0110	+86	**1**010 1010	–86
0010 1111 0101	+757	**1**101 0000 1011	–757

The 2s complement is obtained by adding 1 to the 1s complement; the 1s complement is obtained by inverting all bits of the number. There is a faster way to obtain the 2s complement of a number: keep the low-order 0s and the first 1 unchanged as the number is scanned from right to left, then invert all remaining bits. An example is shown below.

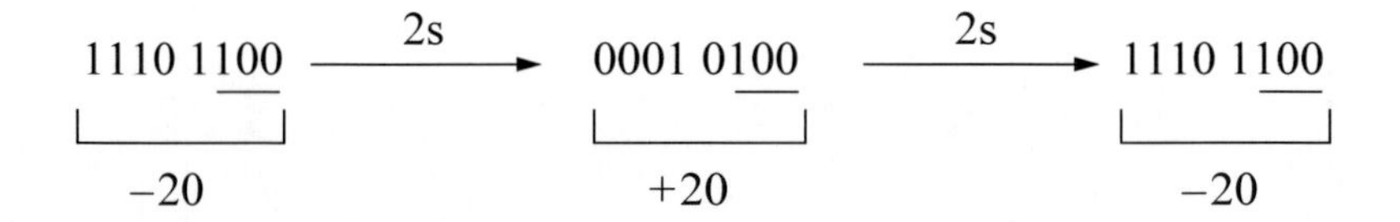

5.1.2 Unsigned Binary Integers

Unsigned integers can occupy a byte (8 bits: 7 through 0), a word (16 bits: 15 through 0), a doubleword (32 bits: 31 through 0), or a quadword (64 bits: 63 through 0). An unsigned integer has the following range:

$$0 \text{ to } 2^n - 1$$

The maximum range for an unsigned byte is $2^8 - 1 = 255$; the range for a word is $2^{16} - 1 = 65{,}535$; the maximum range for a doubleword is $2^{32} - 1 = 4{,}294{,}967{,}295$; and the maximum range for a quadword is $2^{64} - 1 = 1.8446744072 \times 10^{19}$. Examples of unsigned binary integers are shown in Table 5.2.

Table 5.2 Examples of Unsigned Binary Integers

Unsigned Integers	Decimal Value	Unsigned Integers	Decimal Value
0111	7	1001	9
0101 0110	86	1010 1010	170
0010 1111 0101	757	1101 0000 1011	3339

5.1.3 Unpacked and Packed BCD Integers

Each BCD digit is an unsigned number with a range from 0 (0000) to 9 (1001); that is, the decimal number is encoded as an equivalent binary number. All numbers 10 through 15 are invalid for BCD, since a radix 10 digit contains only the decimal numbers 0 through 9.

If the BCD number is *unpacked*, then only the low-order bits of each byte contain a valid decimal number; the high-order bits of each byte can be an indeterminate value for addition or subtraction operations, but must be 0000 for multiplication and division operations.

For *packed* BCD numbers, both the low-order and the high-order half of a byte contain valid decimal numbers. In this case, the digit in the high-order half of the byte is the most significant number. BCD digits can be packed into a format consisting of ten bytes that represent a signed decimal number, as shown in Figure 5.1. The low-order digit is D0 and the high-order digit is D17. The sign bit is bit 79, where a 0 indicates a positive number and a 1 indicates a negative number. Bits 78 through 72 are irrelevant and can be classified as don't care bits.

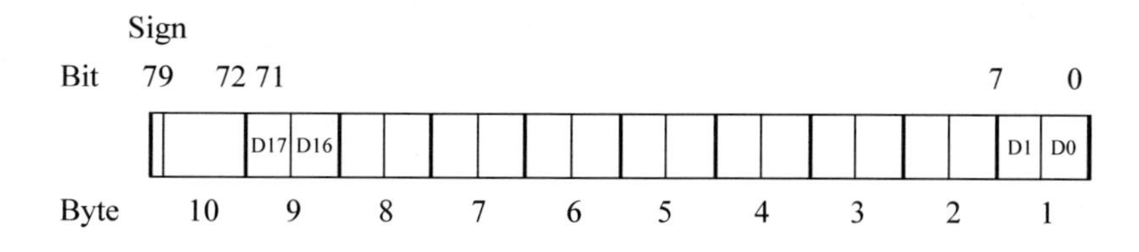

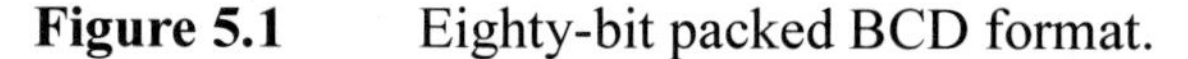
Figure 5.1 Eighty-bit packed BCD format.

Positive and negative BCD numbers with the same absolute value are differentiated only by the sign bit — all decimal digits are identical for both positive and negative numbers. This is in contrast to radix 2 numbers for the diminished-radix complement and the radix complement number representations, where the bit configurations are different for positive and negative numbers. However, the 80-bit packed BCD format is similar to the sign-magnitude number representation, where only the sign bit is different.

5.1.4 Floating-Point Numbers

Floating-point numbers consist of the following three fields: a sign bit, s; an exponent, e; and a fraction, f. These parts represent a number that is obtained by multiplying the fraction, f, by a radix, r, raised to the power of the exponent, e, as shown in Equation 5.1 for the number A, where f and e are signed fixed-point numbers, and r is the radix (or base).

$$A = f \times r^e \tag{5.1}$$

The exponent is also referred to as the *characteristic*; the fraction is also referred to as the *significand* or *mantissa*. Although the fraction can be represented in sign-magnitude, diminished-radix complement, or radix complement, the fraction is predominantly expressed in a true magnitude (unsigned) representation.

Figure 5.2 shows the format for 32-bit single-precision and 64-bit double-precision floating-point numbers. The single-precision format consists of a sign bit that indicates the sign of the number, an 8-bit signed exponent, and a 23-bit fraction. The double-precision format consists of a sign bit, an 11-bit signed exponent, and a 52-bit fraction. Fractions in the IEEE format are normalized; that is, the leftmost significand bit is a 1. Since there will always be a 1 to the immediate right of the radix point, the 1 bit is not explicitly shown — it is an *implied 1*.

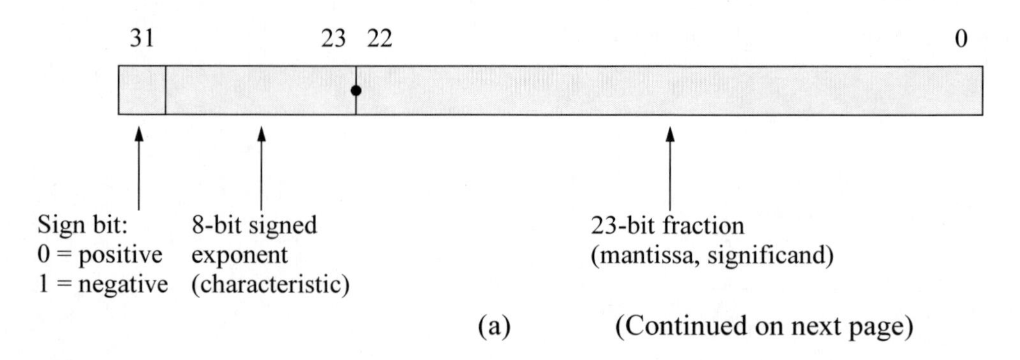

(Continued on next page)

Figure 5.2 IEEE floating-point formats: (a) 32-bit single precision and (b) 64-bit double precision.

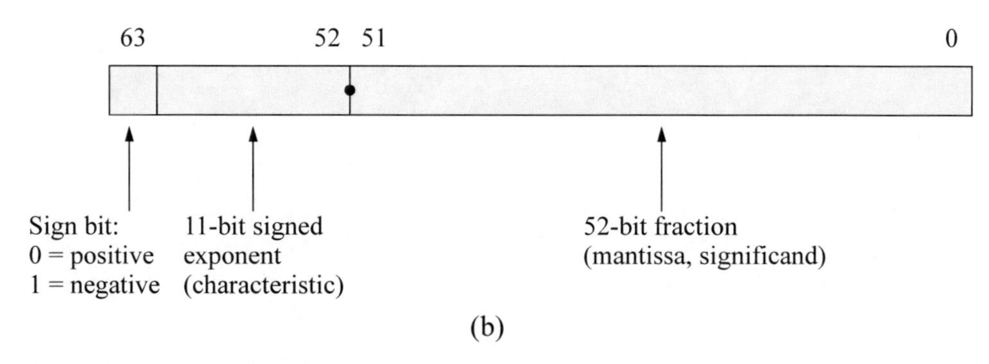

(b)

Figure 5.2 (Continued)

The exponents of the X86 floating-point architecture are initially *unbiased* numbers. An unbiased exponent can be either a positive or a negative integer. During the addition or subtraction of two floating-point numbers, the exponents are compared and the fraction with the smaller exponent is shifted right by an amount equal to the difference of the two exponents. The comparison is simplified by using *biased* exponents; that is, by adding a positive *bias constant* to each exponent during the formation of the numbers. This bias constant has a value that is equal to the most positive exponent and makes all exponents positive numbers.

For example, if the exponents are represented by n bits, then the bias is $2^{n-1} - 1$. For $n = 4$, the most positive number is 0111 (+7). Therefore, all biased exponents are of the form shown in Equation 5.2.

$$e_{\text{biased}} = e_{\text{unbiased}} + 2^{n-1} - 1 \tag{5.2}$$

The advantage of using biased exponents is that they are easier to compare without having to consider the signs of the exponents. The main reason for biasing is to determine the correct alignment of the fractions by aligning the radix points, and to determine the number of bits to shift a fraction in order to obtain proper alignment. An additional advantage is that the smallest exponent contains only zeroes; therefore, the floating-point depiction of the number zero is an exponent of zero with a fraction of zero. If the biased exponent has a maximum value (255) and the fraction is nonzero, then this is interpreted as *Not a Number* (NaN), which is the result of zero divided by zero or the square root of −1.

5.2 Move Instructions

This section presents some of the basic move instructions involving data transfer. These include register-to-register, immediate-data-to-register, immediate-data-to-memory, memory-to-register, and register-to-memory. Also included are moves with sign extension, moves with zero extension, and conditional moves that move data to a destination depending on the state of a flag. Data transfer instructions have the following general syntax:

```
MOV  destination, source
```

The *source* is a general-purpose register, a segment register, immediate data, or a memory operand. The *destination* is a general-purpose register, a segment register, or a memory location. Debug registers can also be used as source or destination registers.

5.2.1 General Move Instructions

Examples of different forms of the MOV instruction are shown below. The general move instruction, MOV, cannot be used for memory-to-memory data transfer. The move string (MOVS) instruction is used for that operation. The MOV instruction also cannot move data from one segment register to another segment register.

Register-to-register	**MOV** EAX, EBX	
Immediate-to-register	**MOV** AX, 1234H	
Direct	**MOV** MEM_ADDR, AX	MEM_ADDR gives the address directly
Register indirect	**MOV** EAX, [EBP]	[EBP] gives the address indirectly
Indexed	**MOV** AX, [DI/SI + displacement]	Displacement may be 0
Based	**MOV** EAX, [EBP/EBX + displacement]	Displacement may be 0
Base plus index	**MOV** [BP/BX + SI/DI], AX	
Base plus index plus displacement	**MOV** [BX/BP + DI/SI + displacement], EAX	

Figure 5.3 shows an example of moving a byte from a memory location, determined by the contents of register EBX, to register AL. This is an indirect addressing mode with an offset of eight. If the operand type is not obvious, then the type can be specified by either BYTE PTR, WORD PTR, DWORD PTR, for example, as shown below, which moves $0C_{16}$ to a memory location specified by the sum of the contents of registers EBX and ESI.

```
MOV  BYTE PTR [EBX + ESI], 0CH
```

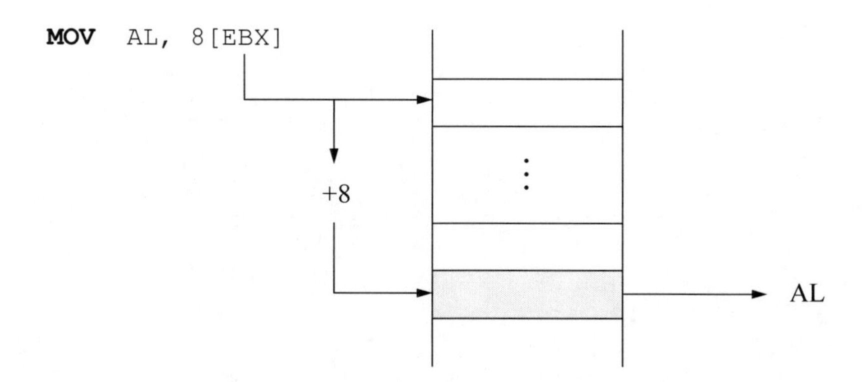

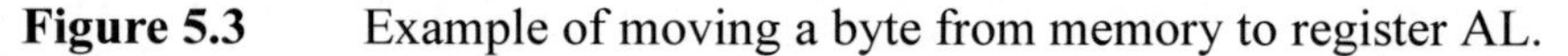
Figure 5.3 Example of moving a byte from memory to register AL.

Structure of an X86 assembly language program Figure 5.4 illustrates a program to interchange two registers using the MOV instruction. This can be accomplished much easier by using the *exchange* (XCHG) instruction, which is presented in a later section. However, since this section introduces the MOV instruction, the program uses the MOV data transfer instruction.

The program is called *swap_bytes* and saved as *swap_bytes.asm*. The **.STACK** is a simplified segment directive that defines the stack segment; in a similar manner, the directives that define the data and code segments are labelled **.DATA** and **.CODE**, respectively. A size value can be appended to these simplified segment directives to specify their respective sizes. These directives generate the appropriate segment statements and the corresponding end segment statements. Prior to the introduction of the simplified segment directives, the segments were defined as shown below.

STSG	**SEGMENT PARA STACK 'STACK'**
	. . .
STSG	**ENDS**
DTSG	**SEGMENT PARA 'DATA'**
	. . .
DTSG	**ENDS**
CDSG	**SEGMENT PARA 'CODE'**
BEGIN	**PROC FAR**
	ASSUME SS:STACK, **DS**:DTSG, **CS**:CDSG
	. . .
BEGIN	**ENDP**
CDSG	**ENDS**
	END BEGIN

```
;swap_bytes.asm
;------------------------------------------------------------
.STACK
;------------------------------------------------------------
.DATA
    TEMP        DB ?
;$ sign is a delimiter meaning end of string
    RSLT        DB 0DH, 0AH, 'BL =   , BH =   $'
;------------------------------------------------------------
.CODE
BEGIN   PROC FAR

;set up pgm ds
      MOV     AX, @DATA     ;place the .DATA addr in ax
      MOV     DS, AX        ;set up data seg addr for this pgm

;assign values to bl and bh
      MOV     BL, 'A'
      MOV     BH, 'B'

;store bl in temp area before swapping
      MOV     TEMP, BL

;swap registers
      MOV     BL, BH
      MOV     BH, TEMP

;move registers to result area for display
      MOV     RSLT + 7, BL
      MOV     RSLT + 15, BH

;print result
      MOV     AH, 09H          ;display string
      MOV     DX, OFFSET RSLT  ;rslt addr must be in dx
      INT     21H              ;a dos interrupt that uses
                               ;a fctn code in ah

BEGIN ENDP
      END     BEGIN            ;start pgm at begin
```

(a)

```
BL = B, BH = A
```

(b)

Figure 5.4 Program to illustrate interchanging the contents of two general-purpose registers BL and BH: (a) the program, and (b) the outputs.

The procedure section of the code begins with the PROC directive and ends with the *end procedure* (ENDP) directive. The code segment, specified by the PROC directive, contains the executable code for the program. The procedure name is BEGIN and must be distinct from any other procedure name. The operand FAR indicates the procedure entry location to begin execution of the program. The ENDP directive ends a procedure and contains the same name as the procedure; the END directive ends the program, and the operand — BEGIN in this case — contains the name of the FAR procedure where program execution is to begin.

The simplified segment directives also include predefined *equates*, such as @code and @data. The @data equate can be used, in conjunction with the MOV instruction, to load the offset of the data segment into register AX by the following statement:

```
MOV  AX, @DATA
```

Register AX is then loaded into the data segment register DS by the following statement:

```
MOV  DS, AX
```

The ASCII characters A and B are then loaded into registers BL and BH, respectively. The contents of registers BL and BH are then interchanged using the temporary storage area in the data segment declared as TEMP, which is defined as a byte location by the define byte(s) directive DB. Once the contents of the registers have been interchanged, they are moved to the RSLT area in the data segment, which is also defined as a field of individual bytes. The control character 0DH places the cursor at the left of the monitor screen; the control character 0AH specifies a line feed, which advances the output by a single line — also called a carriage return.

The control characters 0DH and 0AH occupy one byte each, where 0DH is at location 0 of the storage area designated by RSLT and 0AH is at location 1 of the storage area designated by RSLT. Register BL is moved to location RSLT + 7, which is two spaces to the right of the first equal sign — every space within the single quotation marks is one byte. Register BH is moved to location RSLT + 15, which is two spaces to the right of the second equal sign. The dollar sign ($) indicates end of string within the single quotation marks.

The statements shown below display the contents of the RSLT field. A function code of 09H (display string) is moved to register AH, which must contain the required function. The offset address of RSLT — relative to the beginning of the data segment — is moved to the required register, DX. The DOS interrupt, INT 21H, is an interrupt that uses a function code in register AH to indicate an operation to be performed — in this case the display string function.

```
MOV  AH, 09H
MOV  DX, OFFSET RSLT
INT  21H
```

The program of Figure 5.4 will be rewritten using an assembly language module that is linked to a C program. The revised program is shown in Figure 5.5. External names in a C program are preceded by an underscore (_) character; for example, *_asm*; that is, in-line assembly language is achieved by the *_asm* command. Since a C program is case sensitive, the assembly language section should use the same case for variable names that are common to both the C module and the assembly language module.

Note that there is no need to declare stack, data, or code segments in the linked program. The C module uses the stack to push and pop data as required. There is also no need to set up the data segment in the DS register. The results of the interchange program do not have to be sent to the data segment in preparation for display. The three instructions to display printing the resulting string are also not required — this is accomplished by the **printf()** function. Linking an assembly language to a C program makes the resulting program easier to read and understand.

```
//swap_bytes.cpp
//swap bytes in two GPRs
#include "stdafx.h"
char main (void)
{
   char temp;
   char rslt1, rslt2;

//switch to assembly
      _asm
      {
         MOV   BL, 'A'
         MOV   BH, 'B'
//swap bytes
         MOV   temp, BL
         MOV   rslt1, BH
         MOV   BH, temp
         MOV   rslt2, BH
      }

//print result
   printf ("BL = %c, BH = %c\n", rslt1, rslt2);
   return 0;
}
```
(a)

```
BL = B, BH = A
Press any key to continue . . . _
```
(b)

Figure 5.5 Assembly language module linked to a C module to interchange two general-purpose registers: (a) the program and (b) the outputs.

5.2.2 Move with Sign/Zero Extension

The *move with sign extension* (MOVSX) instruction moves the source operand into a destination general-purpose register, then extends the sign bit into the high-order bits of the destination. The sign bit is extended into a 16-bit or a 32-bit destination. The MOVSX instruction is used with signed operands.

The *move with zero extension* (MOVZX) instruction is similar to the MOVSX instruction, except that zeroes are extended in the destination instead of the sign bit. The syntax for the MOVSX and the MOVZX instructions is shown below.

```
MOVSX/MOVZX  register, register/memory/immediate
```

Figure 5.6 contains a program that illustrates both the MOVSX instruction and the MOVZX instruction. An immediate operand, 0F5H, is loaded into register AL, which is sign-extended into register EAX to yield a result of FFFFFFF5H. An immediate, 0F5H, is loaded into register BL, which is zero-extended into register EBX to yield a result of F5H (000000F5).

```
//movsx_movzx2.cpp
//move with sign/zero extension
#include "stdafx.h"
int main (void)

{
   int rslt1, rslt2;

//switch to assembly
      _asm
      {
         MOV    AL, 0F5H
         MOVSX  EAX, AL
         MOV    rslt1, EAX

         MOV    BL, 0F5H
         MOVZX  EBX, AL
         MOV    rslt2, EBX
      }
   printf ("Result = %X, %X\n", rslt1, rslt2);
   return 0;
}
```
(a)

```
Result = FFFFFFF5, F5
Press any key to continue . . . _
```
(b)

Figure 5.6 Program to illustrate the MOVSX and the MOVZX instructions: (a) the program and (b) the outputs.

5.2.3 Conditional Move

The *conditional move* (CMOVcc) instructions execute move operations based on the state of certain flags — condition codes (cc) — in the EFLAGS register. The condition codes are appended to the right of the CMOV instruction; for example, the instruction CMOVC will execute if the state of the carry flag is 1 (CF = 1).

The conditional move instructions move data from a source operand to a destination location; that is, from memory to a general-purpose register (GPR) or from one GPR to another GPR. Conditional moves for 8-bit registers are not supported. If the specified condition is true, then the move operation is performed. If the condition is false, then the move operation is not executed and the instruction following the CMOVcc instruction is executed. The conditional move function is similar to the *if* construct, in which a branch takes place if a condition is true. The conditional move instructions are shown in Table 5.3 for both unsigned and signed operations.

Table 5.3 Conditional Move Instructions

Mnemonic	Flags	Description
	Unsigned	
CMOVA	(CF or ZF) = 0	Above
CMOVAE	CF = 0	Above or equal
CMOVNC	CF = 0	No carry
CMOVB	CF = 1	Below
CMOVC	CF = 1	Carry
CMOVBE	(CF or ZF) = 1	Below or equal
CMOVE	ZF = 1	Equal
CMOVNE	ZF = 0	Not equal
CMOVP	PF = 1	Parity even
CMOVNP	PF = 0	Parity odd
	Signed	
CMOVGE	(SF xor OF) = 0	Greater than or equal
CMOVL	(SF xor OF) = 1	Less than
CMOVLE	[(SF xor OF) or ZF] = 1	Less than or equal
CMOVO	OF = 1	Overflow
CMOVNO	OF = 0	No overflow
CMOVS	SF = 1	Sign is negative
CMOVNS	SF = 0	Sign is positive

Compare instruction Since the conditional move instructions depend on the state of flags resulting from the execution of certain instructions, the compare instruction will be introduced at this time in order to generate flags for a conditional move instruction. One example will suffice to illustrate the operation of a conditional move instruction. The syntax for the compare instruction is shown below.

```
CMP    first source operand, second source operand
```

The comparison is achieved by subtracting the second source operand from the first source operand and setting the appropriate flags. The comparison is that of the first source operand to the second source operand. The operands are unchanged after the compare operation, but the following flags are set according to the result of the compare operation: auxiliary carry or adjust flag (AF), the carry flag (CF), the overflow flag (OF), the parity flag (PF), the sign flag (SF), and the zero flag (ZF).

Figure 5.7 shows a program that illustrates the application of the conditional move instruction CMOVAE if the unsigned operand in register EAX is above or equal to the unsigned operand in register EBX. The purpose of the program is to print the larger of two operands or one operand if the two are equal. Two integers are entered by the user and stored in memory locations *x* and *y*.

```
//mov_cond.cpp
//uses cmovae (above or equal); cf = 0.
//If user-entered x is unsigned above or equal to
//user-entered y, then print x; otherwise, print y.
//If integers are equal, then print x

#include "stdafx.h"

int main (void)
{
//define and initialize variables
   int   x, y, rslt;

   printf ("Enter two integers: \n");
   scanf ("%d %d", &x, &y);

//switch to assembly
      _asm
      {
         MOV    EAX, x
         MOV    EBX, y                //continued on next page
```

(a)

Figure 5.7 Program to illustrate the use of the conditional move instruction CMOVAE: (a) the program and (b) the outputs.

```
//to move ebx to result if conditional move fails
         MOV    EDX, EBX

         CMP    EAX, EBX    //set flags

//if eax >= ebx, move eax to rslt;
//otherwise, move ebx to result
         CMOVAE EDX, EAX
         MOV     rslt, EDX
      }

   printf ("Result = %d\n\n", rslt);
   return 0;
}
```

```
Enter two integers:
15 10
Result = 15

Press any key to continue . . . _
----------------------------------------------------------------
Enter two integers:
10 15
Result = 15

Press any key to continue . . . _
----------------------------------------------------------------
Enter two integers:
10 10
Result = 10

Press any key to continue . . . _
```

(b)

Figure 5.7 (Continued)

5.3 Load Effective Address

The *load effective address* (LEA) instruction loads the offset (effective address) of a memory address (source operand) within a memory segment and stores it in a general-purpose register. No flags are affected by this instruction. The address calculation of the LEA instruction is similar to the calculation performed by the MOV instruction, but the *address* of the source operand is stored in the destination, not the contents of the source operand. The LEA instruction can also be used for unsigned integer arithmetic. The syntax for the LEA instruction is shown below.

```
LEA    destination, source
```

A comparable operation to the LEA instruction is a MOV with offset operation and the equate construct that were utilized in the program of Figure 5.4, as shown below, both of which move an offset address to a destination register.

```
MOV    DX, OFFSET RSLT
MOV    AX, @DATA
```

An example that illustrates one use of the LEA instruction is shown in Figure 5.8, which accesses an array element. The address of *array* in the data segment is loaded into register BX and register SI is cleared using the immediate addressing mode — SI will be used to index into the array. Then the word at array [0] is stored in register AX. The register addressing mode is used for the destination (AX); the base-index addressing mode is used for the source [BX][SI]. To index through the array, simply increment register SI by the appropriate amount.

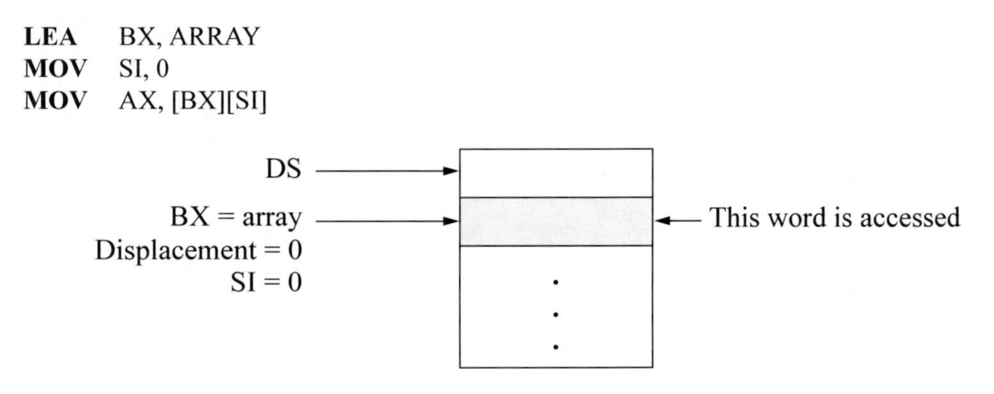

Figure 5.8 One use for the LEA instruction is to access an array element.

5.4 Load Segment Registers

This section describes some of the instructions that load far pointers; for example, *load far pointer* using DS (LDS), *load far pointer* using SS (LSS), *load far pointer* using ES (LES), *load far pointer* using FS (LFS), and *load far pointer* using GS (LGS). These instructions load a far pointer (segment selector and offset), that points to a memory location, into a segment register and a general-purpose register. The 16-bit segment selector component of the far pointer is stored in the segment register specified in the operation code of the instruction; the offset is stored in the specified general-purpose register.

The syntax for a load far pointer instruction is shown below, where destination is any 16-bit or 32-bit general-purpose register and source is a 32-bit or 48-bit memory location.

$$\mathbf{L}_{seg} \quad \texttt{destination, source}$$

An example using the extra segment (ES) is shown in Figure 5.9 for the following instruction:

```
LES    ESI, ptr1
```

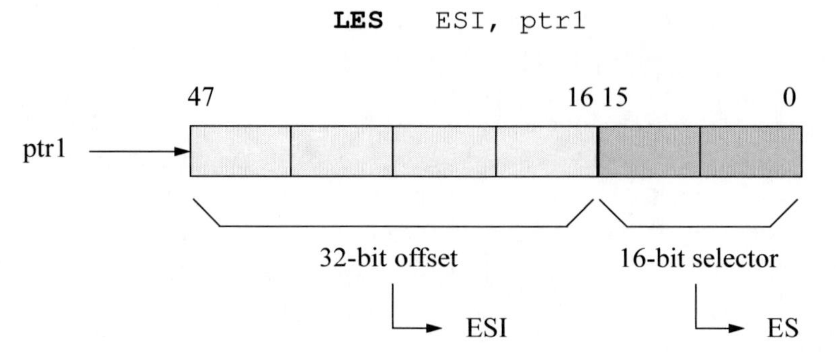

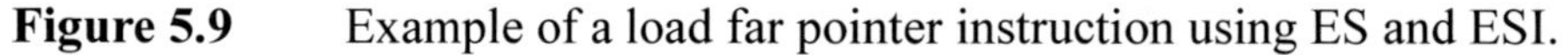
Figure 5.9 Example of a load far pointer instruction using ES and ESI.

An example of a load far pointer instruction using the data segment (DS) register and the source index (SI) register is shown in Figure 5.10. This example sets SI for string operations; SI contains the offset of the string.

```
LDS    SI, label
```

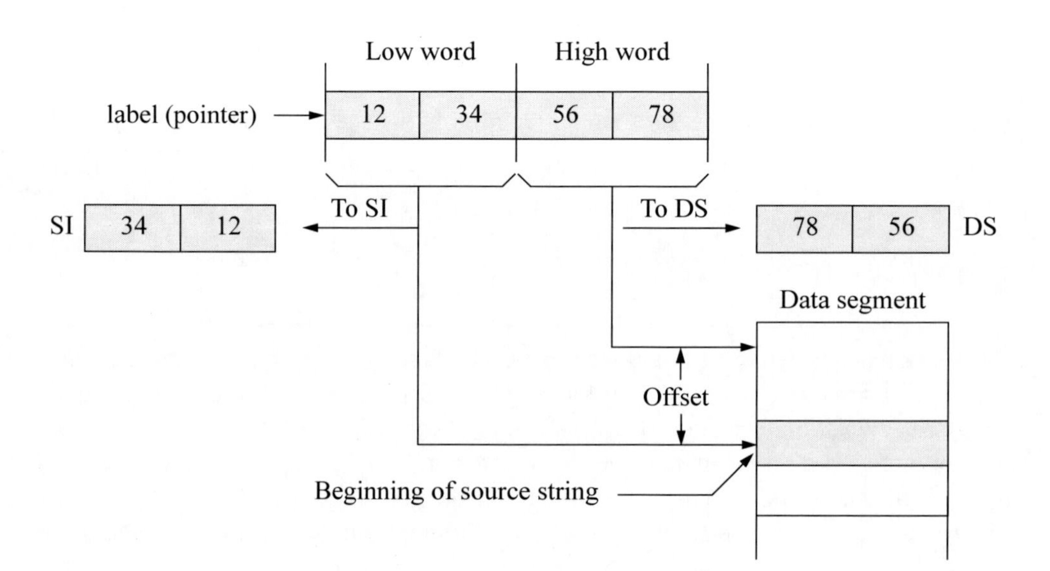

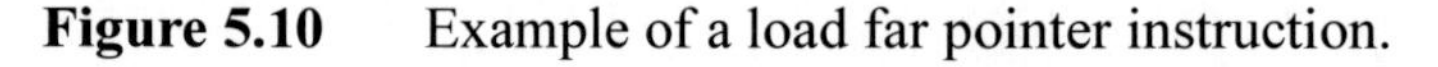
Figure 5.10 Example of a load far pointer instruction.

5.5 Exchange Instructions

This section presents four different interchange-type instructions. One instruction exchanges the contents of a register — or memory location — with a register (XCHG); another reverses the order of bytes in a register (BSWAP); another exchanges the contents of two operands, then stores the sum in the destination (XADD); and another compares the contents of a specific register with the contents of a register — or memory location — then performs a move operation depending on the results of the comparison (CMPXCHG).

5.5.1 Exchange

The *exchange* (XCHG) instruction exchanges the contents of the source and destination operands. The operands can be two general-purpose registers (GPRs) or a GPR and a memory location. This is a simpler and more expeditious method of exchanging two operands than the method using the MOV instruction of Figure 5.4 — no temporary storage location is required to hold one of the operands.

When an XCHG instruction is executed, the LOCK prefix asserts the processor's LOCK# signal. In a multiprocessor environment, the LOCK prefix ensures that the processor maintains uninterrupted use of any shared memory. This prevents other threads from accessing the memory location while the instruction is executing. The LOCK prefix is applicable only to certain instructions and is asserted for the duration of the instruction that follows the LOCK prefix; that is, the LOCK prefix prepends the instruction to which it applies.

The syntax for the XCHG instruction is shown below, where the source and destination operands can be a register or a memory location, but cannot both be memory locations. The flags are not affected by the XCHG instruction.

```
XCHG   destination, source
XCHG   register/memory, register/memory
```

There are several versions of assemblers for use with X86 assembly language programming. The various assemblers include the following: Microsoft Macro-Assembler (MASM), Netwide Assembler (NASM), Lazy Assembler (LZASM), NewBasic Assembler (NBASM), Flat Assembler (FASM), CodeX Assembler, and TMA Macro Assembler, among others. This book uses a version of a flat assembler.

Figure 5.11 illustrates the parameter list array in the data segment that is used to store the keyboard input data. Figure 5.12 shows an example of the XCHG instruction to swap two registers. The name PARLST (parameter list) in the data segment is the name of a one-dimensional array that is labelled as a byte array and accepts input data from the keyboard. The first element of the array, PARLST [0], is called MAXLEN, which defines the maximum number of input characters — in this example, five is the maximum number of allowable characters.

The second array element, PARLST [1], is called ACTLEN, which stores the actual number of characters entered from the keyboard. The third element of the array,

PARLST [2], contains the beginning of the operand field (OPFLD) where the operands from the keyboard are stored — the last byte in the OPFLD is the *Enter* (carriage return) character (↵).

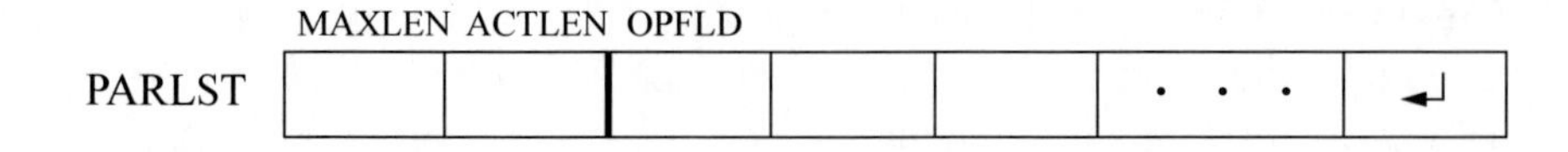

Figure 5.11 Parameter list one-dimensional array in which the keyboard input data are stored.

Following the PARLST array in the data segment is a prompt (PRMPT) for the user to enter two characters. The 0DH byte is the carriage return character; the 0AH byte is the line feed character. A string to be displayed is enclosed in single quotation marks and is terminated by a dollar sign ($), which indicates *end of string*. The code to display the prompt is shown below. The INT 21H is an operating system interrupt that uses a function code in register AH to specify an operation to be performed. The function code 09H is a display string routine. The address of the string to be displayed (PRMPT) is placed in register DX by the LEA instruction.

```
MOV   AH, 09H
LEA   DX, PRMPT
INT   21H
```

The code to receive the keyboard input data is similar, except that the function code 0AH is used to indicate that the function is a buffered keyboard input. The data are stored in the PARLST one-dimensional array beginning at location OPFLD.

The label RSLT1 displays the characters before they are exchanged. The first byte (B) of the string is location RSLT1 [2]. Each character, including spaces, is one byte; therefore, the first character entered from the keyboard is placed at location RSLT1 [20] — one space past the equal sign. In a similar manner, after the characters are exchanged, they are placed in the RSLT2 area of the data segment. The ENDP directive indicates the end of the procedure named BEGIN. The END directive indicates the end of the program.

```
;xchg_2_characters.asm
.STACK

.DATA
PARLST  LABEL BYTE
MAXLEN  DB 5
ACTLEN  DB ?
OPFLD   DB 5 DUP(?)                //continued on next page
```

(a)

Figure 5.12 Example to illustrate the use of the XCHG instruction: (a) the program and (b) the outputs.

```
PRMPT DB 0DH, 0AH, 'Enter two characters: $'
RSLT1 DB 0DH, 0AH, 'Before exchange =      $'
RSLT2 DB 0DH, 0AH, 'After exchange =       $'

.CODE
BEGIN PROC FAR
;set up pgm ds
      MOV     AX, @DATA
      MOV     DS, AX

;read prompt
      MOV     AH, 09H         ;display string
      LEA     DX, PRMPT       ;load addr of prmpt
      INT     21H             ;dos interrupt

;kybd rtn to enter characters
      MOV     AH, 0AH         ;buffered kybd input
      LEA     DX, PARLST      ;load addr of parlst
      INT     21H             ;dos interrupt

;store characters from opfld to bl and bh
      MOV     BL, OPFLD       ;get 1st char, store in bl
      MOV     BH, OPFLD + 1   ;get 2nd char, store in bh

;display original characters
      MOV     RSLT1 + 20, BL
      MOV     RSLT1 + 22, BH

      MOV     AH, 09H
      LEA     DX, RSLT1
      INT     21H

;exchange characters
      XCHG    BL, BH

;display swapped characters
      MOV     RSLT2 + 19, BL
      MOV     RSLT2 + 21, BH

      MOV     AH, 09H
      LEA     DX, RSLT2
      INT     21H

BEGIN ENDP
      END     BEGIN
                                 //continued on next page
```

Figure 5.12 (Continued)

```
Enter two characters: 12
Before exchange = 1 2
After exchange = 2 1
------------------------------------------------------------
Enter two characters: ab
Before exchange = a b
After exchange = b a              (b)
```

Figure 5.12 (Continued)

Figure 5.13 shows a similar program in which two integers are exchanged by an X86 assembly language program that is linked to a C program. Note the simplicity of the code — there is no need to specify a stack segment, a data segment, or a code segment. There is also no need to display the original integers — they are automatically displayed by the **printf ( )** function when they are entered. Two integers are entered by the user and are stored in locations *x* and *y*. The integers are then moved to registers EAX and EBX, where they are exchanged by the instruction XCHG EAX, EBX, then displayed by the **printf ( )** function; thus, there is also no need to precisely position the results of the operation for display.

```
//xchg_2_numb.cpp
//exchange two user-entered integers
#include "stdafx.h"
int main (void)
{
   intx, y, rslt1, rslt2;      //define variables
   printf ("Enter two integers: ");
   scanf ("%d %d", &x, &y);

//switch to assembly
      _asm
      {
         MOV    EAX, x
         MOV    EBX, y
         XCHG   EAX, EBX
         MOV    rslt1, EAX
         MOV    rslt2, EBX
      }
   printf ("Result = %d %d\n\n", rslt1, rslt2);
   return 0;
}                              (a)     //continued on next page
```

Figure 5.13 Exchange two integers by linking as X86 assembly language program to a C program: (a) the program and (b) the outputs.

```
Enter two integers: 11 22
Result = 22 11

Press any key to continue . . . _
------------------------------------------------------------
Enter two integers: 200 400
Result = 400 200

Press any key to continue . . . _
```
(b)

Figure 5.13 (Continued)

5.5.2 Byte Swap

The *byte swap* (BSWAP) instruction reverses the order of the bytes in a 32-bit or a 64-bit general-purpose register. For a 32-bit register, bit positions 0 through 7 are swapped with bit positions 24 through 31 and bit positions 8 through 15 are swapped with bit positions 16 through 23, as shown in Figure 5.14. This effectively converts a little endian format to a big endian format and a big endian format to a little endian format. The syntax for the byte swap instruction is shown below, where the *destination register* is a 32-bit or a 64-bit general-purpose register.

```
BSWAP   destination register
```

The byte swap instruction is not defined for 16-bit registers; to swap bytes in a 16-bit register, the XCHG instruction should be used. Figure 5.15 shows a program to illustrate the BSWAP instruction for a 32-bit general-purpose register, EAX. A user-entered integer is stored in location *x* as uppercase hexadecimal characters. The character is then moved to register EAX where the byte swap operation takes place. The result of the operation is then moved to location *rslt* to be displayed.

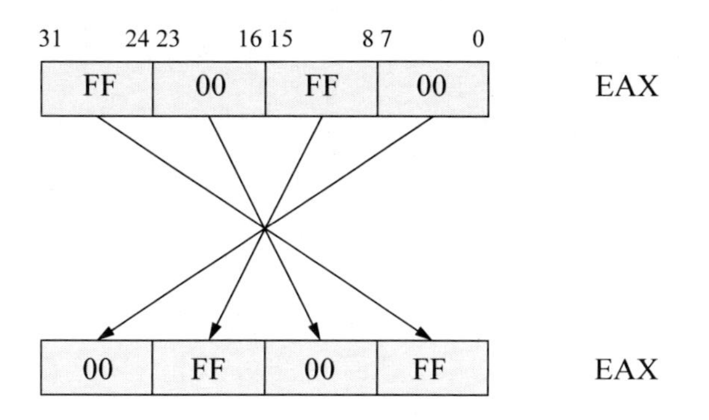

Figure 5.14 Diagram to illustrate the byte swap instruction.

```
//byte_swap.cpp
//swap bytes to convert between little-endian and big-endian

#include "stdafx.h"

int main (void)
{
   //define variables
   int   x, rslt;

   printf ("Enter an integer: \n");
   scanf ("%X", &x);

//switch to assembly
      _asm
      {
         MOV   EAX, x
         BSWAP EAX
         MOV   rslt, EAX
      }

   printf ("\nResult = %X\n\n", rslt);

   return 0;
}
```

(a)

```
Enter an integer:
FF00FF00

Result = FF00FF

Press any key to continue . . . _
------------------------------------------------------------
Enter an integer:
11221122

Result = 22112211

Press any key to continue . . . _
```

(b)

//continued on next page

Figure 5.15 Program to illustrate the byte swap operation: (a) the program and (b) the outputs.

```
Enter an integer:
12345678

Result = 78563412

Press any key to continue . . . _
-------------------------------------------------------------
Enter an integer:
-1

Result = FFFFFFFF

Press any key to continue . . . _
-------------------------------------------------------------
Enter an integer:
aaaabbbb

Result = BBBBAAAA

Press any key to continue . . . _
```

Figure 5.15 (Continued)

5.5.3 Exchange and Add

The *exchange and add* (XADD) instruction exchanges the destination operand with the source operand, then stores the sum of the two operands in the destination location. The destination operand can be in a register or memory location; the source operand is in a general-purpose register. The syntax for the XADD instruction is shown below.

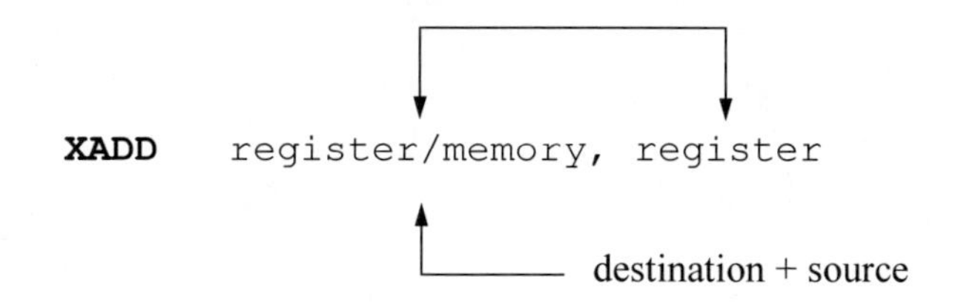

The register and memory operands can be 8-bit, 16-bit, 32-bit, or 64-bit operands. The XADD instruction can also utilize the LOCK prefix to ensure that the processor maintains uninterrupted use of any shared memory until the instruction has completed execution. Figure 5.16 shows a linked C and assembly language program that demonstrates the use of the XADD instruction using the EAX and EBX general-purpose registers.

Two integers are entered from the keyboard and stored in locations *x* and *y*, then moved to registers EAX and EBX, respectively. The XADD instruction is then executed which swaps the contents of the two registers and obtains their sum, which is stored in the destination register EAX. The program then moves the sum to location *rslt1* and the original value of register EAX to *rslt2*. The corresponding results are then displayed. The flags in the EFLAGS register specify the result of the addition operation.

```
//xchg_add.cpp
//exchange and add two general-purpose registers

#include "stdafx.h"

int main (void)
{
//define variables
   int   x, y, rslt1, rslt2;

   printf ("Enter two integers: \n");
   scanf ("%d %d", &x, &y);

//switch to assembly
      _asm
      {
         MOV    EAX, x
         MOV    EBX, y

         XADD   EAX, EBX        //swap EAX and EBX
                                //sum is stored in EAX (dst)
         MOV    rslt1, EAX      //move sum to rslt1
         MOV    rslt2, EBX      //move original EAX to rslt2
      }

   printf ("\nSum = %d\nOriginal EAX = %d\n\n",
      rslt1, rslt2);

   return 0;
}
                                      //continued on next page
```

(a)

Figure 5.16 A linked C and assembly program that demonstrates the use of the exchange and add (XADD) instruction using the EAX and EBX general-purpose registers: (a) the program and (b) the outputs.

```
Enter two integers:
40 25

Sum = 65
Original EAX = 40

Press any key to continue . . . _
------------------------------------------------------------
Enter two integers:
672 538

Sum = 1210
Original EAX = 672

Press any key to continue . . . _
```

(b)

Figure 5.16 (Continued)

5.5.4 Compare and Exchange

Since the primary focus of this book is to present the *code segment* for X86 assembly language programming, it is of lesser importance to include the stack segment and the data segment in each program. Therefore, most of the programs will be structured as an assembly language module embedded in a C program. Also, the methods to obtain keyboard input data and to display the results of an assembly language program are simpler when using the C functions of **scanf ( )** and **printf ( )**.

This section presents the *compare and exchange* (CMPXCHG) instruction, which compares the value in the accumulator with the value in the destination operand. If the two operands are equal, then the source operand is stored in the destination location — a register or a memory location. If the two operands are not equal, then the destination operand is stored in the accumulator. The syntax for the CMPXCHG instruction is shown below.

```
CMPXCHG    register/memory, register
```

The flags in the EFLAGS register indicate the result of the operation that is obtained after subtracting the destination operand from the contents of the accumulator. The CMPXCHG instruction can be combined with the processor's LOCK prefix. In a multiprocessor environment, the LOCK prefix ensures that the processor maintains exclusive use of shared memory, thus preventing other threads from accessing the memory location while the instruction is executing. This is referred to as an *atomic operation*, which is used to maintain synchronization and to avoid race conditions in a multiprocessor environment. Many instructions, such as the XCHG and XADD instructions, among others, always assert the LOCK# signal whether the LOCK prefix is present or not.

The diagram shown in Figure 5.17 graphically portrays the operation of the CMPXCHG instruction using EAX as the accumulator register, EBX as the destination register, and EDX as the source register. Figure 5.18 shows a program to illustrate the use of the instruction CMPXCHG using the registers shown in Figure 5.17.

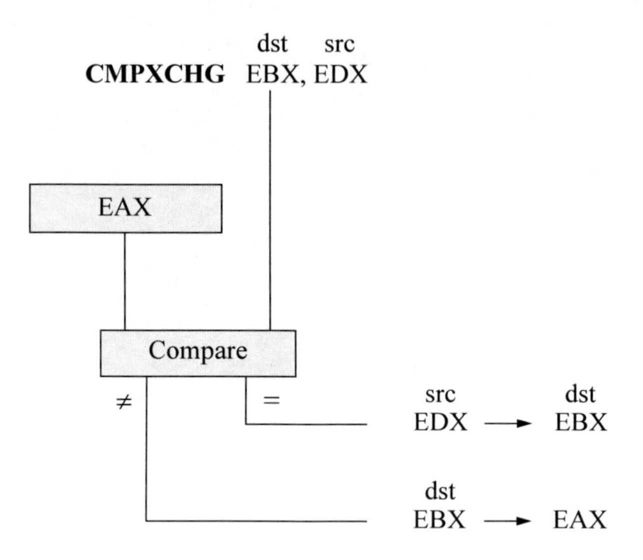

Figure 5.17 Graphical representation of the CMPXCHG instruction.

```
//comp_xchg.cpp
//compare and exchange registers based
//on the results of the comparison

#include "stdafx.h"

int main (void)
{
   //define variables
   int   eax_reg, ebx_reg, edx_reg, equal, not_equal;

   printf ("Enter integers for EAX, EBX_dst, EDX_src: \n");
   scanf ("%d %d %d", &eax_reg, &ebx_reg, &edx_reg);
```

(a) //continued on next page

Figure 5.18 Program to illustrate the use of the CMPXCHG instruction: (a) the program and (b) the outputs.

```
//switch to assembly
      _asm
      {
         MOV      EAX, eax_reg
         MOV      EBX, ebx_reg   //EBX is dst register
         MOV      EDX, edx_reg   //EDX is src register

         CMPXCHG  EBX, EDX       //if EAX = EBX, EDX --> EBX
                                 //if EAX != EBX, EBX --> EAX

         MOV      equal, EBX
         MOV      not_equal, EAX
      }

   printf ("\nEBX = %d\nEAX = %d\n\n", equal, not_equal);

   return 0;
}
```

```
Enter integers for EAX, EBX_dst, EDX_src:
130 130 140

EBX = 140
EAX = 130

Press any key to continue . . . _
------------------------------------------------------------
Enter integers for EAX, EBX_dst, EDX_src:
120 100 150

EBX = 100
EAX = 100

Press any key to continue . . . _
------------------------------------------------------------
Enter integers for EAX, EBX_dst, EDX_src:
-1 4294967295 20              //4294967295 is -1 (FFFFFFFF)

EBX = 20
EAX = -1

Press any key to continue . . . _
------------------------------------------------------------
```

(b) //continued on next page

Figure 5.18 (Continued)

```
Enter integers for EAX, EBX_dst, EDX_src:
4294967295 -1 30

EBX = 30
EAX = -1

Press any key to continue . . . _
-------------------------------------------------------------
Enter integers for EAX, EBX_dst, EDX_src:
4294967295 20 55

EBX = 20
EAX = 20

Press any key to continue . . . _
```

Figure 5.18 (Continued)

5.6 Translate

The *translate* (XLAT or XLATB) instructions use the contents of register AL as an index into a translation table in memory that contains the translated byte of data. The data at the memory location addressed by the index is then stored in AL. The index value in register AL is treated as an unsigned integer, which is added to the contents of the base register (E)BX to obtain the base address of the translation table in the data segment, as shown in Figure 5.19 using register BX as the base register.

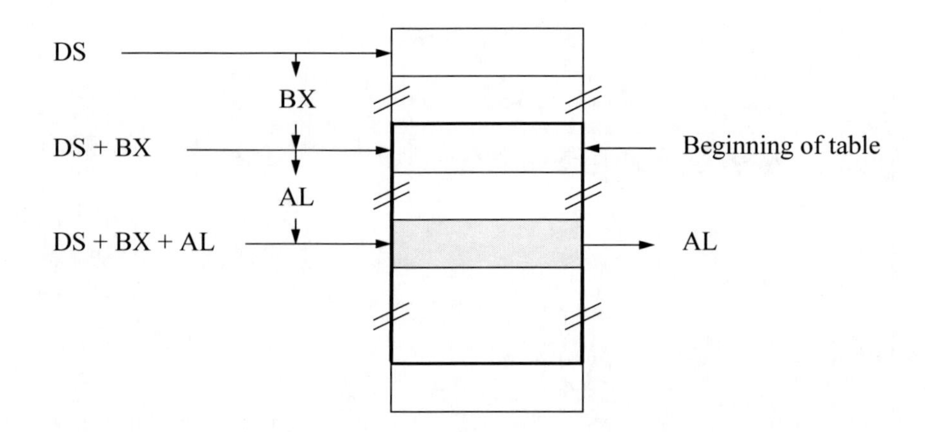

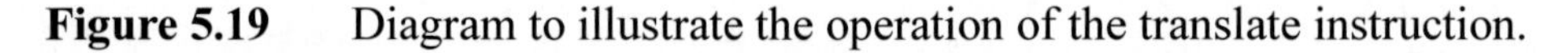
Figure 5.19 Diagram to illustrate the operation of the translate instruction.

The XLAT instruction permits the base address of the translation table to be specified with a symbol. The base address must be explicitly stored in registers DS:(E)BX before execution of the XLAT instruction. The XLATB instruction, however, assumes that the DS:(E)BX registers contain the base address of the translation table. This is often referred to as the *short form* of the translation instruction. The translation instructions are often used to translate from one code to another code; for example, from the ASCII code to the EBCDIC code that is used in the IBM mainframes.

An assembly language program will be coded to translate a 4-bit binary number to the corresponding 4-bit Gray code number using the XLATB instruction. The binary code and Gray code for four bits are shown in Table 5.4.

The Gray code is an nonweighted code that has the characteristic whereby only one bit changes between adjacent code words. The Gray code belongs to a class of cyclic codes called *reflective codes*, as can be seen in Table 5.4. Notice in the first four rows, that g_0 reflects across the reflecting axis; that is, g_0 in rows 2 and 3 is the mirror image of g_0 in rows 0 and 1. In the same manner, g_1 and g_0 reflect across the reflecting axis drawn under row 3. Thus, rows 4 through 7 reflect the state of rows 0 through 3 for g_1 and g_0. The same is true for g_2, g_1, and g_0 relative to rows 8 through 15 and rows 0 through 7.

Table 5.4 Binary 8421 Code and the Gray Code

	Binary Code				Gray Code				
Row	b_3	b_2	b_1	b_0	g_3	g_2	g_1	g_0	
0	0	0	0	0	0	0	0	0	
1	0	0	0	1	0	0	0	1	
2	0	0	1	0	0	0	1	1	← g_0 is reflected
3	0	0	1	1	0	0	1	0	
4	0	1	0	0	0	1	1	0	← g_1 and g_0 are reflected
5	0	1	0	1	0	1	1	1	
6	0	1	1	0	0	1	0	1	
7	0	1	1	1	0	1	0	0	
8	1	0	0	0	1	1	0	0	← g_2, g_1, and g_0 are reflected
9	1	0	0	1	1	1	0	1	
10	1	0	1	0	1	1	1	1	
11	1	0	1	1	1	1	1	0	
12	1	1	0	0	1	0	1	0	
13	1	1	0	1	1	0	1	1	
14	1	1	1	0	1	0	0	1	
15	1	1	1	1	1	0	0	0	

Since the data in the translation table is one byte, the translation table is $2^8 = 256$ bytes in length. Figure 5.20 shows an assembly language program to demonstrate the use of the XLATB instruction, which is a one-byte instruction. This program translates a binary code of four bits to the corresponding Gray code of four bits. An ASCII character is entered that corresponds to a 4-bit binary number, in the form of a hexadecimal digit, that is used to generate a 4-bit Gray code number 0 – 9, A – F. This is only one of several methods to perform the conversion. In a later chapter, a different method will be used that is more appropriate — 1s and 0s are entered as the binary number — but that method requires the use of instructions that have not yet been discussed. This method achieves the translation using the binary-to-Gray conversion algorithm.

In Figure 5.20, the translation table, XLTBL, is defined where the first 16 bytes represent the Gray code numbers that were translated from the corresponding binary numbers. The remaining 240 bytes are declared as zeroes (ASCII 30H). The usual sequence of instructions is utilized: set up the program's data segment; load the address of the translation table into register BX; read the prompt; and enter the keyboard character.

Then the character that was entered from the keyboard and placed into the operand field (OPFLD) is moved to register AL as the index, where the ASCII bias is removed. For example, if the number 5 (0101) is entered, then this is represented as 35H in ASCII — 0011 0101 in binary — but the binary number 0000 0101 is required as the index. Therefore, the high-order four bits — the ASCII bias — must be removed. This is accomplished by ANDing the 0011 0101 with 0000 1111 to produce 0000 0101, the required number to access the sixth entry in the translation table to yield the corresponding Gray code number of 0000 0111. This number can then be printed on the monitor screen as a 7, which is 37H in ASCII. The AND instruction is presented in more detail in Chapter 9, together with the other logic operations.

The translation instruction, XLATB, is then executed, which places the translated data in register AL. The contents of AL are then moved to the result area (RSLT) where the Gray code is displayed by the DOS interrupt INT 21H.

```
;translate.asm

;Enter an ascii character that corresponds to a 4-bit
;hexadecimal number that is used to generate a 4-bit
;Gray code number 0 - 9, A - F

.STACK

.DATA
PARLST   LABEL BYTE
MAXLEN   DB 5
ACTLEN   DB ?
OPFLD    DB 5 DUP(0)          (a)         //continued on next page
```

Figure 5.20 Program to illustrate using the XLATB instruction to convert binary code to Gray code: (a) the program and (b) the outputs.

```
;define ascii translation table of 256 bytes
XLTBL   DB 30H, 31H, 33H, 32H   ;0, 1, 3, 2
        DB 36H, 37H, 35H, 34H   ;6, 7, 5, 4
        DB 43H, 44H, 46H, 45H   ;12(C), 13(D), 15(F), 14(E)
        DB 41H, 42H, 39H, 38H   ;10(A), 11(B), 9, 8
        DB 240 DUP(30H)         ;0

PRMPT   DB 0DH, 0AH, 'Enter an ascii character: $'
RSLT    DB 0DH, 0AH, 'Gray =       $'

.CODE
BEGIN   PROC FAR

;set up pgm ds
        MOV     AX, @DATA
        MOV     DS, AX

;put addr of translation table xltbl in bx
        LEA     BX, XLTBL

;read prompt
        MOV     AH, 09H     ;display string
        LEA     DX, PRMPT   ;load addr of prmpt
        INT     21H         ;dos interrupt

;kybd rtn to enter characters
        MOV     AH, 0AH     ;buffered kybd input
        LEA     DX, PARLST  ;load addr of parlst
        INT     21H         ;dos interrupt

;store number from opfld in al and translate
        MOV     AL, OPFLD
        AND     AL, 0FH     ;remove ascii bias

        XLATB               ;translate to ascii

        MOV     RSLT+9, AL  ;move gray number to rslt

;display gray number
        MOV     AH, 09H     ;display string
        LEA     DX, RSLT    ;load addr of rslt field
        INT     21H         ;dos interrupt

BEGIN   ENDP
        END     BEGIN
                                //continued on next page
```

Figure 5.20 (Continued)

```
Enter an ascii character: 3
Gray = 2
------------------------------------------------------------
Enter an ascii character: 5
Gray = 7
------------------------------------------------------------
Enter an ascii character: 8
Gray = C
------------------------------------------------------------
Enter an ascii character: 9
Gray = D                          (b)
```

Figure 5.20 (Continued)

The conversion from binary to ASCII can also be accomplished by a simple C program, as shown in Figure 5.21. An integer array, labelled *gray*, is a 256-byte array in which the first 16 bytes contain the Gray code for the equivalent binary code — the remaining 240 bytes are initialized to zero by the C compiler.

A user-entered integer specifies the decimal equivalent of a binary number that is stored in location *x*. For example, if 11_{10} (1011) is entered, then the corresponding Gray code is E (1110). The Gray code is printed as specified by the *gray[x]* entry in the array *gray[256]*.

```
//translate.cpp
//Convert from binary to gray code.
#include "stdafx.h"
int main (void)
{
//define array for gray code
   int gray[256] = {0x00, 0x01, 0x03, 0x02,
                    0x06, 0x07, 0x05, 0x04,
                    0x0c, 0x0d, 0x0f, 0x0e,
                    0x0a, 0x0b, 0x09, 0x08};
   int x, rslt;
   printf ("Enter an integer 0-15 that represents
            a binary number: \n");
   scanf ("%d", &x);
   printf ("Binary = %d, Gray = %x\n\n", x, gray[x]);
   return 0;
}                          (a)       //continued on next page
```

Figure 5.21 C program to convert from binary to Gray code: (a) the program and (b) the outputs.

```
Enter an integer 0-15 that represents a binary number:
13
Binary = 13, Gray = b

Press any key to continue . . . _
----------------------------------------------------------------
Enter an integer 0-15 that represents a binary number:
2
Binary = 2, Gray = 3

Press any key to continue . . . _
----------------------------------------------------------------
Enter an integer 0-15 that represents a binary number:
15
Binary = 15, Gray = 8

Press any key to continue . . . _
----------------------------------------------------------------
Enter an integer 0-15 that represents a binary number:
10
Binary = 10, Gray = f

Press any key to continue . . . _
----------------------------------------------------------------
Enter an integer 0-15 that represents a binary number:
11
Binary = 11, Gray = e

Press any key to continue . . . _
```

(b)

Figure 5.21 (Continued)

5.7 Conversion Instructions

This section presents the following conversion instructions: *convert byte to word* (CBW), *convert word to doubleword* (CWD), *convert word to doubleword extended* (CWDE), and *convert doubleword to quadword* (CDQ). Each of these conversion instructions requires no operands and they double the size of the implied source register by extending the sign bit into the implied destination extension register. The flags are not affected by any of these instructions.

These are similar in function to the two move instructions discussed in Section 5.2.2. The *move with sign extension* (MOVSX) moves the source operand into a destination general-purpose register, then extends the sign bit into the high-order bits of the destination. The *move with zero extension* (MOVZX) instruction extends zeroes into the destination. Figure 5.22 shows the four conversion instructions together with a graphical illustration of their function.

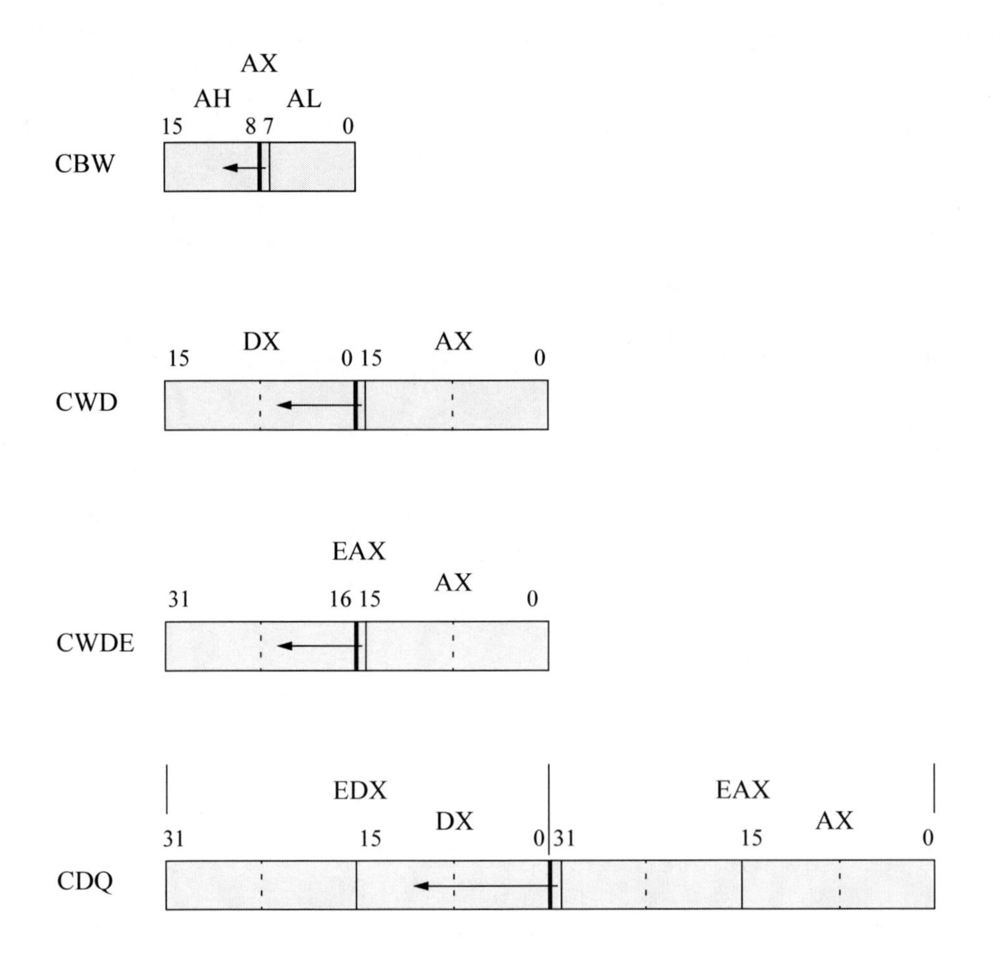

Figure 5.22 Conversion instructions and their function.

Convert byte to word (CBW) The CBW instruction copies the sign bit (bit 7) of register AL into every bit position of register AH, effectively increasing the size of the implied source register, AL, from a byte to a word. The CBW instruction is designed for use when the operand-size attribute is 16 bits. During a division operation, a $2n$-bit dividend is divided by an n-bit divisor; therefore, the CBW instruction can be used to produce a word dividend from a byte dividend prior to a byte division operation.

Convert word to doubleword (CWD) The CWD instruction copies the sign bit (bit 15) of register AX into every bit position of register DX, thereby doubling the size of the implied source operand register. Thus, the CWD instruction uses the DX:AX register pair as the destination operand. During a division operation, the CWD instruction can be used to produce a doubleword dividend from a word dividend before the divide operation commences.

Convert word to doubleword extended (CWDE) The CWDE instruction copies the sign bit (bit 15) of register AX into the high-order 16 bits of register EAX. The CWDE instruction is used when the operand-size attribute is 32 bits. The CWDE instruction differs from the CWD instruction, in that the CWDE instruction uses register EAX as the destination register; whereas, the CWD instruction extends the sign bit in register AX throughout register DX.

Convert doubleword to quadword (CDQ) The CDQ instruction copies the sign bit (bit 31) of register EAX into every bit position of register EDX, thereby doubling the size of the implied source operand register. Thus, the CDQ instruction uses the EDX:EAX register pair as the destination operand. The CDQ instruction can be used to produce a quadword dividend from a doubleword dividend before the division operation begins. Figure 5.23 shows an assembly language module embedded in a C program that illustrates the use of the CDQ instruction. An integer is entered from the keyboard and stored in location *eax_reg*, then copied to general-purpose register EAX. The conversion instruction CDQ is then executed and the results are displayed by the **printf ( )** function.

```
//convert.cpp
//convert doubleword to quadword (CDQ)
#include "stdafx.h"
int main (void)
{
//define variables
   int   eax_reg, edx_reg;
   printf ("Enter an integer: \n");
   scanf ("%x", &eax_reg);

//switch to assembly
      _asm
      {
         MOV    EAX, eax_reg
         CDQ
         MOV    eax_reg, EAX
         MOV    edx_reg, EDX
      }
   printf ("\nEDX = %X, EAX = %X\n\n", edx_reg, eax_reg);
   return 0;
}                                   (a)       //continued on next page
```

Figure 5.23 Program to illustrate using the conversion instruction CDQ: (a) the program and (b) the outputs.

```
Enter an integer:
F0F0F0F0

EDX = FFFFFFFF, EAX = F0F0F0F0

Press any key to continue . . . _
-------------------------------------------------------------
Enter an integer:
ffffffff

EDX = FFFFFFFF, EAX = FFFFFFFF

Press any key to continue . . . _
-------------------------------------------------------------
Enter an integer:
3F

EDX = 0, EAX = 3F

Press any key to continue . . . _
-------------------------------------------------------------
Enter an integer:
63

EDX = 0, EAX = 63

Press any key to continue . . . _
-------------------------------------------------------------
Enter an integer:
80000000

EDX = FFFFFFFF, EAX = 80000000

Press any key to continue . . . _
-------------------------------------------------------------
Enter an integer:
7fffffff

EDX = 0, EAX = 7FFFFFFF

Press any key to continue . . . _
```

(b)

Figure 5.23 (Continued)

5.8 Problems

5.1 Let A and B be two binary integers in 2s complement representation, where A = 1011 0001 and B = 1110 0100. A' and B' are the 1s complement of A and B, respectively. Determine the following: $A' + 1$ and $B' + 1$.

5.2 The 1s complement (diminished-radix complement) can be obtained by subtracting all bits of the number from 1. The 2s complement (radix complement) is obtained by adding 1 to the 1s complement, as shown below for the binary number 1110 (–2), which is in 2s complement representation.

1–1 = 0 1–1 = 0 1–1 = 0 1–0 = 1 + 1 = 0010 (+2)

The following number is in radix complement representation for radix 16: $F8B6_{16}$. Perform the operation of radix complementation (16s complementation) on the number.

5.3 The sign-magnitude notation for a positive number is represented by the equation shown below, where the sign bit of 0 indicates a positive number.

$$A = (0\ a_{n-2}a_{n-3}\ \ldots\ a_1a_0)_r$$

The sign-magnitude notation for a negative number is represented by the equation shown below, where the sign bit is the radix minus 1. In sign-magnitude notation, the positive version differs from the negative version only in the sign digit position. The magnitude portion $a_{n-2}a_{n-3}\ \ldots\ a_1a_0$ is identical for both positive and negative numbers of the same absolute value.

$$A' = [(r-1)\ a_{n-2}a_{n-3}\ \ldots\ a_1a_0]_r$$

The numbers shown below are in sign-magnitude representation for radix 2. Convert the numbers to 2s complement representation with the same numerical value using eight bits.

	Sign-magnitude	2s Complement
(a)	**0**111 1111	
(b)	**1**111 1111	
(c)	**0**000 1111	
(d)	**1**000 1111	
(e)	**1**001 0000	

5.4 Convert the positive 2s complement numbers shown below to negative numbers in 2s complement representation.

$0000\ 1100\ 0011_2$
$0101\ 0101\ 0101_2$

5.5 Determine the bias constant for an 8-bit floating-point exponent.

5.6 Write a program using C and assembly language that shows the application of the *conditional move below or equal* (CMOVBE) instruction for unsigned operands.

5.7 Write a program using C and assembly language that shows the application of the *conditional move greater than or equal* (CMOVGE) instruction for signed operands.

5.8 Write a program using C and assembly language that shows the application of the *conditional move less than or equal* (CMOVLE) instruction for signed operands.

5.9 Write a program using C and assembly language that shows the application of the *conditional move no parity* (CMOVNP) instruction for unsigned operands. The compare instruction subtracts the second source operand from the first source operand and sets the parity flag accordingly. If the low-order byte of the difference contains an odd number of 1s, then the parity flag equals zero; otherwise, the parity flag equals one.

5.10 Write a program using C and assembly language that shows the application of the *conditional move sign* (CMOVS) instruction for signed operands. The compare instruction subtracts the second source operand from the first source operand and the state of the sign flag is set to either 0 or 1, depending on the sign of the result. If the sign of the difference is negative (SF = 1), then the move operation is executed.

5.11 Use the *exchange and add* (XADD) instruction for two 32-bit user-entered hexadecimal integers. Display the results in lowercase hexadecimal.

5.12 Use the *compare and exchange* (CMPXCHG) instruction for three 32-bit user-entered hexadecimal integers. Display the results in lowercase hexadecimal.

5.13 Write a C program to convert a decimal number to the square of the number. Use an array to translate the decimal numbers 0 through 15 to their corresponding squares.

6

Branching and Looping Instructions

This chapter describes program transfer control using branching and looping instructions. These instructions transfer control to a section of the program that does not immediately follow the current instruction. The transfer may be a *backward transfer* to a section of code that was previously executed or a *forward transfer* to a section of code that follows the current instruction.

There are three basic types of addresses (or pointers) provided by the assembler: a short address, a near address, and a far address. A *short address* (or short jump) transfers control to an address that is located –128 to +127 bytes from the current location (the current EIP value). A *near address* (or near jump) transfers control to an address within the current code segment that has a 16-bit or 32-bit offset — also referred to as an effective address. A *far address* (or far jump) transfers control to an address obtained from a 16-bit segment selector and a 16-bit or 32-bit offset. Far addresses (pointers) are used to transfer control to an address outside the current code segment. Near and far pointers are shown in Figure 6.1 for non-64-bit mode.

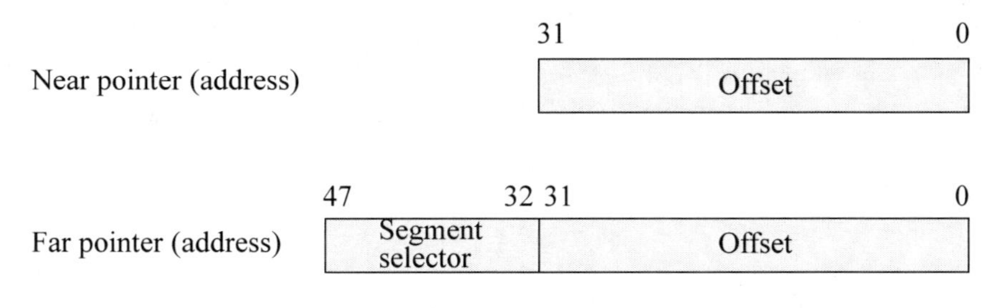

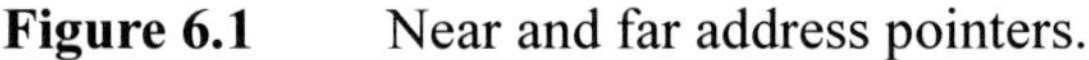
Figure 6.1 Near and far address pointers.

6.1 Branching Instructions

There are various types of branch instructions, including unconditional jumps, conditional jumps, and *if* structures. A jump instruction alters the sequence of flow through the program by branching — or jumping — to another point in the program. A branch can be executed or not executed depending on certain conditions, such as the state of flags in the EFLAGS register. If a branch is not taken, then the program executes the instruction that follows the branch instruction. If a branch is taken, then the program jumps to a new location in the program; that is, an instruction in a different location in memory.

Branch instructions are relative to the current instruction pointer (E)IP, which points to a particular instruction in the current code segment. The combination of the segment selector and the offset is the logical address of the next instruction to be executed. Branching is facilitated by means of a *branch target buffer* (BTB), which is a set-associative cache that is used to predict whether a branch will be performed — refer to Chapter 2 for a review of a set-associative cache. The BTB cache entries contain the address of the branch instruction in conjunction with the branch target address. The cache contains information regarding the branch history; that is, about recently encountered branch instructions.

When a branch instruction is encountered, the BTB is examined to determine if there is an entry for the branch instruction. If there is a miss, the branch is assumed to have not been executed; if there is a hit, the branch history is checked to determine if the branch should be taken. If the branch is predicted to have failed, then program execution continues with the next sequential instruction; otherwise, a branch to the target address is performed.

6.1.1 Unconditional Jump Instruction

The *unconditional jump* (JMP) instruction advances the (E)IP register forward or backward a specific number of instructions. It transfers control to a destination address and provides no return address. The destination address may be in the current code segment or outside the current segment and can be a relative address or an absolute address. If the transfer is to an address within the current code segment, then this is referred to as an *intrasegment* transfer, which may be direct or indirect. If the transfer is to an address outside the current code segment, then this is referred to as an *intersegment* transfer, which may be direct or indirect.

A *relative address* is obtained by adding a signed offset to the (E)IP register to generate a near pointer. The signed offset (or displacement) allows for either a forward jump or a backward jump. An *absolute address* is measured from the base of the current code segment and is stored in the (E)IP register as an offset in the code segment (near pointer), or the address is obtained using a segment selector stored in the CS register and an offset in the (E)IP register. The syntax for the JMP instruction is shown below.

```
JMP    short/near/far address
```

A *short jump* is an intrasegment jump and has a 1-byte displacement that is added to the RIP register then sign-extended to 64 bits — the EIP register becomes the RIP register in 64-bit mode. The jump range for a 1-byte displacement is –128 to +127 from the current value of the (E)IP register. When a short jump is specified as a relative offset, this is represented by label and is encoded as a signed displacement that is relative to the current value of the (E)IP register.

A *near jump* is an intrasegment jump whose branch target address is either an absolute offset or a relative offset that is relative to the next instruction. In 64-bit mode, the relative near jump has a displacement of 32 bits that is added to the RIP register then sign-extended to 64 bits.

A *far jump* is an intersegment jump. The branch target address can specify an absolute address directly by using a pointer in the instruction, where the segment and offset are encoded in the instruction; or the branch target address can specify an indirect address from a memory location, where the low-order part of the memory location is placed in the (E)IP register and the high-order part of the memory location is placed in the CS register. All JMP instructions also clear the instruction prefetch queue.

Figure 6.2 shows a simple assembly language program that illustrates the JMP instruction. Registers BL and BH are assigned the values of A (41H) and B (42H), respectively. Before they can be changed to AL = BL = 00 by two MOV instructions, however, a jump instruction is executed to a location in the program that moves the original contents of the two registers to the RSLT area in the data segment to be displayed.

```
;uncond_jump.asm
.STACK

.DATA
RSLT   DB 0DH, 0AH, 'After jump: BL =  , BH =    $'

.CODE
BEGIN     PROC FAR
;set up pgm ds
          MOV     AX, @DATA
          MOV     DS, AX

;assign values to bl and bh
          MOV     BL, 'A'
          MOV     BH, 'B'
          JMP     JMP_DST

          MOV     BL, 00
          MOV     BH, 00            //continued on next page
```

(a)

Figure 6.2 Program to illustrate a JMP instruction: (a) the program and (b) the outputs.

```
;move bl and bh to rslt area in ds
JMP_DST: MOV      RSLT + 19, BL
         MOV      RSLT + 28, BH
;display result
         MOV      AH, 09H
         LEA      DX, RSLT
         INT      21H

BEGIN    ENDP
         END      BEGIN
```

```
After jump: BL = A, BH = B     (b)
```

Figure 6.2 (Continued)

The main reason for this program is to introduce the operation codes for the JMP instruction together with the addressing mode byte and the branch target address for a jump instruction. There are four different 1-byte operation codes for a jump instruction, depending on whether it is a short jump, a near jump, or a far jump. The operation code for an intrasegment *short jump* is EBH (1110 1011). The operation codes for *near jumps* are E9H (1110 1001) and FFH (1111 1111). The E9H operation code is intrasegment direct and can be used with a displacement that is relative to the next instruction or with the RIP register with a 32-bit displacement sign extended to 64 bits. The FFH operation code is used with indirect addressing that is either intrasegment indirect or intersegment indirect.

The operation codes for a *far jump* are EAH (1110 1010) or FFH (1111 1111). The EAH operation code is used for an intersegment direct operation with an absolute address given the instruction. The FFH operation code is used with absolute addressing obtained indirectly from memory, where the memory location is partitioned into two parts: one for the CS register and one for the (E)IP register.

Table 6.1 shows the machine codes for the instruction segment shown below. The jump instruction in this program is a short intrasegment direct jump with a 1-byte displacement. Therefore, the jump range is from +127 bytes to –128 bytes from the updated (E)IP register.

```
                    . . .
           JMP      JMP_DST

           MOV      BL, 00
           MOV      BH, 00

;move bl and bh to rslt area in ds
JMP_DST:   MOV      RSLT + 19, BL
                    . . .
```

Assume that the CS register is 0723H and that the IP register is 0009H. Then the real (physical) address is obtained as shown below, where the CS register is shifted left four bits and the IP register is right-aligned for the addition operation.

CS register:	0723**0**
IP register (offset): +)	**0**0009
Physical address	07239

Table 6.1 Machine Codes for the Program Segment of Figure 6.2

Physical Address	Machine Code	Symbolic Instruction	Machine Instruction
		. . .	
07239	EB	**JMP** JMP_DST	**JMP** 0FH
0723A	04		
0723B	B3	**MOV** BL, 00	**MOV** BL, 00H
0723C	00		
0723D	B7	**MOV** BH, 00	**MOV** BH, 00H
0723E	00		
0723F	88	JMP_DST: **MOV** RSLT + 19, BL	**MOV** [00013H], BL
07240	1E		
07241	13		
07242	00		
		. . .	

The machine code for the jump instruction is EBH, which is an intrasegment direct short jump. The second byte is a displacement of 04H, which results in a jump of four bytes past the updated IP register — the updated IP register is 000BH. The resulting target address for the jump instruction is

```
CS × 16 (07230H) + right-aligned IP (0000BH) + 04H = 0723FH
```

This is shown in the machine jump instruction as 0FH, which represents the IP register contents of the jump target address — 0723FH.

The instruction format for the move instruction that moves register BL to the *rslt* area (`MOV  RSLT + 19, BL`) is as follows:

Operation code	Addressing mode	Displacement low and high	
1 0 0 0 1 0 d w	mod reg r/m		

If $d = 0$ in the operation code, then the source operand is a register; if $d = 1$, then the destination operand is a register. If $w = 0$, then the width is a byte; otherwise, the width is a word. For the move instruction, the source is register BL; therefore, $d = 0$. Since register BL is a byte operand, therefore, $w = 0$, resulting in an operation code of 1000 1000 (88H).

An *addressing mode byte* may also be used with an instruction. When used, it is contiguous to the operation code and consists of three fields: a 2-bit *mod* field that is used to differentiate between register addressing and memory addressing and is defined in Table 6.2; a 3-bit *reg* field that determines the size of the operands and is shown in Table 6.3; and a 3-bit *r/m* (*register/memory*) field that is used in conjunction with the *mod* field to determine the addressing mode and is shown in Table 6.4.

The format for the addressing mode byte is shown below for the target move instruction with a bit configuration of 0001 1110 (1EH). The *mod* code of 00 indicates that there is no displacement; the *reg* code of 011 specifies register BL when $w = 0$; and the *r/m* field of 110 indicates a direct addressing mode when the *mod* = 00.

mod		reg			r/m		
0	0	0	1	1	1	1	0

Table 6.2 Definition of the *mod* Bits

mod	Definition
0 0	No displacement unless *r/m* = 110, then there is displacement high and low
0 1	There is displacement (offset) low sign-extended to 16 bits
1 0	There is displacement (offset) high and low
1 1	The *r/m* field specifies a register

Table 6.3 Definition of the *reg* Bits

reg Bits	$w = 0$ Reg 8-Bit	$w = 1$ Reg 16/32 Bit	Segment Register
0 0 0	AL	AX/EAX	ES
0 0 1	CL	CX/ECX	CS
0 1 0	DL	DX/EDX	SS
0 1 1	BL	BX/EBX	DS
1 0 0	AH	SP/ESP	FS
1 0 1	CH	BP/EBP	GS
1 1 0	DH	SI/ESI	
1 1 1	BH	DI/EDI	

Table 6.4 Definition of the *r/m* Bits

r/m Bits	*mod* = 00	*mod* = 01 or 10	*mod* = 11 *w* = 0	*mod* = 11 *w* = 1
0 0 0	BX + SI	BX + SI + displacement	AL	AX
0 0 1	BX + DI	BX + DI + displacement	CL	CX
0 1 0	BP + SI	BP + SI + displacement	DL	DX
0 1 1	BP + DI	BP + DI + displacement	BL	BX
1 0 0	SI	SI + displacement	AH	SP
1 0 1	DI	DI + displacement	CH	BP
1 1 0	Direct	BP + displacement	DH	SI
1 1 1	BX	BX + displacement	BH	DI

6.1.2 Compare Instruction

The *compare* (CMP) was introduced in Chapter 5, but more details are presented in this section. The compare instruction compares two integer operands. It subtracts the second source operand from the first source operand and sets the status flags in the EFLAGS register accordingly. Both operands remain unchanged after the subtract operation, whereas a *subtract* (SUB) operation replaces the destination operand with the difference. The compare instruction is used primarily in conjunction with the *jump on condition* (Jcc) instruction, which performs a branch operation based on the state of the flags resulting from a compare instruction. The syntax for the compare instruction is shown below.

```
CMP    register/memory, immediate/register/memory
```

The compare instruction can be used to compare an immediate operand of 8 bits, 16 bits, or 32 bits with the accumulator register. For example,

```
CMP    AL, 4AH
CMP    AX, 1B36H
CMP    EAX, 45AC7B89H
```

It can also be used to compare an immediate operand with a memory operand or a register. For example,

```
CMP    [BX], 12ADH
CMP    BX, 2244H
```

It can also be used to compare registers of the same width. For example,

```
CMP    EBX, EDX
```

Table 6.5 shows the results of subtracting second source operands from first source operands and the resulting flags, using unsigned operands. The adjust flag, or auxiliary carry, (AF) is set if there is a carry out of or a borrow into the low-order four bits; the carry flag (CF) is set if there is a carry out of or a borrow into the high-order bit position of an arithmetic operation; otherwise it is reset. When subtracting by adding the 2s complement of the subtrahend, the state of the resulting AF and CF flags is inverted.

The overflow flag (OF) is set if the result of an operation is too large or too small to be contained in the destination operand; that is, the number is out of range for the size of the result. The parity flag (PF) is set if the low-order byte of the result has an even number of 1s; otherwise it is reset. The sign flag (SF) is set to the value of the high-order bit position. The zero flag (ZF) is set if the result is zero; otherwise it is reset.

Table 6.5 Compare Operations and the Resulting Flags

First Source Operand	Second Source Operand	Resulting Difference	OF	SF	ZF	AF	PF	CF	
7AH	7AH	00H	0	0	1	0	1	0	op1 = op2
6CH	3FH	2DH	0	0	0	1	1	0	op1 > op2
B5H	CDH	E8H	0	1	0	1	1	1	op1 < op2
15H	32H	E3H	0	1	0	0	0	1	op1 < op2
63H	B3H	B0H	1	1	0	0	0	1	op1 < op2

The second example (6CH – 3FH) will be examined in more detail using binary subtraction; that is, the difference will be obtained by adding the 2s complement of the subtrahend to the minuend. The state of the AF flag is reversed from 0 to 1. The state of the carry flag (CF) is reversed from 1 to 0. There is no overflow (OF), because the signs of the operands are different (minuend = 0, 2s complemented subtrahend = 1). The parity flag (PF) is 1, because there are an even number of 1s in the resulting byte (or low-order byte). The sign flag (SF) is 0, because the high-order bit is 0, which is the sign bit for signed operands. The zero flag (ZF) is 0, because the result is nonzero.

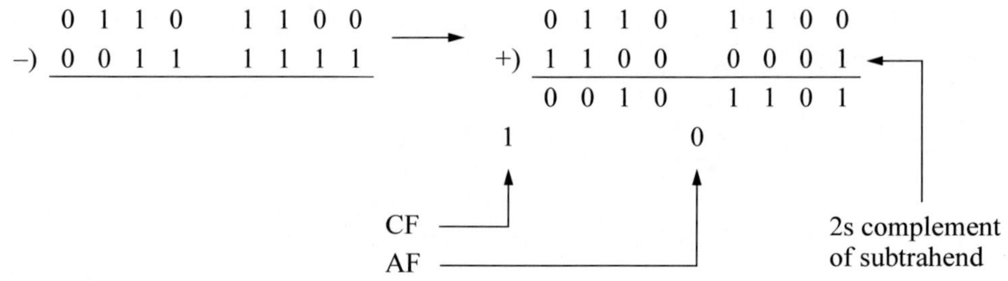

```
    0 1 1 0   1 1 0 0   ---->      0 1 1 0   1 1 0 0
-)  0 0 1 1   1 1 1 1          +)  1 1 0 0   0 0 0 1  <-- 2s complement
    ---------------------          -------------------     of subtrahend
                                   0 0 1 0   1 1 0 1
                               1           0
                               ^           ^
                        CF ----+           |
                        AF ----------------+
```

The low-order byte of the EFLAGS register is shown below, which will be used in conjunction with the program shown in Figure 6.3 — an assembly language module linked to a C program. Bit positions 1, 3, and 5 are reserved.

7	6	5	4	3	2	1	0
SF	ZF	0	AF	0	PF	1	CF

The *flags* variable in the program is declared as type **char**, because only the low-order byte in the EFLAGS register is being used. The user enters two hexadecimal characters, which are stored in **int** variables *x* and *y*. In the assembly language module, *x* and *y* are moved to registers EAX and EBX, respectively. The variables are then compared, resulting in the flags being generated.

A new instruction will now be introduced in order to display the flags. This is the *load AH from flags* (LAHF) instruction, which has a 1-byte operation code. The LAHF instruction copies the low-order byte of the EFLAGS register into the corresponding bit positions of the AH register, but does not alter the flags in the EFLAGS register. The contents of register AH are then moved to the *flags* variable for display.

```
//cmp_flags_gen.cpp
//compare two operands and generate the status flags
#include "stdafx.h"
int main (void)
{
   int x, y;            //define variables
   char flags;

   printf ("Enter two hexadecimal integers: \n");
   scanf ("%X %X", &x, &y);

//switch to assembly
      _asm
      {
         MOV   EAX, x
         MOV   EBX, y
         CMP   EAX, EBX
         LAHF
         MOV   flags, AH
      }

   printf ("\nOpnd1 = %d\nOpnd2 = %d\nFlags = %X\n\n",
               x, y, flags);
   return 0;
}
```

(a) //continued on next page

Figure 6.3 Program to illustrate the CMP instruction and the resulting status flags: (a) the program and (b) the outputs.

```
Enter two hexadecimal integers:
7A 7A

Opnd1 = 122
Opnd2 = 122          //Difference is 0 (00H); Opnd1 = Opnd2
Flags = 46           //ZF = 1, PF = 1

Press any key to continue . . . _
--------------------------------------------------------------
Enter two hexadecimal integers:
6C 3F

Opnd1 = 108
Opnd2 = 63           //Difference is 45 (2DH); Opnd1 > Opnd2
Flags = 16           //AF = 1, PF = 1

Press any key to continue . . . _
--------------------------------------------------------------
Enter two hexadecimal integers:
B5 CD

Opnd1 = 181
Opnd2 = 205          //Difference is -24 (E8H); Opnd1 < Opnd2
Flags = 97           //SF = 1, AF = 1, PF = 1, CF = 1

Press any key to continue . . . _
--------------------------------------------------------------
Enter two hexadecimal integers:
15 32

Opnd1 = 21
Opnd2 = 50           //Difference is -29 (E3H); Opnd1 < Opnd2
Flags = 83           //SF = 1, CF = 1

Press any key to continue . . . _
```

(b)

Figure 6.3 (Continued)

In Figure 6.3(b), the first pair of hexadecimal operands are $7A_{16}$ (122_{10}) and $7A_{16}$ (122_{10}), which results in a difference of zero (0000 0000) after subtracting the second source operand from the first source operand, to yield a flag byte of 0100 0110. Thus, ZF = 1, indicating a result of all zeroes; and PF = 1 in order to maintain odd parity for the byte; that is, there are an even number of 1s in the difference.

The second pair of hexadecimal operands are $6C_{16}$ (108_{10}) and $3F_{16}$ (63_{10}), which results in a difference of 45 (0010 1101), to yield a flag byte of 0001 0110. Thus, AF

= 1, indicating a borrow from bit 4 of the minuend; and PF = 1, because there are an even number of 1s in the difference.

The third pair of operands are $B5_{16}$ (181_{10}) and CD_{16} (205_{10}), which results in a difference of –24 (1110 1000), to yield a flag byte of 1001 0111. Thus, SF = 1, because bit 7 is a 1; AF = 1, indicating a borrow from bit 4 of the minuend; PF =1, indicating an even number of 1s in the difference; and CF = 1, indicating a borrow into bit 7 of the minuend.

The fourth pair of operands are 15_{16} (21_{10}) and 32_{16} (50_{10}), resulting in a difference of –29 (1110 0011), to yield a flag byte of 1000 0011. Thus, SF = 1, because bit 7 is a 1; and CF = 1, indicating a borrow into bit 7 of the minuend.

Another instruction that operates on the EFLAGS register and the AH register is the *store AH into flags* (SAHF) instruction. The SAHF instruction copies register AH into the low-order byte of the EFLAGS register.

6.1.3 Conditional Jump Instructions

The *conditional jump/transfer* (Jcc) instructions transfer control to a destination instruction in the same code segment if certain condition codes (cc) are set — the condition is specified in the instruction mnemonic. If the condition is not met, then program execution continues with the next instruction that follows the Jcc instruction. The conditional jump instructions are partitioned into three groups: those that are used with unsigned integers, those that are used with signed integers, and those that are used irrespective of the sign of the operands, as shown in Table 6.6. There are usually two mnemonics associated with each instruction that give alternative names for the instruction; this may facilitate easier understanding of the operation of the instruction.

Table 6.6 Conditional Jump Instructions

Mnemonic	Description	Flags Examined
	Unsigned Conditional Jumps	
JA/JNBE	Jump if above or not below/equal	(CF or ZF) = 0
JAE/JNB	Jump if above or equal/not below	CF = 0
JB/JNAE	Jump if below or not above/equal	CF = 1
JBE/JNA	Jump if below/equal or not above	(CF or ZF) = 1
JE/JZ	Jump if equal or zero — unsigned/signed	ZF = 1
JNE/JNZ	Jump if not equal/not zero — unsigned/signed	ZF = 0
	Signed Conditional Jumps	
JG/JNLE	Jump if greater or not less/equal	[(SF xor OF) or ZF] = 0
JGE/JNL	Jump if greater/equal or not less	(SF xor OF) = 0
JL/JNGE	Jump if less or not greater/equal	(SF xor OF) = 1
JLE/JNG	Jump if less/equal or not greater	[(SF xor OF) or ZF] = 1

(Continued on next page)

Table 6.6 Conditional Jump Instructions

Mnemonic	Description	Flags Examined
	Other Conditional Jumps	
JCXZ	Jump if CX = 0	Register CX = 0
JECXZ	Jump if ECX = 0	Register ECX = 0
JC	Jump if carry	CF = 1
JNC	Jump if no carry	CF = 0
JNO	Jump if no overflow	OF = 0
JO	Jump if overflow	OF = 1
JNP/JPO	Jump if parity is odd	PF = 0
JP/JPE	Jump if parity is even	PF = 1
JNS	Jump if sign bit = 0	SF = 0
JS	Jump if sign bit = 1	SF = 1

As can be seen from Table 6.6, conditional jump instructions that use the words *above* or *equal* in their descriptions refer to unsigned operands; conditional jump instructions that use the words *greater* or *less* in their descriptions refer to signed operands. There is no return address for the conditional jump instructions and they do not support far jumps; that is, jumps to other code segments. The destination address is a signed offset that is relative to the contents of the (E)IP register, and thus, resides in the current code segment.

The conditional jump instructions examine the state of one or more flags in the EFLAGS register to determine if a jump should take place. The flags that are inspected are SF, ZF, AF, PF, and CF. Examples will now be presented that illustrate the operation of select conditional jump instructions.

Jump if above (JA) for unsigned numbers The JA instruction will cause a jump to occur if the following equation is true:

```
(CF or ZF) = 0
```

Example 6.1 Let register AL = 1111 1011 (251) and register BL = 0000 0111 (7), then compare the two operands by adding the 2s complement of the subtrahend to the minuend.

```
        1 1 1 1     1 0 1 1
+)      1 1 1 1     1 0 0 1
       -------------------------
        1 1 1 1     0 1 0 0
   1(0)         1(0)
```

Since [(CF = 0) or (ZF = 0)] = 0, therefore, the jump will occur, because the number 251 is above the number 7.

Example 6.2 Let register AL = 1111 1011 (251) and register BL = 1111 1011 (251), then compare the two operands by adding the 2s complement of the subtrahend to the minuend.

```
            1  1  1  1      1  0  1  1
+)          0  0  0  0      0  1  0  1
           ------------------------------
            0  0  0  0      0  0  0  0
      1(0)           1(0)
```

Since [(CF = 0) or (ZF = 1)] = 1, therefore, the jump will not occur, because the numbers are equal; that is, the number 251 is not above the number 251.

Example 6.3 Let register AL = 0000 1001 (9) and register BL = 0001 1100 (28), then compare the two operands by adding the 2s complement of the subtrahend to the minuend.

```
            0  0  0  0      1  0  0  1
+)          1  1  1  0      0  1  0  0
           ------------------------------
            1  1  1  0      1  1  0  1
      0(1)           0(1)
```

Since [(CF = 1) or (ZF = 0)] = 1, therefore, the jump will not occur, because the number 9 is not above the number 28.

Figure 6.4 shows a program to illustrate the *jump if above* (JA) instruction for unsigned integers. The operands, *opnd1* and *opnd2*, are moved to registers EAX and EBX, respectively, then compared. If *opnd1* is above *opnd2*, then a conditional jump occurs to the label ABOVE, whose address is equal to the contents of register (E)IP plus a signed displacement. A **printf ( )** function then displays that operand1 is above operand2. If operand1 is not above operand2, then an unconditional jump occurs to the label BELOW where the **printf ( )** function displays that operand1 is not above (below or equal) operand2.

The **goto** instruction simply bypasses the second **printf ( )** function if operand1 is above operand2; that is, it is equivalent to an unconditional jump in assembly language. The **goto** statement is rarely used in C programming, because other forms of control make the program easier to follow, such as the **while** loop, the **if** and **if ... else** constructs, and the **for** loop. Too many **goto** statements make the program difficult to follow and to understand, causing the code to become what some programmers refer to as *spaghetti code*. The **goto** instruction, therefore, should be used sparingly.

```
//jump_above.cpp
//illustrate JA for unsigned operands

#include "stdafx.h"

int main (void)
{
//define variables
   int opnd1, opnd2;

   printf ("Enter two unsigned integers: \n");
   scanf ("%d %d", &opnd1, &opnd2);

//switch to assembly
      _asm
      {
         MOV   EAX, opnd1
         MOV   EBX, opnd2

         CMP   EAX, EBX
         JA    ABOVE
         JMP   BELOW
      }

ABOVE:
   printf ("Opnd1 is above opnd2 \n\n");
   goto end;

BELOW:
   printf ("Opnd1 is below/equal opnd2 \n\n");

end:
   return 0;
}
```

(a)

```
Enter two unsigned integers:
226 227
Opnd1 is below/equal opnd2

Press any key to continue . . . _
```

(b) //continued on next page

Figure 6.4 Program to illustrate the use of the JA instruction: (a) the program and (b) the outputs.

```
Enter two unsigned integers:
350 330
Opnd1 is above opnd2

Press any key to continue . . . _
------------------------------------------------------------
Enter two unsigned integers:
670 670
Opnd1 is below/equal opnd2

Press any key to continue . . . _
```

Figure 6.4 (Continued)

Jump if greater (JG) for signed numbers The JG instruction will cause a jump to occur if the following equation is true:

```
([SF xor OF) or ZF] = 0
```

Example 6.4 Let register AL = 0001 0111 (+23) and register BL = 0000 1011 (+11), then compare the two operands by adding the 2s complement of the subtrahend to the minuend.

```
          0 0 0 1     0 1 1 1
+)        1 1 1 1     0 1 0 1
          -------------------
          0 0 0 0     1 1 0 0
      1(0)        0(1)
```

Since {[(SF = 0) xor (OF = 0)] or ZF = 0} = 0, therefore, the jump will occur, because the number +23 is greater than the number +11.

Example 6.5 Let register AL = 0001 1101 (+29) and register BL = 0001 1101 (+29), then compare the two operands by adding the 2s complement of the subtrahend to the minuend.

```
          0 0 0 1     1 1 0 1
+)        1 1 1 0     0 0 1 1
          -------------------
          0 0 0 0     0 0 0 0
      1(0)        1(0)
```

Since {[(SF = 0) xor (OF = 0)] or ZF = 1} = 1, therefore, the jump will not occur, because the numbers are equal; that is, the number +29 is not greater than the number +29.

Example 6.6 Let register AL = 0011 0111 (+55) and register BL = 1001 1011, a negative number (–101), then compare the two operands by adding the 2s complement of the subtrahend to the minuend.

```
          0 0 1 1       0 1 1 1
+)        0 1 1 0       0 1 0 1
          -------------------------
          1 0 0 1       1 1 0 0
     0(1)          0(1)
```

Since {[(SF = 1) xor (OF = 1)] or ZF = 0} = 0, therefore, the jump will occur, because +55 is greater than –101.

Example 6.7 Let registers AL and BL both contain identical negative operands: 1101 1011 (–37), then compare the two operands by adding the 2s complement of the subtrahend to the minuend.

```
          1 1 0 1       1 0 1 1
+)        0 0 1 0       0 1 0 1
          -------------------------
          0 0 0 0       0 0 0 0
     1(0)          1(0)
```

Since {[(SF = 0) xor (OF = 0)] or ZF = 1} = 1, therefore, the jump will not occur, because both operands are equal; that is, the number –37 is not greater than the number –37.

Example 6.8 Let registers AL and BL both contain negative operands: AL = 1110 1100 (–20) and BL = 1110 1001 (–23), then compare the two operands by adding the 2s complement of the subtrahend to the minuend.

```
          1 1 1 0       1 1 0 0
+)        0 0 0 1       0 1 1 1
          -------------------------
          0 0 0 0       0 0 1 1
     1(0)          1(0)
```

Since {[(SF = 0) xor (OF = 0)] or ZF = 0} = 0, therefore, the jump will occur, because –20 is greater than –23.

Figure 6.5 shows a program to illustrate the *jump if greater* (JG) instruction for signed integers. The operands, *opnd1* and *opnd2*, are moved to registers EAX and EBX, respectively, then compared. If *opnd1* is greater than *opnd2*, then a conditional jump occurs to the label GREATER, whose address is equal to the contents of register (E)IP plus a signed displacement. A **printf ( )** function then displays that operand1 is greater than operand2. If operand1 is not greater than operand2, then an unconditional jump occurs to the label LESS, where the **printf ( )** function displays that operand1 is not greater than (less or equal) operand2.

```
//jump_greater.cpp
//illustrate JG for signed operands

#include "stdafx.h"
int main (void)
{
//define variables
   int opnd1, opnd2;

   printf ("Enter two signed integers: \n");
   scanf ("%d %d", &opnd1, &opnd2);

//switch to assembly
      _asm
      {
         MOV   EAX, opnd1
         MOV   EBX, opnd2

         CMP   EAX, EBX
         JG    GREATER
         JMP   LESS
      }

GREATER:
   printf ("Opnd1 is greater than opnd2 \n\n");
   goto end;

LESS:
   printf ("Opnd1 is less/equal opnd2 \n\n");

end:
   return 0;
}                          (a)          //continued on next page
```

Figure 6.5 Program to illustrate the use of the JG instruction: (a) the program and (b) the outputs.

```
Enter two signed integers:
-29 -30
Opnd1 is greater than opnd2

Press any key to continue . . . _
------------------------------------------------------------
Enter two signed integers:
+1 -1
Opnd1 is greater than opnd2

Press any key to continue . . . _
------------------------------------------------------------
Enter two signed integers:
+450 +450
Opnd1 is less/equal opnd2

Press any key to continue . . . _
------------------------------------------------------------
Enter two signed integers:
+780 +779
Opnd1 is greater than opnd2

Press any key to continue . . . _
------------------------------------------------------------
Enter two signed integers:
+425 +426
Opnd1 is less/equal opnd2

Press any key to continue . . . _    (b)
```

Figure 6.5 (Continued)

Jump if carry (JC) The JC instruction will cause a jump to occur if the following equation is true:

$$CF = 1$$

Example 6.9 Let register AL = 0001 1101 (+29) and register BL = 0011 1100 (+60). Then compare the two operands by subtracting the subtrahend from the minuend. This results in a carry out of the high-order bit position and produces a jump.

```
          0 0 0 1     1 1 0 1
    +)    1 1 0 0     0 1 0 0
         ------------------------
          1 1 1 0     0 0 0 1
      0(1)        1(0)
```

Example 6.10 Let register AL = 0011 0111 (+55) and register BL = 1001 1011, a negative number (–101), then compare the two operands by subtracting the subtrahend from the minuend. This results in a carry out of the high-order bit position and produces a jump.

```
            0 0 1 1      0 1 1 1
+)          0 1 1 0      0 1 0 1
            -------------------
            1 0 0 1      1 1 0 0
      0(1)         0(1)
```

Example 6.11 Let registers AL and BL both contain negative operands: AL = 1110 1100 (–20) and BL = 1110 1001 (–23), then compare the two operands by subtracting the subtrahend from the minuend. This results in no carry out of the high-order bit position; therefore, a jump is not executed.

```
            1 1 1 0      1 1 0 0
+)          0 0 0 1      0 1 1 1
            -------------------
            0 0 0 0      0 0 1 1
      1(0)         1(0)
```

Example 6.12 Let registers AL and BL both contain identical positive operands: 0010 1000 (+40), then compare the two operands by subtracting the subtrahend from the minuend. This results in no carry out of the high-order bit position; therefore, a jump is not executed.

```
            0 0 1 0      1 0 0 0
+)          1 1 0 1      1 0 0 0
            -------------------
            0 0 0 0      0 0 0 0
      1(0)         1(0)
```

Figure 6.6 shows a program that illustrates the *jump if carry* (JC) instruction. The operands, *opnd1* and *opnd2*, are moved to registers EAX and EBX, respectively, then compared. A conditional jump will occur if there is a carry out of the high-order bit position, in which case the program jumps to the label CARRY, whose address is equal to the contents of register (E)IP plus a signed displacement. A **printf ( )** function then displays that a carry has occurred. If there is no carry, then an unconditional jump occurs to the label NO_CARRY where the **printf ( )** function displays that there was no carry.

```
//jump_carry.cpp
//illustrate JC for unsigned/signed operands

#include "stdafx.h"

int main (void)
{
//define variables
   int opnd1, opnd2;

   printf ("Enter two unsigned/signed integers: \n");
   scanf ("%d %d", &opnd1, &opnd2);

//switch to assembly
      _asm
      {
         MOV   EAX, opnd1
         MOV   EBX, opnd2

         CMP   EAX, EBX
         JC    CARRY
         JMP   NO_CARRY
      }

CARRY:
   printf ("A carry was generated \n\n");
   goto end;

NO_CARRY:
   printf ("No carry was generated \n\n");

end:
   return 0;
}
```

(a)

```
Enter two unsigned/signed integers:
+29 +60
A carry was generated

Press any key to continue . . . _
```

//continued on next page

(b)

Figure 6.6 Program to illustrate the use of the JC instruction: (a) the program and (b) the outputs.

```
Enter two unsigned/signed integers:
+55 -101
A carry was generated

Press any key to continue . . . _
--------------------------------------------------------------
Enter two unsigned/signed integers:
-20 -23
No carry was generated

Press any key to continue . . . _
--------------------------------------------------------------
Enter two unsigned/signed integers:
+40 +40
No carry was generated

Press any key to continue . . . _
```

Figure 6.6 (Continued)

6.2 Looping Instructions

There are two categories of software loop instructions: an unconditional loop and conditional loops. Both categories use the (E)CX register as a loop control counter; the counter determines the number of times that the loop will be executed. All loop instructions decrement the count in the (E)CX register by one each time the loop instruction is decoded. If the count is nonzero, then the loop operation is executed; if the count is zero then the loop operation is not executed and program control is transferred to the instruction that immediately follows the loop instruction. The loop instructions do not change the state of the flags in the EFLAGS register.

If the loop instruction is executed, then the destination address is relative to the contents of the (E)IP register and is characterized as a short jump; that is, within –128 bytes to +127 bytes of the current value in the (E)IP register. If the count in the (E)CX register is zero when the loop instruction is initially decoded, then the counter is decremented to a value of 2^{16} = FFFFH if register CX is used or to a value of 2^{32} = FFFFFFFFH if register ECX is used. In order to avoid this situation, the *jump if CX register is 0* (JCXZ) or the *jump if ECX register is 0* (JECXZ) should be used.

6.2.1 Unconditional Loop

An *unconditional loop* (LOOP) instruction transfers control to another instruction in the specified range as indicated by a label. The label at the destination address is terminated by a colon, which indicates an instruction within the current code segment. The label name in the LOOP instruction, however, does not have a colon. If the LOOP instruction does not generate a transfer, then the instruction immediately following the LOOP instruction is executed more quickly than if a transfer occurred. This is because fewer clock cycles are required, since there is no address calculation to determine the destination address — the (E)IP register is simply incremented.

The program segment shown below illustrates an unconditional LOOP instruction to transfer control to an instruction with a label specified as NXT_NUM. The program in the loop performs a calculation on numbers in the loop. The body of the loop is executed 20 times.

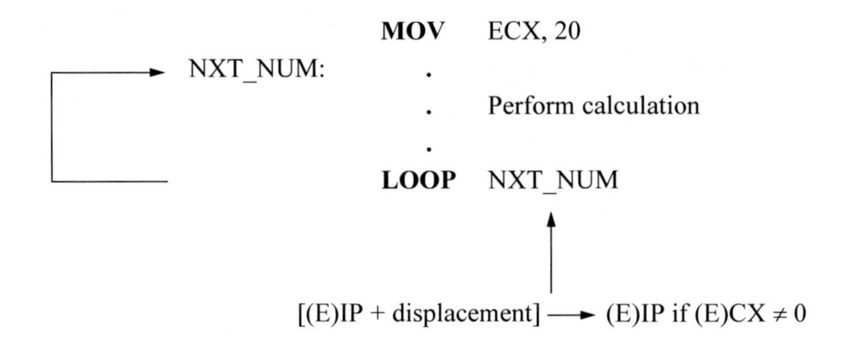

Figure 6.7 shows an assembly language program embedded in a C program to illustrate using the unconditional loop to double the value of a user-entered integer six times. The instruction shown below adds the contents of register EAX to register EAX, effectively doubling the value in EAX.

```
ADD     EAX, EAX
```

```
//loop_dbl.cpp
//program to illustrate the loop instruction.
//an integer is entered, then doubled with
//each iteration of the loop (6 times)

#include "stdafx.h"

int main (void)
{
//define variables
   int   x, rslt;
```

(a) //continued on next page

Figure 6.7 Program to illustrate using the LOOP instruction to double a user-entered integer six times: (a) the program and (b) the outputs.

```
   printf ("Enter a decimal integer: \n");
   scanf ("%d", &x);

//switch to assembly
      _asm
      {
         MOV   EAX, x
         MOV   ECX, 6

DBL:     ADD   EAX,  EAX    //double the value in EAX
         MOV   rslt, EAX    //move EAX to rslt
         LOOP  DBL          //loop to DBL if ECX is not zero,
                            //otherwise print result
      }

   printf ("Result = %d\n\n", rslt);

   return 0;
}
```

```
Enter a decimal integer:
1
Result = 64

Press any key to continue . . . _
------------------------------------------------------------
Enter a decimal integer:
2
Result = 128

Press any key to continue . . . _
------------------------------------------------------------
Enter a decimal integer:
5
Result = 320

Press any key to continue . . . _
------------------------------------------------------------
Enter a decimal integer:
25
Result = 1600

Press any key to continue . . . _
```

(b)

Figure 6.7 (Continued)

6.2.2 Conditional Loops

The conditional loop instructions are *loop while equal/zero* (LOOPE/LOOPZ) and *loop while not equal/not zero* (LOOPNE/LOOPNZ). The LOOPE and LOOPZ instructions are different mnemonics that refer to the same instruction and they repeat the loop — a short jump — if the (E)CX register is nonzero and the ZF flag is equal to 1. Otherwise, the instruction immediately following the loop instruction is executed. The ZF flag is set by a previous instruction.

The LOOPNE and LOOPNZ are different mnemonics that refer to the same instruction and repeat the loop — a short jump — if (E)CX is nonzero and the ZF flag is equal to 0. Otherwise, the instruction immediately following the loop instruction is executed. The ZF is reset by a previous instruction.

The conditional loop instructions use the count in the (E)CX register to determine the number of times to execute the loop; the count in the (E)CX register is decremented by one for each iteration. None of the flags are affected by these conditional loop instructions.

Figure 6.8 shows an embedded assembly language program illustrating the use of the LOOPNE instruction. Although the program for the loop instruction is relatively simple, it illustrates the principle of how it can be used for certain applications. In the program of Figure 6.8, the count in register ECX is initially set to a value of 20, register EAX is set to a value of 10, and register EBX is set to a value of 1. The conditional loop repeats while register ECX $\neq 0$ and the zero flag ZF $= 0$.

A value of 1 is added to register EBX with each iteration of the loop. After nine iterations, the values in registers EAX and EBX are equal. Therefore, even though the count in register ECX is nonzero (11_{10}), the ZF flag is set to a value of 1, which results in the termination of the loop. The sequence of iterations through the loop is shown below.

Iteration	EAX	EBX	ZF (CMP EAX, EBX)	ECX
0	10	1	0	20
1	10	2	0	19
2	10	3	0	18
3	10	4	0	17
4	10	5	0	16
5	10	6	0	15
6	10	7	0	14
7	10	8	0	13
8	10	9	0	12
9	10	10	1	11

When the program exits the loop, the value in register ECX is moved to the *rslt* variable. The **printf ( )** function then displays the equality of registers EAX and EBX and displays the resulting value (11_{10}) in register ECX. The count in register ECX does not decrement to zero because the ZF flag was set at iteration 9, resulting in the termination of the loop.

```
//loopne.cpp
//program to illustrate the
//loop while not equal (LOOPNE) instruction.

#include "stdafx.h"

int main (void)
{
//define variable
   int rslt;

//switch to assembly
         _asm
         {
            MOV        ECX, 20
            MOV        EAX, 10
            MOV        EBX, 1

LP1:        ADD        EBX, 1       //add 1 to EBX
            CMP        EAX, EBX     //ZF = 0 for 9 iterations
            LOOPNE     LP1          //loop if ECX ≠ 0 and ZF = 0,
            MOV        rslt, ECX    //otherwise print result
         }

   printf ("EAX = EBX, ECX = %d\n\n", rslt);

   return 0;
}
```

(a)

```
EAX = EBX, ECX = 11

Press any key to continue . . . _
```

(b)

Figure 6.8 Program to illustrate the function of the LOOPNE instruction: (a) the program and (b) the outputs.

6.2.3 Implementing While Loops

The **while** loop (or **while** statement) is one of many constructs in the C programming language that operate in a looping manner. The **while** loop executes a statement or block of statements as long as a test expression is true (nonzero). The statements that are controlled by the **while** statement loop repeatedly until the expression becomes false (0). Although the **while** statement is not part of the X86 assembly language instruction set, it can be implemented by using a series of standard assembly language instructions that simulate the structure of a **while** loop.

Figure 6.9 shows an assembly language program module embedded in a C program. The **while** structure is contained within the WHILE and END_WHILE labels, which act as delimiters for the assembly language program. The user enters a number in the range of 1 through 9; the number is then moved to register EAX prior to entering the loop, which begins at the WHILE label.

A comparison is made with the contents of register EAX and the number 10 to determine if the value in register EAX is greater than 10. This is done by the following instruction:

```
CMP     EAX, 10
```

If register EAX is greater than 10, then the loop is exited by the following instruction:

```
JG     END_WHILE
```

A value of 1 is added to register EAX with each iteration of the loop until the value in EAX is greater than 10. The final contents (11_{10}) of register EAX are then printed.

```
//while_loop_asm.cpp
//program to illustrate implementing
//a while loop using C and assembly language

#include "stdafx.h"

int main (void)
{
//define variables
   int int1, rslt;

   printf ("Enter an integer between 1 and 9: \n");
   scanf ("%d", &int1);
                                      //continued on next page
```

(a)

Figure 6.9 An assembly language module embedded in a C program to simulate a **while** loop structure: (a) the program and (b) the outputs.

```
//switch to assembly
      _asm
      {
         MOV    EAX, int1
WHILE:
         CMP    EAX, 10        //is EAX > 10?
         JG     END_WHILE      //if EAX > 10, end the while loop
         ADD    EAX, 1         //if EAX !> 10, add 1 to EAX
         MOV    rslt, EAX
         JMP    WHILE          //do another iteration of the loop
      }

END_WHILE:
   printf ("EAX = %d\n\n", rslt);

   return 0;
}
```

```
Enter an integer between 1 and 9:
1
EAX = 11

Press any key to continue . . . _
----------------------------------------------------------------
Enter an integer between 1 and 9:
5
EAX = 11

Press any key to continue . . . _
----------------------------------------------------------------
Enter an integer between 1 and 9:
10
EAX = 11

Press any key to continue . . . _
```

(b)

Figure 6.9 (Continued)

A similar program will now be written entirely in assembly language using the number 9 as an upper limit, as shown in Figure 6.10. This program displays the value of register BL for each iteration of the WHILE loop. The **.STACK** directive is a simplified way to define the stack segment — the default stack size is 1,024 bytes, but can be changed as required.

The data segment is defined by the **.DATA** directive and includes a one-dimensional parameter list array defined as a byte array, labelled PARLST, that is used to store the keyboard input data. The first element of the array, PARLST [0], is called MAXLEN, which defines the maximum number of input character bytes (DB) — in this example, ten is the maximum number of allowable characters, although only one character is needed.

The second array element, PARLST [1], is defined as a byte (DB) called ACTLEN, which stores the actual number of characters entered from the keyboard. The third element of the array, PARLST [2], contains the beginning of the operand field (OPFLD) where the operands from the keyboard are stored — in this example, ten bytes are specified. The last byte in the OPFLD is the *Enter* character (carriage return ↵).

A prompt field follows the parameter list, labelled PRMPT; this field prompts the user to enter an integer. The final field in the data segment is the result (RSLT) field where the results of the program are stored.

In a similar manner, the code segment is declared by the directive **.CODE**. A size value can be appended to these simplified segment directives to specify their respective sizes. These directives generate the appropriate segment statements and the corresponding end segment statements.

The address of the data segment (@DATA) is moved into the DS register. Then the prompt is displayed by the following instructions:

```
MOV   AH, 09H
LEA   DX, PRMPT
INT   21H
```

The MOV instruction places 09H, which is the *display string* function, in the required register AH for the interrupt function call, INT 21H. The LEA instruction places the address of the prompt for the display area in the required register DX. The INT instruction is the function call to execute the function code in register AH.

The next sequence of instructions is similar to those just presented, except that the function (0AH) in register AH is for a buffered keyboard input, which places the keyboard characters in the OPFLD area of the parameter list (PARLST). Data entered from the keyboard is stored in the OPFLD area as American Standard Code for Information Interchange (ASCII) characters. The number that is stored in the OPFLD is then moved to register BL. The address of the display area, RSLT + 11, to store the character from the first iteration of the loop is stored in register DI, which is used as a destination index.

The first instruction of the WHILE loop moves the user-entered number in register BL to the location specified by the contents of register DI ([DI]), where the brackets indicate *the contents of.* Since ASCII characters are being compared, the contents of register BL are compared to 39H — the upper limit. If the contents of register BL are greater than 9 (39H), then the program jumps to the END_WHILE label and exits the WHILE loop.

If the contents of register BL are not greater than 9, then a value of 1 is added to register BL and the destination address in register DI is incremented by 3, which provides for spacing between successive numbers. Then the next iteration of the loop occurs with an unconditional jump to the label WHILE. The result area (RSLT), containing all of the numbers from the user-entered number to the number 9, is then displayed.

```
PAGE 66, 80
TITLE while_loop.asm
;-----------------------------------------------------------

.STACK

;-----------------------------------------------------------

.DATA
PARLST    LABEL BYTE
MAXLEN    DB 10
ACTLEN    DB ?
OPFLD     DB 10 DUP(?)

PRMPT     DB 0DH, 0AH, 'Enter a number between 1 and 9: $'
RSLT      DB 0DH, 0AH, 'Result =                          $'

;-----------------------------------------------------------

.CODE
BEGIN     PROC FAR

;set up pgm ds
          MOV     AX, @DATA         ;put addr of data seg in ax
          MOV     DS, AX            ;put addr in ds

;read prompt
          MOV     AH, 09H           ;display string
          LEA     DX, PRMPT         ;put addr of prompt in dx
          INT     21H               ;dos interrupt

;kybd rtn to enter number
          MOV     AH, 0AH           ;buffered kybd input
          LEA     DX, PARLST        ;load addr of parlst
          INT     21H               ;dos interrupt

;store number in bl
          MOV     BL, OPFLD         ;store ascii number in bl
          LEA     DI, RSLT + 11     ;set up destination addr
                          (a)       //continued on next page
```

Figure 6.10 Assembly language program to illustrate the implementation of a **while** loop: (a) the program and (b) the outputs.

```
;implement the while loop
WHILE:
        MOV   [DI], BL        ;move bl to where di points
        CMP   BL, 39H         ;is bl > 9 (39 ascii)?
        JG    END_WHILE       ;if bl > 9, end the while loop
        ADD   BL, 1           ;if bl !> 9, add 1 to bl
        ADD   DI, 3           ;obtain next destination addr
        JMP   WHILE           ;repeat while loop

END_WHILE:
;display the values of bl
        MOV   AH, 09H         ;display string
        LEA   DX, RSLT        ;load addr of rslt field
        INT   21H             ;dos interrupt

BEGIN   ENDP
        END     BEGIN         ;start pgm at begin
```

```
Enter a number between 1 and 9: 1
Result = 1  2  3  4  5  6  7  8  9  :
--------------------------------------------------------------
Enter a number between 1 and 9: 5
Result = 5  6  7  8  9  :
--------------------------------------------------------------
Enter a number between 1 and 9: 7
Result = 7  8  9  :
--------------------------------------------------------------
Enter a number between 1 and 9: 9
Result = 9  :
```

(b)

Figure 6.10 (Continued)

6.2.4 Implementing for Loops

Another common looping technique is the **for** loop, which is used in the C programming language and in the Verilog Hardware Description Language (HDL). The **for** loop repeats a statement or block of statements a specific number of times. This is different than the **while** loop, which repeats the loop as long as a certain condition is met. When the **for** loop has completed the final loop, the program exits the loop and transfers control to the first statement following the block of statements.

The **for** loop contains three parts:

1. An *initial* condition to assign a value to the counter E(CX) as a control variable to determine the number of iterations for the loop. This is executed once before the beginning of the loop.

2. A *test* condition to determine when the loop terminates. This is a jump on condition instruction; for example, a *jump if greater than* (JG) expression that is executed before the procedural statements of the loop to determine if the loop should execute. The loop is repeated as long as the expression is true. If the expression is false, the loop terminates and the activity flow proceeds to the next statement in the module.

3. An *assignment* to modify the control variable, usually an increment or a decrement. This assignment is executed after each execution of the loop and before the next test to terminate the loop. The syntax for a **for** loop is shown below.

```
for (initial assignment for the counter; test instruction;
                    increment/decrement)
```

Although the **for** statement is not part of the X86 assembly language instruction set, it can be implemented by using a series of standard assembly language instructions that simulate the structure of a **for** loop.

Figure 6.11 shows an assembly language module embedded in a C program. The **for** structure is contained within the FOR and END_FOR labels, which act as delimiters for the assembly language program. The user enters a number in the range of 1 through 9 that is stored in an **int** variable called *count*. The number is then moved to register ECX and acts as a loop control variable prior to entering the loop, which begins at the FOR label.

The FOR loop performs an operation on register EAX by incrementing EAX by 1 during each iteration of the loop. The count in register ECX in incremented by 1 near the end of the FOR loop, then checked if it is within range at the beginning of the loop.

```
//for_loop_asm.cpp
//program to illustrate implementing
//a for loop using assembly language
#include "stdafx.h"

int main (void)
```

(a) //continued on next page

Figure 6.11 An assembly language module embedded in a C program to simulate a **for** loop structure: (a) the program and (b) the outputs.

```
{
//define variables
   int count, rslt_ecx, rslt_eax;

   printf ("Enter an integer between 1 and 9: \n");
   scanf ("%d", &count);

//switch to assembly
      _asm
      {
         MOV   EAX, 0
         MOV   ECX, count

FOR:
         CMP   ECX, 10       //is ECX > 10?
         JG    END_FOR       //if ECX > 10, end the for loop

//if ECX !> 10, add 1 to EAX and ECX
         ADD   EAX, 1
         MOV   rslt_eax, EAX
         ADD   ECX, 1
         MOV   rslt_ecx, ECX

         JMP   FOR           //do another iteration of the loop
      }

END_FOR:
   printf ("ECX = %d\nEAX = %d\n\n", rslt_ecx, rslt_eax);

   return 0;
}
```

```
Enter an integer between 1 and 9:
0
ECX = 11
EAX = 11

Press any key to continue . . . _
------------------------------------------------------------
Enter an integer between 1 and 9:
1
ECX = 11
EAX = 10

Press any key to continue . . . _     //continued on next page
```

(b)

Figure 6.11 (Continued)

```
Enter an integer between 1 and 9:
3
ECX = 11
EAX = 8

Press any key to continue. . . _
-------------------------------------------------------------
Enter an integer between 1 and 9:
7
ECX = 11
EAX = 4

Press any key to continue. . . _
-------------------------------------------------------------
Enter an integer between 1 and 9:
9
ECX = 11
EAX = 2

Press any key to continue. . . _
-------------------------------------------------------------
Enter an integer between 1 and 9:
10
ECX = 11
EAX = 1

Press any key to continue. . . _
```

Figure 6.11 (Continued)

The sequence of iterations through the loop is shown below for register ECX and register EAX.

Iteration	ECX	EAX
0	3	0
1	4	1
2	5	2
3	6	3
4	7	4
5	8	5
6	9	6
7	10	7
8	11	8

A similar program will now be written entirely in assembly language using the number 9 as an upper limit, as shown in Figure 6.12. This program displays the value of the count in register CL and the value of the data in register BL for each iteration of the FOR loop. The result areas, RSLT1 and RSLT2, contain all of the numbers from the user-entered number to the number 9 in register CL and corresponding data in register BL. The two result areas are displayed using two separate display routines.

```
PAGE 66, 80
TITLE for_loop.asm

;-------------------------------------------------------

.STACK

;-------------------------------------------------------

.DATA
PARLST    LABEL BYTE
MAXLEN    DB 10
ACTLEN    DB ?
OPFLD     DB 10 DUP(?)

PRMPT     DB 0DH, 0AH, 'Enter a number between 1 and 9: $'
RSLT1     DB 0DH, 0AH, 'CL =                                $'
RSLT2     DB 0DH, 0AH, 'BL =                                $'

;-------------------------------------------------------

.CODE
BEGIN     PROC FAR

;set up pgm ds
          MOV   AX, @DATA        ;put addr of data seg in ax
          MOV   DS, AX           ;put addr in ds

;read prompt
          MOV   AH, 09H          ;display string
          LEA   DX, PRMPT        ;put addr of prompt in dx
          INT   21H              ;dos interrupt

;kybd rtn to enter number (count)
          MOV   AH, 0AH          ;buffered kybd input
          LEA   DX, PARLST       ;load addr of parlst
          INT   21H              ;dos interrupt
```

(a) //continued on next page

Figure 6.12 Assembly language program to illustrate the implementation of a **for** loop: (a) the program and (b) the outputs.

```
;store count in cl
        MOV     CL, OPFLD       ;get ascii number, store in cl
        MOV     BL, 30H         ;initialize bl to ascii 0
        LEA     DI, RSLT1 + 7   ;set up dst addr for cl
        LEA     SI, RSLT2 + 7   ;set up dst addr for bl

;implement the for loop
FOR:
        MOV     [DI], CL        ;move cl to where di points
        CMP     CL, 39H         ;is cl > 9 (39 ascii)?
        JG      END_FOR         ;if cl > 9, end the for loop

        ADD     CL, 1           ;if cl !> 9, add 1 to cl
        ADD     DI, 3           ;obtain next dst addr for cl

        MOV     [SI], BL         ;move bl to where si points

        ADD     BL, 1           ;add 1 to bl
        ADD     SI, 3           ;obtain next dst addr for bl
        JMP     FOR             ;repeat for loop

END_FOR:
;display the values of cl
        MOV     AH, 09H         ;display string
        LEA     DX, RSLT1       ;load addr of rslt1 (cl) field
        INT     21H             ;dos interrupt

;display the values of bl
        MOV     AH, 09H         ;display string
        LEA     DX, RSLT2       ;load addr of rslt2 (bl) field
        INT     21H             ;dos interrupt

BEGIN   ENDP
        END     BEGIN           ;start pgm at begin
```

```
Enter a number between 1 and 9: 1
CL = 1   2   3   4   5   6   7   8   9   :
BL = 0   1   2   3   4   5   6   7   8
-----------------------------------------------------------
Enter a number between 1 and 9: 3
CL = 3   4   5   6   7   8   9   :
BL = 0   1   2   3   4   5   6
-----------------------------------------------------------
Enter a number between 1 and 9: 7
CL = 7   8   9   :
BL = 0   1   2            (b)          //continued on next page
```

Figure 6.12 (Continued)

```
Enter a number between 1 and 9: 9
CL = 9   :
BL = 0
```

Figure 6.12 (Continued)

6.3 Problems

6.1 Define a *near* jump, a *short* jump, and a *far* jump. All are unconditional jump instructions.

6.2 Determine if an overflow occurs for the operation shown below. The operands are signed numbers in 2s complement representation.

```
     1 0 1 1   1 0 0 1
 –)  0 1 0 0   0 1 1 0
     ---------------
```

6.3 Determine if an overflow occurs for the operation shown below. The operands are signed numbers in 2s complement representation.

```
     1 1 1 1   1 1 1 1
 +)  1 1 1 1   1 1 1 1
     ---------------
```

6.4 Show that no overflow occurs for the operations shown below. The operands are in 2s complement representation.

(a) 1101 0011 + 1100 1110
(b) 0110 1101 – 0110 0011
(c) 1111 1111 – 1111 1111
(d) 1111 1111 + 1111 1111

6.5 Determine the state of the flags for the following operation, where register AX = 73B4H and register BX = 6ACDH:

```
CMP   AX, BX
```

6.6 Determine the state of the flags for the following operation, where register AL = F3H and register BL = 72H:

```
CMP   AL, BL
```

6.7 Determine whether the conditional jump instructions shown below will cause a jump to DEST.

(a) 004FH + 200D **JS** DEST
(b) FF38H + 200D **JZ** DEST

6.8 Let AL = 1110 0000B and BL = 1100 0000B. Determine whether the conditional jump instructions shown below will cause a jump to DEST.

(a)
```
CMP     AL, BL
JA      DEST
```

(b)
```
CMP     AL, BL
JG      DEST
```

6.9 Let AL = 1111 1000B and BL = 1111 0000B. Determine whether the conditional jump instruction shown below will cause a jump to LBL1.

```
CMP     AL, BL
JG      LBL1
```

6.10 Let AX = 067CH. Determine whether the conditional jump instruction shown below will cause a jump to BRANCH_ADDR.

```
CMP     AX, 1660D
JNE     BRANCH_ADDR
```

6.11 Let AX = –405D. Determine whether the conditional jump instruction shown below will cause a jump to BRANCH_ADDR.

```
CMP     AX, FE6CH
JGE     BRANCH_ADDR
```

6.12 Let AX = 7768H and BX = 9BCAH. Determine if the conditional jump instruction shown below will cause a jump to DEST for the operation AX + BX.

```
JPO     DEST
```

6.13 Determine the number of times that the following program segment executes the body of the loop:

```
            MOV         CX, -1
LP1:                .
                    .
            LOOP        LP1
```

6.14 Determine the number of times that the following program segment executes the body of the loop:

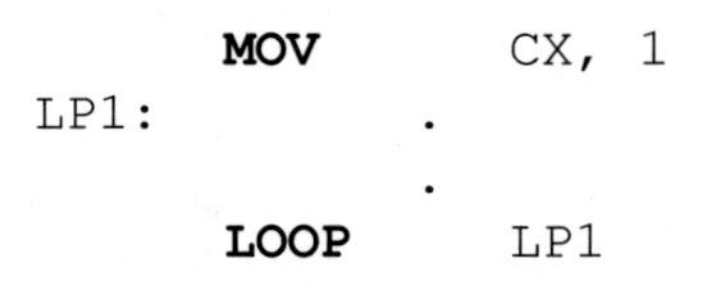

```
        MOV     CX, 1
LP1:            .
                .
        LOOP    LP1
```

6.15 Determine the number of times that the following program segment executes the body of the loop:

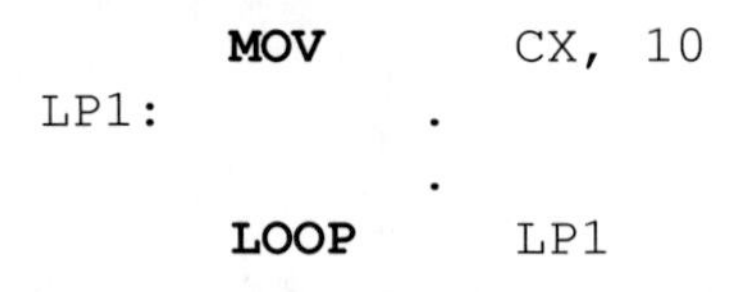

```
        MOV     CX, 10
LP1:            .
                .
        LOOP    LP1
```

6.16 Determine the number of times that the following program segment executes the body of the loop:

```
        MOV     CX, 0
LP1:            .
                .
        LOOP    LP1
```

6.17 Given the program shown below, determine the contents of register EAX after the program has finished execution.

```
//loop_add.cpp
//determine the contents of register EAX
//after execution of the program

#include "stdafx.h"

int main (void)

{
//define variable
   int rslt;

                                    //continued on next page
```

```
//switch to assembly
      _asm
      {
         MOV   CX, 4
         MOV   EAX, 2
LP1:     ADD   AX, 3
         ADD   AL, 4
         LOOP  LP1
         MOV   rslt, EAX
      }

   printf ("EAX = %X\n\n", rslt);

   return 0;
}
```

7

Stack Operations

The stack is a one-dimensional data structure located in contiguous locations of memory that is used for the temporary storage of data. It is one of the segments in a segmented memory model and is called the *stack segment*, in which the base address of the stack is contained in the *stack segment* (SS) register. A stack can have a maximum size of four gigabytes. A general-purpose register (GPR) called the *stack pointer* (E)SP contains the address of the current *top of stack*. A stack is comparable to a stack of trays in a cafeteria in which each tray is stacked on top of the last tray; the tray on the top of the stack is the tray that is normally removed first.

Generally, only the word at the top of stack — pointed to by the stack pointer — is accessed; however, another GPR can also be used to access elements within the stack. This is the *base pointer* (E)BP register, which contains an offset into the stack segment to reference data or parameters that were passed to a called program via the stack. The default segment for the (E)BP register is the stack segment. Since the element at the top of the stack is removed first, stacks are referred to as a *last-in-first-out* (LIFO) queue.

A data element is placed on top of the stack by a PUSH instruction; a data element is removed from the top of the stack by a POP instruction. A stack builds toward lower addresses. When a data item is pushed onto the stack, the (E)SP register is first decremented, then the data item is stored at the new top of stack. When a data item is popped off the stack, it is first stored in the destination address, then the (E)SP register is incremented to point to the new top of stack. The operand size determines the amount that the stack pointer is incremented or decremented. For example, if the operand size is 16 bits, then the SP register is incremented/decremented by 2; if the operand size is 32 bits, then the ESP register is incremented/decremented by 4.

7.1 Stack Structure

Figure 7.1 shows a drawing of a stack structure where the operand size is 16 bits; therefore, the SP register is used as the stack pointer. The structure is similar for an operand size of 32 bits where the items stored on the stack are doublewords; in this case, the stack pointer is the ESP register. In Figure 7.1, a PUSH operation first decrements SP by 2, then stores the operand at the new stack top, as shown below, where the colon (:) indicates concatenation.

PUSH

```
SP ← (SP - 2)
SP:(SP + 1) ← [source operand]
```

A POP operation first stores the operand in the destination then increments SP by 2, as shown below.

POP

```
destination ← [SP:(SP + 1)]
SP ← (SP + 2)
```

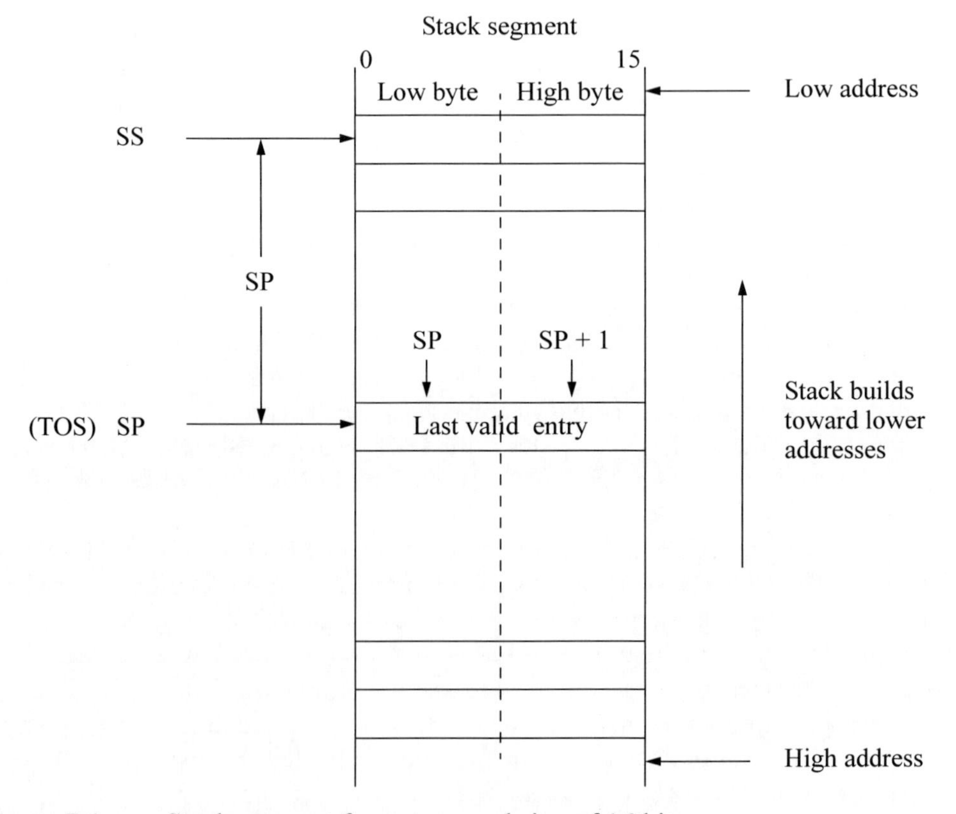

Figure 7.1 Stack segment for an operand size of 16 bits.

Figure 7.2 shows an example of pushing and popping 16-registers using the stack configuration of Figure 7.1. Prior to the operations, the general-purpose register, AX, is assigned a value of 1234H; the stack segment register, SS, is assigned a value of 1050H; and the stack pointer register, SP, is assigned a value of 0008H. Figure 7.2(a) shows the initialization of the stack and registers. Figure 7.2(b) shows the PUSH AX operation; Figure 7.2(c) shows the POP AX operation; and Figure 7.2(d) shows the POP BX operation. The stack pointer is incremented or decremented accordingly for each operation. Flags are not affected by the PUSH and POP operations.

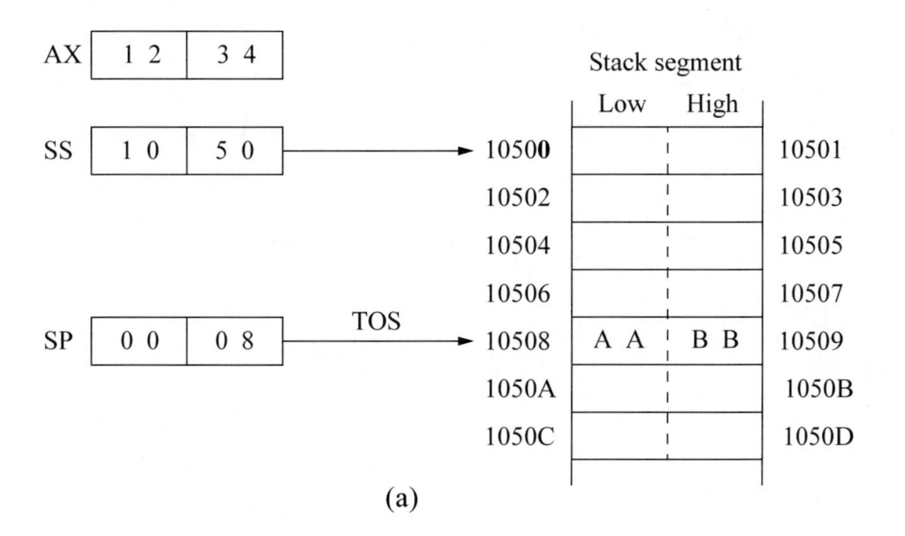

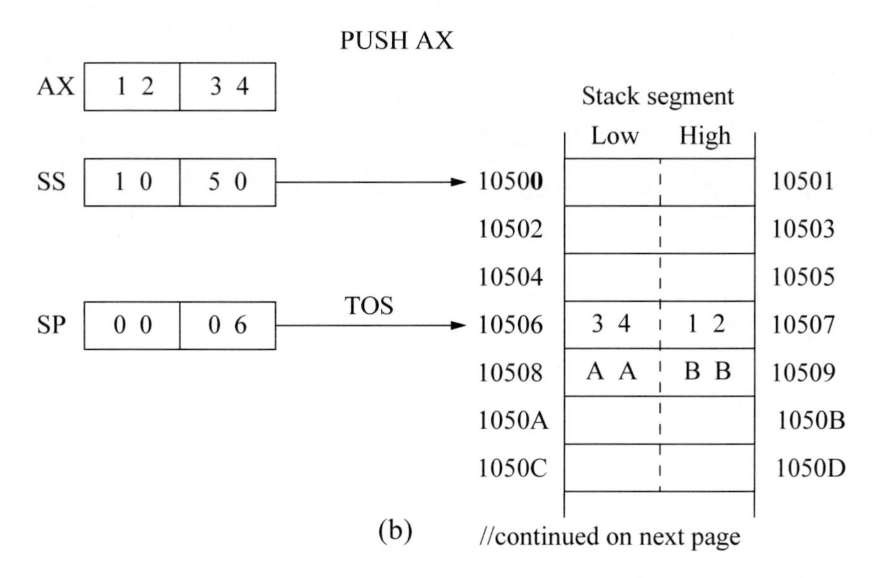

//continued on next page

Figure 7.2 Diagram showing stack operations for PUSH and POP instructions involving the general-purpose registers AX and BX: (a) initialization, (b) PUSH AX, (c) POP AX, and (d) POP BX.

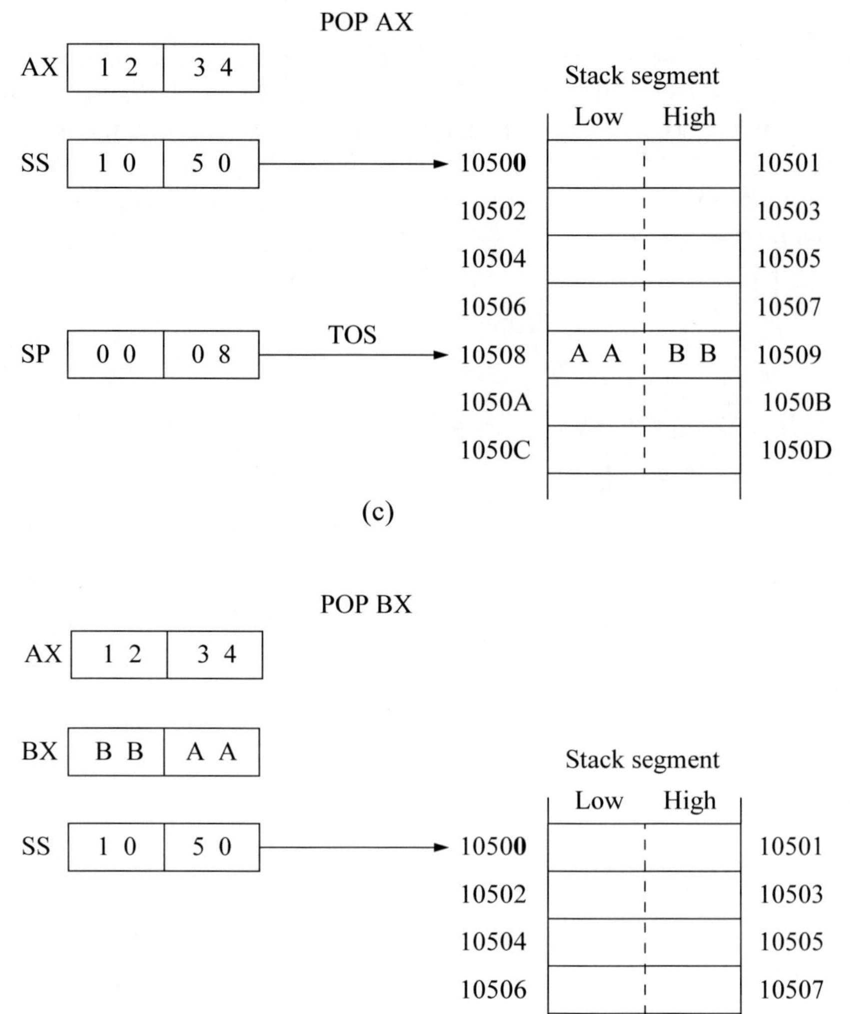

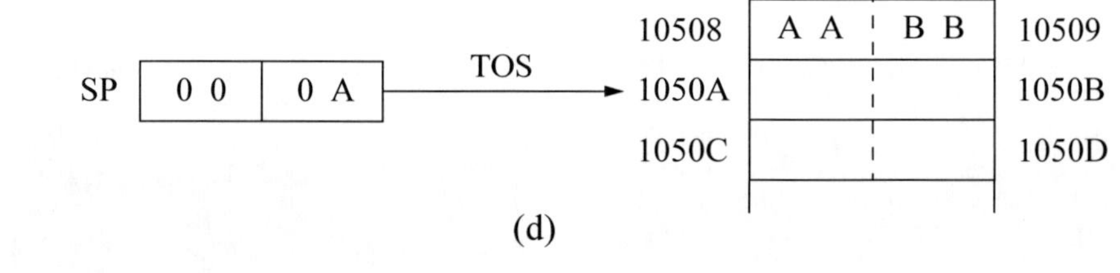

Figure 7.2 (Continued)

Unlike the CS register, the SS register and the (E)SP register can be initialized explicitly. This allows multiple stacks to be used with the capability to switch between the stacks. This is a useful technique for multitasking systems, where each task has a separate stack. One method of initializing a stack is shown in Figure 7.3 to initialize the SS register and the SP register.

The memory model is defined as SMALL by the **.MODEL** directive, which defines a code segment of ≤ 64 kilobytes and a data segment of ≤ 64 kilobytes. The stack is defined as 64 bytes by the **.STACK** directive — the default stack size is 1,024 bytes. Initially, (E)SP points to the doubleword/word that is one location higher than the highest address in the stack.

The name TOS in the stack segment is characterized by **LABEL**, which is a directive used to define the type attribute of a name — in this case the name TOS is defined as type WORD and aligns the stack on a word boundary. The syntax for the LABEL directive is shown below.

```
NAME    LABEL    type-specifier
```

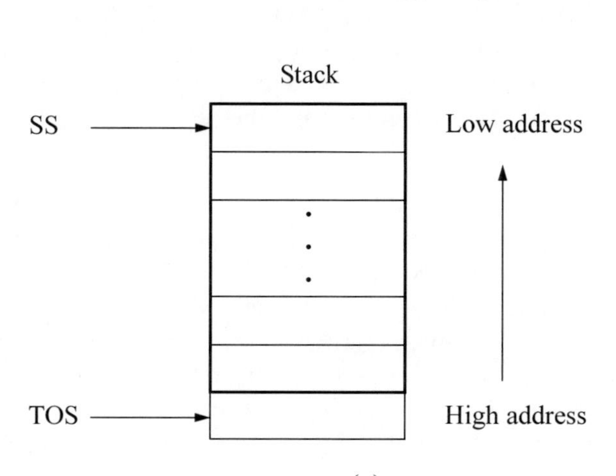

(a)

```
.MODEL    SMALL
;--------------------------------------------------------------
.STACK    64
TOS    LABEL     WORD
;--------------------------------------------------------------
.DATA       . . .
;--------------------------------------------------------------
.CODE
BEGIN       PROC FAR
            . . .
            MOV    AX, @STACK
            MOV    SS, AX               ;initialize SS

            MOV    SP, OFFSET  TOS      ;initialize SP
            . . .
BEGIN       ENDP
            END    BEGIN
```

(b)

Figure 7.3 One technique to initialize the SS register and the SP register: (a) the stack structure and (b) the program snippet.

Figure 7.3 established the SS register and the SP register using the MOV instruction. They can also be established in a simpler more direct method by using the *load far pointer using SS* (LSS) instruction. The LSS instruction loads a far pointer, containing a segment selector and offset, from the source operand in memory into the SS register and the (E)SP register in one operation. The syntax for the LSS instruction is shown below for legacy mode, with different operand-size attributes.

```
LSS   register 16 bits, memory 16:16 bits
LSS   register 32 bits, memory 16:32 bits
```

The operation code (LSS) and the destination specify the stack segment and the general-purpose register (E)SP. The 16-bit segment selector from the source operand is loaded into the stack segment (SS) register and the 32-bit or 16-bit offset is loaded into the stack pointer (E)SP register as specified by the destination operand. The following line of code establishes the SS register and the SP register.

```
LSS   SP, [SRC_OPND]
```

The stack pointer (SP) register should point to a 16-bit word or a 32-bit doubleword boundary for correct alignment, as specified by the stack segment. The PUSH instruction can push memory operands, immediate operands, general-purpose registers, and segment registers onto the stack. The POP instruction stores the word or doubleword to a general-purpose register, a segment register, or a memory location. When an operand is popped off the stack and stored in the destination, it is simply copied to the destination, not removed from the stack until it is overwritten.

If an attempt is made to push an operand onto a full stack, then an error is indicated; likewise, if an attempt is made to pop an empty stack, then an error is indicated. Most operating systems monitor the stack boundaries to detect stack overflow and stack underflow in order to maintain system reliability.

However, if necessary, a safe push and a safe pop can be implemented in software. Assume that the boundary addresses for a stack are 00986 to 10500. If (E)SP is ≤ 00986, and a push operation is attempted, then a stack overflow is indicated, because the stack is full. Similarly, if (E)SP is > 10500, and a pop operation is attempted, then a stack underflow is indicated, because the stack is empty. The code snippets shown check for stack overflow and stack underflow.

```
Stack overflow:     CMP   00986, ESP
                    JLE   FULL_ERR
                    NEXT_INSTR


Stack underflow:    CMP   10500, ESP
                    JG    EMPTY_ERR
                    NEXT_INSTR
```

Figure 7.4 shows an assembly language module embedded in a C program to illustrate stack utilization to perform addition operations on user-entered integers. This is a rudimentary application of a stack, but adequate to describe the use of a stack. The user enters three integers, *int1*, *int2*, and *int3*, which are moved to registers EAX, EBX, and ECX, respectively. The program then adds *int1* and *int2* and stores the sum in register EAX, then moves the sum to location *sum_12* to be displayed later.

Then *int1* and *int3* are added; the sum is stored in register EAX and location *sum_13*. Then *int2* and *int3* are added; the sum is stored in register EBX, and location *sum_23*. Register EBX is then pushed onto the stack. Integer *int1* is moved to register EAX, register EBX (*int2* + *int3*) is popped off the stack and added to register EAX, which contains *int1*, to produce the sum of the three integers: *int1*, *int2*, and *int3*. The program then prints the following sums: *int1* + *int2*, *int1* + *int3*, *int2* + *int3*, and *int1* + *int2* + *int3*.

```
//push_pop_add.cpp
//use the stack to add three integers

#include "stdafx.h"
int main (void)
{
//define variables
   int int1, int2, int3;
   int sum_12, sum_13, sum_23, sum_123;

   printf ("Enter three integers: \n");
   scanf ("%d %d %d", &int1, &int2, &int3);

//switch to assembly
      _asm
      {
         MOV   EAX, int1
         MOV   EBX, int2
         MOV   ECX, int3

         ADD   EAX, EBX     ;int1 + int2 -> EAX
         MOV   sum_12, EAX ;move int1 + int2 to result area

         MOV   EAX, int1    ;move int1 to EAX
         ADD   EAX, ECX     ;int1 + int3 -> EAX
         MOV   sum_13, EAX ;move int1 + int3 to result area

                      (a)        //continued on next page
```

Figure 7.4 Program to illustrate using a stack to add three integers: (a) the program and (b) the outputs.

```
		ADD		EBX, ECX	;int2 + int3 -> EBX
		MOV		sum_23, EBX ;move int2 + int3 to result area
		PUSH	EBX			;save int2 + int3 on stack

		MOV		EAX, int1	;move int1 to EAX
		POP		EBX			;pop int2 + int3
		ADD		EAX, EBX	;int1 + int2 + int3 -> EAX
		MOV		sum_123, EAX;move int1 + int2 + int3 to result
	}

	printf ("\nint1 + int2 = %d\n
			int1 + int3 = %d\n
			int2 + int3 = %d\n
			int1 + int2 + int3 = %d\n\n",
			sum_12, sum_13, sum_23, sum_123);

	return 0;
}
```

```
Enter three integers:
0 0 0

int1 + int2 = 0
int1 + int3 = 0
int2 + int3 = 0
int1 + int2 + int3 = 0

Press any key to continue . . . _
------------------------------------------------------------
Enter three integers:
1 1 1

int1 + int2 = 2
int1 + int3 = 2
int2 + int3 = 2
int1 + int2 + int3 = 3

Press any key to continue . . . _
------------------------------------------------------------
```

(b) //continued on next page

Figure 7.4 (Continued)

```
Enter three integers:
2 2 2

int1 + int2 = 4
int1 + int3 = 4
int2 + int3 = 4
int1 + int2 + int3 = 6

Press any key to continue . . . _
------------------------------------------------------------
Enter three integers:
2 10 25

int1 + int2 = 12
int1 + int3 = 27
int2 + int3 = 35
int1 + int2 + int3 = 37

Press any key to continue . . . _
------------------------------------------------------------
Enter three integers:
10 20 30

int1 + int2 = 30
int1 + int3 = 40
int2 + int3 = 50
int1 + int2 + int3 = 60

Press any key to continue . . . _
------------------------------------------------------------
Enter three integers:
25 50 100

int1 + int2 = 75
int1 + int3 = 125
int2 + int3 = 150
int1 + int2 + int3 = 175

Press any key to continue . . . _
------------------------------------------------------------
Enter three integers:
250 125 600

int1 + int2 = 375
int1 + int3 = 850
int2 + int3 = 725
int1 + int2 + int3 = 975

Press any key to continue . . . _
```

Figure 7.4 (Continued)

7.2 Additional Push Instructions

This section describes additional push instructions, such as PUSHA, PUSHAD, PUSHF, and PUSHFD. These instructions push specific source operands onto the stack. The stack pointer, (E)SP register, operates in the same manner as the regular PUSH operation; that is, it decrements an explicit number of bytes before the push operation, depending on the size of the operand being pushed onto the stack.

Push all 16-bit general-purpose registers (PUSHA) The PUSHA instruction is used to push all general-purpose registers (GPRs) onto the stack. If the operand-size attribute is 16, then the registers are pushed onto the stack in the following order: AX, CX, DX, BX, SP, BP, SI, and DI. Note that the value in SP that is pushed onto the stack is the value in SP before the PUSHA instruction is executed.

When a procedure call is executed, the processor does not automatically save the GPRs onto the stack prior to executing the call. The PUSHA instruction is one way that the calling/called procedure can save all the GPRs by executing a single instruction. The called procedure can also save only the specific registers that will be used before executing the procedure; the registers are then restored to their original values prior to returning to the calling program. No flags are affected by the PUSHA instruction.

Push all 32-bit general-purpose registers (PUSHAD) The instruction that pushes all 32-bit GPRs onto the stack is the PUSHAD instruction. If the operand-size attribute is 32 bits, then the registers are pushed onto the stack in the following order: EAX, ECX, EDX, EBX, ESP, EBP, ESI, and EDI. Note that the value in ESP that is pushed onto the stack is the value in ESP before the PUSHAD instruction is executed. The PUSHA and PUSHAD instructions have the same operation code and their operation is identical except for the registers that are pushed onto the stack, which are a function of the operand-size attribute.

When a procedure is called, saving the GPRs is identical as described under the PUSHA heading. It should be noted that some X86 assemblers may cause the operand size to be 16 bits for a PUSHA instruction or to be 32 bits when a PUSHAD instruction is encountered. Other X86 assemblers use the current value of the operand-size attribute. No flags are affected by the PUSHAD instruction.

Push 16 low-order flags (PUSHF) This instruction pushes the FLAGS register — the low-order 16 bits of the EFLAGS register — onto the stack if the operand-size attribute is 16 bits, then decrements the stack pointer, SP, by two. The FLAGS register contains the status flags (OF, SF, ZF, AP, PF, and CF), the control flag (DF), and four system flags (NT, IOPL, IF, and TF). No flags are affected by the PUSHF instruction.

Push all 32 flags (PUSHFD) This instruction pushes the 32-bit EFLAGS register onto the stack if the operand-size attribute is 32 bits, then decrements the stack pointer, ESP, by four. The EFLAGS register contains the status flags (OF, SF, ZF, AP, PF, and CF), the control flag (DF), and the ten system flags (ID, VIP, VIF, AC, VM, RF, NT, IOPL, IF, and TF).

Some X86 assemblers may cause the operand size to be 16 bits for a PUSHF instruction or to be 32 bits when a PUSHFD instruction is encountered. Other X86 assemblers use the current value of the operand-size attribute. The PUSHF and PUSHFD instructions have the same operation code and their operation is identical except for the registers — FLAGS or EFLAGS — that are pushed onto the stack, which are a function of the operand-size attribute. No flags are affected by the PUSHFD instruction. Certain flags in the EFLAGS register can be modified by specific instructions; for example, *set carry flag* (STC), *clear carry flag* (CLC), and *complement carry flag* (CMC).

7.3 Additional Pop Instructions

This section describes additional pop instructions, such as POPA, POPAD, POPF, and POPFD. The stack pointer, (E)SP, register operates in an identical manner as for the regular POP operation; that is, it increments an explicit number of bytes after the pop operation depending on the size of the operand being popped from the stack.

Pop all 16-bit general-purpose registers (POPA) The POPA instruction is used to pop all general-purpose registers off the stack. If the operand-size attribute is 16, then the word registers are popped off the stack in the following order and stored in their respective registers: DI, SI, BP, the pushed SP is ignored, BX, DX, CX, and AX. Note that the value in SP is discarded, because if SP were popped off the stack, then the value in the current SP register would be changed.

The POPA instruction is one way that the called/calling procedure can restore all the GPRs by executing a single instruction. The called procedure can also restore only the specific registers that were used in the procedure; the registers are then restored to their original values prior to returning to the calling program. The POPA instruction effectively reverses the operation of the PUSHA instruction. No flags are affected by the POPA instruction.

Pop all 32-bit general-purpose registers (POPAD) The POPAD instruction pops all 32-bit general-purpose registers off the stack. If the operand-size attribute is 32 bits, then the registers are popped off the stack in the following order: EDI, ESI, EBP, the pushed ESP is ignored, EBX, EDX, ECX, and EAX. Note that the value in ESP is discarded for the same reason that the value of SP was discarded for the POPA instruction. The POPA and POPAD instructions have the same operation code and their operation is identical except for the registers that are popped off the stack, which are a function of the operand-size attribute.

Some X86 assemblers may cause the operand size to be 16 bits for a POPA instruction or to be 32 bits when a POPAD instruction is encountered. Other X86 assemblers use the current value of the operand-size attribute. No flags are affected by the POPAD instruction.

Pop 16 low-order flags (POPF) The POPF instruction is used when the operand-size attribute is 16. It pops the top word from the stack into the low-order 16 bits of the EFLAGS register, then increments the SP register by two to point to the new stack top. The POPF instruction, in conjunction with the PUSHF instruction, allows a procedure to save the calling program's flags and then to restore the flags. The following flag bits are affected: OF, DF, TF, SF, ZF, AF, PF, and CF. The IOPL flag and the IF flag are also affected depending on the privilege level.

Pop all 32 flags (POPFD) The POPFD instruction pops a doubleword off the stack into the EFLAGS register if the operand-size attribute is 32 bits, then increments the ESP stack pointer by four. The POPF and POPFD instructions have the same operation code and their operation is identical except for the registers that are loaded from the stack — FLAGS or EFLAGS — which are a function of the operand-size attribute. Some X86 assemblers may cause the operand size to be 16 bits for a POPF instruction or to be 32 bits when a POPFD instruction is encountered. Other X86 assemblers use the current value of the operand-size attribute.

7.4 Problems

7.1 Given DS = 2800H, BX = 0400H, SP = 1000H, SS = 2F00H, and memory location 28400H = A020H, find the real (physical) data address of the source operand and the real address of the stack top when the PUSH [BX] instruction is executed. Show the contents of the stack top in memory and determine the new contents of the stack pointer SP.

7.2 Given DS = FF00H, SI = 0008H, SP = 0FEAH, SS = 2F00H, and memory location 2FFEAH = 3BC5H, find the real (physical) data address of the destination operand and the real address of the stack top when the POP [SI] instruction is executed. Show the contents of the stack top in memory and determine the new contents of the stack pointer SP.

7.3 Determine the result of each instruction for the following program segment:

```
PUSH EBP
MOV  EBP, ESP
PUSH EAX
PUSH EBX
PUSH ECX
          . . .
              //continued on next page
```

```
            . . .
MOV   EAX, [EBP - 12]
MOV   EBX, [EBP - 8]
MOV   ECX, [EBP - 4]
            . . .
ADD   ESP, 12
POP   EBP
```

7.4 Determine the result of each instruction for the following program segment:

```
PUSH  EAX
PUSH  EBX
PUSH  ECX
PUSH  EBP
MOV   EBP, ESP
            . . .
MOV   EAX, [EBP + 4]
MOV   EBX, [EBP + 8]
MOV   ECX, [EBP + 12]
            . . .
POP   EBP
ADD   ESP, 12
```

7.5 The partial contents of a stack are shown below before execution of the program segment listed below. Determine the contents of the stack after the program has been executed and indicate the new top of stack.

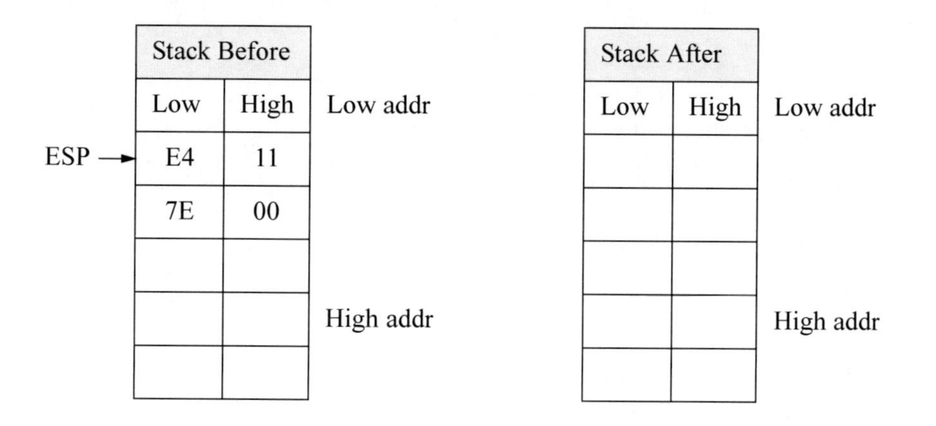

```
POP     BX
MOV     AH, BH
ADD     AH, BL
MOV     BH, AH
PUSH    BX
```

7.6 Why will a PUSH AL instruction cause an error message to be displayed?

7.7 Assume that a stack contains three parameters and that SP points to the third parameter, as shown in the diagram below. Determine how each parameter can be accessed if the following program segment is executed:

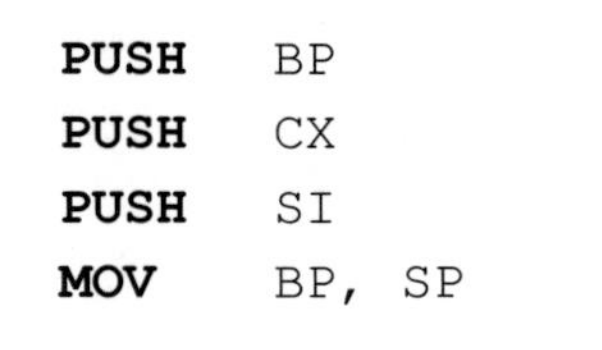

```
PUSH    BP
PUSH    CX
PUSH    SI
MOV     BP, SP
```

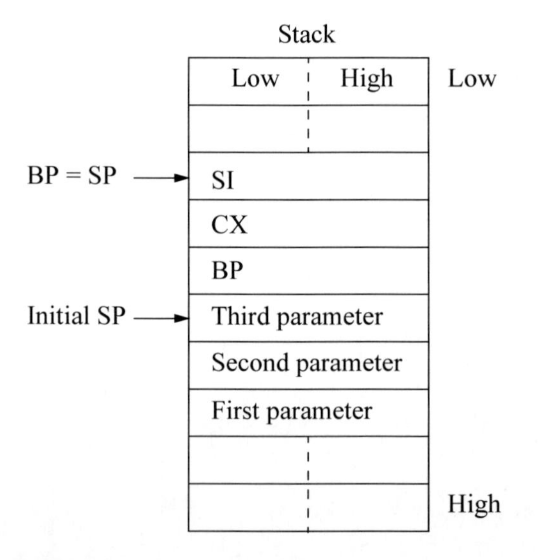

7.8 Write an assembly language program — using the PUSH and POP instructions — that adds four decimal integers, then displays their sum. Embed the assembly module in a C program. The decimal integers are entered from the keyboard.

7.9 Write an assembly language program — using the PUSH and POP instructions — that adds five hexadecimal integers, then displays their sum. Embed the assembly module in a C program. The hexadecimal integers are entered from the keyboard. Enter hexadecimal integers that range from one character to eight characters. Display the sum as upper-case hexadecimal characters.

8

Logical, Bit, Shift, and Rotate Instructions

The logical operations of AND, OR, exclusive-OR, NOT, and TEST are presented in this chapter; these instructions execute the Boolean equivalent of the corresponding operations in digital logic circuits. Also covered are the instructions that operate on single bits, such as *bit test* (BT), *bit test and set* (BTS), *bit test and reset* (BTR), *bit test and complement* (BTC), *bit scan forward* (BSF), and *bit scan reverse* (BSR).

Shift instructions are also presented that perform logical or arithmetic left or right shifts on bytes, words, or doublewords. The shift instructions are *shift arithmetic left* (SAL), *shift logical left* (SHL), *shift arithmetic right* (SAR), *shift logical right* (SHR), *double precision shift left* (SHLD), and *double precision shift right* (SHRD). The rotate instructions include *rotate left* (ROL), *rotate right* (ROR), *rotate through carry left* (RCL), and *rotate through carry right* (RCR). Also covered is the *set byte on condition* (SETcc) instruction.

8.1 Logical AND Instruction

The AND instruction performs the bitwise AND operation of two operands — the first operand (destination) and the second operand (source) and stores the result in the destination operand. The source operand can be located in a general-purpose register, in memory, or can be an immediate operand. The destination operand can be in a general-purpose register or a memory location; however, both operands cannot be in memory. The syntax for the logical AND instruction is shown below.

```
AND    register/memory, register/memory/immediate
```

The truth table for the AND function is shown in Table 8.1 and corresponds to the Boolean product. The variables x_1, x_2, and z_1 represent a bit in the destination operand, a bit in the source operand, and the resulting bit in the destination operand, respectively. The AND instruction can be used to mask off particular bits; that is, to reset certain bits, because any bit that is ANDed with a 0 bit results in a 0 bit. The affected flags are the parity flag (PF), the sign flag (SF), and the zero flag (ZF) — the overflow flag (OF) and the carry flag (CF) are set to 0. Examples of the AND function are shown in Table 8.2(a) and Table 8.2(b). Table 8.2(b) masks off the high-order four bits.

Table 8.1 Truth Table for the AND Function

AND		
x_1	x_2	z_1
0	0	0
0	1	0
1	0	0
1	1	1

Table 8.2 Examples of the AND Function for Eight Bits

AND																
0	1	1	0	1	1	0	1		1	1	1	1	1	1	1	1
0	0	1	1	1	1	1	0		0	0	0	0	1	0	1	0
0	0	1	0	1	1	0	0		0	0	0	0	1	0	1	0
(a)									(b)							

Figure 8.1 shows an assembly language module embedded in a C program that illustrates using the AND function to mask off certain bits in a hexadecimal number that is entered from the keyboard. The number is moved to the EAX register, then the AND instruction resets the high-order 28 bits; that is, bits 31 through 4 by the instruction shown below.

```
AND    EAX, 0000000FH
```

The program checks to determine if the modified number is in the range 00000002H through 00000005H. If the number is within the range, then the number is displayed. If the number is not in the specified range, then an error message is displayed.

```
//and_mask.cpp
//illustrates the use of the AND function
//to mask off certain bits

#include "stdafx.h"

int main (void)
{
//define variables
   int hex1, rslt;

   printf ("Enter an 8-digit hex number in the range
            xxxxxxx2 -- xxxxxxx5: \n");
   scanf ("%X", &hex1);

//switch to assembly
      _asm
      {
         MOV   EAX, hex1
         AND   EAX, 0000000FH
         CMP   EAX, 00000005H
         JA    ERR

         CMP   EAX, 00000002H
         JB    ERR

         MOV   rslt, EAX
         JMP   NO_ERR
      }

ERR:  printf ("\nNumber out of range\n\n");
      goto end;

NO_ERR:
   printf ("\nResult = %X\n\n", rslt);

end:
   return 0;
}
```

(a) //continued on next page

Figure 8.1 Program to illustrate using the AND instruction to mask off certain bits: (a) the program and (b) the outputs.

```
Enter an 8-digit hex number in the range xxxxxxx2 -- xxxxxxx5:
ABCDEF12

Result = 2

Press any key to continue . . . _
-------------------------------------------------------------
Enter an 8-digit hex number in the range xxxxxxx2 -- xxxxxxx5:
12AB34C3

Result = 3

Press any key to continue . . . _
-------------------------------------------------------------
Enter an 8-digit hex number in the range xxxxxxx2 -- xxxxxxx5:
1A2B3C44

Result = 4

Press any key to continue . . . _
-------------------------------------------------------------
Enter an 8-digit hex number in the range xxxxxxx2 -- xxxxxxx5:
12AB34C5

Result = 5

Press any key to continue . . . _
-------------------------------------------------------------
Enter an 8-digit hex number in the range xxxxxxx2 -- xxxxxxx5:
ABCDEF11

Number out of range

Press any key to continue . . . _
-------------------------------------------------------------
Enter an 8-digit hex number in the range xxxxxxx2 -- xxxxxxx5:
12345676

Number out of range

Press any key to continue . . . _
```

(b)

Figure 8.1 (Continued)

Figure 8.2 contains an assembly language module embedded in a C program that illustrates the logical AND function of two operands. The program displays the result of the AND operation together with the resulting flags. The **unsigned** type modifier is applied to the **char** data type. An unsigned character has a range from 0 to 255; a signed character, **char**, has a range from –128 to +127.

The two operands, *opnd1* and *opnd2*, are moved to registers AL and BL, respectively, prior to the AND operation. The *load AH from flags* (LAHF) instruction stores the low-order byte (shown below) of the EFLAGS register in the AH register. Bit positions 1, 3, and 5 are reserved. The flags and result are then moved to the *flags* variable and the *rslt* variable to be displayed.

7	6	5	4	3	2	1	0
SF	ZF	0	AF	0	PF	1	CF

```
//logical_and2.cpp
//illustrates the logical AND function
#include "stdafx.h"
int main (void)
{
//define variables
   unsigned char and1, and2, rslt, flags;

   printf ("Enter two 2-digit hexadecimal characters: \n");
   scanf ("%X %X", &and1, &and2);

//switch to assembly
      _asm
      {
         MOV   AL, and1
         MOV   BL, and2
         AND   BL, AL
         LAHF
         MOV   flags, AH
         MOV   rslt, BL
      }

   printf ("\nOpnd1 = %X\nOpnd2 = %X\n", and1, and2);
   printf ("\nAND result = %X\n\n\nFlags = %X\n\n",
               rslt, flags);
   return 0;
}
```

(a) //continued on next page

Figure 8.2 Program to illustrate the logical AND operation and resulting flags: (a) the program and (b) the outputs.

```
Enter two 2-digit hexadecimal characters:
87 CA

Opnd1 = 87         //1 0 0 0 0 1 1 1
Opnd2 = CA         //1 1 0 0 1 0 1 0

AND result = 82    //1 0 0 0 0 0 1 0

                   //SF ZF 0  AF 0  PF 1  CF
Flags = 86         //1  0  0  0  0  1  1  0

Press any key to continue . . . _
------------------------------------------------------------
Enter two 2-digit hexadecimal characters:
47 C3

Opnd1 = 47         //0 1 0 0 0 1 1 1
Opnd2 = C3         //1 1 0 0 0 0 1 1

AND result = 43    //0 1 0 0 0 0 1 1

                   //SF ZF 0  AF 0  PF 1  CF
Flags = 2          //0  0  0  0  0  0  1  0

Press any key to continue . . . _
------------------------------------------------------------
Enter two 2-digit hexadecimal characters:
FF FF

Opnd1 = FF         //1 1 1 1 1 1 1 1
Opnd2 = FF         //1 1 1 1 1 1 1 1

AND result = FF    //1 1 1 1 1 1 1 1

                   //SF ZF 0  AF 0  PF 1  CF
Flags = 86         //1  0  0  0  0  1  1  0

Press any key to continue . . . _
------------------------------------------------------------
```

//continued on next page

(b)

Figure 8.2 (Continued)

```
Enter two 2-digit hexadecimal characters:
FF 00

Opnd1 = FF          //1 1 1 1 1 1 1 1
Opnd2 = 0           //0 0 0 0 0 0 0 0

AND result = 0      //0 0 0 0 0 0 0 0

                    //SF ZF 0  AF 0  PF 1  CF
Flags = 46          //0  1  0  0  0  1  1  0

Press any key to continue . . . _
------------------------------------------------------------
Enter two 2-digit hexadecimal characters:
EC C3

Opnd1 = EC          //1 1 1 0 1 1 0 0
Opnd2 = C3          //1 1 0 0 0 0 1 1

AND result = C0     //1 1 0 0 0 0 0 0

                    //SF ZF 0  AF 0  PF 1  CF
Flags = 86          //1  0  0  0  0  1  1  0

Press any key to continue . . . _
```

Figure 8.2 (Continued)

8.2 Logical Inclusive-OR Instruction

The OR instruction performs the bitwise OR operation of two operands — the first operand (destination) and the second operand (source) and stores the result in the destination operand. The source operand can be located in a general-purpose register, in memory, or can be an immediate operand. The destination operand can be in a general-purpose register or a memory location; however, both operands cannot be in memory. The syntax for the logical OR instruction is shown below.

```
OR    register/memory, register/memory/immediate
```

The truth table for the OR function is shown in Table 8.3 and corresponds to the Boolean sum. The variables x_1, x_2, and z_1 represent a bit in the destination operand, a bit in the source operand, and the resulting bit in the destination operand, respectively. The OR instruction can be used to set particular bits; that is, to set certain bits to a value of 1, because any bit that is ORed with a 1 bit results in a 1 bit. The affected flags are the parity flag (PF), the sign flag (SF), and the zero flag (ZF) — the overflow flag (OF) and the carry flag (CF) are set to 0. Examples of the OR function are shown in Table 8.4(a) and Table 8.4(b). Table 8.4(b) sets all the bits to a value of 1 regardless of their original value.

Table 8.3 Truth Table for the OR Function

OR		
x_1	x_2	z_1
0	0	0
0	1	1
1	0	1
1	1	1

Table 8.4 Examples of the OR Function for Eight Bits

(a)								(b)							
0	1	1	0	0	1	0	1	0	0	1	1	0	0	1	0
0	0	1	0	1	0	0	1	1	1	1	1	1	1	1	1
0	1	1	0	1	1	0	1	1	1	1	1	1	1	1	1

Figure 8.3 contains an assembly language module embedded in a C program that illustrates the logical OR function of two operands. The program displays the result of the OR operation together with the resulting flags.

```
//logical_or.cpp
//illustrates the logical OR function

#include "stdafx.h"

int main (void)
{
//define variables
   unsigned char or1, or2, rslt, flags;
```

(a) //continued on next page

Figure 8.3 Program to illustrate the logical OR operation and resulting flags: (a) the program and (b) the outputs.

```
    printf ("Enter two 2-digit hexadecimal characters: \n");
    scanf ("%X %X", &or1, &or2);

//switch to assembly
        _asm
        {
            MOV    AL, or1
            MOV    BL, or2
            OR     BL, AL
            LAHF
            MOV    flags, AH
            MOV    rslt, BL
        }

    printf ("\nOpnd1 = %X\nOpnd2 = %X\n", or1, or2);
    printf ("\nOR result = %X\n\n\nFlags = %X\n\n",
                        rslt, flags);
    return 0;
}
```

```
Enter two 2-digit hexadecimal characters:
F0 0F

Opnd1 = F0         //1  1  1  1  0  0  0  0
Opnd2 = F          //0  0  0  0  1  1  1  1

OR result = FF     //1  1  1  1  1  1  1  1

                   //SF ZF 0  AF 0  PF 1  CF
Flags = 86         //1  0  0  0  0  1  1  0

Press any key to continue . . . _
------------------------------------------------------------
Enter two 2-digit hexadecimal characters:
69 37

Opnd1 = 69         //0  1  1  0  1  0  0  1
Opnd2 = 37         //0  0  1  1  0  1  1  1

OR result = 7F     //0  1  1  1  1  1  1  1

                   //SF ZF 0  AF 0  PF 1  CF
Flags = 2          //0  0  0  0  0  0  1  0

Press any key to continue . . . _    (b) //continued on next page
```

Figure 8.3 (Continued)

```
Enter two 2-digit hexadecimal characters:
BD AE

Opnd1 = BD          //1 0 1 1 1 1 0 1
Opnd2 = AE          //1 0 1 0 1 1 1 0

OR result = BF      //1 0 1 1 1 1 1 1

                    //SF ZF 0  AF 0  PF 1  CF
Flags = 82          //1  0  0  0  0  0  1  0

Press any key to continue . . . _
-----------------------------------------------------------
Enter two 2-digit hexadecimal characters:
4D EC

Opnd1 = 4D          //0 1 0 0 1 1 0 1
Opnd2 = EC          //1 1 1 0 1 1 0 0

OR result = ED      //1 1 1 0 1 1 0 1

                    //SF ZF 0  AF 0  PF 1  CF
Flags = 86          //1  0  0  0  0  1  1  0

Press any key to continue . . . _
```

Figure 8.3 (Continued)

8.3 Logical Exclusive-OR Instruction

The XOR instruction performs the bitwise exclusive-OR operation of two operands — the first operand (destination) and the second operand (source) and stores the result in the destination operand. The source operand can be located in a general-purpose register, in memory, or can be an immediate operand. The destination operand can be in a general-purpose register or a memory location; however, both operands cannot be in memory. The syntax for the logical XOR instruction is shown below.

```
XOR    register/memory, register/memory/immediate
```

The truth table for the exclusive-OR function is shown in Table 8.5. The variables x_1, x_2, and z_1 represent a bit in the destination operand, a bit in the source operand, and the resulting bit in the destination operand, respectively. The XOR instruction can be used to invert select bits; that is, to change a bit with a value of 1 to a value of 0 and vice versa, because any bit that is exclusive-ORed with a 1 bit will invert the value of the bit. The affected flags are the parity flag (PF), the sign flag (SF), and the zero flag (ZF) — the overflow flag (OF) and the carry flag (CF) are set to 0. Examples of the exclusive-OR function are shown in Table 8.6(a) and Table 8.6(b). Table 8.6(b) inverts all bits regardless of their original value.

Table 8.5 Truth Table for the Exclusive-OR Function

XOR		
x_1	x_2	z_1
0	0	0
0	1	1
1	0	1
1	1	0

Table 8.6 Examples of the XOR Function for Eight Bits

XOR															
1	1	0	1	0	0	1	0	1	0	0	1	0	1	1	0
1	0	1	1	1	0	1	1	1	1	1	1	1	1	1	1
0	1	1	0	1	0	0	1	0	1	1	0	1	0	0	1
(a)								(b)							

There is no exclusive-NOR instruction; however, this function can be achieved by using the XOR operation in conjunction with the NOT operation (Section 8.4) to invert the result. Figure 8.4 contains an assembly language module embedded in a C program that illustrates the logical XOR function of two operands. The program displays the result of the XOR operation together with the resulting flags.

```
//logical_xor.cpp
//illustrates the logical xor function

#include "stdafx.h"
int main (void)
{
//define variables
   unsigned char xor1, xor2, rslt, flags;
                                              //continued on next page
```

(a)

Figure 8.4 Program to illustrate the logical XOR operation and resulting flags: (a) the program and (b) the outputs.

```
   printf ("Enter two 2-digit hexadecimal characters: \n");
   scanf ("%X %X", &xor1, &xor2);

//switch to assembly
      _asm
      {
         MOV   AL, xor1
         MOV   BL, xor2
         XOR   BL, AL
         LAHF
         MOV   flags, AH
         MOV   rslt, BL
      }
   printf ("\nOpnd1 = %X\nOpnd2 = %X\n", xor1, xor2);
   printf ("\nXOR result = %X\n\n\nFlags = %X\n\n",
               rslt, flags);
   return 0;
}
```

```
Enter two 2-digit hexadecimal characters:
5A 99

Opnd1 = 5A         //0 1 0 1 1 0 1 0
Opnd2 = 99         //1 0 0 1 1 0 0 1

XOR result = C3    //1 1 0 0 0 0 1 1

                   //SF ZF 0  AF 0  PF 1  CF
Flags = 86         //1  0  0  0  0  1  1  0

Press any key to continue . . . _
------------------------------------------------------------
Enter two 2-digit hexadecimal characters:
F0 0F

Opnd1 = F0         //1 1 1 1 0 0 0 0
Opnd2 = F          //0 0 0 0 1 1 1 1

XOR result = FF    //1 1 1 1 1 1 1 1

                   //SF ZF 0  AF 0  PF 1  CF
Flags = 86         //1  0  0  0  0  1  1  0

Press any key to continue . . . _
```

(b) //continued on next page

Figure 8.4 (Continued)

```
Enter two 2-digit hexadecimal characters:
AB CD

Opnd1 = AB        //1 0 1 0 1 0 1 1
Opnd2 = CD        //1 1 0 0 1 1 0 1

XOR result = 66   //0 1 1 0 0 1 1 0

                  //SF ZF 0  AF 0  PF 1  CF
Flags = 6         //0  0  0  0  0  1  1  0

Press any key to continue . . . _
------------------------------------------------------------
Enter two 2-digit hexadecimal characters:
7E 43

Opnd1 = 7E        //0 1 1 1 1 1 1 0
Opnd2 = 43        //0 1 0 0 0 0 1 1

XOR result = 3D   //0 0 1 1 1 1 0 1

                  //SF ZF 0  AF 0  PF 1  CF
Flags = 2         //0  0  0  0  0  0  1  0

Press any key to continue . . . _
------------------------------------------------------------
Enter two 2-digit hexadecimal characters:
D7 D7

Opnd1 = D7        //1 1 0 1 0 1 1 1
Opnd2 = D7        //1 1 0 1 0 1 1 1

XOR result = 0    //0 0 0 0 0 0 0 0

                  //SF ZF 0  AF 0  PF 1  CF
Flags = 46        //0  1  0  0  0  1  1  0

Press any key to continue . . . _
```

Figure 8.4 (Continued)

8.4 Logical NOT Instruction — 1s Complement

The NOT instruction performs a bitwise 1s complement operation on the destination operand and stores the result in the destination operand. The destination can be a general-purpose register or a memory location. The NOT instruction inverts each bit in the destination operand and is also referred to as the *diminished-radix complement* (or the *r – 1 complement*). The syntax for the NOT instruction is shown below.

```
NOT    register/memory
```

Operands that are represented in the 1s complement notation are signed numbers with the following range:

$$-(2^{n-1} - 1) \text{ to } +(2^{n-1} - 1)$$

The truth table for the NOT function is shown in Table 8.7. The variables x_1 and z_1 represent bits in the destination operand and the resulting bit in the destination operand, respectively. No flags are affected by the NOT instruction.

Table 8.7 Truth Table of the NOT Function

NOT	
x_1	z_1
0	1
1	0

The binary number 1101_2 will be 1s complemented. The number has a decimal value of –2. To obtain the 1s complement, subtract each digit in turn from 1 (the highest number in the radix), as shown below — or for radix 2, simply invert each bit. Therefore, the 1s complement of 1101_2 is 0010_2, which has a decimal value of +2.

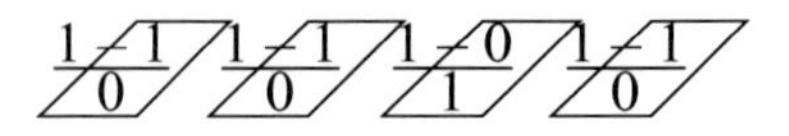

To verify the operation, add the negative and positive numbers to obtain 1111_2, which is zero in 1s complement notation; 0000_2 is also zero in 1s complement.

$$\begin{array}{rcccc} & 1 & 1 & 0 & 1 \\ +) & 0 & 0 & 1 & 0 \\ \hline & 1 & 1 & 1 & 1 \end{array}$$

8.5 NEG Instruction — 2s Complement

The NEG instruction performs a 2s complement operation on the destination operand and stores the result in the destination operand. The destination can be a general-purpose register or a memory location. The NEG instruction subtracts the destination operand from zero, effectively changing the sign of the number while maintaining the same absolute value.

The value of a positive number in 2s complement representation is obtained in the usual manner by adding the values of the 1 bits by the weight of their respective positions. The value of a negative number in 2s complement representation is obtained by adding the values of the 0 bits by weight in their respective positions and then adding a value of one to the result.

Another method of obtaining the 2s complement of a number is to form the 1s complement of the number, then add one to the result. Both methods are shown Figure 8.5 using 4-bit operands. Figure 8.5(a) subtracts the operand from zero; Figure 8.5(b) adds one to the 1s complement.

A third method to generate the radix complement for a radix 2 number is to keep the low-order 0s and the first 1 unchanged, then complement (invert) the remaining high-order bits.

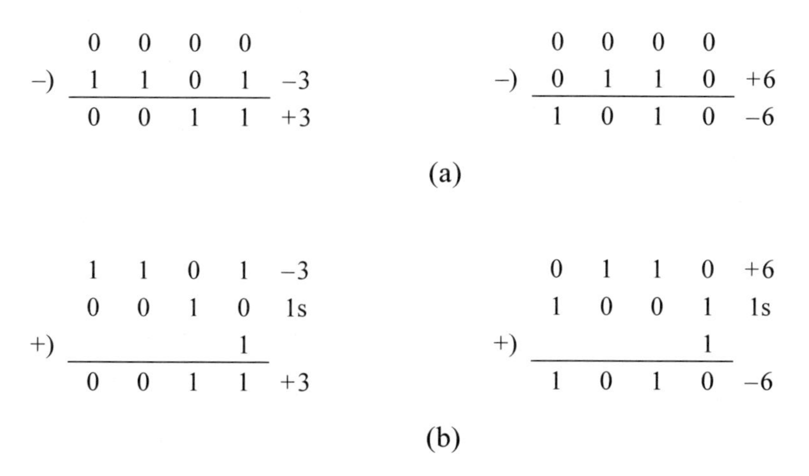

Figure 8.5 Methods to obtain the 2s complement (negation) of a number: (a) subtract from zero and (b) add 1 to the 1s complement.

The NEG instruction is also referred to as the *radix complement* (or the *r complement*). The syntax for the NEG instruction is shown below.

```
NEG    register/memory
```

A integer has the following range in 2s complement representation, where the number zero is considered to be positive:

$$-(2^{n-1}) \text{ to } +(2^{n-1}-1)$$

A positive number A is represented as

$$A = (0\ a_{n-2}a_{n-3}\ \cdots\ a_1a_0)_2$$

and a negative number is represented as

$$(A')_{+1} = \{[(2-1)\ a_{n-2}'a_{n-3}'\ \cdots\ a_1'a_0'] + 1\}_2$$

where A' is the diminished-radix complement. Thus, the radix complement is obtained by adding 1 to the diminished-radix complement; that is, $(r-1)+1=r$.

The flags affected by the NEG instruction are the overflow flag (OF), the sign flag (SF), the zero flag (ZF), the auxiliary carry flag (AF), the parity flag (PF), and the carry flag (CF) — if the operand is zero, then CF = 0. Figure 8.6 shows an assembly language module embedded in a C program that illustrates both the NOT instruction and the NEG instruction and the associated flags for the NEG instruction.

```
//logical_not_neg.cpp
//illustrates the logical not and neg functions
#include "stdafx.h"
int main (void)
{
//define variables
   unsigned char not1, neg1, rslt_not, rslt_neg, flags_neg;

   printf ("Enter two 2-digit hexadecimal characters: \n");
   scanf ("%X %X", &not1, &neg1);

//switch to assembly
      _asm
      {
         MOV   AL, not1
         NOT   AL              ;form 1s complement of opnd1
         MOV   rslt_not, AL
                          (a)     //continued on next page
```

Figure 8.6 Program to illustrate the logical NOT and NEG operations and resulting flags for the NEG function: (a) the program and (b) the outputs.

```
        MOV   BL, neg1
        NEG   BL              ;form 2s complement of opnd2
        LAHF                  ;load AH from flags register
        MOV   flags_neg, AH
        MOV   rslt_neg, BL
    }

    printf ("\nNOT opnd = %X\nResult NOT = %X\n",
            not1, rslt_not);
    printf ("\nNEG opnd = %X\nResult NEG = %X\n\n\n
            NEG flags = %X\n\n\n",
            neg1, rslt_neg, flags_neg);

    return 0;
}
```

```
Enter two 2-digit hexadecimal characters:
1F 3A

NOT opnd = 1F         //0 0 0 1 1 1 1 1
Result NOT = E0       //1 1 1 0 0 0 0 0

NEG opnd = 3A         //0 0 1 1 1 0 1 0
Result NEG = C6       //1 1 0 0 0 1 1 0

                      //SF ZF 0  AF 0  PF 1  CF
NEG flags = 97        //1  0  0  1  0  1  1  1

Press any key to continue . . . _
---------------------------------------------------------
Enter two 2-digit hexadecimal characters:
E6 DB

NOT opnd = E6         //1 1 1 0 0 1 1 0
Result NOT = 19       //0 0 0 1 1 0 0 1

NEG opnd = DB         //1 1 0 1 1 0 1 1
Result NEG = 25       //0 0 1 0 0 1 0 1

                      //SF ZF 0  AF 0  PF 1  CF
NEG flags = 13        //0  0  0  1  0  0  1  1

Press any key to continue . . . _
---------------------------------------------------------
```

(b) //continued on next page

Figure 8.6 (Continued)

```
Enter two 2-digit hexadecimal characters:
00 FF

NOT opnd = 0            //0 0 0 0 0 0 0 0
Result NOT = FF         //1 1 1 1 1 1 1 1

NEG opnd = FF           //1 1 1 1 1 1 1 1
Result NEG = 1          //0 0 0 0 0 0 0 1

                        //SF ZF 0  AF 0  PF 1  CF
NEG flags = 13          //0  0  0  1  0  0  1  1

Press any key to continue . . . _
-------------------------------------------------------------
Enter two 2-digit hexadecimal characters:
FF 00

NOT opnd = FF           //1 1 1 1 1 1 1 1
Result NOT = 0          //0 0 0 0 0 0 0 0

NEG opnd = 0            //0 0 0 0 0 0 0 0
Result NEG = 0          //0 0 0 0 0 0 0 0

                        //SF ZF 0  AF 0  PF 1  CF
NEG flags = 46          //0  1  0  0  0  1  1  0

Press any key to continue . . . _
```

Figure 8.6 (Continued)

To obtain the NOT function of an operand, the bits are simply inverted. The NEG results for the operands shown in Figure 8.6(b) are determined by the calculations shown in Figure 8.7, which subtracts the operands from a value of zero. It was previously stated that the NEG instruction subtracts the destination operand from zero, effectively changing the sign of the number while maintaining the same absolute value. The rules for subtraction in radix 2 are as follows:

0 – 0 =	0
0 – 1 =	1 with a borrow from the next higher-order minuend
1 – 0 =	1
1 – 1 =	0

	0	0	0	0		0	0	0	0		
–)	0	0	1	1		1	0	1	0	3AH	+58
	1	1	0	0		0	1	1	0	C6H	–58
1					1						
CF					AF						

SF	ZF	0	AF	0	PF	1	CF
1	0	0	1	0	1	1	1

	0	0	0	0		0	0	0	0		
–)	1	1	0	1		1	0	1	1	DBH	–37
	0	0	1	0		0	1	0	1	25H	+37
1					1						
CF					AF						

SF	ZF	0	AF	0	PF	1	CF
0	0	0	1	0	0	1	1

	0	0	0	0		0	0	0	0		
–)	1	1	1	1		1	1	1	1	FFH	–1
	0	0	0	0		0	0	0	1	01H	+1
1					1						
CF					AF						

SF	ZF	0	AF	0	PF	1	CF
0	0	0	1	0	0	1	1

	0	0	0	0		0	0	0	0		
–)	0	0	0	0		0	0	0	0	00H	0
	0	0	0	0		0	0	0	0	00H	0
0					0						
CF					AF						

SF	ZF	0	AF	0	PF	1	CF
0	1	0	0	0	1	1	0

Figure 8.7 Calculations for the negation operations of the operands shown in Figure 8.6(b).

8.6 TEST and Set Byte on Condition Instructions

The TEST instruction and the *set byte on condition* (SETcc) instruction can be used to alter the program flow depending on the state of the EFLAGS register. The TEST instruction can be followed by a *jump on condition* (Jcc) instruction.

8.6.1 TEST Instruction

The TEST instruction performs the logical AND operation of two operands on a bit-by-bit basis. The result of the operation affects the sign flag (SF), the zero flag (ZF), and the parity flag (PF) in a manner similar to an AND instruction; however, the destination and source operands are not changed. The result of the AND operation is discarded after execution of the TEST instruction. The syntax for the TEST instruction is shown below.

```
TEST   register/memory, register/immediate
```

If any identical bit positions in the two operands both contains 1s, then the zero flag is reset. The state of the EFLAGS register can then be tested by using the *jump on condition* (Jcc) instruction, the *loop on condition* (LOOPcc), or the *set byte on condition* (SETcc) instruction. For example, the low-order four bits in register AX can be tested to determine if any bits are nonzero, as shown below. The jump will occur if the zero flag is reset (ZF = 0).

```
TEST  AX, 0000000000001111
JNZ   NON_ZERO
```

8.6.2 Set Byte on Condition (SETcc) Instruction

The *set byte on condition* (SETcc) instruction sets the destination byte operand to a value of 0 or 1 depending on the state of certain flags in the EFLAGS register. The syntax for the SETcc instruction is shown below.

```
SETcc   register/memory
```

The affected flags are similar to those of the *jump on condition* (Jcc) instruction described in Chapter 6. An example to determine if register EAX is greater than register EBX is shown below. If register EAX is greater than register EBX, then register CH is set to a value of 00000001.

```
CMP   EAX, EBX
SETG  CH
```

8.7 Bit Test Instructions

The bit test instructions operate on a single bit and are used to scan the bits in an operand and then perform an operation on the selected bit. The operand that contains the bit to be tested is specified by the destination operand (first operand — or bit base), which can be in a general-purpose register of a memory location. The location of the bit to be tested is stipulated by the source operand (bit offset). The location of the bit in the bit string is specified as an offset from bit 0 of the string. The selected bit is then stored in the carry flag (CF). No other flags are affected.

If the destination operand indicates a register, then the instruction assumes modulo-16, modulo-32, or modulo-64 of the source operand, depending on the size of the operand and the mode. Operands that are 64 bits can be used only in 64-bit mode. If the destination operand is a memory location, then this specifies the address of a byte in memory that contains the operand on which the bit test instruction is to be executed. The syntax for the bit test instructions is shown below.

```
BT/S/R/C    register/memory, register/immediate
```

There are four bit test instructions that will be discussed in this section: *bit test* (BT), *bit test and set* (BTS), *bit test and reset* (BTR), and *bit test and complement* (BTC). Figure 8.8 shows a general diagram that illustrates the four bit test instructions.

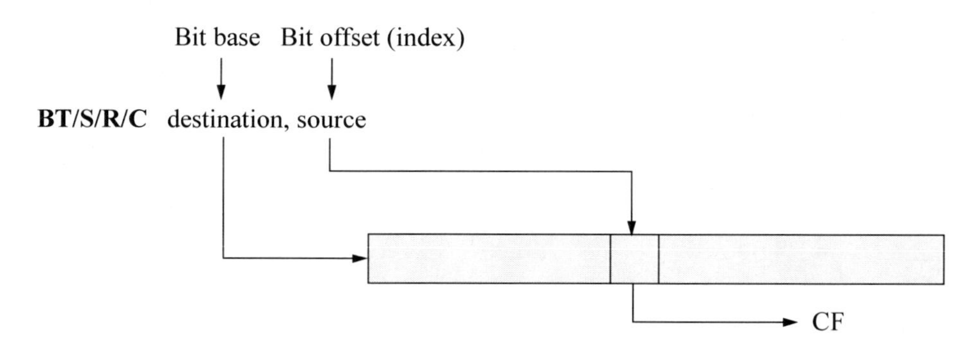

Figure 8.8 Illustration of the bit test instructions.

8.7.1 Bit Test (BT) Instruction

Figure 8.9 shows a drawing for the *bit test* (BT) instruction using a general-purpose register and an immediate operand. The base offset is modulo-32, because register EAX is 32 bits. If the immediate value is greater than 31, then it is modulo-32. For example, if the offset were 198, then bit 6 would be selected.

The BT instruction copies the value of the selected bit within a bit string to the CF flag; the sequence of bits in a general-purpose register or a memory location are numbered from the low-order bit to the high-order bit. If the destination operand is in memory, then the processor can access two bytes beginning with a 16-bit operand address or four bytes beginning with a 32-bit operand address.

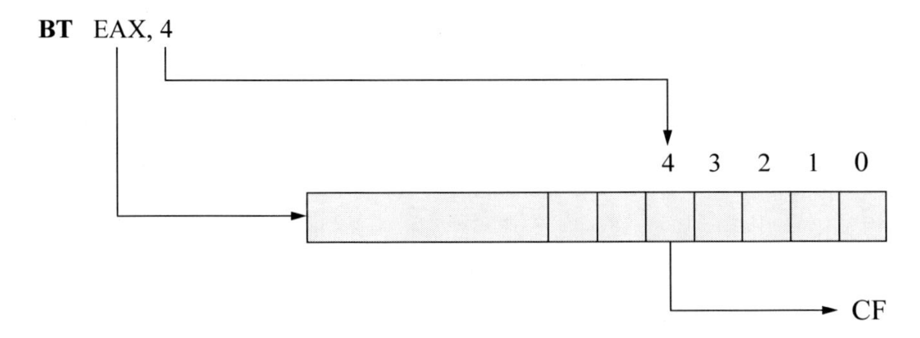

Figure 8.9 Bit test instruction for a general-purpose register with an immediate operand.

8.7.2 Bit Test and Set (BTS) Instruction

The *bit test and set* (BTS) instruction stores the selected bit in the carry flag (CF), then sets the selected bit in the bit string (destination) to a value of 1. The destination operand (bit base) can be a general-purpose register or a memory location, both of which can be 16 bits, 32 bits, or 64 bits. The source operand (bit offset — or index) can be a register or an immediate operand. If the index is an immediate operand, then the selected bit can range from 0 to 31 or greater, depending on the mode. The general comments and figures for the BT instruction also apply to the BTS instruction.

8.7.3 Bit Test and Reset (BTR) Instruction

The *bit test and reset* (BTR) instruction selects a single bit in a bit string specified by the destination operand (bit base) and stores the selected bit in the carry flag (CF), then resets the selected bit in the bit string (destination) to a value of 0. The selected bit is determined by the source operand (bit offset — or index). The destination operand can be a general-purpose register or a memory location, both of which can be 16 bits, 32 bits, or 64 bits. The source operand can be a register or an immediate operand. If the index is an immediate operand, then the selected bit can range from 0 to 31 or greater, depending on the mode. The general comments and figures for the BT instruction also apply to the BTR instruction.

8.7.4 Bit Test and Complement (BTC) Instruction

The *bit test and complement* (BTC) instruction selects a single bit in a bit string specified by the destination operand (bit base) and stores the selected bit in the carry flag (CF), then complements (inverts) the selected bit in the bit string (destination). The selected bit is determined by the source operand (bit offset — or index). The destination operand can be a general-purpose register or a memory location, both of which can be 16 bits, 32 bits, or 64 bits. The source operand can be a register or an immediate operand. If the index is an immediate operand, then the selected bit can range from 0 to 31 or greater, depending on the mode. The general comments and figures for the BT instruction also apply to the BTC instruction.

Figure 8.10 shows an assembly language module embedded in a C program that illustrates the four bit test instructions using immediate data as the bit offset. An 8-character hexadecimal number is entered from the keyboard and stored in the variable *bt_opnd* to be used as the bit test operand. The results of the bit test operations are displayed as: *bt_opnd* for the BT instruction, *bts_rslt_ebx* for the BTS instruction, *btr_rslt_ebx* for the BTR instruction, *btc_rslt_ebx* for the BTC instruction, and *btc_rslt_mod32* for the BTC instruction in which the bit index is greater than 31.

The flags are also displayed for each bit test instruction as: *bt_flags* for the BT instruction, *bts_flags* for the BTS instruction, *btr_flags* for the BTR instruction, *btc_flags* for the BTC instruction, and *btc_mod32_flags* for the BTC instruction in which the bit index is greater than 31. Since the carry flag (CF) is the only flag affected by the bit test instructions, the state of the other flags is irrelevant.

The LAHF instruction is executed for each bit test instruction; register AH is then stored in the appropriate flags variable. The second BTC instruction has a bit offset (index) value of 198. This establishes the selected bit as bit 6, because 198 modulo-32 is 6.

In the first pass, the hexadecimal number of FFFFFF4B (shown below) is entered from the keyboard and stored in the *bt_opnd* variable. Then the BT instruction moves bit 4 (0) to the carry flag and leaves the original bit 4 unchanged.

31	...	23	22	21	20	...	17	16	...	7	6	5	4	3	2	1	0
1		1	1	1	1	1	1	1	1	0	1	0	0	1	0	1	1

The BTS instruction operates on bit 5 (0) of the *bt_opnd* variable, moves bit 5 to the carry flag, then sets the original bit 5, giving the four bits 4 through 7 a value of 6_{16}. The BTR instruction operates on bit 16 (1) of the *bt_opnd* variable, moves bit 16 to the carry flag, then resets the original bit 16, giving the four bits 16 through 19 a value of E_{16} (14_{10}). The first BTC instruction operates on bit 20 (1) of the *bt_opnd* variable, moves bit 20 to the carry flag, then complements bit 20 to yield a value of 0, giving the four bits 20 through 23 a value of E_{16} (14_{10}). The second BTC instruction operates on bit 6 (198 modulo-32) of the *bt_opnd* variable, moves bit 6 to the carry flag, then complements bit 6 to yield a value of 0, giving the four bits 4 through 7 a value of 0. A similar sequence occurs for the two remaining user-entered hexadecimal numbers: 11111111 and 77777777.

```
//bit_test_instr.cpp
//illustrate the operation of the bit test
//instructions BT, BTS, BTR, BTC using
//immediate data for the bit offset (index)

#include "stdafx.h"

int main (void)
{
//define variables
   int bt_opnd, bts_rslt_ebx, btr_rslt_ebx, btc_rslt_ebx,
         btc_rslt_mod32;
   unsigned char bt_flags, bts_flags, btr_flags, btc_flags,
         btc_mod32_flags;

   printf ("Enter an 8-character hexadecimal number: \n");
   scanf ("%X", &bt_opnd);

//switch to assembly
      _asm
      {
//BT instruction
         MOV   EBX, bt_opnd
         BT    EBX, 4
         LAHF
         MOV   bt_flags, AH

//BTS instruction
         BTS   EBX, 5
         LAHF
         MOV   bts_flags, AH
         MOV   bts_rslt_ebx, EBX

//BTR instruction
         MOV   EBX, bt_opnd
         BTR   EBX, 16
         LAHF
         MOV   btr_flags, AH
         MOV   btr_rslt_ebx, EBX
                                 //continued on next page
```

(a)

Figure 8.10 Program to illustrate the bit test instructions: BT, BTS, BTR, and BTC using immediate data as the bit offset: (a) the program and (b) the outputs.

```
//BTC instruction
        MOV    EBX, bt_opnd
        BTC    EBX, 20
        LAHF
        MOV    btc_flags, AH
        MOV    btc_rslt_ebx, EBX

        MOV    EBX, bt_opnd
        BTC    EBX, 198       ;198 modulo-32 = bit 6
        LAHF
        MOV    btc_mod32_flags, AH
        MOV    btc_rslt_mod32, EBX
}


   printf ("\nbt_opnd = %X bit 4\nbt_flags = %X\n\n",
            bt_opnd, bt_flags);


   printf ("bts_opnd = %X bit 5\nbts_flags = %X\n
            rslt_ebx = %X\n\n", bt_opnd, bts_flags,
            bts_rslt_ebx);


   printf ("btr_opnd = %X bit 16\nbtr_flags = %X\n
         rslt_ebx = %X\n\n", bt_opnd, btr_flags,
            btr_rslt_ebx);


   printf ("btc_opnd = %X bit 20\nbtc_flags = %X\n
         rslt_ebx = %X\n\n", bt_opnd, btc_flags,
         btc_rslt_ebx);


   printf ("btc_opnd = %X bit 198(6)\n
         btc_mod32_flags = %X\nrslt_ebx = %X\n\n",
         bt_opnd, btc_mod32_flags, btc_rslt_mod32);

   return 0;
}

                                    //continued on next page
```

Figure 8.10 (Continued)

```
Enter an 8-character hexadecimal number:
FFFFFF4B

bt_opnd = FFFFFF4B bit 4
bt_flags = 46

bts_opnd = FFFFFF4B bit 5
bts_flags = 46
rslt_ebx = FFFFFF6B

btr_opnd = FFFFFF4B bit 16
btr_flags = 47
rslt_ebx = FFFEFF4B

btc_opnd = FFFFFF4B bit 20
btc_flags = 47
rslt_ebx = FFEFFF4B

btc_opnd = FFFFFF4B bit 198(6)
btc_mod32_flags = 47
rslt_ebx = FFFFFF0B

Press any key to continue . . . _
-------------------------------------------------------------
Enter an 8-character hexadecimal number:
11111111

bt_opnd = 11111111 bit 4
bt_flags = 47

bts_opnd = 11111111 bit 5
bts_flags = 46
rslt_ebx = 11111131

btr_opnd = 11111111 bit 16
btr_flags = 47
rslt_ebx = 11101111

btc_opnd = 11111111 bit 20
btc_flags = 47
rslt_ebx = 11011111

btc_opnd = 11111111 bit 198(6)
btc_mod32_flags = 46
rslt_ebx = 11111151

Press any key to continue . . . _ (b) //continued on next page
```

Figure 8.10 (Continued)

```
Enter an 8-character hexadecimal number:
77777777

bt_opnd = 77777777 bit 4
bt_flags = 47

bts_opnd = 77777777 bit 5
bts_flags = 47
rslt_ebx = 77777777

btr_opnd = 77777777 bit 16
btr_flags = 47
rslt_ebx = 77767777

btc_opnd = 77777777 bit 20
btc_flags = 47
rslt_ebx = 77677777

btc_opnd = 77777777 bit 198(6)
btc_mod32_flags = 47
rslt_ebx = 77777737

Press any key to continue . . . _
```

Figure 8.10 (Continued)

Figure 8.11 shows an assembly language module embedded in a C program that illustrates the four bit test instructions using data in register EDX as the bit offset. An 8-character hexadecimal number is entered from the keyboard and stored in the variable *bt_opnd* to be used as the bit test operand. Then a hexadecimal number is entered as the bit offset and stored in the variable *offset_reg*, which is later stored in register EDX.

This program is similar to the program shown in Figure 8.10, except that the bit offset is in a register rather than specified as immediate data. The hexadecimal number that is entered as the bit offset applies to *all* of the bit test instructions. The second set of characters shown in Figure 8.11(b) is hexadecimal FFFFFF4B for the bit test operand and hexadecimal 10 (16_{10}) for the bit offset. Bit offsets greater than 31 are not shown in this example.

The results of the four bit test operations are displayed together with the corresponding flags in a manner identical to that of Figure 8.10. Unlike the program of Figure 8.10 however, the *bit test and set* (BTS) instruction does not reinitialize the bit test operand in register EBX, because the contents were not changed by the *bit test* (BT) instruction.

```
//bit_test_instr_reg.cpp
//illustrates the operation of the bit test
//instructions BT, BTS, BTR, BTC using a
//register as the bit offset (index).
//the first hex number entered is the operand;
//the second hex number entered is the offset in a reg
#include "stdafx.h"
int main (void)
{
//define variables
   int bt_opnd, offset_reg, bts_rslt_ebx, btr_rslt_ebx,
         btc_rslt_ebx;
   unsigned char bt_flags, bts_flags, btr_flags, btc_flags;

   printf ("Enter two hexadecimal numbers: \n");
   scanf ("%X %X", &bt_opnd, &offset_reg);

//switch to assembly
      _asm
      {
//BT instruction
         MOV   EBX, bt_opnd
         MOV   EDX, offset_reg
         BT    EBX, EDX
         LAHF
         MOV   bt_flags, AH

//BTS instruction
         BTS   EBX, EDX
         LAHF
         MOV   bts_flags, AH
         MOV   bts_rslt_ebx, EBX

//BTR instruction
         MOV   EBX, bt_opnd
         MOV   EDX, offset_reg
         BTR   EBX, EDX
         LAHF
         MOV   btr_flags, AH
         MOV   btr_rslt_ebx, EBX
```

(a) //continued on next page

Figure 8.11 Program to illustrate the bit test instructions: BT, BTS, BTR, and BTC using the contents of register EDX as the bit offset: (a) the program and (b) the outputs.

```
//BTC instruction
         MOV    EBX, bt_opnd
         MOV    EDX, offset_reg
         BTC    EBX, EDX
         LAHF
         MOV    btc_flags, AH
         MOV    btc_rslt_ebx, EBX
      }
   printf ("\nbt_opnd = %X\nbt_flags = %X\n\n",
            bt_opnd, bt_flags);

   printf ("bts_opnd = %X\nbts_flags = %X\n
            rslt_ebx = %X\n\n", bt_opnd, bts_flags,
            bts_rslt_ebx);

   printf ("btr_opnd = %X\nbtr_flags = %X\n
            rslt_ebx = %X\n\n", bt_opnd, btr_flags,
            btr_rslt_ebx);

   printf ("btc_opnd = %X\nbtc_flags = %X\n
            rslt_ebx = %X\n\n", bt_opnd, btc_flags,
            btc_rslt_ebx);

   return 0;
}
```

```
Enter two hexadecimal numbers:
FFFFFF4B 4

bt_opnd = FFFFFF4B
bt_flags = 46

bts_opnd = FFFFFF4B
bts_flags = 46
rslt_ebx = FFFFFF5B

btr_opnd = FFFFFF4B
btr_flags = 46
rslt_ebx = FFFFFF4B

btc_opnd = FFFFFF4B
btc_flags = 46
rslt_ebx = FFFFFF5B

Press any key to continue . . . _ (b)  //continued on next page
```

Figure 8.11 (Continued)

```
Enter two hexadecimal numbers:
FFFFFF4B 10

bt_opnd = FFFFFF4B
bt_flags = 47

bts_opnd = FFFFFF4B
bts_flags = 47
rslt_ebx = FFFFFF4B

btr_opnd = FFFFFF4B
btr_flags = 47
rslt_ebx = FFFEFF4B

btc_opnd = FFFFFF4B
btc_flags = 47
rslt_ebx = FFFEFF4B

Press any key to continue . . . _
```

Figure 8.11 (Continued)

8.8 Bit Scan Instructions

There are two bit scan instructions: *bit scan forward* (BSF) and *bit scan reverse* (BSR). These instructions scan the contents of a register or memory location to determine the location of the first 1 bit in the operand. If the scanned operand contains all zeroes, the zero flag (ZF) is set to 1; if the scanned operand contains at least one 1 bit, then the ZF flag is reset — the other flags are undefined. The syntax for the bit scan instructions is shown below.

```
BSF/R    register, register/memory
```

8.8.1 Bit Scan Forward (BSF) Instruction

The BSF instruction scans the source operand in a register or memory location to determine the location of the first 1 bit with reference to bit 0 of the operand. The operation scans the source operand from the low-order bit to the high-order bit. Scanning stops when the first low-order 1 bit is encountered. If a 1 bit is found, then its location is specified as the bit index (offset) and is stored in the destination register. The bit index is an unsigned offset referenced from bit 0. If the source operand is zero, then the ZF flag is set to 1 and the contents of the destination register are undefined. Figure 8.12 illustrates a BSF operation using register EAX as the destination register.

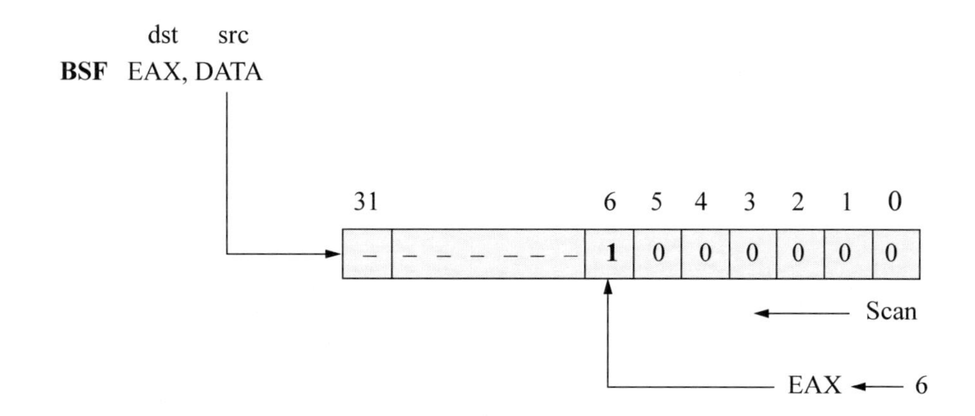

Figure 8.12 Illustration of a bit scan forward (BSF) operation.

Figure 8.13 shows an assembly language module embedded in a C program that illustrates the operation of the BSF instruction. An 8-digit hexadecimal number is entered from the keyboard and moved to register EBX. Since the MOV instruction does not affect the flags, the operand in register EBX is first compared to zero to determine if there are any 1s in the operand. Then a conditional jump if zero (JZ) instruction is executed — the jump will occur if the ZF flag is set, indicating no 1s in the operand. If the jump does not occur, then register EBX is scanned from bit 0 to bit 31 to locate the position of the first 1 bit. The operand and the location of the first low-order 1 bit are then displayed.

```
//bit_scan_fwd.cpp
//illustrates the operation of the BSF instruction

#include "stdafx.h"

int main (void)
{
//define variables
   int bsf_src_opnd, bsf_rslt;
```

(a) //continued on next page

Figure 8.13 Illustrates the operation of the BSF instruction: (a) the program and (b) the outputs.

```
   printf ("Enter an 8-digit hexadecimal number: \n");
   scanf ("%X", &bsf_src_opnd);

//switch to assembly
      _asm
      {
         MOV   EBX, bsf_src_opnd
//the MOV instr does not affect the flags;
//therefore, the state of the flags is unknown
         CMP   EBX, 0
         JZ    NO_ONES_BSF
         BSF   EAX, EBX
         MOV   bsf_rslt, EAX
      }

   printf ("\nBSF bit is %d\n\n", bsf_rslt);
   goto  end;

NO_ONES_BSF:
   printf ("\nThe BSF operand had no 1s \n\n");

end:
   return 0;
}
```

```
Enter an 8-digit hexadecimal number:
00A07000

BSF bit is 12

Press any key to continue . . . _
------------------------------------------------------------
Enter an 8-digit hexadecimal number:
00000001

BSF bit is 0

Press any key to continue . . . _
------------------------------------------------------------
                                     //continued on next page
```

(b)

Figure 8.13 (Continued)

```
Enter an 8-digit hexadecimal number:
80000000

BSF bit is 31

Press any key to continue . . . _
------------------------------------------------------------
Enter an 8-digit hexadecimal number:
0DC06000

BSF bit is 13

Press any key to continue . . . _
------------------------------------------------------------
Enter an 8-digit hexadecimal number:
FF800000

BSF bit is 23

Press any key to continue . . . _
------------------------------------------------------------
Enter an 8-digit hexadecimal number:
00000000

The BSF operand has no 1s

Press any key to continue . . . _
```

Figure 8.13 (Continued)

8.8.2 Bit Scan Reverse (BSR) Instruction

The BSR instruction scans the source operand in a register or memory location to determine the location of the first 1 bit with reference to bit 0 of the operand. The operation scans the source operand from the high-order bit to the low-order bit. Scanning stops when the first high-order 1 bit is encountered. If a 1 bit is found, then its location is specified as the bit index (offset) and is stored in the destination register. The bit index is an unsigned offset referenced from bit 0. If the source operand is zero, then the ZF flag is set to 1 and the contents of the destination register are undefined. Figure 8.12 also applies to the BSR instruction, except that scanning is done from left to right — from bit 31 to bit 0.

Figure 8.14 shows an assembly language module embedded in a C program that illustrates the operation of the BSR instruction. An 8-digit hexadecimal number is entered from the keyboard and is scanned from bit 31 to bit 0 to locate the position of the first 1 bit. The operand and the location of the first high-order 1 bit are displayed.

```
//bit_scan_rev.cpp
//illustrates the operation of the BSR instruction

#include "stdafx.h"

int main (void)
{
//define variables
   int bsr_src_opnd, bsr_rslt;

   printf ("Enter an 8-digit hexadecimal number: \n");
   scanf ("%X", &bsr_src_opnd);


//switch to assembly
      _asm
      {
         MOV   EBX, bsr_src_opnd
//the MOV instr does not affect the flags;
//therefore, the state of the flags is unknown
         CMP   EBX, 0
         JZ    NO_ONES_BSR
         BSR   EAX, EBX
         MOV   bsr_rslt, EAX
      }


   printf ("\nBSR bit is %d\n\n", bsr_rslt);
   goto  end;


NO_ONES_BSR:
   printf ("\nThe BSR operand had no 1s \n\n");

end:
   return 0;
}
```

//continued on next page

(a)

Figure 8.14 Illustrates the operation of the BSR instruction: (a) the program and (b) the outputs.

```
Enter an 8-digit hexadecimal number:
F0000000

BSR bit is 31

Press any key to continue . . . _
------------------------------------------------------------
Enter an 8-digit hexadecimal number:
00000001

BSR bit is 0

Press any key to continue . . . _
------------------------------------------------------------
Enter an 8-digit hexadecimal number:
0000065C

BSR bit is 10

Press any key to continue . . . _
------------------------------------------------------------
Enter an 8-digit hexadecimal number:
01116ED0

BSR bit is 24

Press any key to continue . . . _
------------------------------------------------------------
Enter an 8-digit hexadecimal number:
00000000

The BSR operand had no 1s

Press any key to continue . . . _
------------------------------------------------------------
Enter an 8-digit hexadecimal number:
00A0B7D0

BSR bit is 23

Press any key to continue . . . _
```

(b)

Figure 8.14 (Continued)

8.9 Shift Instructions

There are six shift instructions: *shift arithmetic left* (SAL), *shift logical left* (SHL), *shift arithmetic right* (SAR), *shift logical right* (SHR), *shift left double* (SHLD), and *shift right double* (SHRD). The SAL and SAR instructions perform arithmetic shifts where the operands are signed numbers in 2s complement representation. The SHL and SHR instructions perform logical shifts on unsigned numbers. The SHLD and SHRD instructions perform shifts from one operand to another operand. The syntax for these instructions is shown below.

```
SAL/R/SHL/R    register/memory, immediate/register CL

SHLD/SHRD   register/memory, register, immediate/register CL
```

8.9.1 Shift Arithmetic Left (SAL) Instruction

The bits in the first operand (destination) are shifted left by the number of bits specified by the second operand (count). The count can be an immediate value of 1, an immediate value specified in a byte, or a count in register CL. The count is masked by $1F_{16}$ (00011111_2) so that the count cannot exceed a count of 31 (modulo-32), making the count range from 0 to 31. If the REX prefix is used in 64-bit mode, then the count is masked to produce a range from 0 to 63. In this mode, the number of general-purpose registers is increased from eight to sixteen and can be extended to 64 bits.

During each shift, the high-order bit of the destination is shifted into the carry flag (CF) and a 0 is shifted into the low-order bit. Since the destination operand is a signed number in 2s complement representation, an overflow occurs if the high-order bit of the destination is not equal to the carry flag. This means that significant bits have been lost in the destination operand.

The flags affected by the SAL instruction are the overflow flag (OF), the sign flag (SF), the zero flag (ZF), the parity flag (PF), and the carry flag (CF). These flags reflect the result of the SAL instruction execution. The overflow flag (OF) is not affected by multiple-bit shifts; it is affected only by single-bit shifts. If the initial count is greater than 1, then the overflow flag (OF) is undefined. The OF flag is reset (0) if the high-order bit in the destination is the same as the CF flag after an initial count of 1 has been performed; if these bits are different, then the OF flag is set (1).

Figure 8.15 shows an assembly language module embedded in a C program that illustrates the use of the SAL instruction utilizing a shift count with an immediate value of 1, a shift count with an immediate byte of 5, and a shift count contained in register CL. The original shift operand is shown in the outputs before execution of the SAL instruction and the resulting shifted operand after execution of the SAL instruction. The flags are also shown that reflect the result of the operation.

```
//shift_arith_left.cpp
//illustrates the operation of the SAL instruction
#include "stdafx.h"
int main (void)
{
//define variables
   int sal_dst_opnd, sal_1_rslt, sal_immed_rslt,
         sal_cl_rslt;
   unsigned char sal_1_flags, sal_immed_flags, sal_cl_flags;

   printf ("Enter a decimal number: \n");
   scanf ("%d", &sal_dst_opnd);

//switch to assembly
      _asm
      {
//shift arithmetic left one bit immediate
         MOV   EBX, sal_dst_opnd
         SAL   EBX, 1
         MOV   sal_1_rslt, EBX
         LAHF
         MOV   sal_1_flags, AH

//shift arithmetic left one byte immediate
         MOV   EBX, sal_dst_opnd
         SAL   EBX, 5
         MOV   sal_immed_rslt, EBX
         LAHF
         MOV   sal_immed_flags, AH

//shift arithmetic left the count in CL
         MOV   EBX, sal_dst_opnd
         MOV   CL, 4
         SAL   EBX, CL
         MOV   sal_cl_rslt, EBX
         LAHF
         MOV   sal_cl_flags, AH
      }
```

(a) //continued on next page

Figure 8.15 Illustrates the operation of the SAL instruction: (a) the program and (b) the outputs.

```
    printf ("\nImmediate shift 1\nBefore shift = %d\n
            After shift = %d\nFlags = %X\n\n",
            sal_dst_opnd, sal_1_rslt, sal_1_flags);

    printf ("Immediate byte shift 5\nBefore shift = %d\n
            After shift = %d\nFlags = %X\n\n",
            sal_dst_opnd, sal_immed_rslt, sal_immed_flags);

    printf ("CL register shift 4\nBefore shift = %d\n
            After shift = %d\nFlags = %X\n\n",
            sal_dst_opnd, sal_cl_rslt, sal_cl_flags);

    return 0;
}
```

```
Enter a decimal number:
32

Immediate shift 1
Before shift = 32
After shift = 64
Flags = 2

Immediate byte shift 5
Before shift = 32
After shift = 1024
Flags = 6

CL register shift 4
Before shift = 32
After shift = 512
Flags = 6

Press any key to continue . . . _
------------------------------------------------------------
Enter a decimal number:
25

Immediate shift 1
Before shift = 25
After shift = 50
Flags = 2
```

(b) //continued on next page

Figure 8.15 (Continued)

```
Immediate byte shift 5
Before shift = 25
After shift = 800
Flags = 2

CL register shift 4
Before shift = 25
After shift = 400
Flags = 6

Press any key to continue . . . _
```

Figure 8.15 (Continued)

8.9.2 Shift Logical Left (SHL) Instruction

The SAL instruction and the SHL instruction are equivalent operations, although the SHL instruction is for unsigned operands. They both shift the destination operand left toward the high-order bit position, which is shifted into the carry flag. Zeroes are shifted into the low-order bit position. The SAL and the SHL instructions are equivalent to a multiply operation: a left shift of one bit multiplies the operand by two; a left shift of three bits multiplies the operand by eight. Operations are shown below using eight bits for both the SAL instruction (2s complement) and the SHL instruction.

SAL:	Initial operand =	0	0	0	1	1	0	1	1	+27
	SAL 1 (×2)	0	0	1	1	0	1	1	0	+54
SAL:	Initial operand =	0	0	0	0	1	1	1	1	+15
	SAL 3 (×8)	0	1	1	1	1	0	0	0	+120
SAL:	Initial operand =	1	1	1	0	0	0	1	1	–29
	SAL 2 (×4)	1	0	0	0	1	1	0	0	–116
SHL:	Initial operand =	0	0	0	0	0	1	1	1	7
	SHL 4 (×16)	0	1	1	1	0	0	0	0	112
SHL:	Initial operand =	0	0	0	1	1	1	1	1	31
	SHL 3 (×8)	1	1	1	1	1	0	0	0	248

8.9.3 Shift Arithmetic Right (SAR) Instruction

The bits in the first operand (destination) are shifted right by the number of bits specified by the second operand (count). The count can be an immediate value of 1, an immediate value specified in a byte, or a count in register CL. The count is masked by $1F_{16}$ (00011111_2) so that the count cannot exceed a count of 31 (modulo-32), making the count range from 0 to 31. If the REX prefix is used in 64-bit mode, then the count is masked to produce a range from 0 to 63. In this mode, the number of general-purpose registers is increased from eight to sixteen and can be extended to 64 bits.

For each bit shift, the low-order bit (0) is shifted into the carry flag. The high-order bit — the sign bit of the destination operand — extends to the right replacing the bits that have been shifted, thereby keeping the sign of the number unchanged.

The flags affected by the SAR instruction are the overflow flag (OF), the sign flag (SF), the zero flag (ZF), the parity flag (PF), and the carry flag (CF). These flags reflect the result of the SAR instruction execution. The overflow flag (OF) is not affected by multiple-bit shifts; it is affected only by single-bit shifts. The OF flag is reset (0) for all 1-bit shifts.

Figure 8.16 shows an assembly language module embedded in a C program that illustrates the use of the SAR instruction utilizing a shift count with an immediate value of 1, a shift count with an immediate byte of 4, and a shift count contained in register CL. The original operand is shown in the outputs before execution of the SAR instruction and the resulting shifted operand after execution of the SAR instruction.

```
//shift_arith_right.cpp
//illustrates the operation of the SAR instruction
#include "stdafx.h"
int main (void)
{
//define variables
   int sar_dst_opnd, sar_1_rslt, sar_immed_rslt,
         sar_cl_rslt;
   printf ("Enter an 8-digit hexadecimal number: \n");
   scanf ("%X", &sar_dst_opnd);

//switch to assembly
      _asm
{
//shift arithmetic right one bit immediate
         MOV    EBX, sar_dst_opnd
         SAR    EBX, 1
         MOV    sar_1_rslt, EBX
                              (a)        //continued on next page
```

Figure 8.16 Illustrates the operation of the SAR instruction: (a) the program and (b) the outputs.

```
//shift arithmetic right one byte immediate
         MOV    EBX, sar_dst_opnd
         SAR    EBX, 4
         MOV    sar_immed_rslt, EBX

//shift arithmetic right the count in CL
         MOV    EBX, sar_dst_opnd
         MOV    CL, 3
         SAR    EBX, CL
         MOV    sar_cl_rslt, EBX
}

   printf ("\nImmediate shift 1\nBefore shift = %X\n
            After shift = %X\n\n", sar_dst_opnd,sar_1_rslt);

   printf ("Immediate byte shift 4\nBefore shift = %X\n
            After shift = %X\n\n", sar_dst_opnd,
            sar_immed_rslt);

   printf ("CL register shift 3\nBefore shift = %X\n
            After shift = %X\n\n", sar_dst_opnd,
            sar_cl_rslt);

   return 0;
}
```

```
Enter an 8-digit hexadecimal number:
FF0000FF

Immediate shift 1
Before shift = FF0000FF
After shift = FF80007F

Immediate byte shift 4
Before shift = FF0000FF
After shift = FFF0000F

CL register shift 3
Before shift = FF0000FF
After shift = FFE0001F

Press any key to continue . . . _
```

(b)

//continued on next page

Figure 8.16 (Continued)

```
Enter an 8-digit hexadecimal number:
F0000677

Immediate shift 1
Before shift = F0000677
After shift = F800033B

Immediate byte shift 4
Before shift = F0000677
After shift = FF000067

CL register shift 3
Before shift = F0000677
After shift = FE0000CE

Press any key to continue . . . _
-----------------------------------------------------------
Enter an 8-digit hexadecimal number:
00EF00FF

Immediate shift 1
Before shift = 00EF00FF
After shift = 77807F

Immediate byte shift 4
Before shift = 00EF00FF
After shift = EF00F

CL register shift 3
Before shift = 00EF00FF
After shift = 1DE01F

Press any key to continue . . . _
```

Figure 8.16 (Continued)

8.9.4 Shift Logical Right (SHR) Instruction

The SHR instruction shifts an unsigned operand right the number of bits specified in the count field of the second operand, which can be an immediate value of 1, an immediate value specified in a byte, or a value in the CL register. During each shift operation, the low-order bit is shifted into the carry flag and a zero is shifted into the high-order bit position. The SAR instruction and the SHR instruction can be used to divide the destination operand by powers of 2. A right shift of one bit divides the operand by two; a right shift of three bits divides the operand by eight. Operations are shown below using eight bits for both the SAR instruction (2s complement) and the SHR instruction.

SAR:	Initial operand =	0	0	0	1	0	1	0	0	+20
	SAR 1 (÷2)	0	0	0	0	1	0	1	0	+10
SAR:	Initial operand =	0	1	1	0	1	0	0	0	+104
	SAR 3 (÷8)	0	0	0	0	1	1	0	1	+13
SAR:	Initial operand =	1	1	1	0	0	1	1	0	–26
	SAR 2 (÷4)	1	1	1	1	1	0	0	1	–7
SAR:	Initial operand =	0	0	0	1	0	1	0	1	+21
	SAR 1 (÷2)	0	0	0	0	1	0	1	0	+10
SAR:	Initial operand =	0	0	0	1	0	1	0	1	+21
	SAR 2 (÷4)	0	0	0	0	0	1	0	1	+5
SHR:	Initial operand =	0	0	1	1	0	0	0	0	48
	SHR 4 (÷16)	0	0	0	0	0	0	1	1	3
SHR:	Initial operand =	1	0	0	1	1	0	0	0	152
	SHR 3 (÷8)	0	0	0	1	0	0	1	1	19

Note the result of the third SAR instruction, which divides –26 by 4. The shift operation leaves an incorrect result; however, the last bit shifted out of the operand — the most significant bit (1) from position 2^1 — is stored in the carry flag. This problem occurs only for negative numbers. A signed division of these two numbers yields a correct result with a quotient and remainder. In the fourth SAR instruction, +21 is divided by 2, which yields a result of 10 with the carry flag (CF) = 1; that is, an answer of 10.5. In the fifth SAR instruction, +21 is divided by 4, which yields a result of 5 with the carry flag (CF) = 0. If the two bits shifted out are considered, then the result would be 0000 0101.01 (+5.25), which is correct.

8.9.5 Shift Left Double (SHLD) Instruction

The SHLD instruction is a 3-operand instruction that is used to shift bits from one operand into another operand. Bits from the second operand (source operand) are shifted left into the first operand (destination operand). The number of bits shifted are specified in a count variable that is either an unsigned immediate integer or a count in register CL. Bits are shifted from the source operand, beginning with the high-order bit position, into the destination operand, beginning with the low-order bit position. The destination operand is either a register or a memory location; the source operand must be a register and is not modified by the instruction. For convenience, the syntax for the SHLD instruction is reproduced below.

```
SHLD   register/memory, register, immediate/register CL
```

Both the destination operand and the source operand must be the same length, either 16 bits or 32 bits. The count in register CL is masked by $1F_{16}$ (00011111_2) so that the count cannot exceed a count of 31, making the count range from 0 to 31. A count that is greater than the size of the operand produces a result that is undefined. If the REX prefix is used in 64-bit mode, then the count is masked to produce a range from 0 to 63. In this mode, the number of general-purpose registers is increased from eight to sixteen and can be extended to 64 bits.

The carry flag (CF) contains the last bit that is shifted out of the destination operand. If the shift count is 1, then the overflow flag (OF) is set (1) if a change in the sign occurred; otherwise, it is reset (0); for counts greater than one, the overflow flag is undefined. A count of zero does not affect the flags.

Figure 8.17 shows a diagram that illustrates the operation of the SHLD instruction for two 32-bit operands. Figure 8.18 shows a code segment that demonstrates the SHLD instruction using two operands from memory that are moved to register EAX (destination) and to register EBX (source). The shift count is an immediate value of seven.

Figure 8.19 shows a numerical example of the SHLD instruction using registers AX (destination) and register BX (source). Register AX is assigned a value of 1234H and register BX is assigned a value of 5678H. The source operand in register BX is saved before the shift operation begins so that it can be restored after the SHLD instruction is finished.

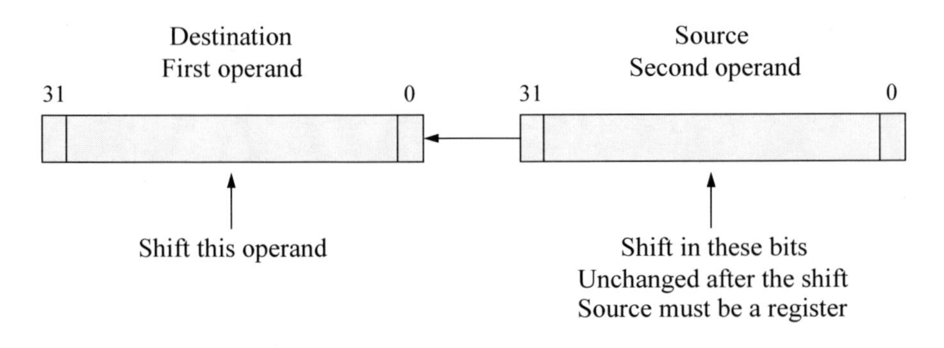

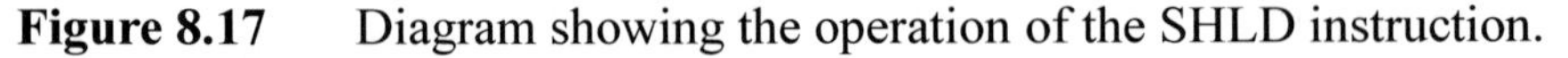
Figure 8.17 Diagram showing the operation of the SHLD instruction.

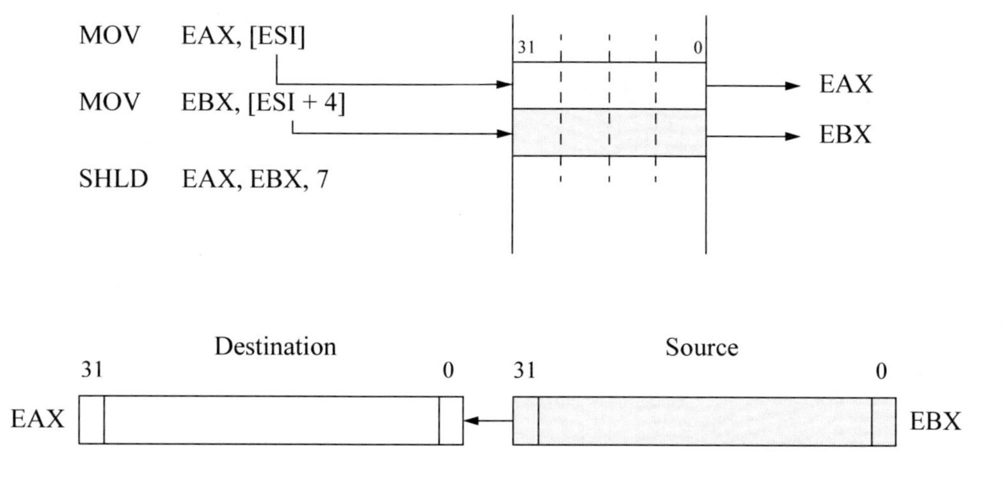

Figure 8.18 Code segment and register initialization of operands from memory for a SHLD instruction.

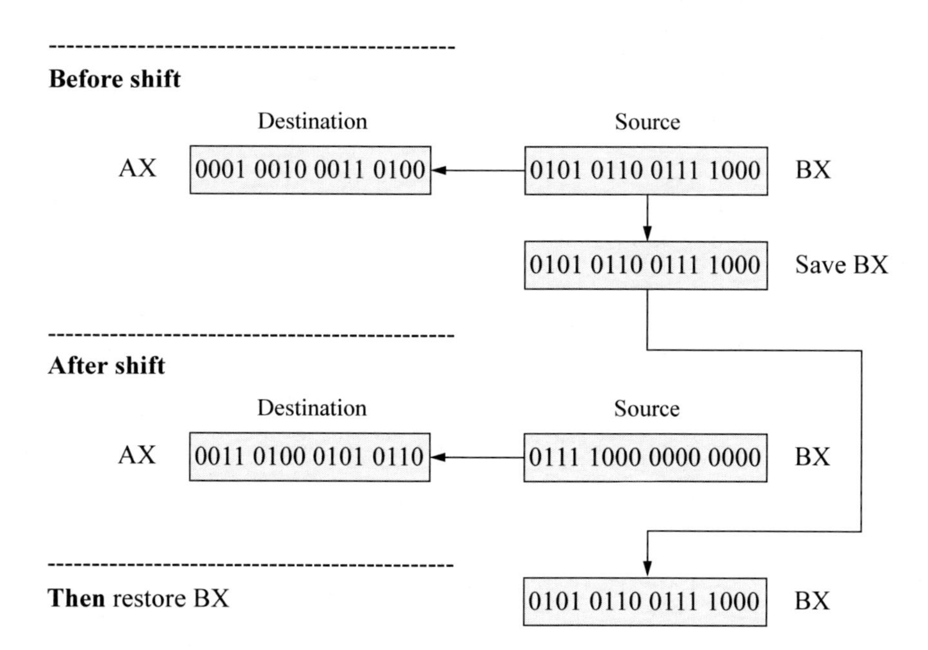

Figure 8.19 Diagram to illustrate a numerical example for a SHLD instruction.

Figure 8.20 shows an assembly language module embedded in a C program that illustrates the application of the SHLD instruction shown in Figure 8.19 using the AX and BX general-purpose registers. The *unsigned short* integer data type is used to characterize the 16-bit unsigned hexadecimal numbers. This data type consists of two bytes with a range from 0 to 65,535.

```
//shift_left_dbl.cpp
//illustrates the use the SHLD instruction

#include "stdafx.h"

int main (void)

{
//define variables
   unsigned short src_opnd, dst_opnd, src_rslt, dst_rslt;

   printf ("Enter two 4-digit hex numbers - src, dst: \n");
   scanf ("%X %X", &src_opnd, &dst_opnd);

//switch to assembly
      _asm
      {
         MOV   BX, src_opnd
         MOV   AX, dst_opnd

         SHLD  AX, BX, 8    ;shift AX:BX left 8 bits

         MOV   src_rslt, BX
         MOV   dst_rslt, AX
      }

   printf ("\nSource result = %X\n
            Destination result = %X\n\n",
            src_rslt, dst_rslt);

   return 0;
}
                               (a)    //continued on next page
```

Figure 8.20 Illustrates the operation of the SHLD instruction: (a) the program and (b) the outputs.

```
Enter two 4-digit hex numbers - src, dst:
1234 5678

Source result = 1234
Destination result = 7812

Press any key to continue . . . _
------------------------------------------------------------
Enter two 4-digit hex numbers - src, dst:
12AB CDEF

Source result = 12AB
Destination result = EF12

Press any key to continue . . . _
```

(b)

Figure 8.20 (Continued)

8.9.6 Shift Right Double (SHRD) Instruction

The SHRD instruction functions in a manner similar to the SHLD instruction, except that the bits are shifted to the right from the low-order bit position of the source operand register to the high-order bit of the destination operand (register/memory). The number of bits shifted is specified in a count variable that is either an unsigned immediate integer or a count in register CL. The syntax for the SHRD instruction is shown below.

```
SHLD   register/memory, register, immediate/register CL
```

Both the destination operand and the source operand must be the same length, either a word (16 bits) or a doubleword (32 bits). The count is masked by $1F_{16}$ (00011111_2) providing a count range from 0 to 31. A count that is greater than the size of the operand produces an undefined result.

If the REX prefix is used in 64-bit mode, then the count is masked to produce a range from 0 to 63. In this mode, the number of general-purpose registers is increased from eight to sixteen and can be extended to 64 bits. The last bit shifted out of the destination operand is stored in the carry flag (CF). The overflow flag (OF) is set if the sign changes for a 1-bit shift; otherwise, the overflow flag is reset. The states of the sign flag (SF), the zero flag (ZF), and the parity flag (PF) are determined by the result of the shift operation. Figure 8.21 shows a diagram that illustrates the operation of the

SHRD instruction for two 32-bit operands. The source operand is unchanged after the shift operation.

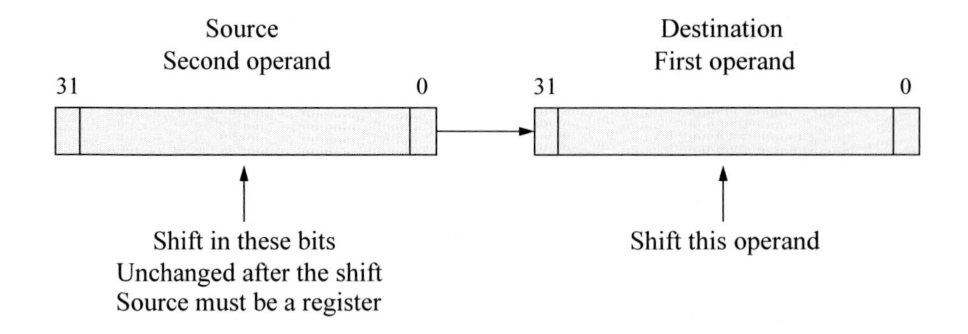

Figure 8.21 Diagram showing the operation of the SHRD instruction.

Figure 8.22 shows an assembly language module embedded in a C program that illustrates the application of the SHRD instruction using the AX and BX general-purpose registers for the source operand and destination operand, respectively. Two 4-digit hexadecimal numbers are entered from the keyboard and displayed together with the source operand result and the destination operand result. The *unsigned short* integer data type is used to characterize the 16-bit unsigned hexadecimal numbers.

```
//shift_right_dbl2.cpp
//illustrates the use the SHRD instruction
#include "stdafx.h"
int main (void)

{
//define variables
   unsigned short src_opnd, dst_opnd, src_rslt, dst_rslt;

   printf ("Enter two 4-digit hex numbers - src, dst: \n");
   scanf ("%X %X", &src_opnd, &dst_opnd);

//switch to assembly
      _asm
      {
         MOV   AX, src_opnd
         MOV   BX, dst_opnd
```

(a) //continued on next page

Figure 8.22 Illustrates the operation of the SHRD instruction: (a) the program and (b) the outputs.

```
        SHRD  BX, AX, 10  ;shift AX:BX right 10 bits

        MOV   src_rslt, AX
        MOV   dst_rslt, BX
    }

  printf ("\nSource result = %X\n
            Destination result = %X\n\n",
            src_rslt, dst_rslt);

  return 0;
}
```

```
Enter two 4-digit hex numbers - src, dst:
1234 5678

Source result = 1234
Destination result = 8D15

Press any key to continue . . . _
------------------------------------------------------------
Enter two 4-digit hex numbers - src, dst:
12AB CDEF

Source result = 12AB
Destination result = AAF3

Press any key to continue . . . _
```

(b)

Figure 8.22 (Continued)

8.10 Rotate Instructions

This section describes the four rotate instructions: *rotate left* (ROL), *rotate through carry left* (RCL), *rotate right* (ROR), and *rotate through carry right* (RCR). The rotate instructions rotate the bits in the destination operand a specific number of bits as stipulated in the count field of the second operand. The count can be a single bit shift, an unsigned integer as an immediate byte, or a count in register CL. The count is masked by $1F_{16}$ (00011111_2) providing a count range from 0 to 31. The shift operations presented in Section 8.9 resulted in an inherent loss of bits; however, the rotate instructions provide a result in which no bits are lost — they are simply shifted in to the other end of the destination operand. The syntax for the rotate instructions is shown below.

```
ROL/RCL/ROR/RCR   register/memory, immediate/register CL
```

8.10.1 Rotate Left (ROL) Instruction

The *rotate left* (ROL) instruction rotates (shifts) the destination bits left toward the high-order bit position of the operand. The high-order bit is shifted into the carry flag and also into the low-order bit position. The carry flag is not included as part of the destination result. The overflow flag (OF) is set for shifts of one bit only and is undefined for all other shift counts. The overflow flag is defined as the exclusive-OR of the carry flag and the high-order bit of the result; that is, a change of sign occurred for the destination operand. Figure 8.23 shows a diagram that illustrates the operation of the ROL instruction. Figure 8.24 provides a numerical example for a ROL instruction using general-purpose register BX with a rotate count of 9.

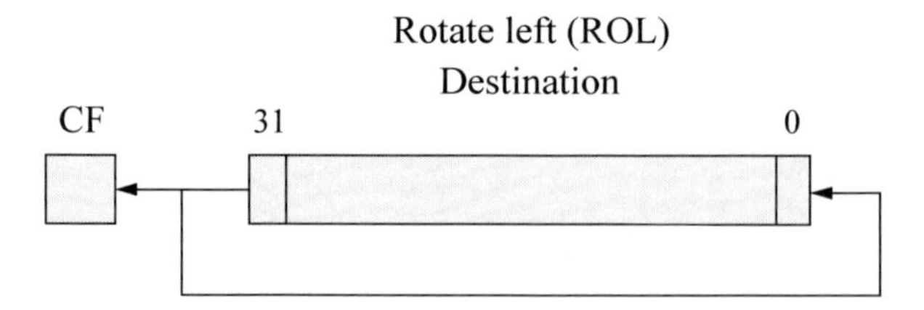

Figure 8.23 Diagram showing the operation of the ROL instruction.

ROL BX, 9 BX =	0 1 0 1	0 0 0 0	1 0 1 0	0 0 0 1
After 1 shift	1 0 1 0	0 0 0 1	0 1 0 0	0 0 1 0
After 2 shifts	0 1 0 0	0 0 1 0	1 0 0 0	0 1 0 1
After 3 shifts	1 0 0 0	0 1 0 1	0 0 0 0	1 0 1 0
After 4 shifts	0 0 0 0	1 0 1 0	0 0 0 1	0 1 0 1
After 5 shifts	0 0 0 1	0 1 0 0	0 0 1 0	1 0 1 0
After 6 shifts	0 0 1 0	1 0 0 0	0 1 0 1	0 1 0 0
After 7 shifts	0 1 0 1	0 0 0 0	1 0 1 0	1 0 0 0
After 8 shifts	1 0 1 0	0 0 0 1	0 1 0 1	0 0 0 0
After 9 shifts	0 1 0 0	0 0 1 0	1 0 1 0	0 0 0 1

Figure 8.24 Numerical example for ROL using register BX with a count of 9.

8.10.2 Rotate through Carry Left (RCL) Instruction

The *rotate through carry left* (RCL) instruction is similar to the ROL instruction except that the carry flag is included in the rotation. The RCL instruction shifts the carry flag into the low-order bit position of the destination operand and shifts the high-order bit of the destination into the carry flag. The overflow flag operates in an identical manner as in the ROL instruction. Figure 8.25 shows a diagram that illustrates the operation of the RCL instruction. Figure 8.26 provides a numerical example for a RCL instruction using general-purpose register EAX with a rotate count of 4 in which the carry flag is initially set to a value of 1. This example depicts one of the user-entered inputs in an assembly language program that is described below.

Figure 8.27 shows an assembly language module embedded in a C program that illustrates the application of the ROL instruction and the RCL instruction using the general-purpose register EAX as the destination operand. An 8-digit hexadecimal number is entered from the keyboard and displayed together with the resulting destination operand. The *set carry flag* (STC) instruction is introduced in the program to set the carry flag to a known state in order to execute the RCL instruction. The STC instruction sets the carry flag in the EFLAGS register to a value of 1. There are two similar instructions: *clear carry flag* (CLC), which resets the carry flag, and the *complement carry flag* (CMC), which changes the state of the carry flag — from 0 to 1 or from 1 to 0.

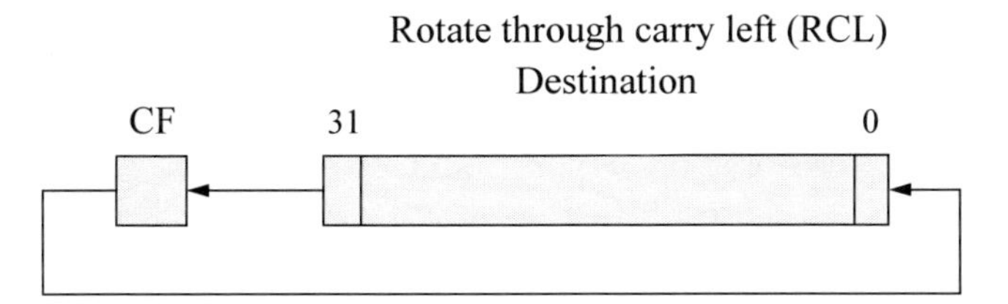

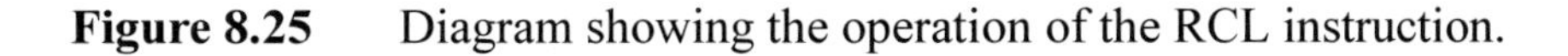

Figure 8.25 Diagram showing the operation of the RCL instruction.

CF	**RCL** EAX, 4								
1	0001	0010	1010	1011	1100	1101	1110	1111	
0	0010	0101	0101	0111	1001	1011	1101	1111	After 1 shift
0	0100	1010	1010	1111	0011	0111	1011	1110	After 2 shifts
0	1001	0101	0101	1110	0110	1111	0111	1100	After 3 shifts
1	0010	1010	1011	1100	1101	1110	1111	1000	After 4 shifts

Figure 8.26 Numerical example for RCL using register EAX with a count of 4.

```
//rotate_left.cpp
//illustrates the operation of the rotate left (ROL)
//and the rotate through carry left (RCL) instructions

#include "stdafx.h"

int main (void)
{
//define variables
   int rotate_opnd, rol_1_rslt, rol_immed_rslt, rol_cl_rslt,
         rcl_rslt;

   printf ("Enter an 8-digit hexadecimal number: \n");
   scanf ("%X", &rotate_opnd);

//switch to assembly
      _asm
      {
//rotate left one bit immediate
         MOV   EAX, rotate_opnd
         ROL   EAX, 1
         MOV   rol_1_rslt, EAX      ;move to result

//rotate left one byte immediate
         MOV   EAX, rotate_opnd
         ROL   EAX, 6
         MOV   rol_immed_rslt, EAX  ;move to result

//rotate left the count in CL
         MOV   EAX, rotate_opnd
         MOV   CL, 3
         ROL   EAX, CL
         MOV   rol_cl_rslt, EAX     ;move to result

//rotate through carry left one byte immediate
         STC
         MOV   EAX, rotate_opnd
         RCL   EAX, 4
         MOV   rcl_rslt, EAX        ;move to result
      }
```

(a) //continued on next page

Figure 8.27 Program to illustrate the application of the ROL and the RCL instructions: (a) the program and (b) the outputs.

```
    printf ("\nImmediate rotate 1\nBefore rotate = %X\n
            After rotate = %X\n\n", rotate_opnd, rol_1_rslt);

    printf ("\nImmediate byte rotate 6\n
            Before rotate = %X\nAfter rotate = %X\n\n",
            rotate_opnd, rol_immed_rslt);

    printf ("\nCL register rotate 3\n
            Before rotate = %X\nAfter rotate = %X\n\n",
            rotate_opnd, rol_cl_rslt);

    printf ("\nRotate through carry 4\n
            Before rotate = %X\nAfter rotate = %X\n\n",
            rotate_opnd, rcl_rslt);

    return 0;
}
```

```
Enter an 8-digit hexadecimal number:
12ABCDEF

Immediate rotate 1
Before rotate = 12ABCDEF
After rotate = 25579BDE

Immediate byte rotate 6
Before rotate = 12ABCDEF
After rotate = AAF37BC4

CL register rotate 3
Before rotate = 12ABCDEF
After rotate = 955E6F78

Rotate through carry 4
Before rotate = 12ABCDEF
After rotate = 2ABCDEF8

Press any key to continue . . . _
------------------------------------------------------------
```

//continued on next page

(b)

Figure 8.27 (Continued)

```
Enter an 8-digit hexadecimal number:
67AB89CD

Immediate rotate 1
Before rotate = 67AB89CD
After rotate = CF57139A

Immediate byte rotate 6
Before rotate = 67AB89CD
After rotate = EAE27359

CL register rotate 3
Before rotate = 67AB89CD
After rotate = 3D5C4E6B

Rotate through carry 4
Before rotate = 67AB89CD
After rotate = 7AB89CDB

Press any key to continue . . . _

------------------------------------------------------------

Enter an 8-digit hexadecimal number:
11111111

Immediate rotate 1
Before rotate = 11111111
After rotate = 22222222

Immediate byte rotate 6
Before rotate = 11111111
After rotate = 44444444

CL register rotate 3
Before rotate = 11111111
After rotate = 88888888

Rotate through carry 4
Before rotate = 11111111
After rotate = 11111118

Press any key to continue . . . _
```

Figure 8.27 (Continued)

8.10.3 Rotate Right (ROR) Instruction

The *rotate right* (ROR) instruction rotates (shifts) the destination bits right toward the low-order bit position of the operand. The low-order bit is shifted into the carry flag and also into the high-order bit position. The carry flag is not included as part of the destination result. The overflow flag (OF) is set for shifts of one bit only and is undefined for all other shift counts. The overflow flag is defined as the exclusive-OR of the two high-order bits of the result; that is, a change of sign occurred for the destination operand. Figure 8.28 shows a diagram that illustrates the operation of the ROR instruction. Figure 8.29 shows a numerical example for a ROR instruction using general-purpose register AX with a rotate count of 5.

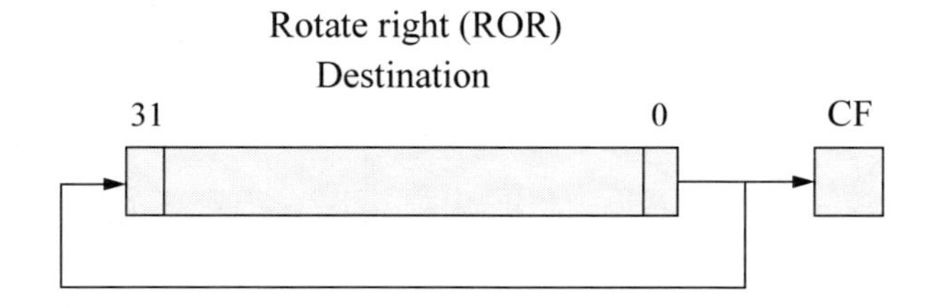

Figure 8.28 Diagram showing the operation of the ROR instruction.

ROR AX, 5 AX =	1 1 0 1	1 1 0 0	0 1 1 0	1 0 0 1
After 1 shift	1 1 1 0	1 1 1 0	0 0 1 1	0 1 0 0
After 2 shifts	0 1 1 1	0 1 1 1	0 0 0 1	1 0 1 0
After 3 shifts	0 0 1 1	1 0 1 1	1 0 0 0	1 1 0 1
After 4 shifts	1 0 0 1	1 1 0 1	1 1 0 0	0 1 1 0
After 5 shifts	0 1 0 0	1 1 1 0	1 1 1 0	0 0 1 1

Figure 8.29 Numerical example for ROR using register AX with a count of 5.

8.10.4 Rotate through Carry Right (RCR) Instruction

The *rotate through carry right* (RCR) instruction is similar to the ROR instruction except that the carry flag is included in the rotation. The RCR instruction shifts the carry flag into the high-order bit position of the destination operand and shifts the low-order bit of the destination into the carry flag. The overflow flag operates in an identical manner as in the ROR instruction. Figure 8.30 shows a diagram that illustrates the operation of the RCR instruction. Figure 8.31 provides a numerical example for a RCR instruction using general-purpose register EAX with a rotate count of 4 in which the carry flag is initially set to a value of 0. This example depicts one of the user-entered inputs in an assembly language program that is described below.

Figure 8.32 shows an assembly language module embedded in a C program that illustrates the application of the ROR instruction and the RCR instruction using the general-purpose register EAX as the destination operand. An 8-digit hexadecimal number is entered from the keyboard and displayed together with the resulting destination operand. The *clear carry flag* (CLC) resets the carry flag in the EFLAGS register to a value of 0 prior to execution of the RCR instruction.

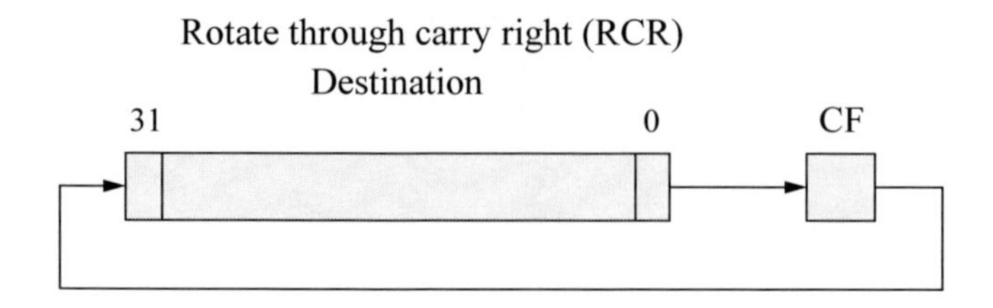

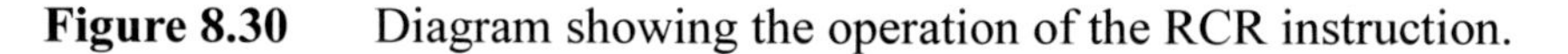

Figure 8.30 Diagram showing the operation of the RCR instruction.

	RCR EAX, 4	CF
	0001 0010 0011 0100 1010 1011 1100 1101	0
After 1 shift	0000 1001 0001 1010 0101 0101 1110 0110	1
After 2 shifts	1000 0100 1000 1101 0010 1010 1111 0011	0
After 3 shifts	0100 0010 0100 0110 1001 0101 0111 1001	1
After 4 shifts	1010 0001 0010 0011 0100 1010 1011 1100	1

Figure 8.31 Numerical example for RCR using register EAX with a count of 4.

```
//rotate_right.cpp
//illustrates the operation of the rotate right (ROR)
//and the rotate through carry right (RCR) instructions
#include "stdafx.h"
int main (void)
{
//define variables
   int rotate_opnd, ror_1_rslt, ror_immed_rslt, ror_cl_rslt,
         rcr_rslt;
```

(a) //continued on next page

Figure 8.32 Program to illustrate the application of the ROR and the RCR instructions: (a) the program and (b) the outputs.

```
   printf ("Enter an 8-digit hexadecimal number: \n");
   scanf ("%X", &rotate_opnd);
//switch to assembly
      _asm
      {
//rotate right one bit immediate
         MOV   EAX, rotate_opnd
         ROR   EAX, 1
         MOV   ror_1_rslt, EAX      ;move to result

//rotate right one byte immediate
         MOV   EAX, rotate_opnd
         ROR   EAX, 6
         MOV   ror_immed_rslt, EAX  ;move to result

//rotate right the count in CL
         MOV   EAX, rotate_opnd
         MOV   CL, 3
         ROR   EAX, CL
         MOV   ror_cl_rslt, EAX     ;move to result

//rotate through carry right one byte immediate
         CLC
         MOV   EAX, rotate_opnd
         RCR   EAX, 4
         MOV   rcr_rslt, EAX        ;move to result
      }
   printf ("\nImmediate rotate 1\nBefore rotate = %X\n
            After rotate = %X\n\n", rotate_opnd, ror_1_rslt);

   printf ("\nImmediate byte rotate 6\nBefore rotate = %X\n
            After rotate = %X\n\n", rotate_opnd,
            ror_immed_rslt);

   printf ("\nCL register rotate 3\nBefore rotate = %X\n
            After rotate = %X\n\n", rotate_opnd,
            ror_cl_rslt);

   printf ("\nRotate through carry 4\nBefore rotate = %X\n
            After rotate = %X\n\n", rotate_opnd, rcr_rslt);
   return 0;
}
                                          //continued on next page
```

Figure 8.32 (Continued)

```
Enter an 8-digit hexadecimal number:
1234ABCD

Immediate rotate 1
Before rotate = 1234ABCD
After rotate = 891A55E6

Immediate byte rotate 6
Before rotate = 1234ABCD
After rotate = 3448D2AF

CL register rotate 3
Before rotate = 1234ABCD
After rotate = A2469579

Rotate through carry 4
Before rotate = 1234ABCD
After rotate = A1234ABC

Press any key to continue . . . _

------------------------------------------------------------

Enter an 8-digit hexadecimal number:
11111111

Immediate rotate 1
Before rotate = 11111111
After rotate = 88888888

Immediate byte rotate 6
Before rotate = 11111111
After rotate = 44444444

CL register rotate 3
Before rotate = 11111111
After rotate = 22222222

Rotate through carry 4
Before rotate = 11111111
After rotate = 21111111

Press any key to continue . . . _
------------------------------------------------------------
                                   //continued on next page
```

(b)

Figure 8.32 (Continued)

```
Enter an 8-digit hexadecimal number:
AB34CD56

Immediate rotate 1
Before rotate = AB34CD56
After rotate = 559A66AB

Immediate byte rotate 6
Before rotate = AB34CD56
After rotate = 5AACD335

CL register rotate 3
Before rotate = AB34CD56
After rotate = D56699AA

Rotate through carry 4
Before rotate = AB34CD56
After rotate = CAB34CD5

Press any key to continue . . . _
```

Figure 8.32 (Continued)

8.11 Problems

8.1 For each of the sections shown below, where the values are in hexadecimal, obtain the resulting values for register AX and the flags. Then write an assembly language module that is embedded in a C program to verify the results and print the results and the flags. When inserting hexadecimal numbers in an assembly language program manually, the first digit must be a number 0 through 9, then the digits A through F, if required, followed by the hexadecimal radix specifier H.

(a) AX = FA75 **AND** AX, 000FH

(b) AX = FA75 **OR** AX, 0FFF0H

(c) AX = FA75 **XOR** AX, 0FFFFH

8.2 Use an XOR instruction to do the following:

(a) Exclusive-OR the memory operand addressed by register BX with the contents of register DX. Save the result in the memory location addressed by BX.

(b) Change the contents of register CL from 53H to 73H.

8.3 Using only logic instructions, write one instruction for each of the following parts that will perform the indicated operations.

(a) Set the low-order four bits of register AX.

(b) Reset the high-order three bits of register AX.

(c) Invert bits 7, 8, and 9 of register AX.

8.4 Write an assembly language program — not embedded in a C program — to perform an AND operation of two single hexadecimal characters that are entered from the keyboard. There is no space between the characters. Display the result of the AND operation and the low-order four bits of the flags register.

8.5 Assume that register AL contains an ASCII code for an uppercase letter. Write a single instruction to change the contents to the corresponding lowercase letter.

8.6 Determine the contents of registers AX, AH, and BL after the following program segment has been executed:

```
MOV     CL, 3
MOV     AX, 7FH
MOV     BX, 0505H
ROL     AX, CL
AND     AH, BH
OR      BL, AL
```

8.7 Write an assembly language module embedded in a C program that executes the bit test instructions BTS, BTR, and BTC. Enter a 4-digit hexadecimal number for the operand and a 2-digit hexadecimal number for the bit offset.

Store the operand in register AX. The program consists of two segments: one segment for an immediate bit offset and one segment for an offset in register BX. Include bit offsets greater than 16 so that the bit offset modulo-16 will generate a bit position in register AX. Display the results of the BTS, BTR, and BTC instructions for both the immediate bit offsets and the bit offset in register BX.

8.8 Write an assembly language module embedded in a C program that uses the *bit scan forward* (BSF) instruction to detect whether the first bit detected is in the low-order half or the high-order half of a 32-bit operand. Do not determine the bit position.

8.9 Repeat problem 8.8 using the *bit scan reverse* (BSR) instruction.

8.10 Determine the contents of register AX after execution of the following program segment:

```
MOV   AX, -15
MOV   CL, 3
SAL   AX, CL
```

8.11 Determine the contents of register AX after execution of the following program segment:

```
MOV   AX, -32,668
MOV   CL, 5
SAR   AX, CL
```

8.12 Determine the contents of register EBX and the flags register after execution of the program segment shown below. Then write an assembly language module embedded in a C program to verify the results.

```
MOV   EBX, 0FFFFA85BH
SAL   EBX, 20
MOV   EBX, 0FFFFA85BH
SAR   EBX, 20
```

8.13 Determine the contents of the destination register EBX, the source register EDX, and the flags register after execution of the program segment shown below and on the following page. Then write an assembly language module embedded in a C program to verify the results.

```
MOV   EBX, 1234ABCDH
MOV   EDX, 0ABCD1234H
```

```
MOV    CL, 16
SHLD   EBX, EDX, CL

MOV    EBX, 1234ABCDH
MOV    EDX, 0ABCD1234H
MOV    CL, 16
SHRD   EBX, EDX, CL
```

8.14 Write an assembly language module embedded in a C program that will multiply and divide a decimal number by 8 using arithmetic shift instructions. When dividing, some numbers will have the fraction truncated.

8.15 Write an assembly language program — not embedded in a C program — that requests one of three 1-digit numbers (1, 2, or 3) to be entered from the keyboard. Determine which number was entered, then display the number. Use the AND instruction to remove the ASCII bias.

8.16 Let register EAX contain AD3E14B5H and the carry flag be set. Determine the contents of register EAX after the following instruction is executed:

```
RCL    EAX, 6
```

8.17 Let register EAX contain 12345678H and the carry flag be set. Determine the contents of register EAX after the following instruction is executed:

```
RCR    EAX, 6
```

8.18 Let register EAX contain 1C78FDA5 and the carry flag be set. Determine the contents of register EAX after the following instruction is executed:

```
RCR    EAX, 6
```

8.19 Write an assembly language module embedded in a C program that illustrates the *shift logical left* (SHL), *shift logical right* (SHR), *rotate left* (ROL), and *rotate right* (ROR) instructions. Enter an 8-digit hexadecimal number for the destination operand with a count of 40 stored in register CL.

8.20 This problem is similar to Problem 8.15. Write an assembly language module embedded in a C program that requests one of three 1-digit numbers (3, 4, or 5) to be entered from the keyboard. Determine which number was entered, then display the number. If an incorrect number is entered, then display that information.

9

Fixed-Point Arithmetic Instructions

In fixed-point operations, the radix point is in a fixed location in the operand. The radix point (or binary point for radix 2) is to the immediate right of the low-order bit for integers, or to the immediate left of the high-order bit for fractions. The operands in a computer can be expressed by any of the following number representations: unsigned, sign-magnitude, diminished-radix complement, or radix complement.

If the numbers are signed, then the sign bit can be extended to the left indefinitely without changing the value of the number. An n-bit signed number A is shown in Equation 9.1, where the leftmost digit a_{n-1} is the sign bit. The sign bit for any radix is 0 for positive numbers and $r-1$ for negative numbers, as shown in Equation 9.2.

$$A = a_{n-1}\ a_{n-2}\ \ldots\ a_1\ a_2 \tag{9.1}$$

$$A = \begin{cases} 0 & \text{for } A \geq 0 \\ r-1 & \text{for } A < 0 \end{cases} \tag{9.2}$$

9.1 Addition

Addition of two binary operands treats both signed and unsigned operands the same — there is no distinction between the two types of numbers during the add operation. The operands for addition are the *augend* and the *addend*, where the addend is added to the augend and the sum replaces the augend in most computers — the addend is unchanged. The rules for radix 2 addition are shown in Table 9.1.

Table 9.1 Rules for Binary Addition

+	0	1
0	0	1
1	1	0_1

(1) 1 + 1 = 0 with a carry to the next higher-order column.

An example of binary addition is shown in Figure 9.1. The subscripted values are the carries from the previous columns. For example, in column 1 the sum of five 1s is 5_{10} (101_2). Therefore, there is a carry of 0 to column 2 and a carry of 1 to column 3. The sum of column 2 is 3_{10} (11_2), which yields a sum of 1 with a carry of 1 to column 3. In column 3, the sum is 6_{10} (110_2), which yields a sum of 0 with a carry of 1 to column 4 and a carry of 1 to column 5. The total sum is 51_{10} (110011_2).

Now consider the rightmost four columns and assume that the operands are in 2s complement representation. Operand 1 is negative because the sign is 1; therefore, the value is –5 (the 0 in column 3 has a value of 4 + 1 = 5). Using the same rationale, operand 2 is also negative with a value of –3. Operand 3 is positive because the sign is 0; therefore, the value is +5. Operand 5 is negative with a value of –1 (count the zeroes by their weight and add one).

	2^5	2^4	2^3	2^2	2^1	2^0	
Column	6	5	4	3	2	1	Radix 10 values
Operand 1	0	0	1	0	1	1	+11(–5)
Operand 2	0	0	1	1	0	1	+13(–3)
Operand 3	0	0	0	1	0	1	+5
Operand 4	0	0	0	1	1	1	+7
Operand 5	0_1	0_{10}	1_1	1_{11}	1_0	1	+15(–1)
Sum =	1	1	0	0	1	1	+51(+3)

Figure 9.1 Example of binary addition.

9.1.1 Add (ADD) Instruction

The *add* (ADD) instruction obtains the sum of the destination operand (the *augend*) and the source operand (the *addend*) and stores the result in the destination operand. The destination operand (first operand) can be a general-purpose register or a memory location. The source operand (second operand) can be a general-purpose register, a memory location, or a signed immediate operand with the sign extended to match the size of the destination operand, if necessary. The syntax for the ADD instruction is shown below.

```
ADD    register/memory, register/memory/immediate
```

The ADD instruction operates on integer operands only and affects the following flags: the overflow flag (OF), the sign flag (SF), the zero flag (ZF), the auxiliary carry flag (AF), the parity flag (PF), and the carry flag (CF).

Overflow occurs when the result of an arithmetic operation (usually addition) exceeds the word size of the machine; that is, the sum is not within the representable range of numbers provided by the number representation. For numbers in 2s complement representation, the range is from -2^{n-1} to $+2^{n-1} - 1$. For two n-bit numbers

$$A = a_{n-1} a_{n-2} a_{n-3} \ldots a_1 a_0$$

$$B = b_{n-1}\, b_{n-2}\, b_{n-3} \ldots b_1 b_0$$

a_{n-1} and b_{n-1} are the sign bits of operands A and B, respectively. Overflow can be detected by either of the following two equations:

$$\text{Overflow} = (a_{n-1} \bullet b_{n-1} \bullet s_{n-1}') + (a_{n-1}' \bullet b_{n-1}' \bullet s_{n-1})$$

$$\text{Overflow} = c_{n-1} \oplus c_{n-2} \tag{9.3}$$

where the symbol "•" is the logical AND operator, the symbol "+" is the logical OR operator, the symbol "⊕" is the logical exclusive-OR operator, the sign bit of the result is s_{n-1}, and the carry bits out of positions $n - 1$ and $n - 2$ are c_{n-1} and c_{n-2}, respectively.

Thus, overflow produces an erroneous sign reversal and is possible only when both operands have the same sign. An overflow cannot occur when adding two operands of different signs, since adding a positive number to a negative number produces a result that falls within the limit specified by the two numbers. Two examples of overflow are shown below: one for positive numbers and one for negative numbers, where both operands are in 2s complement representation.

$A =$	0 0 1 1	1 1 1 0		+62
+) $B =$	0 1 1 0	0 0 1 1		+99
Sum =	1 0 1 0	0 0 0 1		+161

$A =$	1 0 1 1	0 1 0 1		–75
+) $B =$	1 1 0 0	1 0 0 0		–56
Sum =	0 1 1 1	1 1 0 1		–131

For the first example, the sum of +161 requires nine bits: 0 1010 0001; therefore, an overflow has occurred. For the second example, the sum of –131 also requires nine bits: 1 0111 1101; therefore, an overflow has occurred.

The sign flag (SF) is set to the value of the high-order bit of the result; the zero flag (ZF) is set if the sum is zero; the auxiliary carry flag (AF) is set if there is a carry from column 2^3 to column 2^4, which occurs in the first example; the parity flag (PF) is set if there are an even number of 1s in the result (low-order byte only); and the carry flag (CF) is set if there is a carry out of the high-order bit positions.

Figure 9.2 shows an assembly language module embedded in a C program that illustrates an ADD instruction in which all of the flags in the low-order byte of the EFLAGS register are set for specific augends and addends.

```
//add_flags2.cpp
//add two operands and display the resulting flags
#include "stdafx.h"
int main (void)
{
//define variables
   unsigned short augend, addend, sum;
   unsigned char flags;

   printf ("Enter two 4-digit hexadecimal numbers: \n");
   scanf ("%X %X", &augend, &addend);

//switch to assembly
      _asm
      {
         MOV    BX, augend
         MOV    AX, addend
                                   //continued on next page
```

(a)

Figure 9.2 Program to illustrate the AND instruction and flag generation: (a) the program and (b) the outputs.

```
        ADD   BX, AX
        MOV   sum, BX
        LAHF
        MOV   flags, AH
     }

   printf ("\nAugend = %X\nAddend = %X\n", augend, addend);
   printf ("\nSum = %X\nFlags = %X\n\n", sum, flags);

   return 0;
}
```

```
Enter two 4-digit hexadecimal numbers:
0100 0100

Augend = 100
Addend = 100

Sum = 200
Flags = 6         //PF

Press any key to continue . . . _
------------------------------------------------------------
Enter two 4-digit hexadecimal numbers:
0200 FE00

Augend = 200
Addend = FE00

Sum = 0
Flags = 47        //ZF, PF, CF

Press any key to continue . . . _
------------------------------------------------------------
Enter two 4-digit hexadecimal numbers:
0FBC A0DE

Augend = FBC
Addend = A0DE

Sum = B09A
Flags = 96        //SF, AF, PF

Press any key to continue . . . _
```

(b) //continued on next page

Figure 9.2 (Continued)

```
Enter two 4-digit hexadecimal numbers:
10A2 6522

Augend = 10A2
Addend = 6522

Sum = 75C4
Flags = 2          //no status flags set (low-order = C4)

Press any key to continue . . . _
```

Figure 9.2 (Continued)

9.1.2 Add with Carry (ADC) Instruction

The *add with carry* (ADC) instruction adds two integers plus the carry flag (CF = 0 or 1). It is typically used when adding two signed or unsigned operands, which can be multiple bytes, words, doublewords, or larger operands — the carry is propagated from one stage to the next stage. The ADC instruction adds the destination operand, the source operand, and the carry flag and stores the sum in the destination operand. The syntax for the ADC instruction is shown below; however, both operands cannot be in memory.

```
ADC    register/memory, register/memory/immediate
```

The value of the carry flag represents a carry from a previous addition of two operands in a multioperand addition. An immediate operand is sign extended to match the size of the destination operand, if necessary. The ADC instruction normally follows an ADD instruction of a multioperand addition. If a REX prefix is used in 64-bit mode, the number of general-purpose registers is increased from eight to sixteen and can be extended to 64 bits.

Figure 9.3 shows a program segment that depicts the ADC instruction to add two 32-bit operands in two sections. First the low-order section is added, then the high-order section is added plus any carry generated from the addition of the low-order section. The operands are shown below in two segments. Figure 9.4 shows how the numbers are stored in memory.

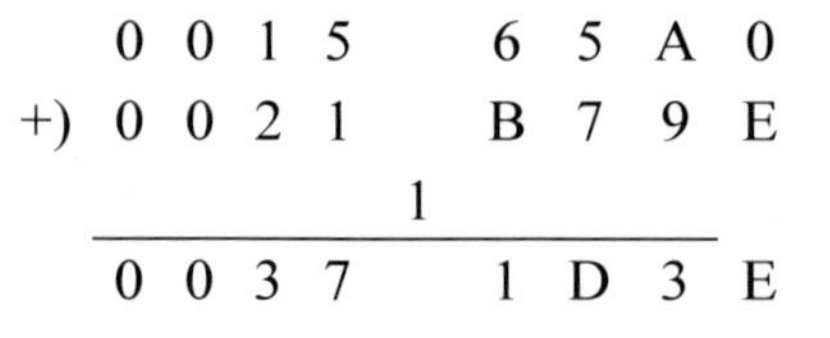

```
                    . . .
DP1         DW 65A0H      ;low word; stored as A065
DP1+2       DW 0015H      ;high word; stored as 1500

DP2         DW B79EH      ;low word; stored as 9EB7
DP+2        DW 0021H      ;high word; stored as 2100

DPSUM       DW ?
DPSUM+2     DW ?
                    . . .
            MOV     AX, DP1
            ADD     AX, DP2
            MOV     DPSUM, AX
            MOV     AX, DP1+2
            ADC     AX, DP2+2
            MOV     DPSUM+2, AX
                    . . .
```

Figure 9.3 Program segment using the ADC instruction.

		+1	+2	+3	Decimal value
DP1	A0	65	15	00	1,402,272
		+1	+2	+3	
DP2	9E	B7	21	00	2,209,694
		+1	+2	+3	
DPSUM	3E	1D	37	00	
Sum =	00	37	1D	3E	3,611,966

Figure 9.4 Values of the numbers as stored in memory.

Figure 9.5 lists an assembly language module embedded in a C program that adds the numbers given in this section. The program adds the two 32-bit operands in two sections: first the low-order segment, then the high-order segment plus the carry from the low-order segment. Both operands are entered as immediate data and the sum is declared as *sum1* and *sum2* for the low-order and the high-order segments, respectively.

```
//dbl_precision.cpp
//illustrates the ADC instruction
//for adding two 32-bit operands

#include "stdafx.h"


int main (void)

{
   unsigned short sum1, sum2;


//switch to assembly
      _asm
      {
         MOV      AX, 65A0H
         MOV      BX, 0B79EH
         ADD      AX, BX
         MOV      sum1, AX


         MOV      AX, 0015H
         MOV      BX, 0021H
         ADC      AX, BX
         MOV      sum2, AX
      }

   printf ("Low sum = %X\nHigh sum = %X\n\n", sum1, sum2);

   return 0;
}
```

(a)

```
Low sum = 1D3E
High sum = 37

Press any key to continue . . . _
```

(b)

Figure 9.5 Program illustrating using the ADC instruction to add two 32-bit operands: (a) the program and (b) the outputs.

Figure 9.6 depicts a program segment illustrating a numerical example of adding two operands: a 32-bit augend and a 16-bit addend using the *add with carry* (ADC) instruction in conjunction with the *convert word to doubleword* (CWD) instruction.

```
          . . .
DP      DW F000H        Stored as 00F0
DP+2    DW 0001H        Stored as 0100
                        Actual augend = 0001F000  (+126,976)
          . . .
MOV     AX, 8000H       Addend value = -32,768
CWD                            DX:AX = FFFF:8000
ADD     AX, DP                    DP = F000
                          ┌───── 1 ←─7000
                          │
MOV     DP, AX            │       DP = 0070
                          │
ADC     DX, DP+2          │       DX = FFFF
                          │     DP+2 = 0001
                          └────► CF =    1
                                  DX = 0001

MOV     DP+2, DX                DP+2 = 0100


Result =                DP = 0070, DP+2 = 0100
Actual value =          0001 7000 (+94,208)
+126,976 + (-32,768) = +94,208
```

Figure 9.6 Program segment to add a 32-bit augend to a 16-bit addend.

Figure 9.7 shows an assembly language module embedded in a C program that adds two unequal length operands using the ADC instruction in combination with the CWD instruction, this time using hexadecimal numbers that are entered from the keyboard. The augend consists of 32 bits, whereas the addend consists of 16 bits. The first operands are those shown in Figure 9.6.

The CWD instruction effectively doubles the size of the operand in the implied general-purpose register AX by extending the sign in AX throughout register DX. Thus, the result obtained by executing the CWD instruction is stored in register pair DX:AX, where the colon signifies concatenation. The 32-bit contents of the addend in registers DX:AX can now be added to the 32-bit augend. No flags are affected by the CWD instruction.

```
//dbl_precision3.cpp
//add a 32-bit operand to a 16-bit operand.
//the CWD and ADC instructions will be used

#include "stdafx.h"

int main (void)
{
//define variables
   unsigned short aug_hi, aug_lo, add_lo, sum_hi, sum_lo;

   printf ("Enter three 4-digit hexadecimal numbers:
            two for augend, one for addend: \n");
   scanf ("%X %X %X", &aug_hi, &aug_lo, &add_lo);

//switch to assembly
      _asm
      {
//obtain low sum
         MOV   AX, add_lo
         CWD                  //dx=ax sign (ext), ax=add_lo
         ADD   AX, aug_lo
         MOV   sum_lo, AX     //flags are not affected by mov

//obtain high sum
         MOV   AX, aug_hi
         ADC   AX, DX         //cf not changed since the add
         MOV   sum_hi, AX
      }

   printf ("\nAugend high = %X, Augend low = %X\n",
               aug_hi, aug_lo);

   printf ("\nAddend low = %X\n", add_lo);

   printf ("\nSum high = %X, Sum low = %X\n\n",
               sum_hi, sum_lo);

   return 0;
}
```

(a) //continued on next page

Figure 9.7 Using the ADC instruction in conjunction with the CWD instruction to add a 32-bit augend to a 16-bit addend: (a) the program and (b) the outputs.

```
Enter three 4-digit hexadecimal numbers: two for augend,
one for addend:
0001 F000 8000    //CF = 1

Augend high = 1, Augend low = F000

Addend low = 8000

Sum high = 1, Sum low = 7000

Press any key to continue . . . _
-------------------------------------------------------------
Enter three 4-digit hexadecimal numbers: two for augend,
one for addend:
ABCD 1234 1234    //CF = 0

Augend high = ABCD, Augend low = 1234

Addend low = 1234

Sum high = ABCD, Sum low = 2468

Press any key to continue . . . _
-------------------------------------------------------------
Enter three 4-digit hexadecimal numbers: two for augend,
one for addend:
1234 5678 F111    //CF = 1

Augend high = 1234, Augend low = 5678

Addend low = F111

Sum high = 1234, Sum low = 4789

Press any key to continue . . . _
-------------------------------------------------------------
Enter three 4-digit hexadecimal numbers: two for augend,
one for addend:
ABCD 0789 1234    //CF = 0

Augend high = ABCD, Augend low = 789

Addend low = 1234

Sum high = ABCD, Sum low = 19BD
                                   //continued on next page
```

(b)

Figure 9.7 (Continued)

```
Enter three 4-digit hexadecimal numbers: two for augend,
one for addend:
CDEF 568D C7AB     //CF = 1

Augend high = CDEF, Augend low = 568D

Addend low = C7AB

Sum high = CDEF, Sum low = 1E38
```

Figure 9.7 (Continued)

9.1.3 Increment by 1 (INC) Instruction

The *increment by 1* (INC) instruction is used primarily to increment counters and other unsigned integers by a value of 1. The INC instruction adds 1 to the destination operand, but does not affect the carry flag (CF). The flags that are affected are the overflow flag (OF), the sign flag (SF), the zero flag (ZF), the auxiliary flag (AF), and the parity flag (PF). The destination operand can be a value in a general-purpose register or a value in a memory location. The syntax for the INC instruction is shown below.

```
INC    register/memory
```

Binary-to-Gray code conversion A sufficient number of instructions have now been presented in order to illustrate the binary-to-Gray code conversion algorithm. A procedure for converting from the binary 8421 code to the Gray code can be formulated. Let an n-bit binary code word be represented as

$$b_{n-1}\, b_{n-2}\, \cdots\, b_1\, b_0$$

and an n-bit Gray code word be represented as

$$g_{n-1}\, g_{n-2}\, \cdots\, g_1\, g_0$$

where b_0 and g_0 are the low-order bits of the binary and Gray codes, respectively. The ith Gray code bit g_i can be obtained from the corresponding binary code word by the following algorithm:

$$\begin{aligned} g_{n-1} &= b_{n-1} \\ g_i &= b_i \oplus b_{i+1} \end{aligned} \tag{9.4}$$

for $0 \leq i \leq n-2$, where the symbol $\oplus$ denotes modulo-2 addition defined as:

$$0 \oplus 0 = 0$$

$$0 \oplus 1 = 1$$

$$1 \oplus 0 = 1$$

$$1 \oplus 1 = 0$$

Table 9.2 shows the relationship between the binary 8421 code and the Gray code. The Gray code belongs to a class of cyclic codes called reflective codes. Notice in the first four rows, that g_0 reflects across the reflecting axis; that is, g_0 in rows 2 and 3 is the mirror image of g_0 in rows 0 and 1. In the same manner, g_1 and g_0 reflect across the reflecting axis drawn under row 3. Thus, rows 4 through 7 reflect the state of rows 0 through 3 for g_1 and g_0. The same is true for g_2, g_1, and g_0 relative to rows 8 through 15 and rows 0 through 7. The Gray code is an unweighted code where only one input changes between adjacent code words.

Table 9.2 Table for Converting from the Binary 8421 Code to the Gray code

	Binary code				Gray code				
Row	b_3	b_2	b_1	b_0	g_3	g_2	g_1	g_0	
0	0	0	0	0	0	0	0	0	
1	0	0	0	1	0	0	0	1	
2	0	0	1	0	0	0	1	1	← g_0 is reflected
3	0	0	1	1	0	0	1	0	
4	0	1	0	0	0	1	1	0	← g_1 and g_0
5	0	1	0	1	0	1	1	1	are reflected
6	0	1	1	0	0	1	0	1	
7	0	1	1	1	0	1	0	0	
8	1	0	0	0	1	1	0	0	← g_2, g_1, and g_0
9	1	0	0	1	1	1	0	1	are reflected
10	1	0	1	0	1	1	1	1	
11	1	0	1	1	1	1	1	0	
12	1	1	0	0	1	0	1	0	
13	1	1	0	1	1	0	1	1	
14	1	1	1	0	1	0	0	1	
15	1	1	1	1	1	0	0	0	

Equation 9.4 indicates that the conversion process can be achieved by repetitive use of the exclusive-OR function. For example, using the algorithm of Equation 9.4, the 4-bit binary code word $b_3\ b_2\ b_1\ b_0 = 1010$ translates to the 4-bit Gray code word $g_3\ g_2\ g_1\ g_0 = 1111$ as follows:

$$g_3 = b_3 \qquad\qquad\qquad = 1$$

$$g_2 = b_2 \oplus b_3 = 0 \oplus 1 = 1$$

$$g_1 = b_1 \oplus b_2 = 1 \oplus 0 = 1$$

$$g_0 = b_0 \oplus b_1 = 0 \oplus 1 = 1$$

An example is shown in Figure 9.8 using the algorithm of Equation 9.4 to translate an 8-bit binary code word $b_7\ b_6\ b_5\ b_4\ b_3\ b_2\ b_1\ b_0 = 10101110$ to an 8-bit Gray code word $g_7\ g_6\ g_5\ g_4\ g_3\ g_2\ g_1\ g_0 = 11111001$ as follows:

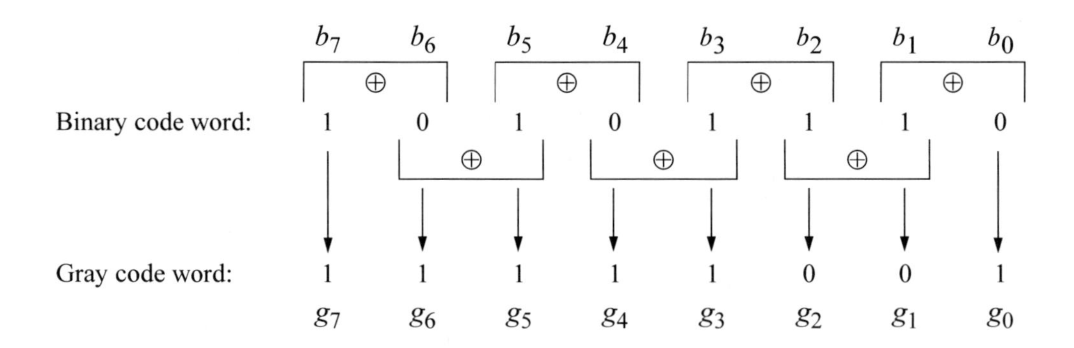

Figure 9.8 Example of translating an 8-bit binary code word to an 8-bit Gray code word.

An assembly language program will now be written to convert a binary code to the corresponding Gray code. Recall that a parameter list (PARLST) is used in the data segment (DTSG) to contain the maximum length (MAXLEN) of the operand field (OPFLD) and the actual length (ACTLEN) of the operand field after data have been entered from the keyboard.

A diagram of the parameter list in the data segment is shown in Figure 9.9 that contains an example of eight binary bits (10101110) that have been entered from the keyboard to be converted to the Gray code. The 1 and 0 bits are entered as ASCII characters and stored in the operand field — numbers have an ASCII bias of 3 (0011). The resulting Gray code is stored in the result area (RSLT) of the data segment and must also be represented as ASCII characters in order to be displayed correctly.

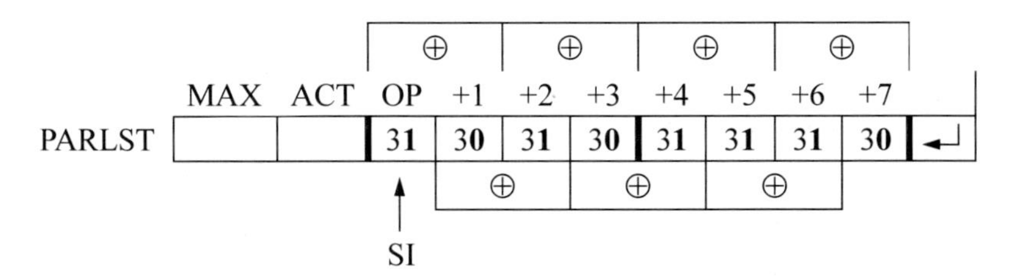

Figure 9.9 An example of the parameter list for the assembly language program of Figure 9.10.

Figure 9.10 contains an assembly language program — not embedded in a C program — that converts binary data to the Gray code. The comments in the code describe the program execution. When performing the exclusive-OR on two ASCII numbers, there is no need to remove the ASCII bias of 3 (0011). For example, $1 \oplus 1 = 0$; that is, 31H ⊕ 31H = 00110001 ⊕ 00110001 = 00000000, which is correct. Before the Gray code number can be displayed, however, the ASCII bias must be restored by performing either the OR operation or the ADD operation of 30H with the resulting Gray code bit. The same reasoning is true for $1 \oplus 0$.

```
PAGE 66, 80
;bin-to-gray2.asm

;-----------------------------------------------
.STACK

;-----------------------------------------------
.DATA
PARLST    LABEL BYTE
MAXLEN    DB    12
ACTLEN    DB    ?
OPFLD     DB    12 DUP(?)
PRMPT     DB    0DH, 0AH, 'Enter 8 binary bits: $'
RSLT      DB    0DH, 0AH, 'Gray code =           $'
;------------------------------------------------
.CODE
BEGIN     PROC  FAR

;set up pgm ds
          MOV   AX, @DATA        ;put addr of data seg in ax
          MOV   DS, AX           ;put addr in ds
```

(a) //continued on next page

Figure 9.10 Program to convert binary data to the corresponding Gray code: (a) the program and (b) the outputs.

```
;read prompt
        MOV   AH, 9             ;display string
        LEA   DX, PRMPT         ;put addr of prompt in dx
        INT   21H               ;dos interrupt

;keyboard request rtn to enter numbers
        MOV   AH, 0AH           ;buffered keyboard input
        LEA   DX, PARLST        ;load addr of parlst
        INT   21H               ;dos interrupt

;set up count for number of bits
        MOV   CX, 7             ;set up count
        LEA   SI, OPFLD         ;set up addr of opfld
        LEA   DI, RSLT + 14     ;set up addr of rslt

;move first char unchanged to rslt
        MOV   AL, [SI]          ;si points to 1st char in opfld
        MOV   [DI], AL          ;di points to the rslt + 14

;set up loop to process characters
NEXT:   MOV   AL, [SI]          ;move char to al
        INC   SI                ;inc si to point to next char
        MOV   AH, [SI]          ;move next char to ah
        XOR   AL, AH            ;perform the xor operation
        ADD   AL, 30H           ;add ascii bias
        INC   DI                ;inc di to next location in rslt
        MOV   [DI], AL          ;move gray code bit to rslt area
        LOOP  NEXT              ;decrement cx. If cx != 0, loop

;display the result
        MOV   AH, 9             ;display string
        LEA   DX, RSLT          ;put addr of rslt in dx
        INT   21H               ;dos interrupt

BEGIN   ENDP
        END   BEGIN
```

```
Enter  8 binary bits: 10101110
Gray code = 11111001
------------------------------------------------------------
Enter  8 binary bits: 01111010
Gray code = 01000111
------------------------------------------------------------
Enter  8 binary bits: 11111111
Gray code = 10000000
------------------------------------------------------------
```

(b) //continued on next page

Figure 9.10 (Continued)

```
Enter  8 binary bits: 01010110
Gray code = 01111101
```

Figure 9.10 (Continued)

9.2 Subtraction

The two operands for subtraction are the minuend and the subtrahend — the subtrahend is subtracted from the minuend according to the rules shown in Table 9.3 for radix 2.

Table 9.3 Truth Table for Subtraction

0 – 0	=	0	
0 – 1	=	1	with a borrow from the next higher-order minuend
1 – 0	=	1	
1 – 1	=	0	

An example is shown in Figure 9.11 using the rules of Table 9.3, in which the subtrahend 0101 0011 (+83) is subtracted from the minuend 0110 0101 (+101), resulting in a difference of 0001 0010 (+18). This is called the paper-and-pencil method, but is not applicable for subtraction in a computer. The difference in column 2^1 is $0 - 1 = 1$ with a borrow from the minuend in column 2^2, which results in a difference of $0 - 0 = 0$ in column 2^2. The same rationale applies to column 2^4 and column 2^5.

	2^7	2^6	2^5	2^4	2^3	2^2	2^1	2^0
Minuend (+101)	0	1	1^1	0	0	1^1	0	1
–) Subtrahend (+83)	0	1	0	1	0	0	1	1
Difference (+18)	0	0	0	1	0	0	1	0

Figure 9.11 Example of the paper-and-pencil method of binary subtraction.

Subtraction is similar to addition, because computers use an adder for the subtraction operation by adding the radix complement of the subtrahend to the minuend. Recall that the *r*s complement is obtained from the $r - 1$ complement by adding 1. For radix 2, the 2s complement is obtained by adding 1 to the 1s complement. Arithmetic processors use an adder to perform subtraction by adding the 2s complement of the subtrahend to the minuend. Let A and B be two n-bit operands, where A is the minuend and B is the subtrahend as follows:

$$A = a_{n-1}\, a_{n-2} \ldots a_1\, a_0$$

$$B = b_{n-1}\, b_{n-2} \ldots b_1\, b_0$$

Therefore, $A - B = A + (B' + 1)$, where B' is the 1s complement of B. Figure 9.11 is repeated as shown in Figure 9.12 using this method.

	2^7	2^6	2^5	2^4	2^3	2^2	2^1	2^0
Minuend (+101)	0	1	1	0	0	1	0	1
+) Subtrahend (–83)	1	0	1	0	1	1	0	1
Difference (+18)	0	0	0	1	0	0	1	0

Figure 9.12 Example of subtraction by adding the radix complement of the subtrahend to the minuend.

When subtracting two operands by adding the 2s complement of the subtrahend, the states of the auxiliary carry flag (AF) and the carry flag (CF) are inverted. For example, a subtraction will be performed on the following operands: minuend 0110 0001 (+97) and subtrahend 0110 0101 (+101) to yield a difference of 1111 1100 (–4). Figure 9.13(a) shows the paper-and-pencil method — including borrows — and Figure 9.13(b) shows the same subtraction accomplished by adding the 2s complement of the subtrahend — including carries.

In Figure 9.13(a), the borrow from column 2^4 represents the auxiliary carry flag (AF = 1) and the borrow from column 2^8 represents the carry flag (CF = 1). Since column 2^4 has a 0 in the minuend, the borrow must come from column 2^5, which leaves a 0 for the minuend in column 2^5. The process repeats for columns 2^5 through column 2^8. In Figure 9.15(b), the auxiliary carry flag and the carry flag are indicated as having a value of 0; however, the true values are 1.

	2^8	2^7	2^6	2^5	2^4	2^3	2^2	2^1	2^0
Minuend (+97)	1	0^1	1^0	1^0	0^1	0^1	0	0	1
–) Subtrahend (+101)		0	1	1	0	0	1	0	1
Difference (–4)		1	1	1	1	1	1	0	0

(a) //continued next page

Figure 9.13 Two methods of subtraction: (a) the paper-and-pencil method and (b) adding the 2s complement of the subtrahend.

	2^8	2^7	2^6	2^5	2^4	2^3	2^2	2^1	2^0
Minuend (+97)		0	1	1	0	0	0	0	1
+) Subtrahend (–101)	0	1	0	0	1_0	1	0_1	1_1	1
Difference (–4)		1	1	1	1	1	1	0	0

(b)

Figure 9.13 (Continued)

Additional examples of subtraction are shown in Figure 9.14 for both positive and negative operands using the radix complement (2s complement) method.

	$A =$	0 0 0 0 1 1 1 1	+15
–)	$B =$	0 1 1 0 0 0 0 0	+96
		↓	
		0 0 0 0 1 1 1 1	+15
+)		1 0 1 0 0 0 0 0	–96
		1 0 1 0 1 1 1 1	–81

	$A =$	1 0 1 1 0 0 0 1	–79
–)	$B =$	1 1 1 0 0 1 0 0	–28
		↓	
		1 0 1 1 0 0 0 1	–79
+)		0 0 0 1 1 1 0 0	+28
		1 1 0 0 1 1 0 1	–51

Figure 9.14 Additional examples of subtraction in which the radix complement of the subtrahend is added to the minuend.

This section presents the following subtract instructions: *subtract* (SUB), *integer subtraction with borrow* (SBB), *decrement by 1* (DEC), and *negate* (NEG).

9.2.1 Subtract (SUB) Instruction

The subtract instruction performs a subtraction on signed (2s complement) or unsigned integer operands that can be bytes, words, or doublewords. The syntax for the subtract operation is shown below. The source operand (second operand — subtrahend) is subtracted from the destination operand (first operand — minuend) and stores the difference in the destination operand.

```
SUB    register/memory, register/memory/immediate
```

The destination operand can be a register or a memory location; the source operand can be a register, a memory location, or an immediate value; however, both operands cannot be in memory. When an immediate operand is used, it is sign-extended to match the size of the destination operand, if necessary.

The following flags are affected: the overflow flag (OF), the sign flag (SF), the zero flag (ZF), the auxiliary carry flag (AF), the parity flag (PF), and the carry flag (CF). If signed operands are utilized, then the overflow flag is set to indicate an overflow condition. The default operand size is 32 bits in 64-bit mode. Using a REX prefix allows access to eight additional registers: R8 through R15.

Figure 9.15 shows an assembly language module embedded in a C program that illustrates the use of the SUB instruction. Two signed or unsigned decimal operands are entered from the keyboard — the first is for the minuend, the second is for the subtrahend. After the subtraction takes place, the resulting difference and flags are displayed. The program is shown in Figure 9.15(a) and the outputs for various operands are shown in Figure 9.15(b).

The results of the first five pairs of operands are easy to understand. The fifth pair of operands has a minuend value of 1879048192_{10} which is $0111\ 0000\ 0000\ 0000\ 0000\ 0000\ 0000\ 0000_2$ and a subtrahend value of 2415919104_{10} which is $1001\ 0000\ 0000\ 0000\ 0000\ 0000\ 0000\ 0000_2$. When the subtrahend is negated it becomes $0111\ 0000\ 0000\ 0000\ 0000\ 0000\ 0000\ 0000_2$, resulting in an overflow condition for a difference of 3758096384_{10} which is $1110\ 0000\ 0000\ 0000\ 0000\ 0000\ 0000\ 0000_2$ as shown below.

	Minuend =	0111	0000	0000	0000	0000	0000	0000	0000
–)	Subtrahend =	1001	0000	0000	0000	0000	0000	0000	0000
					↓				
		0111	0000	0000	0000	0000	0000	0000	0000
+)		0111	0000	0000	0000	0000	0000	0000	0000
		1110	0000	0000	0000	0000	0000	0000	0000

It can be clearly seen that an overflow has occurred, because the signs of the operands and result satisfy the following equation: $(a_{n-1}' \bullet b_{n-1}' \bullet s_{n-1})$. The sign flag has a value of 1; the parity flag has a value of 1, because the low-order byte has an even number of 1s; and the carry flag is 1, because 0 – 1 = 1 with a borrow from the next higher-order bit of the minuend (refer to Table 9.3).

```
//subtract.cpp
//subtraction of two signed or unsigned operands
//and show the resulting flags
#include "stdafx.h"

int main (void)
{
//define variables
   int minuend, subtrahend, difference;
   unsigned short flags;

   printf ("Enter two signed or unsigned decimal integers:
            \n");
   scanf ("%d %d", &minuend, &subtrahend);

//switch to assembly
      _asm
      {
         MOV   EAX, minuend
         MOV   EBX, subtrahend
         SUB   EAX, EBX
         MOV   difference, EAX

         PUSHF
         POP   AX
         MOV   flags, AX
      }

   printf ("\nMinuend = %d\nSubtrahend = %d\n",
            minuend, subtrahend);
   printf ("\nDifference = %d\nFlags = %X\n\n",
            difference, flags);

   return 0;
}
```

(a) //continued on next page

Figure 9.15 Program to illustrate the use of the SUB instruction: (a) the program and (b) the outputs.

```
Enter two signed or unsigned decimal integers:
46 20

Minuend = 46
Subtrahend = 20

Difference = 26
Flags = 202          //IF

Press any key to continue . . . _
-------------------------------------------------------------
Enter two signed or unsigned decimal integers:
42 -30

Minuend = 42
Subtrahend = -30

Difference = 72
Flags = 207          //IF, PF, CF

Press any key to continue . . . _
-------------------------------------------------------------
Enter two signed or unsigned decimal integers:
-250 250

Minuend = -250
Subtrahend = 250

Difference = -500
Flags = 296          //IF, SF, AF, PF

Press any key to continue . . . _
-------------------------------------------------------------
Enter two signed or unsigned decimal integers:
-86 -86

Minuend = -86
Subtrahend = -86

Difference = 0
Flags = 246          //IF, ZF, PF

Press any key to continue . . . _
-------------------------------------------------------------
```

(b) //continued on next page

Figure 9.15 (Continued)

```
Enter two signed or unsigned decimal integers:
28672 -28672

Minuend = 28672
Subtrahend = -28672

Difference = 57344
Flags = 207            //IF, PF, CF

Press any key to continue . . . _
------------------------------------------------------------
Enter two signed or unsigned decimal integers:
1879048192 2415919104

Minuend = 1879048192
Subtrahend = -1879048912

Difference = -536870912
Flags = A87            //OF, IF, SF, PF, CF

Press any key to continue . . . _
```

Figure 9.15 (Continued)

9.2.2 Integer Subtraction with Borrow (SBB) Instruction

The SBB instruction subtracts the source operand (second operand) from the destination operand (first operand) and subtracts 1 if the carry flag is set — the carry flag represents a borrow from a previous subtraction. This operation is accomplished by adding the source operand and the carry flag, then subtracting the sum from the destination operand. The difference is stored in the destination operand.

The SBB instruction is used to subtract multiple bytes or words and is preceded by a SUB instruction. The destination operand can be a register or a memory location; the source operand can be a register, an immediate operand, or a memory location. However, both operands cannot be memory locations. The syntax for the SBB instruction is shown below.

```
SBB    register/memory, register/immediate/memory
```

If an immediate operand is used, the sign is extended to match the size of the destination operand. The SBB instruction is used for both signed and unsigned operands. The overflow flag (OF) is used in conjunction with signed operands; the carry flag (CF) is used for unsigned operands. Using an REX prefix allows access to eight supplementary registers: R8 through R15.

The affected flags resulting from the SBB instruction are the overflow flag (OF), the sign flag (SF), the zero flag (ZF), the auxiliary carry flag (AF), the parity flag (PF), and the carry flag (CF).

9.2.3 Decrement by 1 (DEC) Instruction

The DEC instruction subtracts 1 from the destination operand, which is an integer operand. The destination operand can be a register or a memory location. The affected flags resulting from the DEC instruction are the overflow flag (OF), the sign flag (SF), the zero flag (ZF), the auxiliary carry flag (AF), and the parity flag (PF); the carry flag (CF) is not affected. If the state of the carry flag is required, then a subtraction with an immediate value of 1 can be executed. The DEC instruction can also access eight additional registers — R8 through R15 — if an REX prefix is utilized. The syntax for the DEC instruction is shown below.

```
DEC    register/memory
```

An assembly language program — not embedded in a C program — will now be presented that obtains the sum of the smallest and largest *n* single-digit numbers that are entered from the keyboard, where *n* can range from 1 to a much larger number, for example 30. The DEC instruction will be used throughout the program. Any program of this size is easier to implement if a flowchart is developed prior to coding the program. There are six main symbols used in drawing flowcharts, as shown in Figure 9.16. Arrows indicate the control flow through the flowchart.

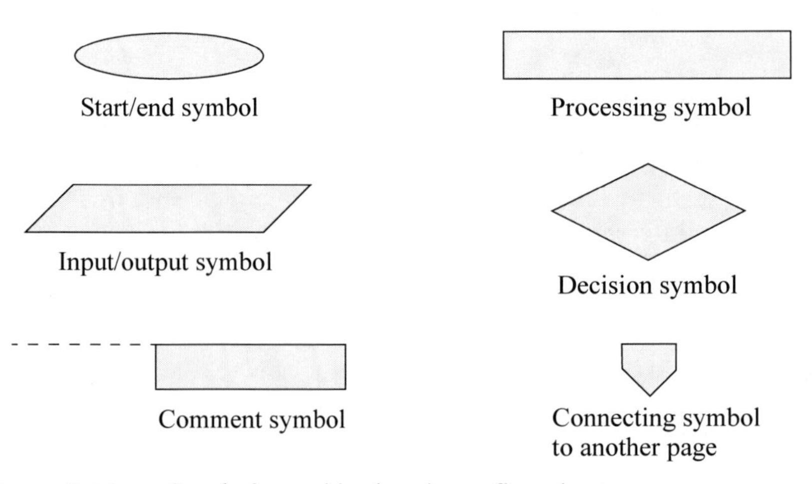

Figure 9.16 Symbols used in drawing a flowchart.

The flowchart for this program is shown in Figure 9.17. The program will be designed directly from the flowchart. The stack, data, and code segments are initialized in the processing symbol following the *start* symbol. The *initialize* symbol sets up the program's data segment (DS). Numbers are entered using the parallelogram labelled *keyboard rtn*. The decision symbol labelled *1 digit entered?* checks to determine if only one number was entered. The remaining symbols are self-explanatory. The assembly language program is shown in Figure 9.18 and follows the sequence presented in the flowchart.

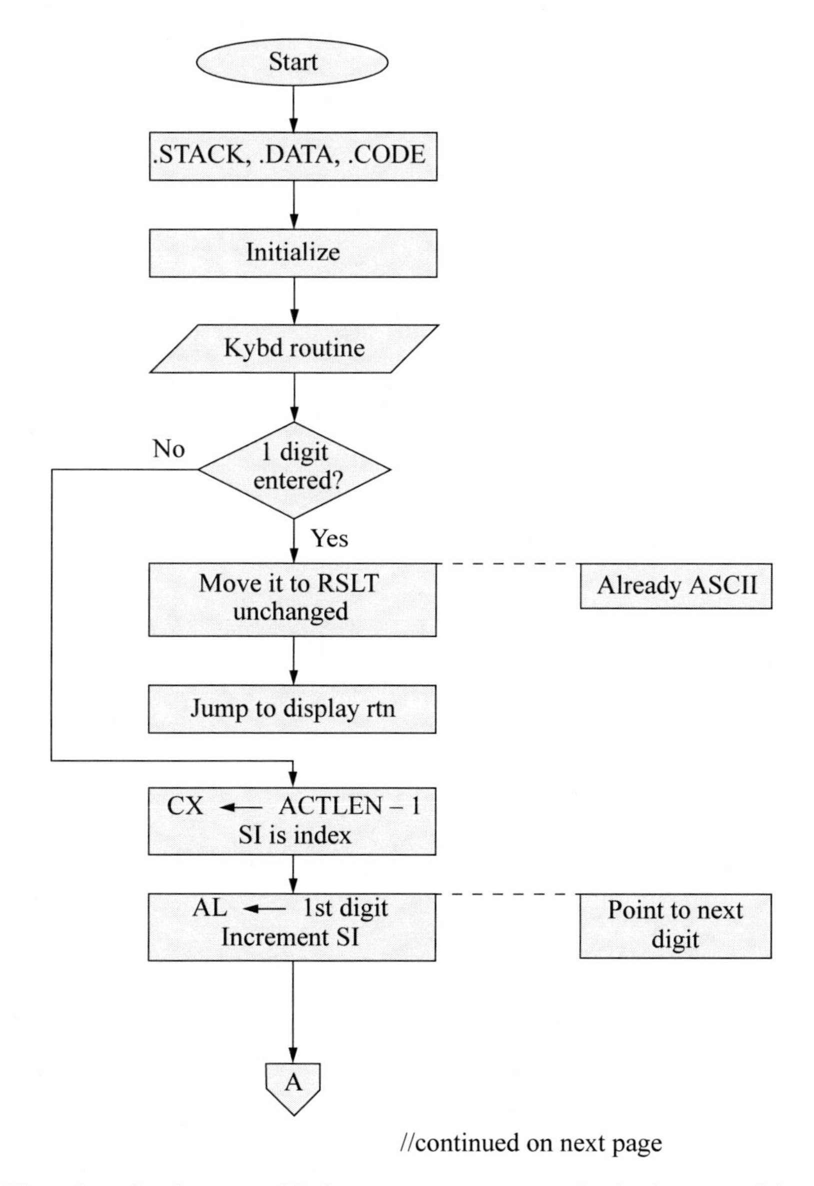

//continued on next page

Figure 9.17 Flowchart for the assembly language program to obtain the sum of the smallest and largest single-digit numbers that are entered from the keyboard.

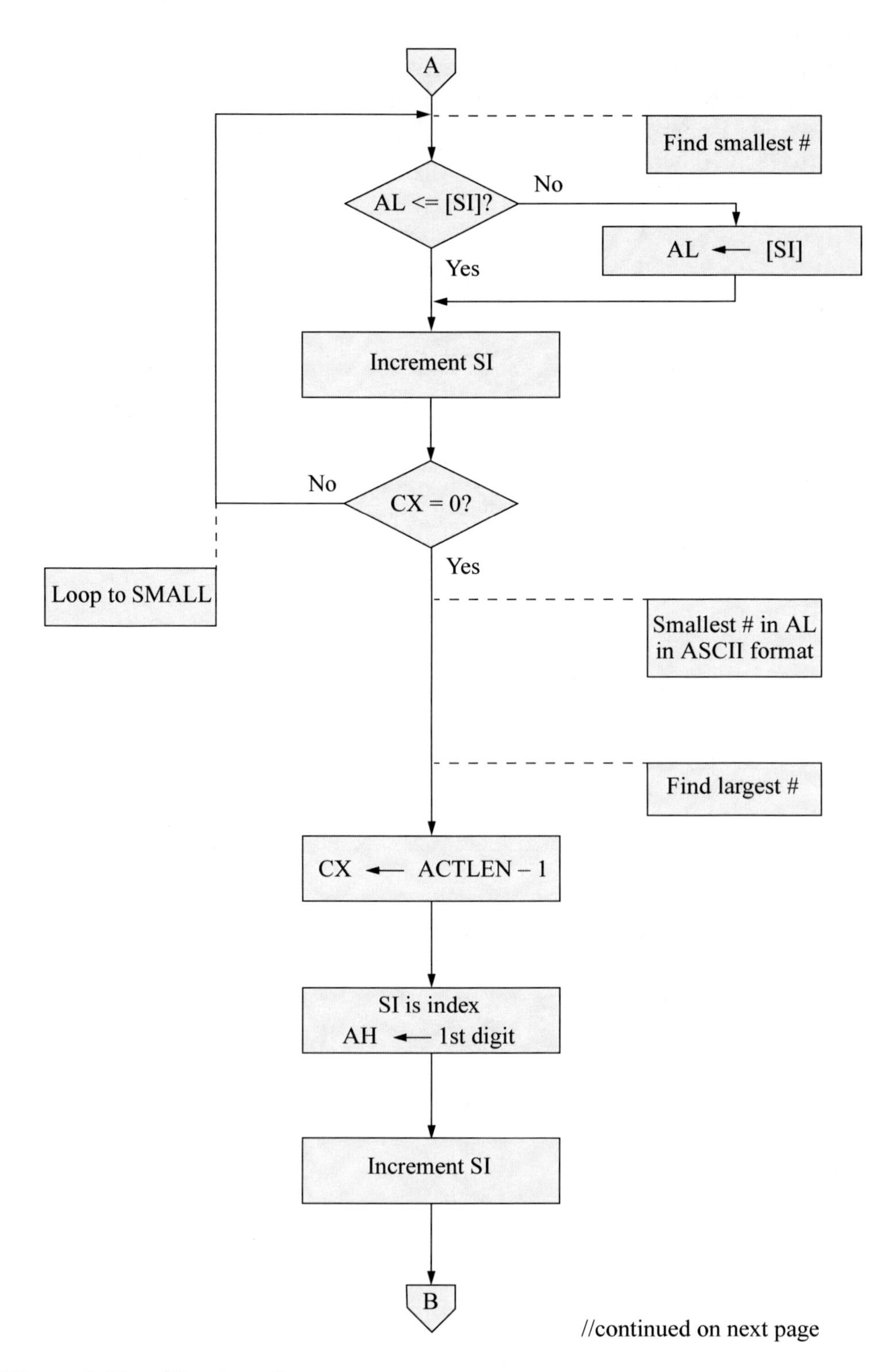

Figure 9.17 (Continued)

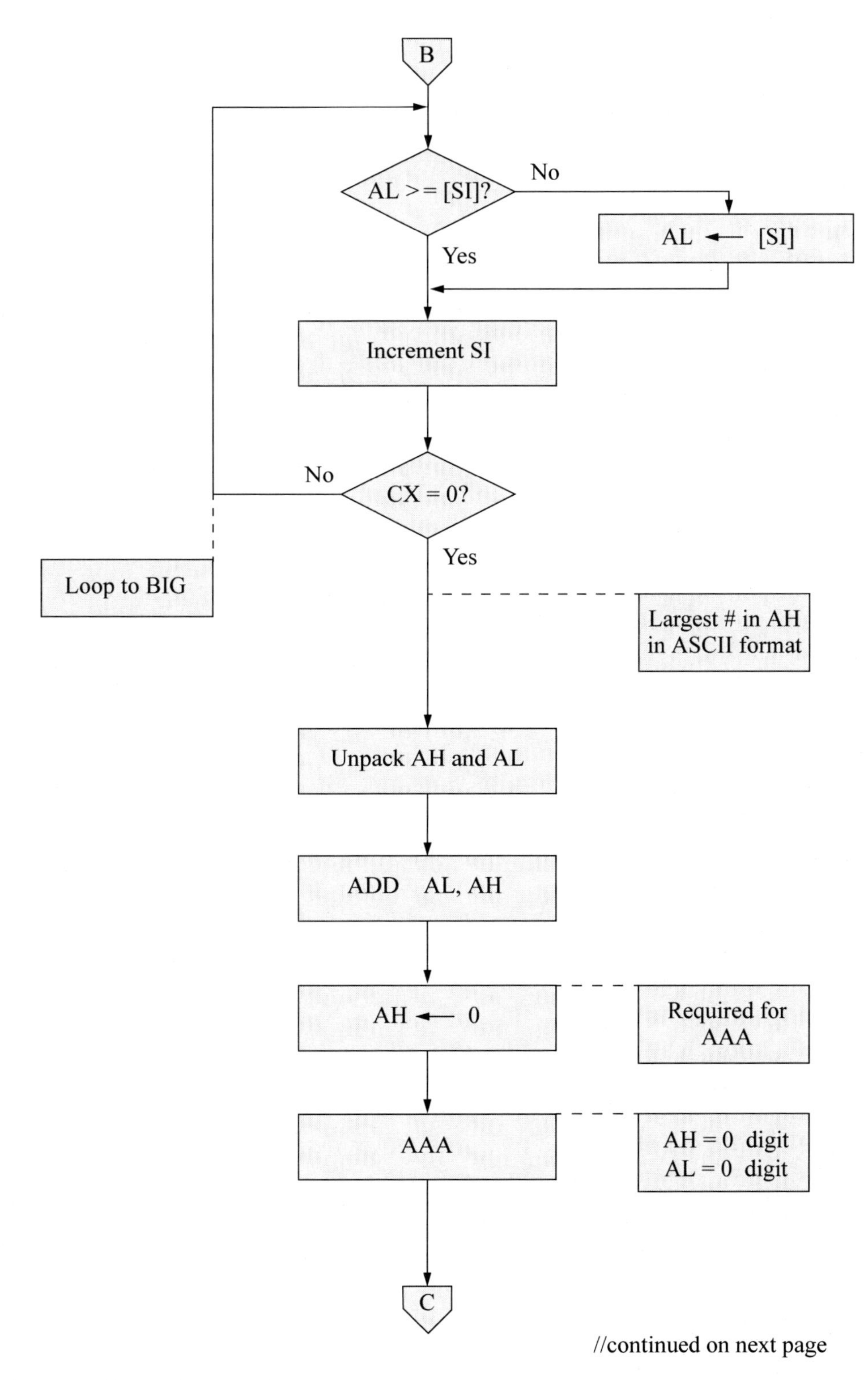

//continued on next page

Figure 9.17 (Continued)

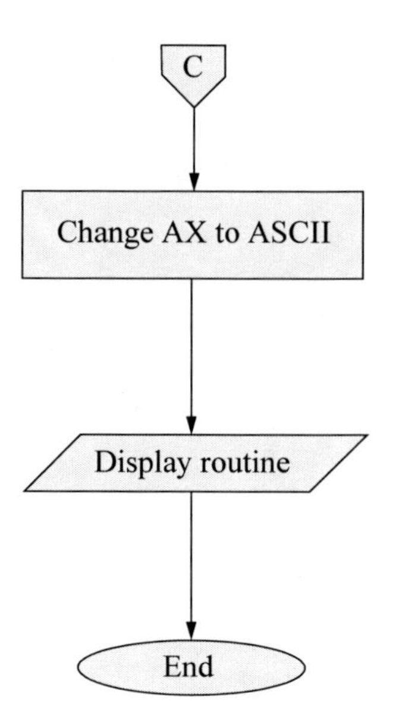

Figure 9.17 (Continued)

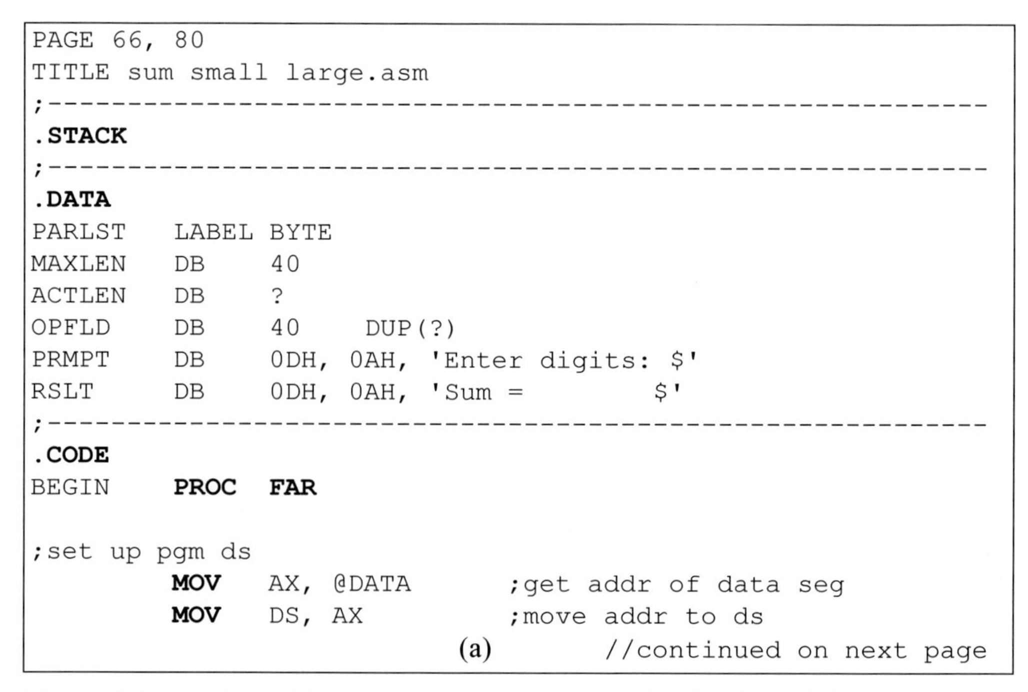

```
PAGE 66, 80
TITLE sum small large.asm
;-------------------------------------------------------------
.STACK
;-------------------------------------------------------------
.DATA
PARLST   LABEL BYTE
MAXLEN   DB    40
ACTLEN   DB    ?
OPFLD    DB    40    DUP(?)
PRMPT    DB    0DH, 0AH, 'Enter digits: $'
RSLT     DB    0DH, 0AH, 'Sum =        $'
;-------------------------------------------------------------
.CODE
BEGIN    PROC  FAR

;set up pgm ds
         MOV   AX, @DATA        ;get addr of data seg
         MOV   DS, AX           ;move addr to ds
```

(a) //continued on next page

Figure 9.18 Assembly language program to obtain the sum of the smallest and largest single-digit numbers that are entered from the keyboard: (a) the program and (b) the outputs.

```
;read prompt
         MOV    AH, 09H          ;display string
         LEA    DX, PRMPT        ;put addr of prompt in dx
         INT    21H              ;dos interrupt

;keyboard request rtn to enter characters
         MOV    AH, 0AH          ;buffered keyboard input
         LEA    DX, PARLST       ;put addr of parlst in dx
         INT    21H              ;dos interrupt

;check if one number entered
         CMP    ACTLEN, 1        ;comp actual #s entered with 1
         JNE    CONT             ;if more than 1, jmp to CONT rtn
         MOV    AL, OPFLD        ;else move the only number to al
         MOV    RSLT + 8, AL     ;move the # to the result area
         JMP    PRINT            ;jump to the print rtn

;if more than one number entered
CONT:    MOV    CL, ACTLEN       ;put qty of #s entered in cl
         MOV    CH, 0            ;clear ch
         DEC    CX               ;decrease count by 1

         LEA    SI, OPFLD        ;load the addr of opfld into si
         MOV    AL, [SI]         ;put the number in al
         INC    SI               ;point to next number in opfld

;find smallest number
SMALL:   CMP    AL, [SI]         ;comp al with # pointed to by si
         JLE    INCR1            ;if # is <=, jump to INCR1 rtn
         MOV    AL, [SI]         ;if !<=, move the larger # to al

INCR1:   INC    SI               ;point to next number in opfld
         LOOP   SMALL            ;find next smallest number

;find largest number
         MOV    CL, ACTLEN       ;put qty of #s entered in cl
         MOV    CH, 0            ;clear ch
         DEC    CX               ;decrease count by 1

         LEA    SI, OPFLD        ;load the addr of opfld into si
         MOV    AH, [SI]         ;put the number in ah
         INC    SI               ;point to next number in opfld

BIG:     CMP    AH, [SI]         ;comp ah with # pointed to by si
         JGE    INCR2            ;if # is >=, jump to INCR2 rtn
         MOV    AH, [SI]         ;if !>=, move smaller # to al
                                    //continued on next page
```

Figure 9.18 (Continued)

```
INCR2:   INC    SI              ;point to next number in opfld
         LOOP   BIG             ;find next largest number


;add smallest and largest
         AND    AL, 0FH         ;remove ascii bias -- unpack
         AND    AH, 0FH         ;remove ascii bias -- unpack
         ADD    AL, AH          ;add the smallest and largest
         MOV    AH, 0           ;required for aaa
         AAA                    ;aaa is presented in chapter 10
         OR     AX, 3030H       ;restore ascii bias
         MOV    RSLT + 8, AH    ;move high-order # to rslt area
         MOV    RSLT + 9, AL    ;move low-order # to rslt area


;display the result
PRINT:   MOV    AH, 09H         ;display string
         LEA    DX, RSLT        ;put addr of rslt in dx
         INT    21H             ;dos interrupt


BEGIN    ENDP
         END    BEGIN
```

```
Enter digits: 7
Sum = 7
------------------------------------------------------------
Enter digits: 26
Sum = 08
------------------------------------------------------------
Enter digits: 739
Sum = 12
------------------------------------------------------------
Enter digits: 0895103
Sum = 09
------------------------------------------------------------
Enter digits: 8769976
Sum = 15
------------------------------------------------------------
Enter digits: 9999999999
Sum = 18
```

(b)

Figure 9.18 (Continued)

9.2.4 Two's Complement Negation (NEG) Instruction

The NEG instruction generates the 2s complement of a number by subtracting the destination integer operand from zero. This procedure changes the sign of the number, but does not alter the absolute value of the number. Numbers that contain a high-order one preceded by all zeroes does not change when it is negated. For example, the number 1000 0000 (–128) will be negated by subtracting the number from zero, as shown below, using the rules of Table 9.3. The 1 bit in column 2^8 represents a borrow from column 2^8.

2^8	2^7	2^6	2^5	2^4	2^3	2^2	2^1	2^0
1	0	0	0	0	0	0	0	0
–)	1	0	0	0	0	0	0	0
	1	0	0	0	0	0	0	0

The destination operand can be in a register or a memory location. If an REX prefix is used in 64-bit mode, the number of general-purpose registers is increased from eight to sixteen and can be extended to 64 bits. The syntax for the NEG instruction is shown below.

```
NEG    register/memory
```

If the destination operand is zero before the NEG instruction is executed, then the carry flag (CF) is reset to zero; otherwise, it is set to one. The remaining flags — OF, SF, ZF, AF, and PF — are set or reset depending on the result of the NEG instruction.

The negation of a number can also be accomplished using a paper-and-pencil method by keeping the low-order 0 bits and the first 1 bit unchanged, then inverting all remaining higher-order bits, as the number is scanned from right to left.

9.3 Multiplication

Fixed-point multiplication is more complex than either addition or subtraction. This section will present techniques and examples to multiply both signed and unsigned operands. The n-bit *multiplicand A* is multiplied by the n-bit *multiplier B* to produce a $2n$-bit *product P*, as shown below.

$$\begin{aligned}\text{Multiplicand:}\quad & A = a_{n-1}\, a_{n-2}\, a_{n-3} \cdots a_1\, a_0 \\ \text{Multiplier:}\quad & B = b_{n-1}\, b_{n-2}\, b_{n-3} \cdots b_1\, b_0 \\ \text{Product:}\quad & P = p_{2n-1}\, p_{2n-2}\, p_{2n-3} \cdots p_1\, p_0\end{aligned}$$

An algorithm for the paper-and-pencil method will be described for multiplying unsigned and signed operands, then examples using these methods will be presented. The algorithm consists of multiplying the multiplicand by the low-order multiplier bit to obtain a partial product. If the multiplier bit is 1, then the multiplicand becomes the partial product; if the multiplier bit is 0, then zeroes become the partial product. For signed multiplication, the sign bit extends left to produce a partial product of $2n$ bits.

The partial product is then shifted left 1 bit position and the multiplicand is multiplied by the next higher-order multiplier bit to obtain a second partial product. Each partial product is shifted left relative to the previous partial product. The process repeats for all remaining multiplier bits, at which time the partial products are added to obtain the product.

For signed multiplication, the sign of the product is positive if both operands have the same sign. If the signs of the operands are different, then the sign of the product is negative. Multiplication of two fixed-point binary numbers is a process of repeated add-shift operations. There are two X86 assembly language instructions that are used for multiplication: *unsigned multiply* (MUL) and *signed multiply* (IMUL).

9.3.1 Unsigned Multiply (MUL) Instruction

The MUL instruction performs a multiplication on two unsigned integer operands: an unsigned multiplicand (destination operand) and an unsigned multiplier (source operand). The implied destination operand is in the accumulator — registers AL, AX, EAX, or RAX, which are a function of the size of the operands. The multiplier source operand can be in a general-purpose register or in a memory location.

The $2n$-bit product is stored in register AX, register pair DX:AX, register pair EDX:EAX, or register pair RDX:RAX. The high-order bits of the product are in registers AH, DX, EDX, or RDX. If the high-order bits of the product are not all zeroes, then the carry flag (CF) and the overflow flag (OF) are set. The syntax is as follows:

```
MUL    register/memory
```

Example 9.1 This example multiplies two 4-bit operands, both with 0s as the high-order bits: multiplicand *a[3:0]* = 0111 (7) and multiplier *b[3:0]* = 0101 (5) to produce a product *p[7:0]* = 0010 0011 (35). A multiplier bit of 1 copies the multiplicand to the low-order four bits of the partial product with zeroes extended to eight bits; a multiplier bit of 0 enters zeroes in the partial product.

Multiplicand A					0	1	1	1	7
Multiplier B				×)	0	1	0	1	5
Partial products	0	0	0	0	0	1	1	1	
	0	0	0	0	0	0	0		
	0	0	0	1	1	1			
	0	0	0	0	0				
Product P	0	0	1	0	0	0	1	1	35

Example 9.2 This example multiplies two 4-bit operands, both with 1s as the high-order bits: multiplicand *a[3:0]* = 1100 (12) and multiplier *b[3:0]* = 1011 (11) to produce a product *p[7:0]* = 1000 0100 (132). A multiplier bit of 1 copies the multiplicand to the low-order four bits of the partial product with zeroes extended to eight bits; a multiplier bit of 0 enters zeroes in the partial product.

Multiplicand *A*					1	1	0	0	12
Multiplier *B*				×)	1	0	1	1	11
Partial Products	0	0	0	0	1	1	0	0	
	0	0	0	1	1	0	0		
	0	0	0	0	0	0			
	0	1	1	0	0				
Product *P*	1	0	0	0	0	1	0	0	132

Figure 9.19 shows a listing for an assembly language module embedded in a C program that multiplies an unsigned multiplicand in register EAX by an unsigned multiplier. The flags are displayed after the multiply operation has been executed. The first set of operands that are entered from the keyboard result in the high-order part of the product being equal to zero; thus, the overflow flag and the carry flag are reset.

The second set of operands result in the high-order part of the product being nonzero; thus, the overflow flag and the carry flag are set. In both cases the parity flag is set, because the low-order byte contains an even number of ones. Although the operands are normally the same length, it is also possible to multiply operands of different lengths, as shown in Figure 9.20 for a doubleword multiplicand and a word multiplier. When the MUL instruction is used to multiply two unsigned operands, the operands are treated as unsigned even though the high-order bit — normally the sign bit — is a 1.

```
//mul.cpp
//illustrates unsigned multiplication and resulting flags
#include "stdafx.h"
int main (void)
{
//define variables
   unsigned long mpcnd, mplyr, prod_unsigned_hi,
                  prod_unsigned_lo;
   unsigned short flags;
   printf ("Enter a hexadecimal mpcnd and mplyr: \n");
   scanf ("%X %X", &mpcnd, &mplyr);
```

(a) //continued on next page

Figure 9.19 Program to illustrate unsigned multiplication and the resulting flags: (a) the program and (b) the outputs.

```
//switch to assembly
      _asm
      {
         MOV   EAX, mpcnd
         MUL   mplyr                ;product in edx:eax
         MOV   prod_unsigned_hi, EDX
         MOV   prod_unsigned_lo, EAX

         PUSHF                      ;push low-order 16 bits
         POP   AX                   ;pop flags to ax
         MOV   flags, AX
      }

   printf ("\nUnsigned product high = %X",
               prod_unsigned_hi);
   printf ("\nUnsigned product low = %X\n",
               prod_unsigned_lo);
   printf ("\nFlags = %X\n\n", flags);

   return 0;
}
```

```
Enter a hexadecimal mpcnd and mplyr:
4567 ABCD

Unsigned product high = 0
Unsigned product low = 2E93607B

Flags = 206             //IF, PF

Press any key to continue . . . _
------------------------------------------------------------
Enter a hexadecimal mpcnd and mplyr:
12345 12345

Unsigned product high = 1
Unsigned product low = 4B65F099

Flags = A07             //OF, IF, PF, CF

Press any key to continue . . . _
```

(b)

Figure 9.19 (Continued)

```
//mul_dblwd_wd.cpp
//multiply a double word by a word
//to generate a product of three words

#include "stdafx.h"

int main (void)
{
//define variables
   unsigned long mpcnd, mplyr, prod_hi, prod_lo;
   unsigned short flags;

   printf ("Enter a hex 8-digit mpcnd and a 4-digit mplyr:
            \n");
   scanf ("%X %X", &mpcnd, &mplyr);

//switch to assembly
      _asm
      {
         MOV   EAX, mpcnd
         MUL   mplyr          ;product in edx:eax
         MOV   prod_hi, EDX
         MOV   prod_lo, EAX

         PUSHF                ;push low-order 16 bits
         POP   AX             ;pop flags
         MOV   flags, AX
      }

   printf ("\nProduct high = %X\nProduct low = %X\n",
               prod_hi, prod_lo);

   printf ("\nFlags = %X\n\n", flags);

   return 0;
}
                                    //continued on next page
```

(a)

Figure 9.20 Program to illustrate multiplying a doubleword multiplicand by a word multiplier: (a) the program and (b) the outputs.

```
Enter a hex 8-digit mpcnd and a 4-digit mplyr:
AAAABBBB CCCC

Product high = 8888
Product low = DA69D04

Flags = A03                //OF, IF, CF
-----------------------------------------------------------------
Enter a hex 8-digit mpcnd and a 4-digit mplyr:
32062521 6400

Product high = 138A
Product low = 6680E400

Flags = A07                //OF, IF, PF, CF
```

(b)

Figure 9.20 (Continued)

9.3.2 Signed Multiply (IMUL) Instruction

The IMUL instruction executes a multiply operation on a signed multiplicand and a signed multiplier. There are three forms for the IMUL instruction, which are a function of the operands being used: one operand, two operands, or three operands. When the IMUL instruction is used to multiply two signed operands, the high-order bit represents the sign bit — 0 for positive operands and 1 for negative operands. For both the MUL instruction and the IMUL instruction, an REX prefix in 64-bit mode increases the number of general-purpose registers from eight to sixteen and can be extended to 64 bits.

One operand This form is equivalent to the syntax used for the MUL instruction; thus, the source operand is in a general-purpose register or in a memory location. The implied destination operand is in the accumulator — registers AL, AX, EAX, or RAX, which are a function of the operand size. The syntax for the one operand IMUL instruction is shown below.

```
IMUL    register/memory
```

Two operands This form multiplies the destination operand (first operand) by the source operand (second operand) and stores the product in the destination operand. The destination operand is a general-purpose register and the source operand is a general-purpose register, a memory location, or an immediate operand. The syntax for the two-operand IMUL instruction is shown below.

```
IMUL    register, register/memory/immediate
```

Three operands This form uses three operands: the destination operand (first operand) and two source operands (second operand and third operand). The destination operand is a general-purpose register; the first source operand can be a general-purpose register or a memory location; the second source operand is an immediate value. The first source operand is multiplied by the second source operand and the product is stored in the destination operand. If an immediate operand is utilized, the sign extends to match the size of the destination operand.

If the high-order half of the product is not the sign extension, then the overflow flag (OF) and the carry flag (CF) are set; otherwise, the flags are reset. When the overflow flag and the carry flag are set, this indicates that the high-order half of the product contains significant bits of the result. The sign flag (SF), the zero flag (ZF), the auxiliary carry flag (AF), and the parity flag (PF) are undefined following execution of the IMUL instruction. An n-bit multiplicand and an n-bit multiplier produce a $2n$-bit product. The syntax for the three-operand IMUL instruction is shown below.

```
IMUL    register, register/memory, immediate
```

An algorithm for the paper-and-pencil method will be described for multiplying signed operands, then examples using this method will be presented. The algorithm is similar to that used for unsigned operands and consists of multiplying the multiplicand by the low-order multiplier bit to obtain a partial product. If the multiplier bit is a 1, then the multiplicand becomes the partial product with the sign extended to match the size of the $2n$-bit product; if the multiplier bit is a 0, then zeroes become the partial product with zeroes extended to match the size of the $2n$-bit product.

The partial product is then shifted left 1 bit position and the multiplicand is multiplied by the next higher-order multiplier bit to obtain a second partial product. Each partial product is shifted left relative to the previous partial product. The process repeats for all remaining multiplier bits, at which time the partial products are added to obtain the product. The sign of the product is positive if both operands have the same sign; the sign of the product is negative if the signs of the operands are different.

Example 9.3 This example multiplies a 4-bit negative multiplicand by a 4-bit negative multiplier to yield an 8-bit positive product. In the paper-and-pencil method, both operands must be positive to obtain the correct positive product. This can be easily accomplished by 2s complementing both operands. If the operands were not negated, then the result would be incorrect due to sign extension in order to generate the partial products. This is not a problem when using the IMUL instruction — the operands do not have to be negated. In this example, the multiplicand is *a[3:0]* = 1100 (–4), the multiplier is *b[3:0]* = 1011 (–5), to generate a product *p[7:0]* = 0001 0100 (+20) when both operands are negated.

Multiplicand A					0	1	0	0	+4
Multiplier B				×)	0	1	0	1	+5
Partial Products	0	0	0	0	0	1	0	0	
	0	0	0	0	0	0	0		
	0	0	0	1	0	0			
	0	0	0	0	0				
Product P	0	0	0	1	0	1	0	0	+20

Example 9.4 This example performs a signed multiplication on two doubleword positive operands: a multiplicand with a hexadecimal value of 228D4A53 and a multiplier with a hexadecimal value of 55C496B2. The multiplication will be performed first using the paper-and-pencil method to display all of the partial products, as shown in Figure 9.21. Then the product will be verified using an assembly language module embedded in a C program in which the operands are entered from the keyboard. The one-operand IMUL form will be used. The three-operand form is assigned as a problem.

The program will also multiply two additional pairs of operands that are entered from the keyboard: a positive multiplicand and a negative multiplier; and a negative multiplicand and a positive multiplier of the same absolute value as the previous values. The product will be the same in both cases. Figure 9.22 illustrates the program for the doubleword multiplication using an assembly language module embedded in a C program.

								2	2	8	D	4	A	5	3
							×)	5	5	C	4	9	6	B	2
0	0	0	0	0	0	0	0	4	5	1	A	9	4	A	6
0	0	0	0	0	0	1	7	C	1	2	3	1	9	1	
0	0	0	0	0	0	C	F	4	F	B	D	F	2		
0	0	0	0	1	3	6	F	7	9	C	E	B			
0	0	0	0	8	A	3	5	2	9	4	C				
0	0	1	9	E	9	F	7	B	E	4					
0	0	A	C	C	2	7	3	9	F						
0	A	C	C	2	7	3	9	F							
	1	2	2	2	3	4	4	4	2	3	2	← Carries			
0	**B**	**9**	**3**	**7**	**2**	**3**	**1**	**4**	**6**	**5**	**6**	**4**	**F**	**B**	**6**

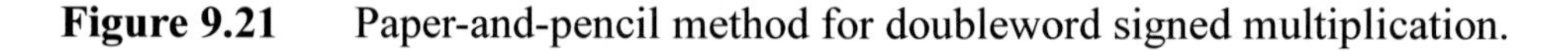

Figure 9.21 Paper-and-pencil method for doubleword signed multiplication.

```
//imul_dblwd_dblwd.cpp
//signed multiply on two doublewords

#include "stdafx.h"

int main (void)
{
//define variables
   long mpcnd, mplyr, prod_hi, prod_lo;

   printf ("Enter two 8-digit hexadecimal numbers: \n");
   scanf ("%X %X", &mpcnd, &mplyr);

//switch to assembly
      _asm
      {
         MOV    EAX, mpcnd      //move 8-digit mpcnd to eax
         IMUL   mplyr           //imul by 8-digit mplyr;
                                //product in edx:eax
         MOV    prod_hi, EDX    //high-order product
                                //in edx to prod_hi
         MOV    prod_lo, EAX    //low-order product
                                //in eax to prod_lo
      }

   printf ("\nProduct high = %X\nProduct low = %X\n\n",
            prod_hi, prod_lo);

   return 0;
}
```

(a)

```
Enter two 8-digit hexadecimal numbers:
228D4A53 55C496B2

Product high = B937231
Product low = 46564FB6

Press any key to continue . . . _
-----------------------------------------------------------
```

(b)

//continued on next page

Figure 9.22 Doubleword multiplication using the one-operand form of IMUL: (a) the program and (b) the outputs.

```
Enter two 8-digit hexadecimal numbers:
228D4A53 85C496B2

Product high = EF80A5CD
Product low = D6564FB6

Press any key to continue . . . _
------------------------------------------------------------
Enter two 8-digit hexadecimal numbers:
85C496B2 228D4A53

Product high = EF80A5CD
Product low = D6564FB6

Press any key to continue . . . _
```

Figure 9.22 (Continued)

This next program performs unsigned (MUL) and signed (IMUL) multiplication on the same operands. All combinations of the high-order bit will be applied for byte operands, as shown below.

Multiplicand Bit 7	Multiplier Bit 7	Unsigned Result	Signed Result
0	0	Positive	Positive
0	1	Positive	Negative
1	0	Positive	Negative
1	1	Positive	Positive

The program is shown in Figure 9.23 as an assembly language module embedded in a C program. Two-digit hexadecimal numbers are entered from the keyboard for both the multiplicand and the multiplier. The multiply instructions are applied, then the results are displayed as unsigned and signed products in decimal notation.

```
//mul_imul.cpp
//demonstrates the use of mul (unsigned numbers)
//and imul (signed numbers) for identical numbers
#include "stdafx.h"
int main (void)
                    (a)         //continued on next page
```

Figure 9.23 Program to illustrate unsigned and signed multiplication for identical numbers: (a) the program and (b) the outputs.

```
{
//define variables
   char mpcnd, mplyr;
   unsigned short prod_unsigned;
   short prod_signed;

   printf ("Enter a 2-digit hex mpcnd and mplyr: \n");
   scanf ("%X %X", &mpcnd, &mplyr);

//switch to assembly
      _asm
      {
//unsigned multiply
         MOV   AL, mpcnd
         MUL   mplyr              ;product in ax
         MOV   prod_unsigned, AX

//signed multiply
         MOV   AL, mpcnd
         IMUL  mplyr              ; product in ax
         MOV   prod_signed, AX
      }

   printf ("\nUnsigned product = %d", prod_unsigned);
   printf ("\nSigned product = %d\n\n", prod_signed);
   return 0;
}
```

```
Enter a 2-digit hex mpcnd and mplyr:
6E 7C

Unsigned product = 13640
Signed product = 13640

Press any key to continue . . . _
------------------------------------------------------------
Enter a 2-digit hex mpcnd and mplyr:
7F 7F

Unsigned product = 16129
Signed product = 16129

Press any key to continue . . . _
```

(b) //continued on next page

Figure 9.23 (Continued)

```
Enter a 2-digit hex mpcnd and mplyr:
7F C7

Unsigned product = 25273
Signed product = -7239

Press any key to continue . . . _
------------------------------------------------------------
Enter a 2-digit hex mpcnd and mplyr:
3D C6

Unsigned product = 12078
Signed product = -3538

Press any key to continue . . . _
------------------------------------------------------------
Enter a 2-digit hex mpcnd and mplyr:
8D 6E

Unsigned product = 15510
Signed product = -12650

Press any key to continue . . . _
------------------------------------------------------------
Enter a 2-digit hex mpcnd and mplyr:
E7 02

Unsigned product = 462
Signed product = -50

Press any key to continue . . . _
------------------------------------------------------------
Enter a 2-digit hex mpcnd and mplyr:
FF FF

Unsigned product = 65025
Signed product = 1

Press any key to continue . . . _
------------------------------------------------------------
Enter a 2-digit hex mpcnd and mplyr:
E1 9A

Unsigned product = 34650
Signed product = 3162

Press any key to continue . . . _
```

Figure 9.23 (Continued)

Multiplication can also be achieved by shifting the operand left a specified number of bits, as shown in Chapter 8. A left shift of one bit position multiplies the operand by two; a left shift of two bit positions multiplies the operand by four; a left shift of three bit positions multiplies the operand by eight, and so forth.

9.4 Division

Division is essentially the inverse of multiplication, where the $2n$-bit dividend corresponds to the $2n$-bit product; the n-bit divisor corresponds to the n-bit multiplicand; and the n-bit quotient corresponds to the n-bit multiplier. The equation that represents this concept is shown below and includes the n-bit remainder as one of the variables.

$$2n\text{-bit dividend} = (n\text{-bit divisor} \times n\text{-bit quotient}) + n\text{-bit remainder}$$

Unlike multiplication, division is not commutative; that is, $A/B \neq B/A$, except when $A = B$, where A and B are the dividend and divisor, respectively. Before introducing the divide instructions, this section presents an example of division using the paper-and-pencil method of sequential shift-subtract/add restoring division. Using this algorithm, the dividend is shifted left one bit position, then the divisor is subtracted from the shifted dividend by adding the 2s complement of the divisor. In general, the operands are as shown below, where A is the $2n$-bit dividend and B is the n-bit divisor. The quotient is Q and the remainder is R, both of which are n bits.

$$A = a_{2n-1}\, a_{2n-2} \ldots a_n\, a_{n-1} \ldots a_1\, a_0$$

$$B = b_{n-1}\, b_{n-2} \ldots b_1\, b_0$$

$$Q = q_{n-1}\, q_{n-2} \ldots q_1\, q_0$$

$$R = r_{n-1}\, r_{n-2} \ldots r_1\, r_0$$

Overflow will occur if the high-order half of the dividend is greater than or equal to the divisor. For example, assume that the high-order half of the dividend is equal to the divisor, as shown below for a dividend of 112 and a divisor of 7, yielding a quotient of 16. The resulting quotient value of 16 cannot be contained in the machine's word size of four bits; therefore, an overflow had occurred. If the high-order half of the dividend is greater than the divisor, then the value of the quotient will be even greater.

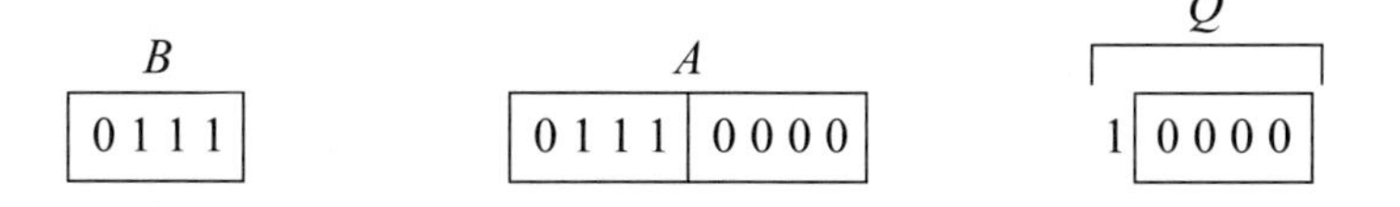

Overflow can be detected by subtracting the divisor from the high-order half of the dividend before the division operation commences. If the result is positive, then an overflow will occur. An overflow is indicated by a divide error exception.

Figure 9.24 illustrates an example of dividing an 8-bit dividend (13) by a 4-bit divisor (5) to yield a 4-bit quotient (2) and a 4-bit remainder (3). Since there are four bits in the divisor, there are four left-shift operations each followed by a subtract operation. The carry-out of the subtraction is placed in the low-order bit position vacated by the left-shifted partial remainder.

Divisor *B* (+5)	*C*		Dividend *A* (+13)	
0 1 0 1	1		0 0 0 0	1 1 0 1
Shift left 1			0 0 0 1	1 0 1 —
Subtract *B*		+)	1 0 1 1	
	0 ←		1 1 0 0	
Restore (Add *B*)		+)	0 1 0 1	
Set *C*	1		0 0 0 1	1 0 1 0
Shift left 1			0 0 1 1	0 1 0 —
Subtract *B*		+)	1 0 1 1	
	0 ←		1 1 1 0	
Restore (Add *B*)		+)	0 1 0 1	
Set *C*	1		0 0 1 1	0 1 0 0
Shift left 1			0 1 1 0	1 0 0 —
Subtract *B*		+)	1 0 1 1	
	1 ←		0 0 0 1	
No Restore (*C* = 1)	1		0 0 0 1	1 0 0 1
Shift left 1			0 0 1 1	0 0 1 —
Subtract *B*		+)	1 0 1 1	
	0 ←		1 1 1 0	
Restore (Add *B*)		+)	0 1 0 1	
			0 0 1 1	0 0 1 0
			Remainder	Quotient

Figure 9.24 Example of the paper-and-pencil method for binary division.

A carry-out of 0 indicates that the difference was greater than the divisor and the partial remainder must be restored by adding the divisor to the partial remainder. A carry-out of 1 indicates that the difference requires no restoration. At the completion of the final cycle, the remainder is contained in the high-order half of the dividend and the quotient is contained in the low-order half of the dividend.

9.4.1 Unsigned Divide (DIV) Instruction

The DIV instruction is one of two divide instructions for fixed-point division; the other is signed divide (IDIV), which is presented in the next section. The DIV instruction divides the unsigned dividend in an implied general-purpose register by the unsigned divisor source operand in a register or memory location. The results are stored in the implied registers. The syntax for the DIV instruction is shown below.

```
DIV    register/memory
```

The locations of the DIV operands are as follows: the dividend registers are AX, DX:AX, EDX:EAX, and RDX:RAX and the size of the corresponding divisors are 8 bits, 16 bits, 32 bits, and 64 bits, respectively; the resulting quotients are stored in registers AL, AX, EAX, and RAX, respectively; the resulting remainders are stored in registers AH, DX, EDX, and RDX, respectively. If an REX prefix is used in 64-bit mode, the number of general-purpose registers is increased from eight to sixteen and can be extended to 64 bits. Results that are not integers are truncated to integers. The above descriptions are illustrated in Figure 9.25 for 8 bit, 16 bit, and 32 bit divisors.

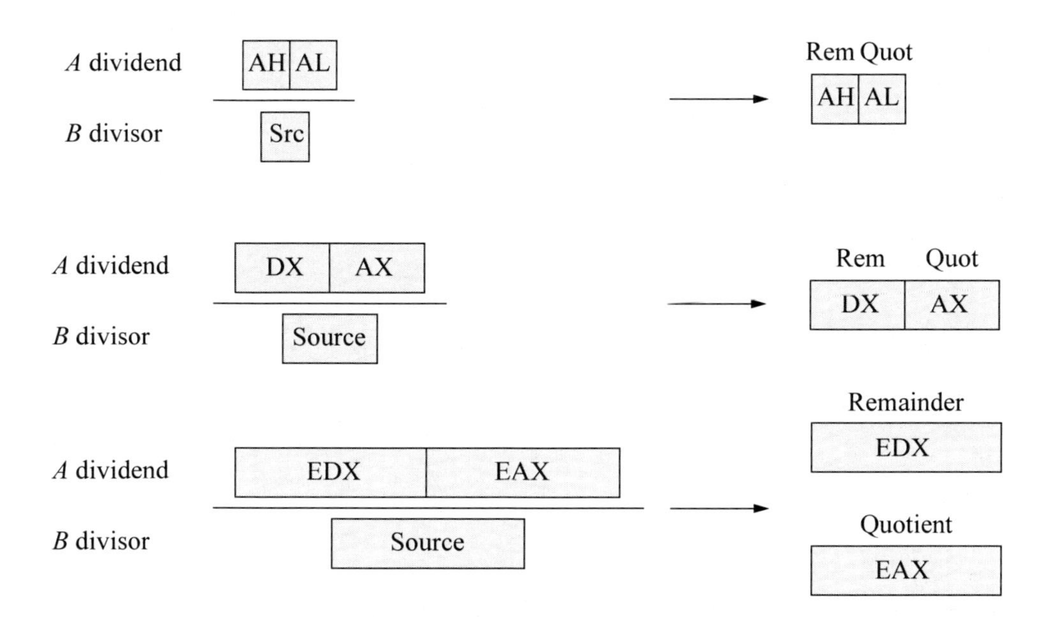

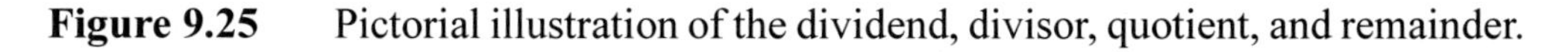
Figure 9.25 Pictorial illustration of the dividend, divisor, quotient, and remainder.

Figure 9.26 shows an assembly language module embedded in a C program that illustrates the DIV instruction. Register pair DX:AX contains the dividend, which is entered from the keyboard as *dvdnd_hi* (DX) and *dvdnd_lo* (AX). The divisor is also entered from the keyboard. The dividend, divisor, quotient, and remainder are displayed after the program has executed. The following flags are undefined: OF, SF, ZF, AF, PF, and CF.

The first set of operands is entered as: dividend = 0100 2050H and divisor = 3500H to yield a quotient = 4D5H and a remainder = 750H. The calculations shown below illustrate the division operation and show the resulting quotient and remainder.

$$\frac{0100\ 2050_{16}}{3500_{16}} \rightarrow \begin{array}{ll} \text{Quotient} = 4D5_{16} & (1237_{10}) \\ \text{Remainder} = 750_{16} & (1872_{10}) \end{array}$$

$$\downarrow \qquad\qquad \downarrow$$

$$\frac{16785488_{10}}{13568_{10}} \rightarrow \begin{array}{l} \text{Quotient} = 1237_{10} \\ \text{Remainder} = 0.137972_{10} \times 13568_{10} = 1872_{10} \end{array}$$

```
//div2.cpp
//unsigned division of two integers
#include "stdafx.h"
int main (void)
{
//define variables
   unsigned short dvdnd_hi, dvdnd_lo, dvsr, quot, rem;

   printf ("Enter a hex dvdnd (hi then lo) and dvsr: \n");
   scanf ("%X %X %X", &dvdnd_hi, &dvdnd_lo, &dvsr);

//switch to assembly
      _asm
      {
         MOV   DX, dvdnd_hi
         MOV   AX, dvdnd_lo
         DIV   dvsr

         MOV   quot, AX
         MOV   rem, DX
      }
```

(a) //continued on next page

Figure 9.26 Unsigned division with the dividend in register pair DX:AX: (a) the program and (b) the outputs.

```
    printf ("\nDividend_hi = %X\nDividend_lo = %X\n",
                dvdnd_hi, dvdnd_lo);
    printf ("\nDivisor = %X\n", dvsr);

    printf ("\nQuotient = %X\nRemainder = %X\n\n", quot, rem);

    return 0;
}
```

```
Enter a hex dvdnd (hi then lo) and dvsr:
0100 2050 3500

Dividend_hi = 100
Dividend_lo = 2050

Divisor = 3500

Quotient = 4D5
Remainder = 750

Press any key to continue . . . _
------------------------------------------------------------
Enter a hex dvdnd (hi then lo) and dvsr:
12AB E543 3500

Dividend_hi = 12AB
Dividend_lo = E543

Divisor = 3500

Quotient = 5A2F
Remainder = 2A43

Press any key to continue . . . _
- - - - - - - - - - - - - - - - - - - - - - - - - - - - - -
```

$12AB\ E543_{16}$ / 3500_{16} → 313255235_{10} / 13568_{10}
313255235_{10} / 13568_{10} = 23087.79739_{10}

Quotient = $5A2F_{16}$ (23087_{10})

Remainder = $2A43_{16}$ → $0.79739_{10} \times 13568_{10} = 10818.98752_{10}$
$= 10819_{10} = 2A43_{16}$

(b)

Figure 9.26 (Continued)

9.4.2 Signed Divide (IDIV) Instruction

The signed divide instruction (IDIV) is similar to the unsigned divide instruction (DIV), except that IDIV operates on signed operands. The IDIV instruction divides the signed dividend in registers AH:AL, DX:AX, EDX:EAX, or RDX:RAX (if an REX prefix is used in 64-bit mode) by the signed source divisor and stores the resulting quotient in the low-order part of the register pairs and stores the remainder in the high-order part of the register pairs.

The source operand can be in a general-purpose register or in a memory location. Results that are not integers are truncated toward zero as integers. The registers shown in Figure 9.25 also apply to the IDIV instruction. The following flags are undefined: OF, SF, ZF, AF, PF, and CF. The syntax for the IDIV instruction is shown below.

```
IDIV    register/memory
```

The remainder has the same sign as the dividend and satisfies the following equation as shown in the examples of Figure 9.27:

$$\text{Dividend} = (\text{Quotient} \times \text{Divisor}) + \text{Remainder}$$

Divisor	Dividend	Quotient	Remainder
4075	50	81	25
4075	–50	–81	25
–4075	50	–81	–25
–4075	–50	81	–25

Figure 9.27 Examples of signed division with resulting quotient and remainder.

Figure 9.28 shows an assembly language module embedded in a C program that illustrates the application of the IDIV instruction. The dividend and divisor are entered from the keyboard as signed decimal integers. All combinations of the signs are presented for the dividend and divisor. The quotient and remainder are displayed after the program has executed.

```
//idiv2.cpp
//signed divide operation
#include "stdafx.h"
int main (void)
```

(a) //continued on next page

Figure 9.28 Example to illustrate signed division: (a) the program and (b) the outputs.

```
{
//define variables
   short dvdnd;
   char dvsr, quot, rem;

   printf ("Enter a signed decimal dvdnd and dvsr: \n");
   scanf ("%d %d", &dvdnd, &dvsr);

//switch to assembly
      _asm
      {
         MOV   AX, dvdnd
         IDIV  dvsr     ;quot in al; rem in ah

         MOV   quot, AL
         MOV   rem, AH
      }
   printf ("\nDividend = %d\nDivisor = %d\n", dvdnd, dvsr);
   printf ("\nQuotient = %d\nRemainder = %d\n\n", quot, rem);
   return 0;
}
```

```
Enter a signed decimal dvdnd and dvsr:
4075 50

Dividend =  4075
Divisor = 50

Quotient = 81
Remainder = 25

Press any key to continue . . . _
------------------------------------------------------------
Enter a signed decimal dvdnd and dvsr:
3500 65

Dividend =  3500
Divisor = 65

Quotient = 53
Remainder = 55

Press any key to continue . . . _    //continued on next page
```

(b)

Figure 9.28 (Continued)

```
Enter a signed decimal dvdnd and dvsr:
4075 -50

Dividend =  4075
Divisor = -50

Quotient = -81
Remainder = 25

Press any key to continue . . . _
------------------------------------------------------------
Enter a signed decimal dvdnd and dvsr:
12288 -96

Dividend =  12288
Divisor = -96

Quotient = -128
Remainder = 0

Press any key to continue . . . _
------------------------------------------------------------
Enter a signed decimal dvdnd and dvsr:
-4075 50

Dividend =  -4075
Divisor = 50

Quotient = -81
Remainder = -25

Press any key to continue . . . _
------------------------------------------------------------
Enter a signed decimal dvdnd and dvsr:
-600 35

Dividend =  -600
Divisor = 35

Quotient = -17
Remainder = -5

Press any key to continue . . . _
------------------------------------------------------------
                                   //continued on next page
```

Figure 9.28 (Continued)

```
Enter a signed decimal dvdnd and dvsr:
-4075 -50

Dividend =  -4075
Divisor = -50

Quotient = 81
Remainder = -25

Press any key to continue . . . _
-----------------------------------------------------------
Enter a signed decimal dvdnd and dvsr:
-100 -25

Dividend =  -100
Divisor = -25

Quotient = 4
Remainder = 0

Press any key to continue . . . _
```

Figure 9.28 (Continued)

Division can also be achieved by shifting the operand right a specified number of bits, as shown in Chapter 8. A right shift of one bit position divides the operand by two; a right shift of two bit positions divides the operand by four; a right shift of three bit positions divides the operand by eight, and so forth. All of the problems in the following section utilize one or more of the instructions presented in this chapter.

9.5 Problems

9.1 Add the following numbers, which are in radix complementation for radix 7.

$$\begin{array}{r} 1\ 2\ 3\ 4_7 \\ 5\ 6\ 1\ 2_7 \\ 3\ 4\ 5\ 6_7 \\ \underline{1\ 2\ 3\ 4_7} \end{array}$$

9.2 Add the following numbers, which are in radix complementation for radix 2.

$$\begin{array}{r} 1\ 1\ 1\ 1_2 \\ 1\ 1\ 1\ 1_2 \\ 1\ 1\ 1\ 1_2 \\ 1\ 1\ 1\ 1_2 \\ \underline{1\ 1\ 1\ 1_2} \end{array}$$

9.3 Indicate whether an overflow occurs for the operation shown below. The numbers are in radix complementation for radix 3.

$$\begin{array}{r} 2\ 1\ 0\ 1\ 2_3 \\ +)\ \underline{0\ 2\ 1\ 2\ 2_3} \end{array}$$

9.4 Write a program segment to negate a 32-bit binary number. The number is in 2s complement representation and resides in register pair DX:AX. Do not use the NEG instruction.

9.5 Let AX = 00 4FH and the word OPND in memory = FF 38H. Determine whether the conditional jump instructions shown below will jump to DEST.

(a)
```
ADD    AX, 200
JS     DEST
```

(b)
```
ADD    OPND, 200
JZ     DEST
```

9.6 Determine the contents of registers AL and BL after the program segment shown below has executed. Then write an assembly language module embedded in a C program to verify the results.

```
        MOV  AL, 00H
        MOV  BL, -5
LOOP1:  ADD  BL, 2
        INC  AL
        ADD  BL, -1
        JNZ  LOOP1
```

9.7 Let AX = 47E7H and BX = 4BED, then add the two registers and obtain the state of the following flags: AF, CF, OF, PF, SF, and ZF.

9.8 Write an assembly language module — not embedded in a C program — that requests two characters to be entered from the keyboard. Display the first character unchanged and display the second character shifted left logically one bit position. The characters can be numbers or special characters in the range 21H to 3FH.

9.9 Write an assembly language program — not embedded in a C program — that removes all nonletters from a string of characters that are entered from the keyboard.

9.10 Write an assembly language module embedded in a C program that adds two 4-digit hexadecimal numbers and displays the sum and the low-order two bytes of the EFLAGS register.

9.11 Write an assembly language module embedded in a C program that uses the *add with carry* (ADC) instruction in the addition of four 4-digit hexadecimal numbers: two for the augend and two for the addend. Display the high-order sum and the low-order sum. Also display the high-order byte and the low-order byte of the EFLAGS register.

9.12 Given the program shown below, determine the contents of the RSLT field after each of the following hexadecimal characters are entered from the keyboard:

123456_{16} $ABCDEF_{16}$ $4a5b6c_{16}$ $UVWXYZ_{16}$

```
;logic_inc.asm
;-------------------------------------------------------
;STACK

;-------------------------------------------------------
.DATA
PARLST  LABEL BYTE
MAXLEN  DB  15
ACTLEN  DB  ?
OPFLD   DB  15  DUP(?)
PRMPT   DB  0DH, 0AH, 'Enter text: $'
RSLT    DB  0DH, 0AH, 'Result =        $'

;-------------------------------------------------------
.CODE
BEGIN     PROC  FAR

;set up pgm ds
          MOV   AX, @DATA
          MOV   DS, AX

;read prompt
          MOV   AH, 09H
          LEA   DX, PRMPT
          INT   21H

;keyboard request rtn to enter characters
          MOV   AH, 0AH
          LEA   DX, PARLST
          INT   21H

;set up addresses
          LEA   SI, OPFLD
          LEA   DI, OPFLD+2
          LEA   BX, RSLT+12
          MOV   CX, 4

;begin main pgm
LP1:      MOV   AL, [SI]
          MOV   AH, [DI]
          XOR   AL, AH
          OR    AL, 30H
          MOV   [BX] ,AL
          INC   SI
          INC   DI
          INC   BX
          LOOP  LP1
                              //continued on next page
```

```
;display result
          MOV    AH, 09H
          LEA    DX, RSLT
          INT    21H

BEGIN     ENDP
          END    BEGIN
```

9.13 Perform a subtract operation on the radix 5 numbers shown below. The procedure is identical to the radix 2 method — the only difference is the radix.

$$\begin{array}{rccccc} & 0 & 4 & 3 & 2 & 1_5 \\ -) & 4 & 3 & 3 & 4 & 0_5 \\ \hline \end{array}$$

9.14 Perform a subtract operation on the two hexadecimal operands shown below.

$$\begin{array}{rcccc} & A & B & C & D_{16} \\ -) & 1 & 2 & 3 & 4_{16} \\ \hline \end{array}$$

9.15 Indicate whether an overflow occurs for the operation shown below. The numbers are in radix complement representation for radix 2.

$$\begin{array}{rcccccccc} & 0 & 1 & 1 & 1 & 1 & 1 & 0 & 0_2 \\ -) & 1 & 1 & 1 & 1 & 1 & 1 & 1 & 1_2 \\ \hline \end{array}$$

9.16 Indicate whether an overflow occurs for the operation shown below. The numbers are in radix complement representation for radix 2.

$$\begin{array}{rcccccccc} & 0 & 1 & 1 & 0 & 1 & 1 & 0 & 0_2 \\ -) & 1 & 0 & 0 & 0 & 1 & 1 & 1 & 0_2 \\ \hline \end{array}$$

9.17 Let register AL = 61H (97_{10}). Then execute the instruction shown below to obtain the difference and the state of the following flags: OF, SF, ZF, AF, PF, and CF.

```
SUB    AL, 65H
```

9.18 Let register DL = F3H and register BH = 72H. Then execute the instruction shown below to obtain the difference and the state of the following flags: OF, SF, ZF, AF, PF, and CF.

```
SUB    DL, BH
```

9.19 Determine the contents of register AL after the program shown below has executed.

```
//decrement.cpp
//program to illustrate the DEC instruction.

#include "stdafx.h"

int main (void)
{

//define variables
   char result;

//switch to assembly
         _asm
         {
            MOV    AL, 3AH
            MOV    CH, 0A9H
            ADD    CH, 06H
            ADD    AL, CH

            NEG    AL
            DEC    AL
            MOV    result, AL
         }

   printf ("\nAL = %X\n\n", result);

   return 0;
}
```

9.20 Show the result of the program listed below when the following digits are entered from the keyboard: 34689000678305. Refer to Chapter 10 for a description of the ASCII Adjust After Addition (AAA) instruction.

```
PAGE 66, 80
;adc_aaa_dec.asm

;-------------------------------------------------------
.STACK

;-------------------------------------------------------
.DATA
PARLST  LABEL   BYTE
MAXLEN  DB  20
ACTLEN  DB  ?
OPFLD   DB  20  DUP(?)
PRMPT   DB  0DH, 0AH, 'Enter 14 single-digit numbers: $'
RSLT    DB  0DH, 0AH, 'Result =                      $'
;OPFLD #s   34689000678305

;-------------------------------------------------------
.CODE
BEGIN    PROC  FAR

;set up pgm ds
         MOV   AX, @DATA
         MOV   DS, AX

;read prompt
         MOV   AH, 09H
         LEA   DX, PRMPT
         INT   21H

;keyboard request rtn to enter characters
         MOV   AH, 0AH
         LEA   DX, PARLST
         INT   21H

;start main pgm
         LEA   SI, OPFLD+5
         LEA   DI, OPFLD+13
         LEA   BX, RSLT+12

         MOV   CX, 6
         CLC

                          //continued on next page
```

```
LP1:    MOV     AH, 0
        MOV     AL, [SI]
        ADC     AL, [DI]
        AAA
        OR      AL, 30H
        MOV     [BX], AL
        DEC     SI
        DEC     DI
        INC     BX
        LOOP    LP1

        MOV     AH, 09H
        LEA     DX, RSLT
        INT     21H

BEGIN   ENDP
        END     BEGIN
```

9.21 Write an assembly language program — not embedded in a C program — to reverse the order and change to uppercase all letters that are entered from the keyboard. The letters that are entered may be lowercase or uppercase.

9.22 Eight digits, either 0 or 1, are entered from the keyboard. Write an assembly language program — not embedded in a C program — to determine the parity of the byte. If there are an even number of 1s in the byte, then a flag bit (PF) is set to a value of 1 maintaining odd parity over all nine bits — the eight bits plus the parity flag; otherwise, the parity flag is reset to 0. Display the parity flag. Draw a flowchart that represents the operation of the odd parity generator. Remember that a 0 and a 1 are represented in ASCII as 30H and 31H, respectively.

9.23 Write an assembly language program — not embedded in a C program — to sort *n* single-digit numbers in ascending numerical order that are entered from the keyboard. A simple exchange sort is sufficient for this program. The technique is slow, but appropriate for sorting small lists. The method is as follows:

1. Search the list to find the smallest number.
2. After finding the smallest number, exchange this number with the first number.
3. Search the list again to find the next smallest number, starting with the second number.
4. Then exchange this (second smallest) number with the second number.
5. Repeat this process, starting with the third number, then the fourth number, and so forth.

9.24 Let register AL = 6CH and register BL = A7H. After execution of

```
IMUL    BL
```

determine the hexadecimal contents of register AX.

9.25 Determine the contents of register pair DX:AX after the following program segment has been executed:

```
OP1    DW     8080H
OP2    DB     0A2H
MOV    AL,    OP2
CBW
MUL    OP1
```

9.26 Determine the results of each of the following operations:

Before	Instruction
AX = FF FFH	MUL AX
AX = FF E4H BX = 04 C2H	MUL BX
AX = 00 17H CX = 00 B2H	IMUL CX
AX = FF E4H BX = 04 C2H	IMUL BX

9.27 Determine the hexadecimal contents of register AX after the following program segment has been executed:

```
       MOV    CX, 3
       MOV    AX, 2
LP1:   MUL    AX
       LOOP   LP1
```

9.28 Write an assembly language module embedded in a C program that multiplies two unsigned (MUL) hexadecimal numbers. Enter numbers from the keyboard that produce a product where the high-order half is all zeroes and where the high-order half is not all zeroes. The overflow flag and the carry flag should be set if the high-order half of the product is not all zeroes. Display the products and the low-order 16 bits of the EFLAGS register.

9.29 Perform a signed multiplication (IMUL) on two doubleword operands: a negative multiplicand with a hexadecimal value of $85C496B2_{16}$ and a positive multiplier with a hexadecimal value of 22441243_{16}. Use the paper-and-pencil method and display all of the partial products. Then verify the product using an assembly language module embedded in a C program in which the operands are entered from the keyboard.

9.30 Write an assembly language module embedded in a C program that uses the three-operand form for signed multiplication. Enter a signed decimal multiplicand from the keyboard and assign an immediate multiplier for the IMUL operation. Display the multiplicand, the multiplier, and the product.

9.31 Use the shift logical left (SHL) instruction in an assembly language module embedded in a C program to multiply an operand. Enter an 8-digit hexadecimal number and a shift amount from the keyboard. Include numbers and shift amounts that produce no overflow and that produce an overflow. Display the shift amount, the preshifted value, the postshifted value, and the accompanying flags.

9.32 Determine the result of the following operation:

Before	Instruction
AX = FE 01H	**DIV** BL
BL = FFH	

9.33 Determine whether the following operands produce an overflow for an unsigned binary divide (DIV) operation:

(a) Dividend = 00011111
Divisor = 0001

(b) Dividend = 00001111
Divisor = 0001

(c) Dividend = 01000001
Divisor = 0011

9.34 Write an assembly language module embedded in a C program to perform unsigned division (DIV) on two decimal integers that are entered from the keyboard. Print the dividend, divisor, quotient, and remainder.

9.35 Write an assembly language program — not embedded in a C program — for a sequence detector that detects the number of times that the four-bit sequence 0111 appears in the keyboard input buffer (OPFLD) when n characters are entered. The variable n has the following range: $1 \leq n \leq 30$.

The bit configuration 0111 may be in the high-order four bits of the keyboard character, the low-order four bits, or both high- and low-order four bits. Display the number of times that the sequence occurs. The DIV instruction can be used in this program. Draw a flowchart prior to designing the program. An example of input data is shown below.

7	5	g	q	T	W
0011 **0111**	0011 0101	0110 **0111**	**0111** 0001	0101 0100	**0111 0111**

0111 occurs five times

9.36 Write an assembly language program — not embedded in a C program — that calculates the surface area of a rectangular solid box. The length, width, and height are single-digit numbers that are entered from the keyboard. The program will use the ADD, MUL, and DIV instructions. Enter at least three sets of numbers and display the corresponding surface area.

9.37 Determine the contents of register pair DX:AX for parts (a) and (b) after execution of the following program segment:

```
MOV   AX, dvdnd
CWD
IDIV  dvsr
```

(a) dvdnd = 059A; dvsr = FFC7

(b) dvdnd = 7C58; dvsr = FFA9

9.38 Write an assembly language module embedded in a C program that evaluates the expression shown below, where $x = 400_{10}$ (0190_{16}), $y = 5_{10}$ (0005_{16}), $z = 1000_{10}$ $(03E8_{16})$, and $v = 4000_{10}$ $(0FA0_{16})$.

$$[v - (x \times y + z - 500)] / x$$

10

Binary-Coded Decimal Arithmetic Instructions

The radix is 10 in the *decimal number system*; therefore, ten digits are used, 0 through 9. The low-value digit is 0 and the high-value digit is $(r - 1) = 9$. The weight assigned to each position of a decimal number is as follows:

$$10^{n-1}\,10^{n-2}\,\ldots\;10^3\,10^2\,10^1\,10^0.\,10^{-1}\,10^{-2}\,10^{-3}\,\ldots\;10^{-m}$$

where the integer and fraction are separated by the radix point (decimal point).

Binary-coded decimal (BCD) instructions operate on decimal numbers that are encoded as 4-bit binary numbers in the 8421 code. For example, the decimal number 576 is encoded in BCD as 0101 0111 0110. BCD numbers have a range of 0 to 9; therefore, any number greater than 9 must be adjusted by adding a value of 6 to the number to yield a valid BCD number. For example, if the result of an operation is 1010, then 0110 is added to this intermediate sum. This yields a value of 0000 with a carry of 1, or 0001 0000 in BCD which is 10 in decimal.

The condition for a correction (adjustment) of an intermediate sum that also produces a carry-out is shown in Equation 10.1. This specifies that a carry-out will be generated whenever bit position b_8 is a 1 in both decades, when bit positions b_8 and b_4 are both 1s, or when bit positions b_8 and b_2 are both 1s, as shown in the examples of Figure 10.1. Table 10.1 lists the ten decimal digits (0 through 9) and the corresponding binary-coded decimal digits, plus BCD numbers of more than one decimal digit.

$$\text{Carry} = c_8 + b_8 b_4 + b_8 b_2 \qquad (10.1)$$

$$
\begin{array}{llcccc|cccc|l}
 & & & & & & b_8 & b_4 & b_2 & b_1 & \\
 & & & & & & 1 & 0 & 0 & 0 & (8) \\
 & & & & & +) & 1 & 0 & 0 & 0 & (8) \\
c_8 = 1 & & & & 1 & \leftarrow & 0 & 0 & 0 & 0 & \leftarrow \text{Intermediate sum} \\
 & & & & \downarrow & +) & 0 & 1 & 1 & 0 & \leftarrow \text{Adjust} \\
 & 0 & 0 & 0 & 1 & & 0 & 1 & 1 & 0 & (16)
\end{array}
$$

$$
\begin{array}{llcccc|cccc|l}
 & & & & & & b_8 & b_4 & b_2 & b_1 & \\
 & & & & & & 0 & 1 & 1 & 1 & (7) \\
 & & & & & +) & 0 & 1 & 0 & 1 & (5) \\
b_8 \,.\, b_4 = 11 & & & & & & 1 & 1 & 0 & 0 & \leftarrow \text{Intermediate sum} \\
 & & & & & +) & 0 & 1 & 1 & 0 & \leftarrow \text{Adjust} \\
 & 0 & 0 & 0 & 1 & \leftarrow & 0 & 0 & 1 & 0 & (12)
\end{array}
$$

$$
\begin{array}{llcccc|cccc|l}
 & & & & & & b_8 & b_4 & b_2 & b_1 & \\
 & & & & & & 0 & 1 & 1 & 0 & (6) \\
 & & & & & +) & 0 & 1 & 0 & 0 & (4) \\
b_8 \,.\, b_2 = 11 & & & & & & 1 & 0 & 1 & 0 & \leftarrow \text{Intermediate sum} \\
 & & & & & +) & 0 & 1 & 1 & 0 & \leftarrow \text{Adjust} \\
 & 0 & 0 & 0 & 1 & \leftarrow & 0 & 0 & 0 & 0 & (10)
\end{array}
$$

Figure 10.1 Examples showing adjustment of the intermediate sum.

Table 10.1 Binary-Coded Decimal Numbers

Decimal	Binary-Coded Decimal
0	0000
1	0001
2	0010
3	0011
4	0100
5	0101
6	0110
7	0111
8	1000
9	1001
10	0001 0000

Continued on next page

Table 10.1 Binary-Coded Decimal Numbers

Decimal	Binary-Coded Decimal
11	0001 0001
12	0001 0010
...	...
124	0001 0010 0100
...	...
365	0011 0110 0101

Figure 10.2 shows an additional example of adding BCD numbers, where the augend and addend both contain three digits each: the augend = 436, the addend = 825, to yield a sum of 1261. The intermediate sum in this example is 1 1100 0110 1011. The sum of the second decade does not require adjustment, because the intermediate sum does not satisfy Equation 10.1. The same is true for the final carry-out.

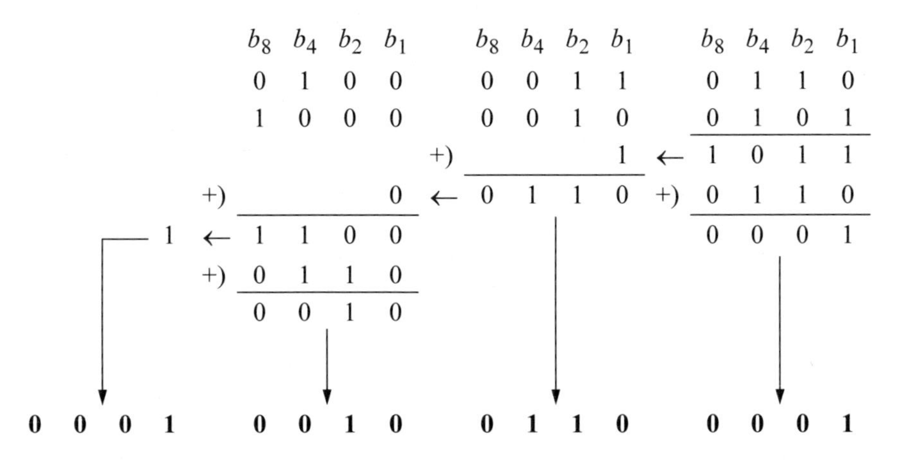

Figure 10.2 Example of adding two BCD operands containing three digits each.

10.1 ASCII Adjust After Addition (AAA) Instruction

The AAA instruction adjusts the result of an addition operation of two unpacked binary-coded decimal (BCD) operands. An unpacked BCD number contains zeroes in the high-order four bits of the byte; the numerical value of the number is contained in the low-order four bits of the byte. General-purpose register AL is the implied source and destination for the AAA instruction.

The sum obtained by the addition of two unpacked BCD or ASCII numbers is stored in register AL. This sum may not be a valid BCD value. The AAA instruction converts the sum in register AL into a valid BCD number and stores the result in bits 3 through 0 of register AL and resets bits 7 through 4 of register AL. If the low-order four bits of the sum in register AL are greater than nine or if the AF flag is set, then six is added to AL and register AH is incremented by 1. In this case, both the auxiliary carry flag (AF) and the carry flag (CF) are set; otherwise, they are reset. The following flags are undefined: OF, SF, ZF, and PF.

Figure 10.3 shows examples of sums in which the result is not a valid BCD number — greater than nine — and a sum in which the AF flag is set. When used in a program, the AAA instruction is preceded by an ADD instruction or an ADC instruction.

AX =	0 0	3 8	ASCII 8
+) BX =	0 0	3 4	ASCII 4
AX =	0 0	6C	Low half of AL > 9
AAA	1	6	
	0 1	0 2	After AAA

(a)

AX =	0 0	3 9	ASCII 9
+) BX =	0 0	$3_1 9$	ASCII 9
AX =	0 0	7 2	AF = 1
AAA	1	6	
	0 1	0 8	After AAA

(b)

Figure 10.3 Examples of using the AAA instruction: (a) where the sum was not a valid BCD number and (b) where the AF flag was set.

Figure 10.4 shows an assembly language module — not embedded in a C program — that illustrates the use of the AAA instruction. Two single-digit numbers are entered from the keyboard and added using the ADD instruction; then the AAA instruction is executed to produce a valid result, if necessary. Numbers that generate a valid BCD sum are entered, also numbers that generate an invalid BCD result, and numbers that generate an auxiliary carry flag (AF). In order to display the sums correctly, the ASCII bias is added to the sums generated by the AAA instruction. A program that adds two 3-digit numbers is left as a problem.

```
;add_aaa.asm
;-------------------------------------------------------------------
.STACK
;-------------------------------------------------------------------
.DATA
PARLST   LABEL BYTE
MAXLEN   DB  5
ACTLEN   DB  ?
OPFLD    DB  5 DUP(?)
PRMPT    DB  0DH, 0AH, 'Enter two single digits: $'
RSLT     DB  0DH, 0AH, 'Sum =      $'

;-------------------------------------------------------------------
.CODE
BEGIN    PROC FAR

;set up pgm ds
         MOV   AX, @DATA      ;get addr of data seg
         MOV   DS, AX         ;move addr to ds

;read prompt
         MOV   AH, 09H        ;display string
         LEA   DX, PRMPT      ;put addr of prompt in dx
         INT   21H            ;dos interrupt

;keyboard request rtn to enter numbers
         MOV   AH, 0AH        ;buffered keyboard input
         LEA   DX, PARLST     ;put addr of parlst in dx
         INT   21H            ;dos interrupt

;get first number
         MOV   AL, OPFLD      ;move first number to al
         MOV   AH, 00H        ;clear ah for the aaa instr

;get second number
         MOV   BL, OPFLD+1    ;move second number to al
         MOV   BH, 00H        ;clear bh for the aaa instr

         ADD   AL, BL         ;add al and bl
         AAA                  ;change al to valid unpacked #
         OR    AX, 3030H      ;restore ascii bias for display
```

(a) //continued on next page

Figure 10.4 Program to illustrate using the AAA instruction: (a) the program and (b) the outputs.

```
          MOV   RSLT+8, AH      ;move high-order digit to rslt
          MOV   RSLT+9, AL      ;move low-order digit to rslt

;display the result
          MOV   AH, 09H         ;display string
          LEA   DX, RSLT        ;put addr of rslt in dx
          INT   21H             ;dos interrupt

BEGIN     ENDP
          END   BEGIN
```

```
Enter two single digits: 84   //generates an invalid BCD #
Sum = 12
--------------------------------------------------------------
Enter two single digits: 99   //AF =1
Sum = 18
--------------------------------------------------------------
Enter two single digits: 97   //AF =1
Sum = 16
--------------------------------------------------------------
Enter two single digits: 78   //generates an invalid BCD #
Sum = 15
--------------------------------------------------------------
Enter two single digits: 43
Sum = 07
--------------------------------------------------------------
Enter two single digits: 45
Sum = 09
```

(b)

Figure 10.4 (Continued)

10.2 Decimal Adjust AL after Addition (DAA) Instruction

The DAA instruction adjusts the sum of two packed BCD integers to generate a packed BCD result. A packed BCD number contains a valid BCD digit in both the high-order four bits and the low-order four bits of a byte. An ADD instruction precedes the DAA instruction. The ADD instruction adds the two packed BCD numbers and stores the sum in register AL. The DAA instruction is then executed and adjusts the sum in register AL to a valid 2-digit packed BCD value, if necessary. Register AL is the implied source and destination register for the DAA instruction.

There are two rules that are used when adding packed BCD numbers regarding the resulting sum.

Rule 1: If the low-order four bits of the sum are greater than nine or if the auxiliary carry flag (AF) is set, then 06H is added to the sum — which is contained in register AL — and the AF flag is set; otherwise, the AF is reset.

Rule 2: If the sum in AL is greater than 9FH or if the carry flag (CF) is set, then 60H is added to AL and the CF flag is set; otherwise, the CF is reset.

For both rules, the overflow flag (OF) is undefined. Examples of adding packed BCD numbers are shown in Figure 10.5.

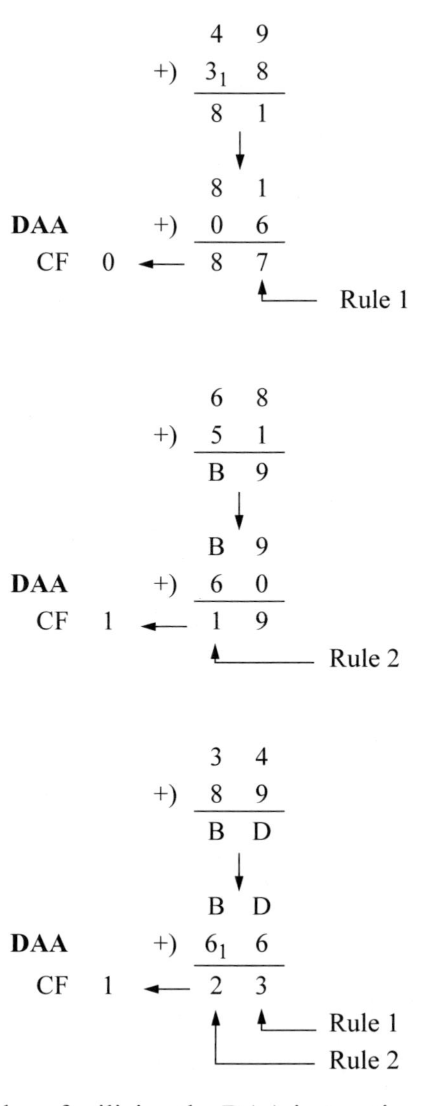

Figure 10.5 Examples of utilizing the DAA instruction.

Figure 10.6 shows an assembly language module embedded in a C program that illustrates the application of the DAA instruction. Two hexadecimal numbers are entered from the keyboard: an augend and an addend. They are added using the ADD instruction, then adjusted using the DAA instruction. The sum is then displayed as a valid 2-digit packed BCD result in register AL. Since the carry flag (CF) and the auxiliary carry flag (AF) are affected, the low-order byte of the EFLAGS register is also displayed — shown below for convenience.

7	6	5	4	3	2	1	0
SF	ZF	0	AF	0	PF	1	CF

```
//daa.cpp
//illustrates the use of the daa instruction
#include "stdafx.h"
int main (void)
{
//define variables
   unsigned char augend, addend, sum, flags;

   printf ("Enter two 2-digit hexadecimal numbers: \n");
   scanf ("%X %X", &augend, &addend);

//switch to assembly
      _asm
      {
//obtain sum
         MOV   AL, augend
         ADD   AL, addend
         DAA
         MOV   sum, AL

//save flags
         PUSHF
         POP   AX
         MOV   flags, AL
      }

   printf ("\nAugend = %X, Addend = %X\n", augend, addend);
   printf ("Sum = %X, Flags = %X\n\n", sum, flags);
   return 0;
}
```

(a) //continued on next page

Figure 10.6 Program to illustrate the use of the DAA instruction: (a) the program and (b) the outputs.

```
Enter two 2-digit hexadecimal numbers:
49 38

Augend = 49, Addend = 38
Sum = 87, Flags = 96         //SF, AF, PF
                             //Sum = 087; CF = 0
Press any key to continue . . . _
-------------------------------------------------------------
Enter two 2-digit hexadecimal numbers:
68 51

Augend = 68, Addend = 51
Sum = 19, Flags = 3          //CF
                             //Sum = 119; CF= 1
Press any key to continue . . . _
-------------------------------------------------------------
Enter two 2-digit hexadecimal numbers:
34 89

Augend = 34, Addend = 89
Sum = 23, Flags = 13         //AF, CF
                             //Sum = 123; CF = 1
Press any key to continue . . . _
-------------------------------------------------------------
Enter two 2-digit hexadecimal numbers:
89 76

Augend = 89, Addend = 76
Sum = 65, Flags = 17         //AF, PF, CF
                             /Sum = 165; CF = 1
Press any key to continue . . . _
-------------------------------------------------------------
Enter two 2-digit hexadecimal numbers:
98 89

Augend = 98, Addend = 89
Sum = 87, Flags = 97         //SF, AF, PF, CF
                             //Sum = 187; CF = 1
Press any key to continue . . . _
-------------------------------------------------------------
Enter two 2-digit hexadecimal numbers:
99 99

Augend = 99, Addend = 99
Sum = 98, Flags = 93         //SF, AF, CF
                             //Sum = 198; CF = 1
Press any key to continue . . . _      (b)
```

Figure 10.6 (Continued)

A similar program is shown in Figure 10.7, this time using only assembly language. The numbers that are entered from the keyboard are stored in the OPFLD input buffer as hexadecimal ASCII characters. If only the numbers 0 through 9 were entered, then only the hexadecimal characters 30 through 39 could be used.

In order to obtain higher-valued hexadecimal numbers as valid packed BCD numbers, the ASCII characters @ through *Y* and ‘ through *y* can be entered. This yields @ = 40 through *Y* = 59 and ‘ = 60 through *y* = 79. Thus, *8y* yields 38 79 and *9q* yields 37 91. The only drawback with this method is that the higher-valued numbers 80 through 89 and 90 through 99 cannot be realized. This presents no problem in illustrating the use of the DAA instruction.

If the higher-valued packed hexadecimal numbers are required, then a simple — but longer — procedure can be employed. Assume that it is required to use the DAA instruction to adjust the sum obtained by adding 99 + 99. The numbers are entered from the keyboard and stored in the OPFLD buffer, as shown below, where the first two digits are altered and changed to a valid packed BCD number. The second pair of digits are similarly altered.

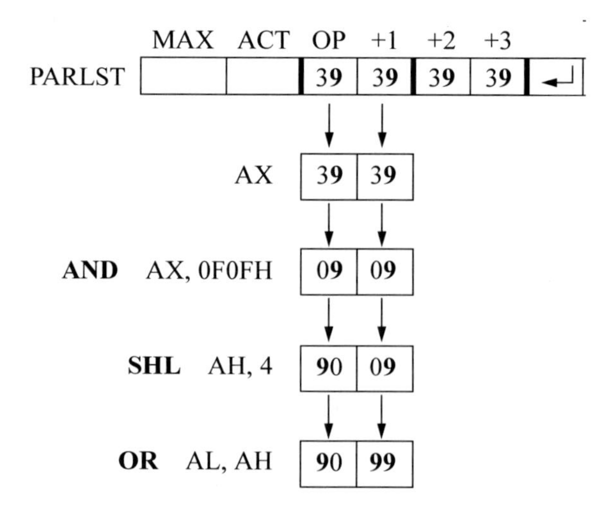


```
;daa.asm
;obtain the sum of two packed bcd operands
;-----------------------------------------------------------------
.STACK
;-----------------------------------------------------------------
.DATA
PARLST    LABEL BYTE
MAXLEN    DB 5
ACTLEN    DB ?
OPFLD     DB 5  DUP(?)          (a)     //continued on next page
```

Figure 10.7 Assembly language program to illustrate using the DAA instruction: (a) the program and (b) the outputs.

```
PRMPT    DB 0DH,0AH,'Enter 2 ascii chars as packed bcd #s: $'
RSLT     DB 0DH, 0AH, 'Result =          $'
FLAGS    DB 0DH, 0AH, 'Flags =     $'

;-----------------------------------------------------------
.CODE
BEGIN    PROC  FAR

;set up pgm ds
         MOV   AX, @DATA       ;get addr of data seg
         MOV   DS, AX          ;move addr to ds

;read prompt
         MOV   AH, 09H         ;display string
         LEA   DX, PRMPT       ;put addr of prompt in dx
         INT   21H             ;dos interrupt

;keyboard request rtn to enter characters
         MOV   AH, 0AH         ;buffered keyboard input
         LEA   DX, PARLST      ;put addr of parlst in dx
         INT   21H             ;dos interrupt

;-----------------------------------------------------------
;get the two numbers
         MOV   AL, OPFLD       ;get first number
         MOV   AH, OPFLD+1     ;get second number

;add the two numbers and adjust sum
         ADD   AL, AH          ;add the two numbers
         DAA                   ;adjust the sum

;-----------------------------------------------------------
;move flags to flag area first (or instr resets flags)
         PUSHF
         POP   BX              ;flags word to bx

         MOV   DL, BL
         AND   DL, 10H         ;isolate af
         SHR   DL, 4           ;shift left logical 4
         OR    DL, 30H         ;add ascii bias
         MOV   FLAGS+11, DL    ;move af to flag area

         MOV   DL, BL
         AND   DL, 01H         ;isolate cf
         OR    DL, 30H         ;add ascii bias
         MOV   FLAGS+12, DL    ;move cf to flag area
                                     //continued on next page
```

Figure 10.7 (Continued)

```
;obtain high-order sum
         MOV    BL, AL
         SHR    BL, 4             ;put # in bits 3-0
         OR     BL, 30H           ;add ascii bias
         MOV    RSLT+11, BL       ;move sum to result area

;obtain low-order sum
         MOV    BL, AL
         AND    BL, 0FH           ;isolate low-order digit
         OR     BL, 30H           ;add ascii bias
         MOV    RSLT+12, BL       ;move sum to result area

;display sum, af, and cf flags
         MOV    AH, 09H           ;display string
         LEA    DX, RSLT          ;put addr of rslt (sum) in dx
         INT    21H               ;dos interrupt

         MOV    AH, 09H           ;display string
         LEA    DX, FLAGS         ;addr of flags (af, cf) in dx
         INT    21H               ;dos interrupt

BEGIN    ENDP
         END    BEGIN
```

```
Enter 2 ascii chars as packed bcd #s: 8y  //38 79
Result = 17
Flags = 11          //AF = 1, CF = 1; Sum = 117
-------------------------------------------------------------
Enter 2 ascii chars as packed bcd #s: 9q  //39 71
Result = 10
Flags = 11          //AF = 1, CF = 1; Sum = 110
-------------------------------------------------------------
Enter 2 ascii chars as packed bcd #s: 2R  //32 52
Result = 84
Flags = 00          //AF = 0, CF = 0; Sum = 084
-------------------------------------------------------------
Enter 2 ascii chars as packed bcd #s: cr  //63 72
Result = 35
Flags = 01          //AF = 0, CF = 1; Sum = 135
-------------------------------------------------------------
Enter 2 ascii chars as packed bcd #s: I8  //49 38
Result = 87
Flags = 10          //AF = 1, CF = 0; Sum = 087
-------------------------------------------------------------
```

(b) //continued on next page

Figure 10.7 (Continued)

```
Enter 2 ascii chars as packed bcd #s: 99  //39 39
Result = 78
Flags = 10         //AF = 1, CF = 0; Sum = 078
----------------------------------------------------------
Enter 2 ascii chars as packed bcd #s: 76  //37 36
Result = 73
Flags = 10         //AF = 1, CF = 0; Sum = 073
----------------------------------------------------------
Enter 2 ascii chars as packed bcd #s: Ec  //45 63
Result = 08
Flags = 01         //AF = 0, CF = 1; Sum = 108
```

Figure 10.7 (Continued)

10.3 ASCII Adjust AL after Subtraction (AAS) Instruction

The AAS instruction adjusts the result of a subtraction operation of two unpacked binary-coded decimal (BCD) operands. The resulting difference is converted to an unpacked BCD value and stored in general-purpose register AL, which is the implied source and destination for the AAS instruction.

This result may not be a valid BCD value. The AAS instruction converts the difference in register AL into a valid BCD number and stores the result in bits 3 through 0 of register AL and resets bits 7 through 4 of register AL. If the low-order four bits of the result in register AL are greater than nine or if the AF flag is set, then six is subtracted from AL and register AH is decremented by 1. In this case, both the auxiliary carry flag (AF) and the carry flag (CF) are set; otherwise, they are reset. The following flags are undefined: OF, SF, ZF, and PF.

When used in a program, the AAS instruction is preceded by a subtract (SUB) instruction or an integer subtraction with borrow (SBB) instruction. Shown below are two examples of subtraction in which the results are a valid and an invalid BCD number before using the AAS instruction.

```
AX = 00 08 - 00 03:   AL = 05; AF = 0, CF = 0
AAS                   AL = 05H; AH = 00H
--------------------------------------------------
AX = 00 06 - 00 08:   AL = -2 (0FEH)
AL & 0FH = 0EH > 9;   AF = 1, CF = 1
AAS                   AL - 6 = -8 (0F8H)
                      AL & 0FH = 08; AH = FFH
```

Figure 10.8 shows an assembly language module embedded in a C program that illustrates using the AAS instruction to adjust the result of subtracting two single-digit numbers. The numbers used in the preceding examples are entered from the keyboard for comparison.

```
//aas.cpp
//illustrates the use of the aas instruction

#include "stdafx.h"

int main (void)

{
//define variables
   unsigned char minuend, subtrahend, result_lo, result_hi;

   printf ("Enter two single-digit numbers:
               minuend -- subtrahend: \n");
   scanf ("%d %d", &minuend, &subtrahend);

//switch to assembly
      _asm
      {
         MOV   AL, minuend
         MOV   BL, subtrahend

         SUB   AL, BL
         AAS

         MOV   result_lo, AL
         MOV   result_hi, AH
      }

   printf ("\nResult low-order = %X\n", result_lo);

   printf ("Result high-order = %X\n\n", result_hi);

   return 0;
}
                                   //continued on next page
```

(a)

Figure 10.8 Program to illustrate using the AAS instruction: (a) the program and (b) the outputs.

```
Enter two single-digit numbers: minuend -- subtrahend:
8 3

Result low-order = 5
Result high-order = 0

Press any key to continue . . . _
------------------------------------------------------------
Enter two single-digit numbers: minuend -- subtrahend:
6 8

Result low-order = 8
Result high-order = FF

Press any key to continue . . . _
------------------------------------------------------------
Enter two single-digit numbers: minuend -- subtrahend:
4 7

Result low-order = 7
Result high-order = FF

Press any key to continue . . . _
------------------------------------------------------------
Enter two single-digit numbers: minuend -- subtrahend:
5 9

Result low-order = 6
Result high-order = FF

Press any key to continue . . . _
------------------------------------------------------------
Enter two single-digit numbers: minuend -- subtrahend:
8 4

Result low-order = 4
Result high-order = 0

Press any key to continue . . . _
------------------------------------------------------------
Enter two single-digit numbers: minuend -- subtrahend:
2 9

Result low-order = 3
Result high-order = FF

Press any key to continue . . . _(b)
```

Figure 10.8 (Continued)

10.4 Decimal Adjust AL after Subtraction (DAS) Instruction

The DAS instruction adjusts the result of a subtraction of two packed BCD operands. The instruction generates a valid packed BCD result and stores the difference in general-purpose register AL, which is the implied source and destination operand. The DAS instruction is preceded by a subtract (SUB) instruction or by an integer subtraction with borrow (SBB) instruction.

There are two rules that are used when subtracting packed BCD numbers regarding the resulting difference.

Rule 1: If the low-order four bits of the difference are greater than nine or if the auxiliary carry flag (AF) is set, then 06H is subtracted from the difference — which is contained in register AL — and the AF flag is set; otherwise, the AF flag is reset.

Rule 2: If the difference in AL is greater than 9FH or if the carry flag (CF) is set, then 60H is subtracted from AL and the CF flag is set; otherwise, the CF flag is reset.

For both rules, the overflow flag (OF) is undefined. Examples of subtracting packed BCD numbers are shown in Figure 10.9 for the following operands:

42 – 23 78 – 94 34 – 89.

			AL		
42 – 23 = 19			4	2	
		+)	D	D	(2s complement of BCD 23)
			1	F	
DAS (Rule 1)		+)	F_1	A	(–6)
	AF = 1		1	9	

			AL		
78 – 94 = –84			7	8	
		+)	6_1	C	(2s complement of BCD 94)
			E	4	
DAS (Rule 2)		+)	A	0	(–60)
	CF = 1		8	4	

//continued on next page

Figure 10.9 Examples of utilizing the DAS instruction.

```
                                        AL
34 – 89 = –45                         3   4
                              +)      7   7   (2s complement of BCD 89)
                                     -------
                                      A   B
DAS (Rule 1)                  +)          A          (–6)
                                     -------
                     AF = 1           A   5
DAS (Rule 2)                  +)      A   0          (–60)
                                     -------
                     CF = 1           4   5
```

Figure 10.9 (Continued)

Figure 10.10 shows an assembly language module embedded in a C program that illustrates the application of the DAS instruction. Two valid packed BCD operands are entered from the keyboard and subtracted using the SUB instruction, then the difference is adjusted using the DAS instruction. The difference after the SUB instruction is displayed together with the result after the DAS instruction is executed. The low-order byte of the EFLAGS register is also displayed. The first three sets of operands are the same as the operands shown in the examples of Figure 10.9.

```
//das.cpp
//illustrates the use of the das instruction

#include "stdafx.h"

int main (void)
{
//define variables
   unsigned char minuend, subtrahend, sub_diff, das_diff,
                  flags;

   printf ("Enter two 2-digit hexadecimal numbers: \n");
   scanf ("%X %X", &minuend, &subtrahend);
```

//continued on next page

(a)

Figure 10.10 Program to illustrate the application of the DAS instruction: (a) the program and (b) the outputs.

```
//switch to assembly
      _asm
      {
//obtain difference
         MOV   AL, minuend
         SUB   AL, subtrahend
         MOV   sub_diff, AL
         DAS
         MOV   das_diff, AL

//save flags
         PUSHF
         POP   AX
         MOV   flags, AL
      }

   printf ("\nMinuend = %X, Subtrahend = %X\n",
               minuend, subtrahend);
   printf ("\nSUB difference = %X, DAS difference = %X\n",
               sub_diff, das_diff);
   printf ("Flags = %X\n\n", flags);
   return 0;
}
```

```
Enter two 2-digit hexadecimal numbers:
42 23

Minuend = 42, Subtrahend = 23

SUB difference = 1F, DAS difference = 19
Flags = 12        //AF

Press any key to continue . . . _
------------------------------------------------------------
Enter two 2-digit hexadecimal numbers:
78 94

Minuend = 78, Subtrahend = 94

SUB difference = E4, DAS difference = 84
Flags = 87        //SF, PF, CF

Press any key to continue . . . _
```

(b) //continued on next page

Figure 10.10 (Continued)

```
Enter two 2-digit hexadecimal numbers:
34 89

Minuend = 34, Subtrahend = 89

SUB difference = AB, DAS difference = 45
Flags = 13        //AF, CF

Press any key to continue . . . _
------------------------------------------------------------
Enter two 2-digit hexadecimal numbers:
68 51

Minuend = 68, Subtrahend = 51

SUB difference = 17, DAS difference = 17
Flags = 6         //PF

Press any key to continue . . . _
------------------------------------------------------------
Enter two 2-digit hexadecimal numbers:
45 87

Minuend = 45, Subtrahend = 87

SUB difference = BE, DAS difference = 58
Flags = 13        //AF, CF

Press any key to continue . . . _
```

Figure 10.10 (Continued)

10.5 ASCII Adjust AX after Multiplication (AAM) Instruction

The AAM instruction adjusts the product in general-purpose register AX resulting from multiplying two valid unpacked BCD operands. The adjustment creates a product of two unpacked BCD values in register AX: the high-order digit is in AH, the low-order digit is in AL. Register AX is the implied source and destination operand for the AAM instruction. The AAM instruction is preceded by a MUL instruction. The affected flags are the sign flag (SF), the zero flag (ZF), and the parity flag (PF), which are set according to the result in register AL. The overflow flag (OF), the auxiliary carry flag (AF), and the carry flag (CF) are undefined.

The AAM instruction divides register AL by 10_{10} (0AH) and stores the quotient in register AH and the remainder in register AL, as shown below. Since single digits have a maximum value of 9, the product has a maximum value of 81_{10} (51H); therefore, only register AL can contain an invalid BCD digit. Figure 10.11 depicts the sequence of operations performed by the MUL and AAM instructions.

```
AH  ←—  AL/0AH   (Quotient)
AL  ←—  AL%0AH   (Remainder, % is modulo)
```

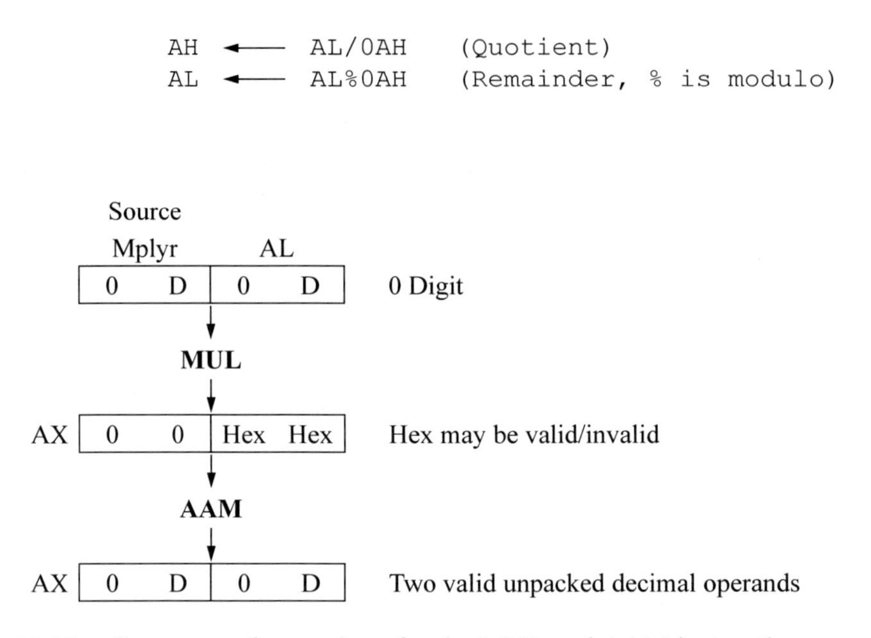

Figure 10.11 Sequence of operations for the MUL and AAM instructions.

Figure 10.12 shows an assembly language module embedded in a C program that illustrates the use of the AAM instruction when preceded by a MUL instruction. In this program, the product in register AX is in the same format as shown in Figure 10.11; that is, two unpacked BCD digits. A one-digit multiplicand and a one-digit multiplier are entered from the keyboard. The resulting product is displayed as a hexadecimal value after the MUL instruction is executed and is displayed as two unpacked BCD digits after the AAM instruction is executed. The flags are also displayed.

```
//aam2.cpp
//illustrates the use of the aam instruction

#include "stdafx.h"
int main (void)                 (a)          //continued on next page
```

Figure 10.12 Program to illustrate the use of the AAM instruction: (a) the program and (b) the outputs.

```
{
//define variables
   unsigned char mpcnd, mplyr, flags;

   unsigned short mul_prod, aam_prod;

   printf ("Enter a 1-digit multiplicand and multiplier:
            \n");

   scanf ("%X %X", &mpcnd, &mplyr);

//switch to assembly
      _asm
      {
//multiply the operands
         MOV   AL, mpcnd
         MUL   mplyr
         MOV   mul_prod, AX

//adjust the result
         AAM
         MOV   aam_prod, AX

//save the flags
         PUSHF
         POP   AX
         MOV   flags, AL
      }

   printf ("\nMultiplicand = %X, Multiplier = %X",
               mpcnd, mplyr);

   printf ("\nMUL product = %X\n", mul_prod);

   printf ("AAM product = %X\n", aam_prod);

   printf ("Flags = %X\n\n", flags);

   return 0;
}
                                        //continued on next page
```

Figure 10.12 (Continued)

```
Enter a 1-digit multiplicand and multiplier:
9 9

Multiplicand = 9, Multiplier = 9
MUL product = 51
AAM product = 801
Flags = 2       //bit 2 is set

Press any key to continue . . . _

------------------------------------------------------------
Enter a 1-digit multiplicand and multiplier:
4 7

Multiplicand = 4, Multiplier = 7
MUL product = 1C
AAM product = 208
Flags = 2       //bit 2 is set

Press any key to continue . . . _

------------------------------------------------------------
Enter a 1-digit multiplicand and multiplier:
9 5

Multiplicand = 9, Multiplier = 5
MUL product = 2D
AAM product = 405
Flags = 6       //PF

Press any key to continue . . . _

------------------------------------------------------------
Enter a 1-digit multiplicand and multiplier:
9 7

Multiplicand = 9, Multiplier = 7
MUL product = 3F
AAM product = 603
Flags = 6       //PF

Press any key to continue . . . _
```

(b)

Figure 10.12 (Continued)

Figure 10.13 shows an assembly language module embedded in a C program that also illustrates the application of the AAM instruction in conjunction with the MUL instruction. This program demonstrates a slightly different version of the same application using the MUL instruction and the AAM instruction that was shown in Figure 10.12; this time the product is displayed as a packed BCD value.

The unpacked product is shown below in register AX, where *D* represents a BCD digit. In order to obtain a packed format in register AL, the unpacked digit in register AH is shifted left logically (SHL) four bit positions. Then register AH is ORed with register AL to obtain the requisite two-digit packed format.

	AH		AL	
AX	0	D	0	D

A one-digit multiplicand and a one-digit multiplier are entered from the keyboard; the resulting product is displayed after the MUL instruction is executed and after the AAM instruction is executed. Although the flags are not usually required, they are also displayed. A program to multiply a multidigit multiplicand by a single-digit multiplier is left as a problem.

```
//aam3.cpp
//illustrates the use of the aam instruction

#include "stdafx.h"
int main (void)
{
//define variables
   unsigned char mpcnd, mplyr, flags, aam_prod;
   unsigned short mul_prod;

   printf ("Enter a 1-digit multiplicand and multiplier:
            \n");
   scanf ("%X %X", &mpcnd, &mplyr);

//switch to assembly
      _asm
      {
//multiply the operands
         MOV   AL, mpcnd
         MUL   mplyr
         MOV   mul_prod, AX
```

(a) //continued on next page

Figure 10.13 Program to illustrate the use of the AAM instruction to obtain a packed BCD product: (a) the program and (b) the outputs.

```
//adjust the result
         AAM

//save the flags; flags are affected by shl later
         PUSHF
         POP   BX
         MOV   flags, BL

//change product from 0d 0d to dd
         SHL   AH, 4
         OR    AL, AH
         MOV   aam_prod, AL
      }

   printf ("\nMultiplicand = %X, Multiplier = %X",
               mpcnd, mplyr);
   printf ("\nMUL product = %X\n", mul_prod);
   printf ("AAM product = %X\n", aam_prod);
   printf ("Flags = %X\n\n", flags);
   return 0;
}
```

```
Enter a 1-digit multiplicand and multiplier:
9 9

Multiplicand = 9, Multiplier = 9
MUL product = 51H
AAM product = 81₁₀
Flags = 2       //bit 2 is set

Press any key to continue . . . _
------------------------------------------------------------
Enter a 1-digit multiplicand and multiplier:
4 7

Multiplicand = 4, Multiplier = 7
MUL product = 1CH
AAM product = 28₁₀
Flags = 2       //bit 2 is set

Press any key to continue . . . _
------------------------------------------------------------
```

(b) //continued on next page

Figure 10.13 (Continued)

```
Enter a 1-digit multiplicand and multiplier:
9 5

Multiplicand = 9, Multiplier = 5
MUL product = 2DH
AAM product = 45₁₀
Flags = 6       //PF

Press any key to continue . . . _
------------------------------------------------------------
Enter a 1-digit multiplicand and multiplier:
9 7

Multiplicand = 9, Multiplier = 7
MUL product = 3FH
AAM product = 63₁₀
Flags = 6       //PF

Press any key to continue . . . _
```

Figure 10.13 (Continued)

Figure 10.14 shows an assembly language program — not embedded in a C program — that uses the MUL instruction in conjunction with the AAM instruction to calculate the area of a triangle, as shown below.

$$\text{Area} = \tfrac{1}{2}\ \text{base} \times \text{height}$$

A single-digit base and a single-digit height are entered from the keyboard and stored in the OPFLD input buffer of the parameter list (PARLST). Since the parameters are entered as ASCII characters, they must first be unpacked — ASCII bias removed — before they are used. If the base value is an odd number, then the base is truncated — rounded down — when it is divided by 2.

```
;area of triangle.asm
;area = 1/2 base x height
;-------------------------------------------------------------
.STACK
;-------------------------------------------------------------
.DATA
PARLST   LABEL BYTE
```

(a) //continued on next page

Figure 10.14 Calculate the area of a triangle using the MUL and AAM instructions: (a) the program and (b) the outputs.

```
MAXLEN    DB 5
ACTLEN    DB ?
OPFLD     DB 5  DUP(?)
PRMPT     DB 0DH, 0AH, 'Enter base and height: $'
RSLT      DB 0DH, 0AH, 'Area of triangle =      $'

;-------------------------------------------------------------
.CODE
BEGIN     PROC FAR

;set up pgm ds
          MOV   AX, @DATA         ;get addr of data seg
          MOV   DS, AX            ;move addr to ds

;read prompt
          MOV   AH, 09H           ;display string
          LEA   DX, PRMPT         ;put addr of prompt in dx
          INT   21H               ;dos interrupt

;keyboard request rtn to enter characters
          MOV   AH, 0AH           ;buffered keyboard input
          LEA   DX, PARLST        ;put addr of parlst in dx
          INT   21H               ;dos interrupt

;get base and height and remove ascii bias
          MOV   BL, OPFLD         ;get 1st digit (base)
          AND   BL, 0FH           ;remove ascii bias
          MOV   AL, OPFLD+1       ;get 1st digit (height)
          AND   AL, 0FH           ;remove ascii bias

;calculate area
          SHR   BL, 1             ;divide base by 2
          MUL   BL                ;mul 1/2 base x height in al
          AAM                     ;convert to unpacked decimal #s
          OR    AX, 3030H         ;add ascii bias

          MOV   RSLT+21, AH       ;move tens to result area
          MOV   RSLT+22, AL       ;move units to result area

;-----------------------------------------------------------
;display area
          MOV   AH, 09H           ;display string
          LEA   DX, RSLT          ;put addr of rslt (area) in dx
          INT   21H               ;dos interrupt

BEGIN     ENDP
          END   BEGIN                //continued on next page
```

Figure 10.14 (Continued)

```
Enter base and height: 67
Area of triangle = 21
------------------------------------------------------------
Enter base and height: 89
Area of triangle = 36
------------------------------------------------------------
Enter base and height: 98
Area of triangle = 32   //1/2 the base truncates; rounds down
------------------------------------------------------------
Enter base and height: 87
Area of triangle = 28
------------------------------------------------------------
Enter base and height: 45
Area of triangle = 10
------------------------------------------------------------
Enter base and height: 44
Area of triangle = 08
```

(b)

Figure 10.14 (Continued)

10.6 ASCII Adjust AX before Division (AAD) Instruction

The AAD instruction converts two unpacked digits in register AX — high-order digit in register AH and low-order digit in register AL — to an equivalent binary value, stores the result in register AL, and resets register AH. The AAD instruction normally precedes a DIV instruction, but can be used independently to convert two unpacked BCD digits in register AX to an equivalent radix 10 value in registers AH and AL, where AH = 00H.

When the value in register AX is then divided by an unpacked BCD value, the unpacked quotient is stored in register AL and the unpacked remainder is stored in register AH. The AAD instruction multiplies register AH by 10_{10} (0AH), then adds the product to the contents of register AL, and stores the result in register AL, as shown below.

```
AL  ←——  (AH × 0AH) + AL
AH  ←——  00H
```

The affected flags are the sign flag (SF), the zero flag (ZF), and the parity flag (PF), which are set or reset according to the result in register AL. The overflow flag (OF), the auxiliary carry flag (AF), and the carry flag (CF) are unaffected.

An example is shown in Figure 10.15 that divides AX = 08 04H by BL = 09H using the AAD instruction and the DIV instruction. The AAD instruction changes AX from AX = 08 04H to AX = 00 84_{10} (00 54H), then the DIV instruction produces the result in AX as AH = 03 (remainder), AL = 09 (quotient).

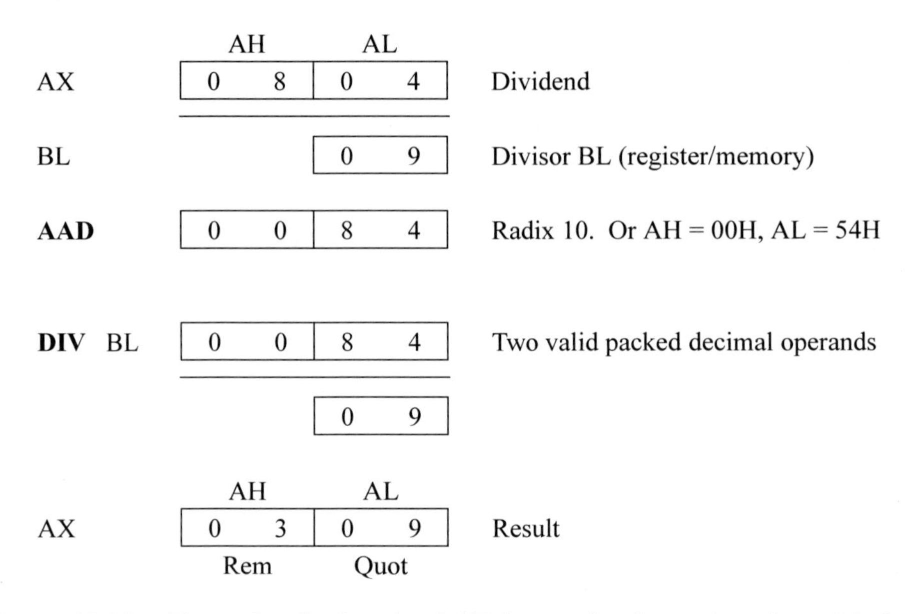

Figure 10.15 Example of using the AAD instruction in conjunction with the DIV instruction.

Figure 10.16 shows an assembly language program — not embedded in a C program — that uses the AAD instruction in conjunction with the DIV instruction to divide a 2-digit unpacked BCD dividend by a 1-digit unpacked BCD divisor. The divide operation is executed with several different operands that are entered from the keyboard.

The results of the AAD instruction are displayed together with the quotient and remainder obtained from the DIV instruction. The results obtained from executing the AAD instruction and the DIV instruction may not necessarily be displayed as numerical ASCII digits — 30 (0) through 39 (9). The ASCII digits can range from 21H (!) to 7EH (~) depending on the results. For example, dividing a dividend of 74 by a divisor of 3 yields a quotient of 24 and a remainder of 2. Executing the AAD instruction for these operands displays a value of *J* (4AH = 74); executing the DIV instruction displays a quotient of 18H (24), and a remainder of 2H (2).

```
;aad.asm
;obtain the quotient and remainder
;of a 2-digit unpacked bcd dividend
;and a 1-digit unpacked bcd divisor
;-----------------------------------------------------------
.STACK

;-----------------------------------------------------------
.DATA
PARLST          LABEL   BYTE
MAXLEN          DB   5
ACTLEN          DB   ?
OPFLD           DB   5    DUP(?)
PRMPT           DB   0DH, 0AH, ' Enter 2-digit dvdnd &
                                   1-digit dvsr: $'
RSLT_AAD        DW   0DH, 0AH, 'After AAD, AX =        $'
RSLT_DIV_QUOT   DB   0DH, 0AH, 'After DIV, Quot =      $'
RSLT_DIV_REM    DB   0DH, 0AH, 'After DIV, Rem =       $'

;-----------------------------------------------------------
.CODE
BEGIN    PROC   FAR

;set up pgm ds
         MOV    AX, @DATA       ;get addr of data seg
         MOV    DS, AX          ;move addr to ds

;read prompt
         MOV    AH, 09H         ;display string
         LEA    DX, PRMPT       ;put addr of prompt in dx
         INT    21H             ;dos interrupt

;keyboard request rtn to enter characters
         MOV    AH, 0AH         ;buffered keyboard input
         LEA    DX, PARLST      ;put addr of parlst in dx
         INT    21H             ;dos interrupt

;-----------------------------------------------------------
;get the two numbers and unpack them
         MOV    AH, OPFLD       ;get dvdnd first digit
         AND    AH, 0FH         ;unpack it
         MOV    AL, OPFLD+1     ;get dvdnd second digit
         AND    AL, 0FH         ;unpack it
         MOV    DL, OPFLD+2     ;get dvsr digit
         AND    DL, 0FH         ;unpack it
```

(a) //continued on next page

Figure 10.16 Assembly language program to illustrate using the AAD instruction in conjunction with the DIV instruction: (a) the program and (b) the outputs.

```
;------------------------------------------------------------
ascii adjust ax before division
         AAD
         MOV   RSLT_AAD+20, AX;move aad value to result area

;------------------------------------------------------------
;divide ax by dl and form quotient and remainder
         DIV   DL              ;quot in al; rem in ah
         MOV   BL, AL          ;move al to bl


         OR    AH, 30H         ;add ascii bias to ah (rem)


         SHR   BL, 4           ;shift right logical 4
         OR    BL, 30H         ;add ascii bias to bl (quot hi)


         AND   AL, 0FH         ;isolate low-order al
         OR    AL, 30H         ;add ascii bias to al (quot lo)

;------------------------------------------------------------
;move quotient and remainder to result area
         MOV   RSLT_DIV_REM+19, AH
         MOV   RSLT_DIV_QUOT+20, BL
         MOV   RSLT_DIV_QUOT+21, AL

;------------------------------------------------------------
;display aad result
         MOV   AH, 09H         ;display string
         LEA   DX, RSLT_AAD
         INT   21H             ;dos interrupt

;------------------------------------------------------------
;display div result
         MOV   AH, 09H         ;display string
         LEA   DX, RSLT_DIV_QUOT
         INT   21H             ;dos interrupt


;display div result
         MOV   AH, 09H         ;display string
         LEA   DX, RSLT_DIV_REM
         INT   21H             ;dos interrupt

;------------------------------------------------------------
BEGIN    ENDP
         END   BEGIN
                                  //continued on next page
```

Figure 10.16 (Continued)

```
Enter 2-digit dvdnd & 1-digit dvsr: 758
After AAD, AX = K        //K = 4BH (75₁₀)
After DIV, Quot = 09H
After DIV, Rem = 3

-----------------------------------------------------------
Enter 2-digit dvdnd & 1-digit dvsr: 743
After AAD, AX = J        //J = 4AH (74₁₀)
After DIV, Quot = 18H    //24₁₀
After DIV, Rem = 2

-----------------------------------------------------------
Enter 2-digit dvdnd & 1-digit dvsr: 934
After AAD, AX = ]        //] = 5DH (93₁₀)
After DIV, Quot = 17H    //23₁₀
After DIV, Rem = 1

-----------------------------------------------------------
Enter 2-digit dvdnd & 1-digit dvsr: 859
After AAD, AX = U        //U = 55H (85₁₀)
After DIV, Quot = 09H
After DIV, Rem = 4

-----------------------------------------------------------
Enter 2-digit dvdnd & 1-digit dvsr: 932
After AAD, AX = ]        //] = 5DH (93₁₀)
After DIV, Quot = 2>     //2> = 2EH (46₁₀)
After DIV, Rem = 1

-----------------------------------------------------------
Enter 2-digit dvdnd & 1-digit dvsr: 849
After AAD, AX = T        //T = 54H (84₁₀)
After DIV, Quot = 09H
After DIV, Rem = 3

-----------------------------------------------------------
Enter 2-digit dvdnd & 1-digit dvsr: 997
After AAD, AX = c        //c = 63H (99₁₀)
After DIV, Quot = 0>     //0> = 0EH (14₁₀)
After DIV, Rem = 1

-----------------------------------------------------------
Enter 2-digit dvdnd & 1-digit dvsr: 465
After AAD, AX = .        //. = 2EH (46₁₀)
After DIV, Quot = 09
After DIV, Rem = 1
```

(b)

Figure 10.16 (Continued)

10.7 Problems

10.1 Show the contents of register AX and the values of the indicated flags after the ADD instruction has been executed and after the AAA instruction has been executed.

```
ADD    AL, CH
AAA
```

	Before	After ADD	After AAA
(a)	AX = 00 04H	AX =	AX =
	CH = 07H	AF =	AF = , CF =
(b)	AX = 05 05H	AX =	AX =
	CH = 07H	AF =	AF = , CF =

10.2 Write an assembly language program — not embedded in a C program — that adds two ASCII numbers that are entered from the keyboard, where the ASCII numbers have three digits each. The first number entered is the augend; the second number entered is the addend. The parameter list (PARLST) format is shown below for convenience. The ASCII numbers shown in the PARLST yield a sum of 1272_{10}. Enter at least four sets of numbers and display the resulting sums.

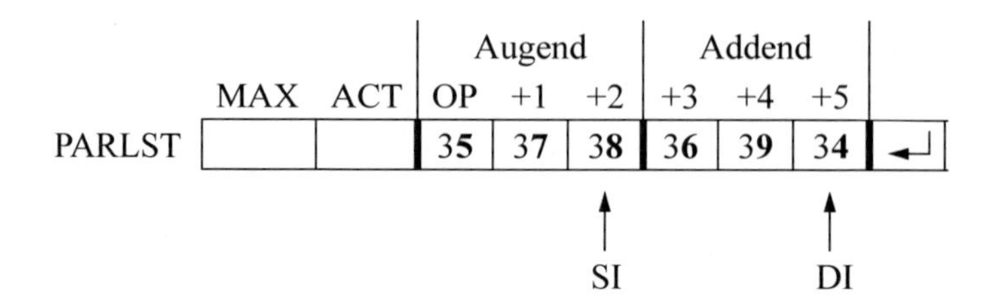

10.3 Let register AX = 05 35H and register BL = 39H. After executing the instructions shown below, determine the contents of AX and the auxiliary carry flag (AF).

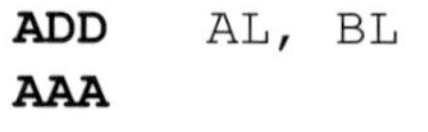

10.4 Write an assembly language program — not embedded in a C program — for a sequence detector that detects the number of times that the four-bit sequence 0110 appears *anywhere* in the keyboard input buffer (OPFLD) when three ASCII characters are entered. This program does not necessarily require BCD instructions.

The bit sequence 0110 may be in the high-order four bits of the keyboard character, the low-order four bits, both high-order and low-order four bits, across the boundary of the high- and low-order bits, or across the boundary of two contiguous bytes. Display the number of times that the sequence occurs. A flowchart may make the program easier to code. Two examples of input data are shown below for *cct* and *F67*.

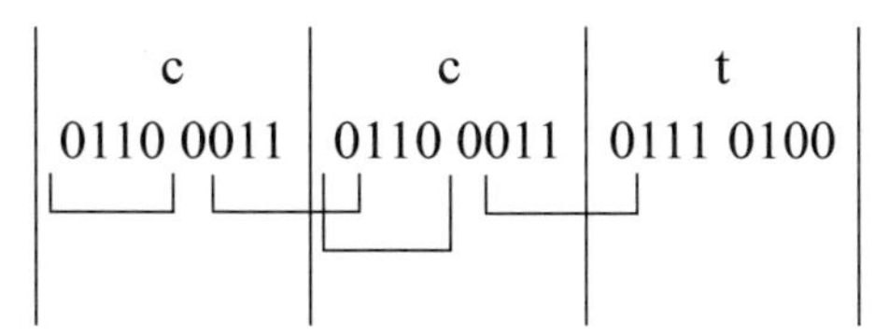

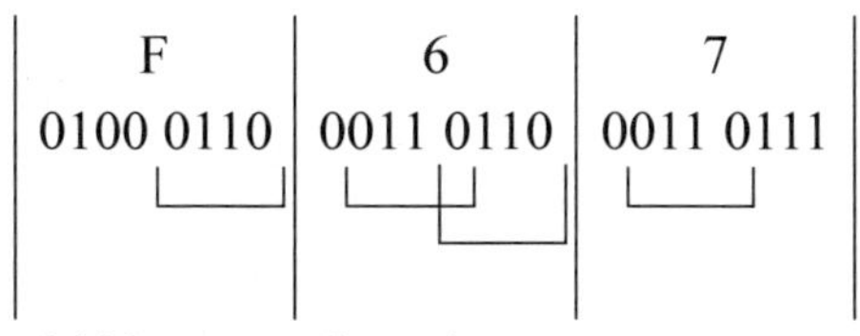

10.5 In each part below, assume that the following instructions are executed:

```
ADD    AL, BL
DAA
```

(a) Give the values of AL after the ADD instruction has been executed, but before the DAA instruction has been executed for the indicated register values shown below.

(b) Give the values of AL, the carry flag (CF), and the auxiliary carry flag (AF) after the DAA instruction has been executed for the indicated register values shown below.

(1) AL = 35H, BL = 48H
(2) AL = 47H, BL = 61H
(3) AL = 75H, BL = 46H

Then write an assembly language module embedded in a C program to verify the results.

10.6 Given the two instructions shown below, determine the ADD sum, the DAA sum, the auxiliary carry flag (AF), the parity flag (PF), and the carry flag (CF) after the instructions have been executed when AL = 89H and BL = 64H.

```
ADD    AL, BL
DAA
```

10.7 Write an assembly language module embedded in a C program that adds two 4-digit hexadecimal numbers. Display the carry flag (CF) and the sum for several different numbers.

10.8 Perform a subtraction and adjustment of the following two operands: 07 – 09. Show the details of the subtraction and adjustment.

10.9 Design an assembly language program — not embedded in a C program — to perform the following subtraction using the AAS instruction: 76 – 48, then display the result.

10.10 Design an assembly language module embedded in a C program to perform subtraction on two unpacked BCD operands using the AAS instruction, then display the result. Enter several different operands.

10.11 Design an assembly language program — not embedded in a C program — that performs subtraction on two unpacked BCD operands using the AAS instruction. Display the results of several different operands. The first two operands are shown below as: 76 (37 36) and 48 (34 38).

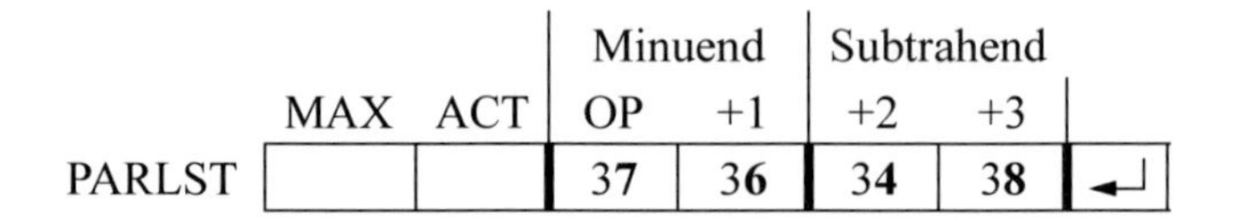

			Minuend		Subtrahend		
	MAX	ACT	OP	+1	+2	+3	
PARLST			37	36	34	38	↵

10.12 Let AL = 00H and BL = 61H. Execute the following two instructions and then show the contents of AL, CF, and AF after the SUB instruction has been executed and after the DAS instruction has been executed:

```
SUB    AL, BL
DAS
```

10.13 Write an assembly language module embedded in a C program that uses the DAS instruction to subtract and adjust two 4-digit hexadecimal numbers. Display the difference and the resulting flags.

10.14 Write an assembly language program — not embedded in a C program — that performs subtraction on two valid packed BCD operands using the DAS instruction. Display the results of several different operands.

10.15 Show the contents of register AX after the MUL instruction has been executed and after the AAM instruction has been executed for AL = 04H, BL = 06H and AL = 05H, BL = 07H.

```
MUL    BL
AAM
```

10.16 Write an assembly language program — not embedded in a C program — that obtains the product of two unpacked single-digit ASCII (BCD) numbers using the MUL and AAM instructions. Display the multiplicand, the multiplier, and the product.

10.17 Write an assembly language program — not embedded in a C program — that multiplies a five-digit multiplicand by a one-digit multiplier using the MUL instruction in combination with the AAM instruction. The ADD instruction and the AAA instruction will also be used in implementing the program.

The five-digit multiplicand and the one-digit multiplier are entered from the keyboard and stored in the OPFLD, as shown below using a multiplicand of 12345 and a multiplier of 6. Enter several different multiplicands and multipliers, then display the operands and the resulting product.

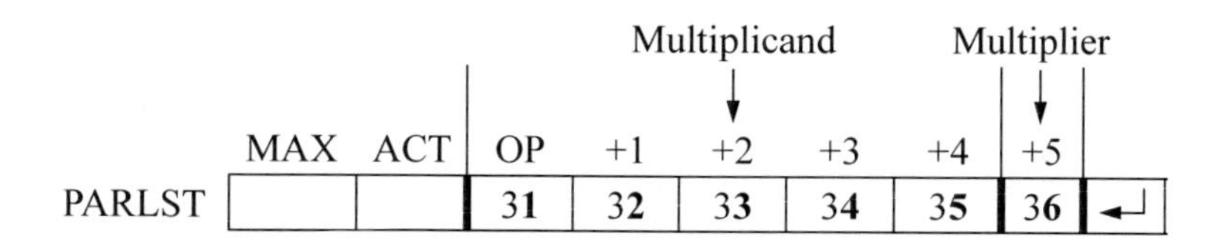

The paper-and-pencil example shown below may help to illustrate the algorithm used in the program.

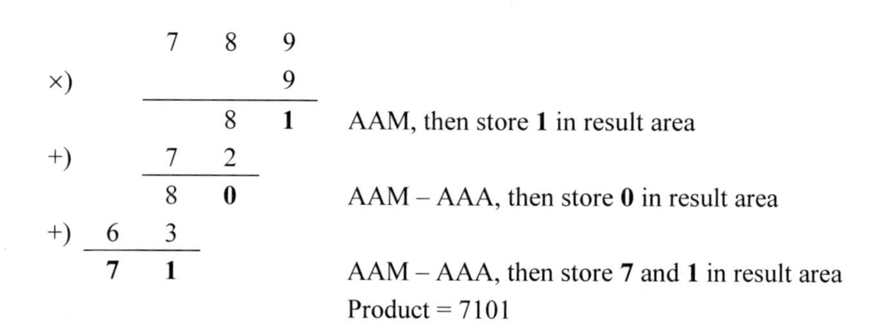

10.18 Write an assembly language program — not embedded in a C program — that calculates the number of hours, minutes, and seconds from a number of seconds that are entered from the keyboard. The number of seconds entered ranges from 10,000 to 60,000. Enter several different 5-digit numbers of seconds. Do not use the LOOP instruction in this program — Problem 10.20 will use the LOOP instruction to reduce the amount of code. This program will use the AAM instruction.

10.19 Repeat Problem 10.18, this time using an assembly language module embedded in a C program. Note the simplicity of this program. There is no requirement to remove the ASCII bias for the numbers that are entered from the keyboard, or to restore the ASCII bias before displaying the hours, minutes, and seconds. It is also not necessary to establish stack or data segments. The range is the same: 10,000 to 60,000 seconds.

10.20 Write an assembly language program to repeat Problem 10.18 — not embedded in a C program — that calculates the number of hours, minutes, and seconds from a number of seconds that are entered from the keyboard. This time use a loop to calculate the sum that is represented by the digits that are entered from the keyboard. The number of seconds entered ranges from 10,000 to 60,000. Enter several different 5-digit numbers of seconds. This program will use the MUL, DIV, and AAM instructions.

10.21 Use the AAD instruction in conjunction with the DIV instruction to divide register AX = 06 04H by register BL = 05H. Show the result of the AAD instruction and the quotient and remainder obtained from the DIV instruction.

10.22 Use the AAD instruction in conjunction with the DIV instruction to divide register AX = 05 09H by register BL = 03H. Show the result of the AAD instruction and the quotient and remainder obtained from the DIV instruction.

10.23 The AAD instruction can be used independently of the DIV instruction for certain applications, as used in the program for this problem. Create a flowchart that demonstrates a method to display one of the twelve months of a calendar from a number that is entered from the keyboard. For example, the number 09 will display the month of September. Then write an assembly language program — not embedded in a C program — that performs this function.

The months are listed in the data segment and each month contains nine letters to correspond to the month with the most letters — September. Enter several month numbers from the keyboard. Other arithmetic instructions will be used in this program in conjunction with the AAD instruction.

11

Floating-Point Arithmetic Instructions

The fixed-point number representation is appropriate for representing numbers with small numerical values that are considered as positive or negative integers; that is, the implied radix point is to the right of the low-order bit. The same algorithms for arithmetic operations can be employed if the implied radix point is to the immediate right of the sign bit, thus representing a signed fraction.

The range for a 16-bit fixed-point number is from (-2^{15}) to $(+2^{15} - 1)$, which is inadequate for some numbers; for example, the following operation:

$$28,400,000,000. \times 0.0000000546$$

This operation can also be written in scientific notation, as follows:

$$(0.284 \times 10^{11}) \times (0.546 \times 10^{-7})$$

where 10 is the *base* and 11 and –7 are the *exponents*. Floating-point notation is equivalent to scientific notation in which the radix point (or binary point) can be made to *float* around the fraction by changing the value of the exponent; thus, the term *floating point*. In contrast, fixed-point numbers have the radix point located in a fixed position, usually to the immediate right of the low-order bit position, indicating an integer.

The base and exponent are called the *scaling factor*, which specify the position of the radix point relative to the *significand digits* (or fraction digits). Common bases are 2 for binary, 10 for decimal, and 16 for hexadecimal. The base in the scaling factor does not have to be explicitly specified in the floating-point number.

11.1 Floating-Point Fundamentals

Floating-point numbers consist of the following three fields: a sign bit, s; an exponent, e; and a fraction, f. These parts represent a number that is obtained by multiplying the fraction, f, by a radix, r, raised to the power of the exponent, e, as shown in Equation 11.1 for the number A, where f and e are signed fixed-point numbers, and r is the radix (or base).

$$A = f \times r^e \tag{11.1}$$

The exponent is also referred to as the *characteristic*; the fraction is also referred to as the *significand* or *mantissa*. Although the fraction can be represented in sign-magnitude, diminished-radix complement, or radix complement, the fraction is predominantly expressed in sign-magnitude representation — sign bit plus fraction.

If the fraction is shifted left k bits, then the exponent is decremented by an amount equal to k; similarly, if the fraction is shifted right k bits, then the exponent is incremented by an amount equal to k. Consider an example in the radix 10 floating-point representation. Let $A = 0.0000074569 \times 10^{+3}$. This number can be rewritten as $A = 0.0000074569 \; +3$ or $A = 0.74569 \; -2$, both with an implied base of 10.

Figure 11.1 shows the format for 32-bit single-precision and 64-bit double-precision floating-point numbers. The single-precision format consists of a sign bit that indicates the sign of the number, an 8-bit signed exponent, and a 23-bit unsigned fraction. The double-precision format consists of a sign bit, an 11-bit signed exponent, and a 52-bit unsigned fraction.

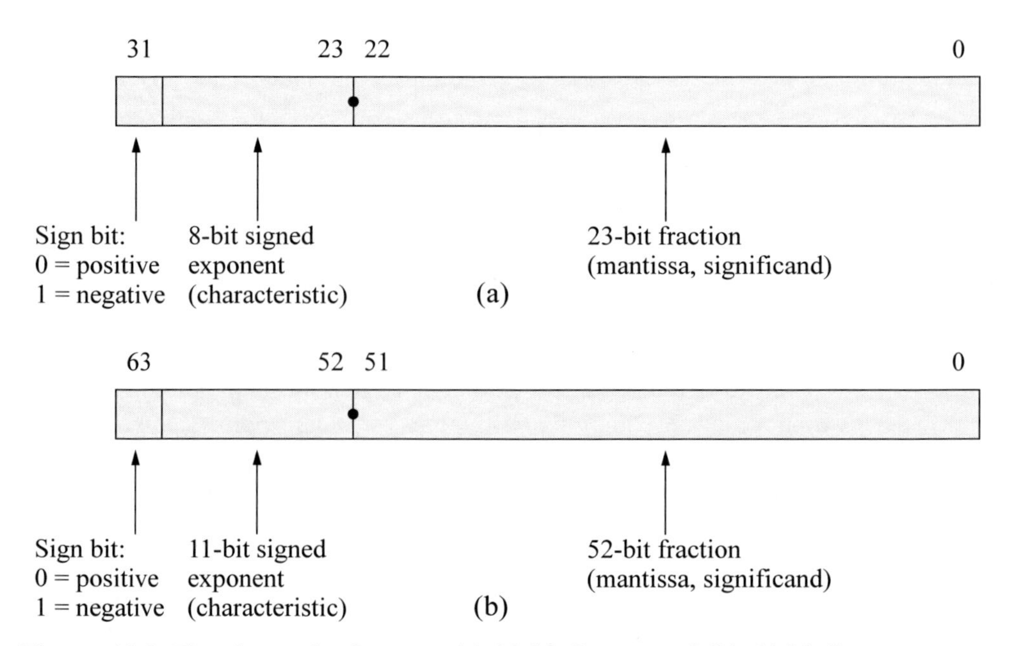

Figure 11.1 Floating-point formats: (a) 32-bit format and (b) 64-bit format.

Figure 11.2(a) shows the eight data registers — called the register stack — used in the floating-point unit (X87 FPU). The stack top ST(0) — also referred to as ST — is register R0 and, like a normal stack, builds toward lower-numbered registers. The register immediately below the stack top is referred to as ST(1); the register immediately below ST(1) is referred to as ST(2), and so forth. When the stack is full, that is, the registers at ST(0) through the register at ST(7) contain valid data, a stack wraparound occurs if an attempt is made to store additional data on the stack. This results in a stack overflow because the unsaved data is overwritten. The stack registers are specified by three bits — 000 through 111 — to reference ST(0) through ST(7). Therefore, ST(*i*) references the *i*th register from the current stack top.

The 16-bit tag register contains a 2-bit tag field for each register that specifies the type of data contained in the corresponding register, as shown in Figure 11.2(b). The values signify whether the data is valid (00); zero (01); a special floating-point number, such as not-a-number (NaN), a value of infinity, a denormal number, or unsupported format (10); or an empty register (11).

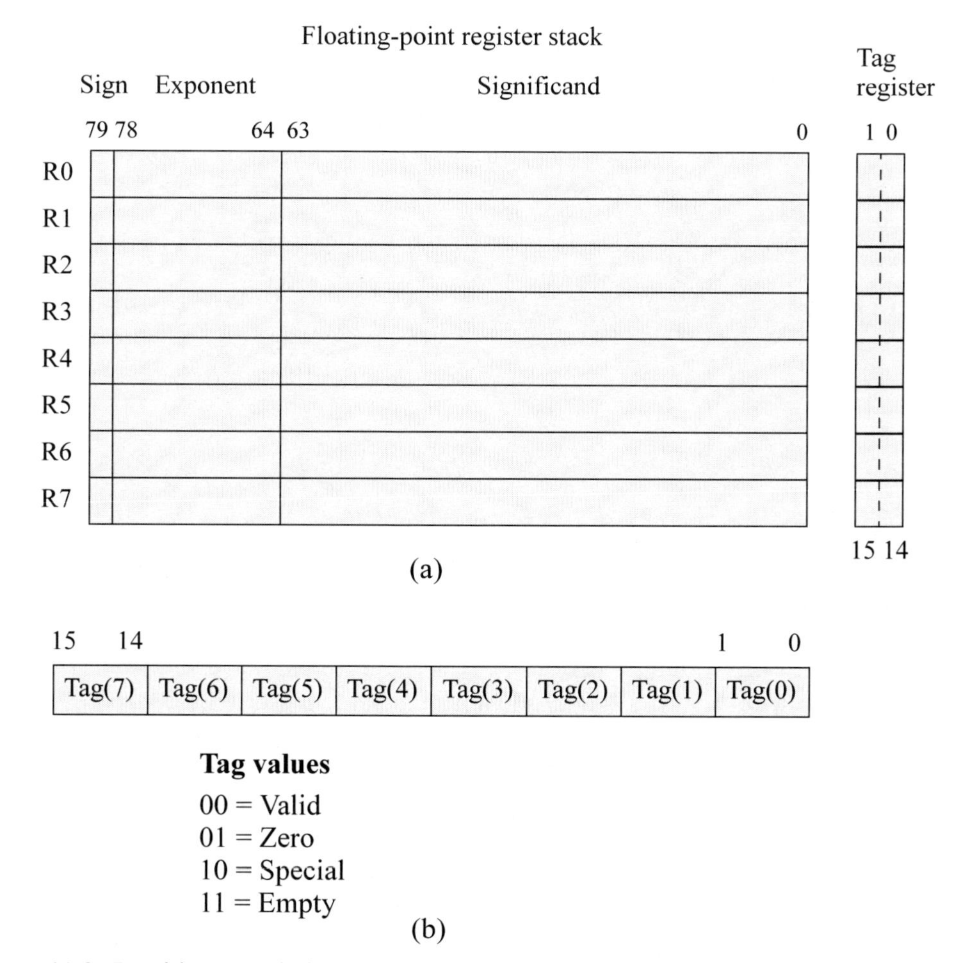

Figure 11.2 Double extended-precision register stack and tag register for the floating-point unit: (a) the register stack and (b) the tag register.

When adding or subtracting floating-point numbers, the exponents are compared and made equal resulting in a right shift of the fraction with the smaller exponent. The comparison is easier if the exponents are unsigned — a simple comparator can be used for the comparison. As the exponents are being formed, a *bias constant* is added to the exponents such that all exponents are positive internally.

For the single-precision format, the bias constant is +127 — also called excess-127; therefore, the biased exponent has a range of

$$0 \leq e_{\text{biased}} \leq 255$$

Fractions in the IEEE format are normalized; that is, the leftmost significant bit is a 1. Figure 11.3 shows unnormalized and normalized numbers in the 32-bit format. Since there will always be a 1 to the immediate right of the radix point, the 1 bit is not explicitly shown — it is an *implied 1.*

	S	Exponent	Fraction
Unnormalized	0	0 0 0 0 0 1 1 1	• 0 0 1 1 x ... x

$+ \ .0011x \ldots x \times 2^7$

	S	Exponent	Fraction
Normalized	0	0 0 0 0 0 1 0 0	• 1 x ... x 0 0 0

$+ \ 1.1x \ldots x000 \times 2^4$

Figure 11.3 Unnormalized and normalized floating-point numbers.

The bias constant has a value that is equal to the most positive exponent. For example, if the exponents are represented by n bits, then the bias is $2^{n-1} - 1$. For n = 4, the most positive number is 0111 (+7). Therefore, all biased exponents are of the form shown in Equation 11.2. The advantage of using biased exponents is that they are easier to compare without having to consider the signs of the exponents. The main reason for biasing is to determine the correct alignment of the fractions by aligning the radix points, and to determine the number of bits to shift a fraction in order to obtain proper alignment.

$$e_{\text{biased}} = e_{\text{unbiased}} + 2^{n-1} - 1 \tag{11.2}$$

11.1.1 Rounding Methods

Rounding deletes one or more low-order bits of the significand and adjusts the retained bits according to a particular rounding technique. The reason for rounding is to reduce the number of bits in the result in order to conform to the size of the significand; that is, in order to be retained within the word size of the machine. Since bits are deleted, this limits the precision of the result.

In some floating-point operations, the result may exceed the number of bits of the significand. For example, rounding can occur when adding two n-bit numbers that result in a sum of $n + 1$ bits. The overflow is handled by shifting the fraction right 1 bit position, resulting in the low-order bit being lost unless it is saved. Rounding attempts to dispose of the extra bits and yet preserve a high degree of accuracy. This section presents three common techniques for rounding that still maintain a high degree of accuracy.

Truncation rounding This method of rounding is also called *chopping*. Truncation deletes extra bits and makes no changes to the retained bits. Aligning fractions during addition or subtraction could result is losing several low-order bits, so there is obviously an error associated with truncation. Assume that the following fraction is to be truncated to four bits:

$$0.b_{-1}\ b_{-2}\ b_{-3}\ b_{-4}\ b_{-5}\ b_{-6}\ b_{-7}\ b_{-8}$$

Then all fractions in the range $0.b_{-1}\ b_{-2}\ b_{-3}\ b_{-4}\ 0000$ to $0.b_{-1}\ b_{-2}\ b_{-3}\ b_{-4}\ 1111$ will be truncated to $0.b_{-1}\ b_{-2}\ b_{-3}\ b_{-4}$. The error ranges from 0 to 0.00001111. In general, the error ranges from 0 to approximately 1 in the low-order position of the retained bits.

Truncation is a fast and easy method for deleting bits resulting from a fraction underflow and requires no additional hardware. There is one disadvantage in that a significant error may result. A fraction underflow can occur when aligning fractions during addition or subtraction when one of the fractions is shifted to the right. Truncation does not round up or round down, but simply deletes a specified number of the low-order significand bits.

Adder-based rounding The result of a floating-point arithmetic operation can be rounded to the nearest number that contains n bits. This method is called *adder-based rounding* and rounds the result to the nearest approximation that contains n bits. The operation is as follows: The bits to be deleted are truncated and a 1 is added to the retained bits if the high-order bit of the deleted bits is a 1. When a 1 is added to the retained bits, the carry is propagated to the higher-order bits. If the addition results in a carry-out of the high-order bit position, then the fraction is shifted right one bit position and the exponent is incremented.

Consider the fraction $0.b_{-1}\ b_{-2}\ b_{-3}\ b_{-4}\ 1\ x\ x\ x$ — where the xs are 0s or 1s — which is to be truncated and rounded to four bits. Using adder-based rounding, this rounds to $0.b_{-1}\ b_{-2}\ b_{-3}\ b_{-4} + 0.0001$ and the resulting fraction is $0.b_{-1}\ b_{-2}\ b_{-3}\ b_{-4}$, where b_{-4} is a 1 or 0. Examples of adder-based rounding are shown in Figure 11.4 in which the fractions are to be rounded to four bits.

Delete

0. 0 1 0 1 | 1 | 0 0 0 | $\times 2^8 = +88$ True value

+) 0. 0 0 0 1

Rounded result 0. 0 1 1 0 $\times 2^8 = +96$ Approach from above

(a)

Delete

0. 0 1 1 1 | 0 | 1 1 1 | $\times 2^8 = +119$ True value

↓

Rounded result 0. 0 1 1 1 $\times 2^8 = +112$ Approach from below

(b)

Figure 11.4 Adder-based rounding examples: (a) a 1 is added to the retained bits and (b) no rounding occurs.

In Figure 11.4(a), the part of the fraction to be deleted for rounding has a value that is greater than or equal to half its maximum value of 15. Therefore, a 1 is added to the retained bits, which results in the true value being approached from above. That is, the part being deleted has a maximum value of 1111 (15), while its actual value is 1000 (8). Since a value of $8 \geq 7.5$, a 1 is added to the retained bits. A similar reasoning is used for Figure 11.4(b); however, the actual value of the part to be deleted is 0111 (7). Since $7 < 7.5$, the low-order four bits are deleted and a 1 is not added to the retained bits, which results in the true value being approached from below.

Adder-based rounding is an unbiased method that generates the nearest approximation to the number being rounded. Although adder-based rounding is obviously a better method of rounding than truncation, additional hardware is required to accommodate the addition cycle, thus adding more delay to the rounding operation.

von Neumann rounding The von Neumann rounding method is also referred to as *jamming* and is similar to truncation. If the bits to be deleted are all zeroes, then the bits are truncated and there is no change to the retained bits. However, if the bits to be deleted are not all zeroes, then the bits are deleted and the low-order bit of the retained bits is set to 1. Thus, when 8-bit fractions are rounded to four bits, fractions in the range

$$0.b_{-1}\, b_{-2}\, b_{-3}\, b_{-4}\, 0001 \text{ to } 0.b_{-1}\, b_{-2}\, b_{-3}\, b_{-4}\, 1111$$

will all be rounded to $0.b_{-1}\, b_{-2}\, b_{-3}\, 1$.

Therefore, the error ranges from approximately −1 to +1 in the low-order bit of the retained bits when

$$0.b_{-1}\, b_{-2}\, b_{-3}\, b_{-4}\, 0001 \text{ is rounded to } 0.b_{-1}\, b_{-2}\, b_{-3}\, 1$$

and when

$$0.b_{-1}\, b_{-2}\, b_{-3}\, b_{-4}\, 1111 \text{ is rounded to } 0.b_{-1}\, b_{-2}\, b_{-3}\, 1$$

Although the error range is larger in von Neumann rounding than with truncation rounding, the individual errors are evenly distributed over the error range. Thus, positive errors will be inclined to offset negative errors for long sequences of floating-point calculations involving rounding. The von Neumann rounding method has the same total bias as adder-based rounding; however, it requires no more time than truncation.

There are over 90 floating-point instructions in the X86 instruction set; therefore, only the most commonly used instructions will be presented in detail. The predominant prefix for the floating-point mnemonics is the letter *F*. The following types of floating-point instructions will be presented: load data instructions, store data instructions, addition instructions, subtraction instructions, multiplication instructions, division instructions, compare instructions, trigonometric instructions, and a select variety of additional instructions.

11.2 Load Data Instructions

This section describes the floating-point instructions that push different types of data onto the register stack. These are classified as data transfer instructions and include the *load floating-point value* (FLD) instruction, several *load constant* instructions, such as FLD1, FLDL2T, FLDL2E, FLDPI, FLDLG2, FLDLN2, and FLDZ (all of which will be described in later sections), the *load X87 FPU control word* (FLDCW) instruction, and the *load X87 FPU environment* (FLDENV) instruction. Also included is the *load integer* (FILD) and the *load binary-coded decimal* (FBLD) instructions.

11.2.1 Load Floating-Point Value (FLD) Instruction

The FLD instruction pushes the contents of the X87 source operand onto the register stack. The source operand can be any of the three floating-point data type formats: single precision, double precision, or double extended precision. Single-precision formats and double-precision formats are automatically converted to the double extended-precision format. The syntax for the FLD instruction is shown below, where FLD ST(*i*) is a register in the register stack.

```
FLD   m32/64/80fp (memory, 32-, 64-, or 80-bit floating-point)
FLD   ST(i)
```

If a stack overflow or a stack underflow results from a floating-point operation, the stack fault flag (SF) — bit 6 of the status word reproduced in Figure 11.5 — is set. When the SF flag is set, the condition code flag C1 (bit 9) is examined. If C1 = 1, a stack overflow has occurred; if C1 = 0, a stack underflow has occurred.

15	14	13	12	11	10	9	8
B	C3	TOS			C2	C1	C0

7	6	5	4	3	2	1	0
ES	SF	PE	UE	OE	ZE	DE	IE

Figure 11.5 Floating-point unit status word format.

A simplified register stack is shown in Figure 11.6. Assume that memory contains the following floating-point values:

```
flp1 = 25.0
flp2 = 10.0
flp3 = 15.0
```

Assume also that the instructions shown below are executed sequentially. The stack will contain the values shown in Figure 11.6 after the instructions have been executed. The fourth instruction will push the contents of ST(2) onto the stack; however, the contents of location ST(2) will not change.

```
FLD    flp1     Figure 11.6(a)
FLD    flp2     Figure 11.6(b)
FLD    flp3     Figure 11.6(c)
FLD    ST(2)    Figure 11.6(d)
```

	(a)		(b)		(c)		(d)	
R0	25.0	ST	10.0	ST	15.0	ST	25.0	ST
R1		ST(1)	25.0	ST(1)	10.0	ST(1)	15.0	ST(1)
R2		ST(2)		ST(2)	25.0	ST(2)	10.0	ST(2)
R3		ST(3)		ST(3)		ST(3)	25.0	ST(3)
R4		ST(4)		ST(4)		ST(4)		ST(4)
R5		ST(5)		ST(5)		ST(5)		ST(5)
R6		ST(6)		ST(6)		ST(6)		ST(6)
R7		ST(7)		ST(7)		ST(7)		ST(7)

Figure 11.6 Simplified register stack.

The floating-point instructions can access only the X87 registers, not the X86 general-purpose registers. Data must be transmitted between processors via memory. All addressing of the register stack is relative to the current top of stack (TOS), which is contained in bits 13 through 11 of the X87 status word. Like a regular stack, a load operation decrements the TOS by one and stores the new data in the new TOS register; this is similar to a PUSH operation. A store operation sends the data that resides in the current TOS register to the destination, then increments the TOS by one; this is similar to a POP operation.

11.2.2 Load Constant Instructions

There are seven *load constant* instructions that push specific values onto the register stack as double extended-precision floating-point values. These are listed below.

FLD1 instruction This instruction pushes +1.0 onto the register stack.

FLDL2T instruction This instruction pushes $\log_2 10$ onto the register stack, where $\log_2 10$ represents the exponent to which the base *2* must be raised to yield 10. The general equation is

$$(\log_b x = y) \equiv (b^y = x)$$

where b is the base. Thus,

$$(\log_2 10 = y) \equiv (2^y = 10)$$

$$\therefore\ 2^{3.333} = 10$$

$$\therefore\ \log_2 10 \approx 3.333$$

FLDL2E instruction This instruction pushes $\log_2 e$ onto the register stack, where $\log_2 e$ represents the exponent to which the base *2* must be raised to yield e, where

$$e \approx 2.71828$$

The notation for the constant e was selected by the mathematician Leonhard Euler because it was the first letter of the word *exponential*. The general equation is

$$(\log_b x = y) \equiv (b^y = x)$$

where b is the base. Thus,

$$(\log_2 e = y) \equiv (2^y = e)$$

$$\therefore\ 2^{1.4425} \approx 2.717914$$

$$\therefore\ \log_2 e \approx 1.4425$$

FLDPI instruction This instruction pushes π onto the register stack, where π is approximately 3.14159.

FLDLG2 instruction This instruction pushes $\log_{10}2$ onto the register stack, where $\log_{10}2$ represents the exponent to which the base 10 must be raised to yield 2. The general equation is

$$(\log_b x = y) \equiv (b^y = x)$$

where b is the base. Thus,

$$(\log_{10}2 = y) \equiv (10^y = 2)$$

$$\therefore\ 10^{0.3} = 2$$

$$\therefore\ \log_{10}2 \approx 0.3$$

FLDLN2 instruction This instruction pushes $\log_e 2$ onto the register stack, where $\log_e 2$ represents the exponent to which the base e must be raised to yield 2 and

$$e \approx 2.71828$$

The general equation is

$$(\log_b x = y) \equiv (b^y = x)$$

where b is the base. Thus,

$$(\log_e 2 = y) \equiv (e^y = 2)$$

$$\therefore\ 2.71828^{0.695} = 2$$

$$\therefore\ \log_e 2 \approx 0.695$$

FLDZ instruction This pushes +0.0 onto the register stack.

11.2.3 Load X87 FPU Control Word (FLDCW) Instruction

This instruction loads a 16-bit source operand from memory into the floating-point unit control word register, which is reproduced in Figure 11.7.

15	14	13	12	11	10	9	8
			IC	RC		PC	

7	6	5	4	3	2	1	0
		PM	UM	OM	ZM	DM	IM

Figure 11.7 Floating-point unit control word register.

The syntax for the FLDCW instruction is shown below, where `m2byte` specifies a 2-byte memory location. This instruction is used to load a new control word from memory in order to modify the existing control word — thus changing the mode of operation of the floating-point unit — or to establish a new control word.

```
FLDCW   m2byte (memory, 2 bytes)
```

11.2.4 Load X87 FPU Environment (FLDENV) Instruction

The FLDENV instruction loads the operating environment into the floating-point registers from memory as 14-byte data or as 28-byte data. The operating environment is loaded into the following registers: control word, status word, tag word, instruction pointer (IP) offset, data pointer offset, and last opcode pointer. The information that is loaded depends on the operating mode of the floating-point unit — protected mode or real mode — and the current operand size attribute, either 16 bits or 32 bits. The syntax for the FLDENV instruction is shown below, where `m14/28byte` specifies a memory operand of 14 bytes or 28 bytes.

```
FLDENV   m14/28byte (memory, 14 or 28 bytes)
```

The FLDENV instruction should use the identical operating mode — protected mode or real mode — as was used with the *store X87 FPU environment* (FSTENV) instruction, which is covered in the next section.

11.2.5 Load Integer (FILD) Instruction

The FILD instruction changes a signed integer source operand in memory to a double extended-precision floating-point number and pushes that value onto the register stack. The format for the integer source operand can be a word, a doubleword, or a quadword. The syntax for the FILD instruction is shown below.

```
FILD   m16int (memory, 16-bit integer)
FILD   m32int (memory, 32-bit integer)
FILD   m64int (memory, 64-bit integer)
```

11.2.6 Load Binary-Coded Decimal (FBLD) Instruction

The FBLD instruction converts a signed 80-bit packed binary-coded decimal (BCD) source operand in memory to a double extended-precision floating-point number and pushes that value onto the register stack. The instruction does not check for invalid digits. The syntax for the FBLD instruction is shown below.

```
FBLD   m80dec (memory, 80 bits decimal)
```

11.3 Store Data Instructions

This section describes the floating-point instructions that store — or store and pop — different types of data on the register stack and other specific registers. These instructions include the *store bcd integer and pop* (FBSTP) instruction, the *store integer* (FIST) instruction, the *store integer and pop* (FISTP) instruction, the *store integer with truncation* (FISTTP) instruction, the *store floating-point value* (FST) instruction, the *store floating-point value and pop* (FSTP) instruction, the *store X87 FPU control word* (FSTCW) instruction, the *store X87 FPU environment* (FSTENV) instruction, and the *store X87 FPU status word* (FSTSW).

11.3.1 Store BCD Integer and Pop (FBSTP) Instruction

The FBSTP instruction converts the value in ST(0) of the register stack to a BCD integer and stores the result in the destination operand located in a 10-byte area in memory. If the stored value is not an integer, then the operand is rounded to an integer using the rounding method specified by bits 11 and 10 of the RC field in the floating-point control word register of Figure 11.7. The register stack is then popped. A pop operation marks the ST(0) register as empty and increments the stack pointer by 1 — bits 13 through 11 (TOS or TOP) of the X87 floating-point status word, as shown in Figure 11.5. The syntax for the FBSTP instruction is shown below, where `m80bcd` is the destination operand of 80 bits in the BCD format.

```
FBSTP   m80bcd (memory, 80 bits)
```

11.3.2 Store Integer (FIST) Instruction

The FIST instruction converts the value in ST(0) of the register stack to a signed integer — rounded if necessary — and stores the result in the destination memory location as a word or doubleword. The syntax for the FIST instruction is shown below, where the destination operand is either a 16-bit integer or a 32-bit integer.

```
FIST   m16int (memory, 16-bit integer)
FIST   m32int (memory, 32-bit integer)
```

11.3.3 Store Integer and Pop (FISTP) Instruction

The FISTP instruction operates identically to the FIST instruction and then pops the register stack. It stores the value in ST(0) into memory as a word, doubleword, or quadword integer. The syntax for the FISTP instruction is shown below, where the destination operand is either a 16-bit integer, a 32-bit integer, or a 64-bit integer.

```
FISTP   m16int (memory, 16-bit integer)
FISTP   m32int (memory, 32-bit integer)
FISTP   m64int (memory, 64-bit integer)
```

11.3.4 Store Integer with Truncation and Pop (FISTTP) Instruction

The FISTTP instruction converts the operand in ST(0) to a signed integer using the truncation rounding method, then stores the result in the destination location and pops the register stack. Truncation deletes extra bits and makes no changes to the retained bits. This method of rounding is also referred to as *chopping*. The syntax for the FISTTP instruction is shown below.

```
FISTTP   m16int (memory, 16-bit integer)
FISTTP   m32int (memory, 32-bit integer)
FISTTP   m64int (memory, 64-bit integer)
```

Figure 11.8 shows an assembly language module embedded in a C program that illustrates the application of the FIST and the FISTTP instructions. Two floating-point numbers are entered from the keyboard and then used by the FIST and FISTTP instructions. The integer results are then displayed.

```
//fist_fisttp.cpp
//program to illustrate using
//FIST store integer and
//FISTTP store integer with truncation and pop

#include "stdafx.h"

int main (void)
{
//define variables
   float fist_num, fisttp_num;
   short fist_rslt, fisttp_rslt;

   printf ("Enter two floating-point numbers: \n");
   scanf ("%f %f", &fist_num, &fisttp_num);
                                    //continued on next page
```

(a)

Figure 11.8 Program to illustrate using the FIST and the FISTTP instructions: (a) the program and (b) the outputs.

```
//switch to assembly
      _asm
      {
         FLD      fist_num    //push first flp number
         FIST     fist_rslt   //convert to integer and store

         FLD      fisttp_num  //push second flp number
         FISTTP   fisttp_rslt //convert to integer, truncate,
                              //store, and pop
      }

//print result
   printf ("\nFIST result = %d\n", fist_rslt);
   printf ("FISTTP result = %d\n\n", fisttp_rslt);

   return 0;
}
```

```
Enter two  floating-point numbers:
43.6789 5.9865

FIST result = 44
FISTTP result = 5

Press any key to continue . . . _
------------------------------------------------------------
Enter two  floating-point numbers:
38.00375 640.76438

FIST result = 38
FISTTP result = 640

Press any key to continue . . . _
------------------------------------------------------------
Enter two  floating-point numbers:
720.475 34.444

FIST result = 720
FISTTP result = 34

Press any key to continue . . . _
------------------------------------------------------------
```

//continued on next page

(b)

Figure 11.8 (Continued)

```
Enter two  floating-point numbers:
1.4456 3.1546

FIST result = 1
FISTTP result = 3

Press any key to continue . . . _
------------------------------------------------------------
Enter two  floating-point numbers:
567.5000 567.5000

FIST result = 568
FISTTP result = 567

Press any key to continue . . . _
```

Figure 11.8 (Continued)

11.3.5 Store Floating-Point Value (FST) Instruction

The FST instruction stores the operand in ST(0) to the destination location, which can be a location in memory or another register in the register stack. If the destination is a memory location, then the operand is converted to the single-precision format or the double-precision format. The syntax for the FST instruction is shown below.

```
FST   m32fp (memory, 32 bits floating-point)
FST   m64fp (memory, 64 bits floating-point)
FST   ST(i) (copy ST(0) to ST(i))
```

Figure 11.9 shows an assembly language module embedded in a C program that illustrates using the FST instruction in conjunction with the *load constant* instructions: FLD1, FLDL2T, FLDL2E, FLDPI, FLDLG2, FLDLN2, and FLDZ. The *load constant* instructions push the appropriate values onto the register stack; the FST instruction stores them in the assigned locations in memory, then the results are displayed.

```
//fld_fst.cpp
//use floating-point load constant instructions and
//floating-point store instructions
#include "stdafx.h"
                                   //continued on next page
```

(a)

Figure 11.9 Program to illustrate using the FST instruction: (a) the program and (b) the outputs.

```
int main (void)
{
//define variables
   float fld1_num, fldl2t_num, fldl2e_num, fldpi_num,
         fldlg2_num, fldln2_num, fldz_num;

//switch to assembly
      _asm
      {
         FLD1                 //push +1.0 onto stack
         FST    fld1_num      //copy ST(0) to fld1_num

         FLDL2T               //push log210 onto stack
         FST    fldl2t_num    //copy ST(0) to fldl2t

         FLDL2E               //push log2e onto stack
         FST    fldl2e_num    //copy ST(0) to fldl2e_num

         FLDPI                //push pi onto stack
         FST    fldpi_num     //copy ST(0) to fldpi_num

         FLDLG2               //push log102 onto stack
         FST    fldlg2_num    //copy ST(0) to fldlg2_num

         FLDLN2               //push loge2 onto stack
         FST    fldln2_num    //copy ST(0) to fldln2_num

         FLDZ                 //push +0.0 onto stack
         FST    fldz_num      //copy ST(0) to fldz_num
      }

//print result
   printf ("FLD1 result = %f\n", fld1_num);
   printf ("FLDL2T result = %f\n", fldl2t_num);
   printf ("FLDL2E result = %f\n", fldl2e_num);
   printf ("FLDPI result = %f\n", fldpi_num);
   printf ("FLDLG2 result = %f\n", fldlg2_num);
   printf ("FLDLN2 result = %f\n", fldln2_num);
   printf ("FLDZ result = %f\n\n", fldz_num);

   return 0;
}
```

Figure 11.9 (Continued)

```
FLD1 result = 1.000000
FLDL2T result = 3.321928
FLDL2E result = 1.442695
FLDPI result = 3.141593
FLDLG2 result = 0.301030
FLDLN2 result = 0.693147
FLDZ result = 0.000000

Press any key to continue . . . _
```

(b)

Figure 11.9 (Continued)

11.3.6 Store Floating-Point Value and Pop (FSTP) Instruction

The FSTP instruction stores the operand in ST(0) to the destination location, which can be a location in memory or another register in the register stack, then pops the register stack. If the destination is a memory location, then the operand is converted to the single-precision format, the double-precision format, or the double extended-precision format. The syntax for the FSTP instruction is shown below.

```
FSTP   m32fp (memory, 32 bits floating-point and pop stack)
FSTP   m64fp (memory, 64 bits floating-point and pop stack)
FSTP   m80fp (memory, 80 bits floating-point and pop stack)
FSTP   ST(i) (copy ST(0) to ST(i) and pop stack)
```

11.3.7 Store X87 FPU Control Word (FSTCW) Instruction

The FSTCW instruction stores the floating-point control word at the memory location specified by the destination location. The FSTCW also resolves any pending unmasked floating-point exceptions before storing the control word. Refer to Section 11.2.3 for the control word format. The syntax for the FSTCW instruction is shown below. There is a second *store X87 FPU control word* (FNSTCW) that does not check for pending unmasked floating-point exceptions. The syntax for the FNSTCW is also shown below.

```
FSTCW  m2byte (memory, two bytes check for exceptions)
FNSTCW m2byte (memory, two bytes do not check for exceptions)
```

11.3.8 Store X87 FPU Environment (FSTENV) Instruction

The FSTENV instruction saves the floating-point unit environment in a memory location as indicated by the destination operand. The operating environment consists of the following registers: control word, status word, tag word, instruction pointer (IP) offset, data pointer offset, and last opcode pointer. The format of the environment depends on the operating mode of the floating-point unit — protected mode or real mode — and the current operand size attribute, either 16 bits or 32 bits. The FSTENV instruction then masks all floating-point exceptions.

The syntax for the FSTENV instruction is shown below. There is a second *store X87 FPU environment* (FNSTENV) that does not check for pending unmasked floating-point exceptions. The syntax for the FNSTENV is also shown below.

```
FSTENV    m14byte (memory, 14 bytes check for exceptions,
                   then mask exceptions)

FSTENV    m28byte (memory, 28 bytes check for exceptions,
                   then mask exceptions)

FNSTENV   m14byte (memory, 14 bytes do not check for
                   exceptions, then mask exceptions)

FNSTENV   m28byte (memory, 28 bytes do not check for
                   exceptions, then mask exceptions)
```

11.3.9 Store X87 FPU Status Word (FSTSW) Instruction

The FSTSW instruction stores the floating-point status word at the memory location specified by the destination location. The destination is a 2-byte memory location or the general-purpose register AX. The FSTSW also resolves any pending unmasked floating-point exceptions before storing the status word.

The syntax for the FSTSW instruction is shown below. There is a second *store X87 FPU status word* (FNSTSW) that does not check for pending unmasked floating-point exceptions. The syntax for the FNSTSW is also shown below.

```
FSTSW    m2byte (memory, two bytes, check for exceptions)

FSTSW    AX (register AX, check for exceptions)


FNSTSW   m2byte (memory, two bytes,
                 do not check for exceptions)

FNSTSW   AX (register AX, do not check for exceptions)
```

11.4 Addition Instructions

The addition of two fractions is identical to the addition algorithm presented in fixed-point addition. If the signs of the operands are the same ($A_{sign} \oplus B_{sign} = 0$), then this is referred to as *true addition* and the fractions are added. True addition corresponds to one of the following conditions:

$$(+A) + (+B)$$
$$(-A) + (-B)$$
$$(+A) - (-B)$$
$$(-A) - (+B)$$

Floating-point addition is defined as shown in Equation 11.3 for two numbers A and B, where $A = f_A \times r^{e_A}$ and $B = f_B \times r^{e_B}$.

$$\begin{aligned} A + B &= (f_A \times r^{e_A}) + (f_B \times r^{e_B}) \\ &= [f_A + (f_B \times r^{-(e_A - e_B)})] \times r^{e_A} \text{ for } e_A > e_B \\ &= [(f_A \times r^{-(e_B - e_A)}) + f_B] \times r^{e_B} \text{ for } e_A \leq e_B \end{aligned} \tag{11.3}$$

The terms $r^{-(e_A - e_B)}$ and $r^{-(e_B - e_A)}$ are *shifting factors* to shift the fraction with the smaller exponent. This is analogous to a divide operation, since $r^{-(e_A - e_B)}$ is equivalent to $1/r^{(e_A - e_B)}$, which is a right shift. For $e_A > e_B$, fraction f_B is shifted right the number of bit positions specified by the absolute value of $| e_A - e_B |$. An example of using the shifting factor for addition is shown in Figure 11.10 for two operands $A = +9.75$ and $B = +3.875$.

The fractions must be properly aligned before addition can take place; therefore, the fraction with the smaller exponent is shifted right and the exponent is adjusted by increasing the exponent by one for each bit position shifted.

Before alignment

$A = f_A \times r^4$

$A = 0 \ . \ 1\ 0\ 0\ 1\ 1\ 1\ 0\ 0 \quad \times 2^4 \qquad +9.75$

$B = f_B \times r^2$

$B = 0 \ . \ 1\ 1\ 1\ 1\ 1\ 0\ 0\ 0 \quad \times 2^2 \qquad +3.875$

Continued on next page

Figure 11.10 Addition of two floating-point numbers.

After alignment

$$A = 0\ .\ 1\ 0\ 0\ 1\ 1\ 1\ 0\ 0 \quad \times 2^4 \qquad +9.75$$

$$B = 0\ .\ 0\ 0\ 1\ 1\ 1\ 1\ 1\ 0 \quad \times 2^4 \qquad +3.875$$

$$A + B = 0\ .\ 1\ 1\ 0\ 1\ 1\ 0\ 1\ 0 \quad \times 2^4 \qquad +13.625$$

Figure 11.10 (Continued)

Figure 11.11 shows an example of floating-point addition when adding $A = +5.75$ and $B = +30.5$, in which the 8-bit fractions are not properly aligned initially and postnormalization is required. *Postnormalization* occurs when the resulting fraction overflows, requiring a right shift of one bit position with a corresponding increment of the exponent. The bit causing the overflow is shifted right into the high-order fraction bit position.

Before alignment

$$A = f_A \times r^3$$

$$A = 0\ .\ 1\ 0\ 1\ 1\ 1\ 0\ 0\ 0 \quad \times 2^3 \qquad +5.75$$

$$B = f_B \times r^5$$

$$B = 0\ .\ 1\ 1\ 1\ 1\ 0\ 1\ 0\ 0 \quad \times 2^5 \qquad +30.5$$

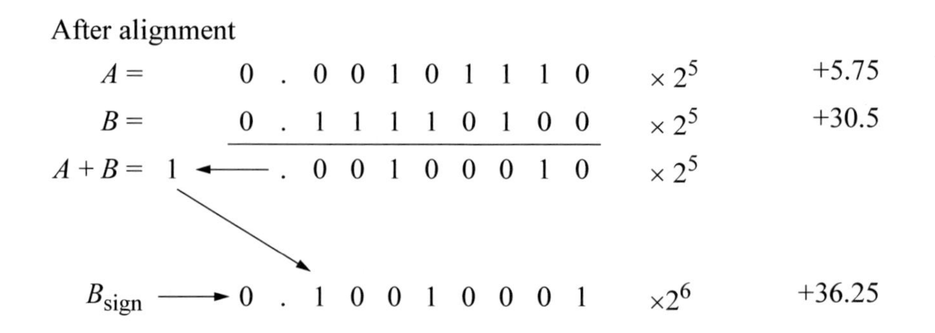

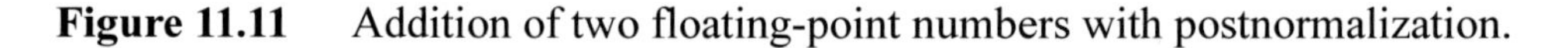

Figure 11.11 Addition of two floating-point numbers with postnormalization.

The alignment and shifting of the fractions is now summarized. Equation 11.3 states that if $e_A > e_B$, then fraction f_A is added to the aligned fraction f_B with the exponent e_A assigned to the resulting sum. The radix points of the two operands must be aligned prior to the addition operation. This is achieved by comparing the relative magnitudes of the two exponents. The fraction with the smaller exponent is then shifted $| e_A - e_B |$ positions to the right.

The augend and addend are then added and the sum is characterized by the larger exponent. A carry-out of the high-order bit position may occur, yielding a result with an absolute value of $1 \leq | result | < 2$ before postnormalization.

11.4.1 Overflow and Underflow

The floating-point addition example of Figure 11.11 generated a carry-out of the high-order bit position, which caused a *fraction overflow*. When adding two numbers with the same sign, the absolute value of the result may be in the following range before postnormalization:

$$1 \leq | result | < 2$$

This indicates that the fraction is in the range of 1.000 . . . 0 to 1.111 . . .1. The overflow can be corrected by shifting the carry-out in concatenation with the fraction one bit position to the right and incrementing the exponent by 1. This operation is shown in Equation 11.4.

$$\begin{aligned} A + B &= (f_A \times r^{e_A}) + (f_B \times r^{e_B}) \\ &= \{[f_A + (f_B \times r^{-(e_A - e_B)})] \times r^{-1}\} \times r^{e_A + 1} \quad \text{for } e_A > e_B \\ &= \{[(f_A \times r^{-(e_B - e_A)}) + f_B] \times r^{-1}\} \times r^{e_B + 1} \quad \text{for } e_A \leq e_B \qquad (11.4) \end{aligned}$$

The term r^{-1} is the shifting factor that shifts the resulting fraction and the carry-out one bit position to the right. For radix 2, the shifting factor is 2^{-1} (or 1/2), which divides the result by 2 by executing a right shift of one bit position. The terms $r^{e_A + 1}$ and $r^{e_B + 1}$ increment the appropriate exponents by 1. Equation 11.3 is similar to Equation 11.4, but does not require a shift operation.

When aligning a fraction by shifting the fraction right and adjusting the exponent, bits may be lost off the right end of the fraction, resulting in a *fraction underflow*. This can be resolved by using a rounding method discussed in Section 11.1.1.

11.4.2 Add Instructions

There are different versions of the *add* instruction. One version, FADD, adds the single-operand floating-point destination operand in ST(0) of the register stack to a 32-bit or a 64-bit source operand in memory and stores the sum in ST(0). For some add instructions, the source operand can be a single-precision floating-point operand, a double-precision floating-point operand, an integer word operand, or an integer doubleword operand. The syntax for the FADD instruction is shown below.

```
FADD   m32fp (memory, 32 bits, floating-point)
FADD   m64fp (memory, 64 bits, floating-point)
```

Another version of the add instruction adds the operand in register ST(0) to the operand in register ST(*i*) and stores the sum in ST(0) as destination or in ST(*i*) as destination depending on syntax of the instruction, as shown below.

```
FADD   ST(0), ST(i) (stores sum in destination ST(0))
FADD   ST(i), ST(0) (stores sum in destination ST(i))
```

Another version of the add instruction, FADDP, is similar to the double-operand version shown above, where the sum is stored in ST(*i*). However, in this version the register stack is popped after the sum is stored. The syntax is shown below.

```
FADDP   ST(i), ST(0)
```

Another version of the add instruction, FADDP, is the no-operand version, which adds the operand in ST(0) to the operand in ST(1) and stores the sum in ST(1), then pops the register stack. The syntax is shown below.

```
FADDP
```

Another version of the add instruction, FIADD, adds the operand in ST(0) of the register stack to a 16-bit or a 32-bit integer source operand in memory and stores the sum in ST(0). The FIADD instruction converts the integer source operand to a double extended-precision floating-point number before adding it to ST(0). The syntax is shown below.

```
FIADD   m16int (memory, 16 bits, integer)
FIADD   m32int (memory, 32 bits, integer)
```

Figure 11.12 shows an assembly language module embedded in a C program that illustrates utilizing different versions of the FADD instruction. The FADD single-operand and the FADD double-operand instructions are used in the program. Two floating-point numbers are entered from the keyboard for use in the program. The *initialize floating-point unit* (FINIT) can be used to initialize the register stack. This instruction does not change the contents of the stack; however, each register is tagged as being empty — the tag register is set to 11_2.

```
//fadd_versions.cpp
//show the use of versions of the FADD instruction

#include "stdafx.h"

int main (void)
{
//define variables
   float flp1_num, flp2_num, flp_rslt1, flp_rslt2, flp_rslt3;

   printf ("Enter two floating-point numbers: \n");
   scanf ("%f %f", &flp1_num, &flp2_num);

//switch to assembly
      _asm
      {
         FLD    flp1_num    //flp1_num -> ST(0)
         FLDPI              //pi -> ST(0), flp1_num -> ST(1)
         FADD   flp2_num    //sum = flp2_num + pi -> ST(0)
         FST    flp_rslt1   //ST(0) sum -> flp_rslt1
                            //fld1_num -> ST(1)

         FADD   ST(0), ST(0)//2 x ST(0) -> ST(0)
         FST    flp_rslt2   //ST(0) -> flp_rslt2

         FADD   ST(0), ST(1)//ST(0) + ST(1) -> ST(0)
         FST    flp_rslt3   //ST(0) -> flp_rslt3
      }

//print result
   printf ("\nflp_rslt1 = %f\n", flp_rslt1);
   printf ("flp_rslt2 = %f\n", flp_rslt2);
   printf ("flp_rslt3 = %f\n\n", flp_rslt3);

   return 0;
}                              //continued on next page
```

(a)

Figure 11.12 Program to illustrate using versions of the FADD instruction: (a) the program and (b) the outputs.

```
Enter two floating-point numbers:
2.4 6.8

flp_rslt1 = 9.941593
flp_rslt2 = 19.883186
flp_rslt3 = 22.283186

Press any key to continue . . . _
```
(b)

Figure 11.12 (Continued)

Figure 11.13 shows the register stack contents for different stages of the program. Figure 11.13(a) shows the result of *flp1_num* (2.4) being pushed onto the register stack. Then the value of *pi* (≈3.141593) is pushed onto the stack as shown in Figure 11.13(b). Next the FADD instruction adds *flp2_num* (6.8) to *pi* and stores the sum (9.941593) in ST(0), as shown in Figure 11.13(c). Then ST(0) is doubled (19.883186), as shown in Figure 11.13(d). Finally, ST(0) is added to ST(1) — 19.883186 + 2.4 = 22.283186 and stored in ST(0).

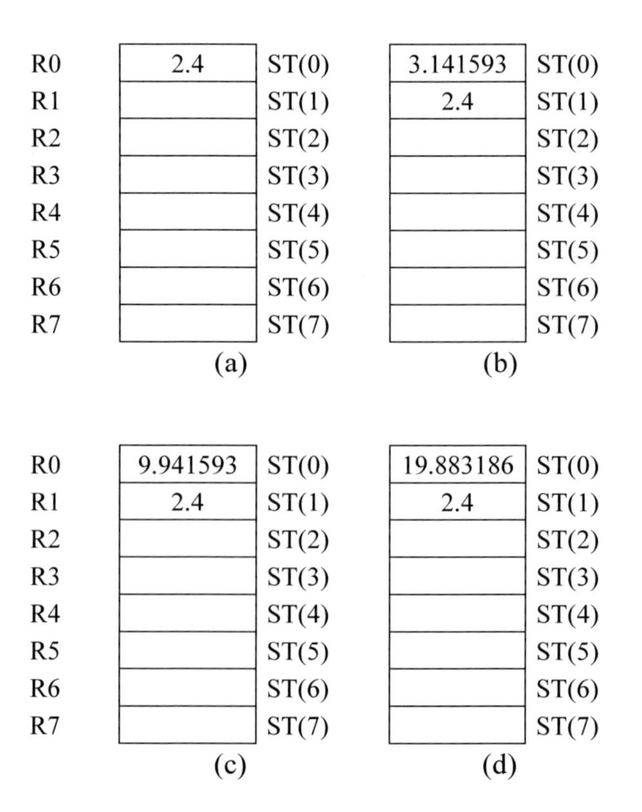

Figure 11.13 Register stack utilization for the program of Figure 11.12.

11.5 Subtraction Instructions

Floating-point subtraction also requires that the fractions be aligned before subtracting. Fraction overflow can also occur in subtraction since subtraction is accomplished by adding the 2s complement of the subtrahend. The subtraction of two fractions is identical to the subtraction algorithm presented in fixed-point addition. If the signs of the operands are the same ($A_{sign} \oplus B_{sign} = 0$) and the operation is subtraction, then this is referred to as *true subtraction* and the fractions are subtracted. If the signs of the operands are different ($A_{sign} \oplus B_{sign} = 1$) and the operation is addition, then this is also specified as *true subtraction*. True subtraction corresponds to one of the following conditions:

$$
\begin{array}{ccc}
(+A) & - & (+B) \\
(-A) & - & (-B) \\
(+A) & + & (-B) \\
(-A) & + & (+B)
\end{array}
$$

As in fixed-point notation, the same hardware can be used for both floating-point addition and subtraction to add or subtract the fractions. All operands will consist of normalized fractions properly aligned with biased exponents. Floating-point subtraction is defined as shown in Equation 11.5 for two numbers A and B, where $A = f_A \times r^{e_A}$ and $B = f_B \times r^{e_B}$.

$$
\begin{aligned}
A - B &= (f_A \times r^{e_A}) - (f_B \times r^{e_B}) \\
&= [f_A - (f_B \times r^{-(e_A - e_B)})] \times r^{e_A} \text{ for } e_A > e_B \\
&= [(f_A \times r^{-(e_B - e_A)}) - f_B] \times r^{e_B} \text{ for } e_A \leq e_B
\end{aligned} \tag{11.5}
$$

The terms $r^{-(e_A - e_B)}$ and $r^{-(e_B - e_A)}$ in Equation 11.5 are analogous to the terms used in floating-point addition. These terms are called *shifting factors* to shift the fraction with the smaller exponent. This is equivalent to a divide operation, since $r^{-(e_A - e_B)}$ is equivalent to $1/r^{(e_A - e_B)}$, which is a right shift. For $e_A > e_B$, fraction f_B is shifted right the number of bits specified by the absolute value of $| e_A - e_B |$. An example of using the shifting factor for subtraction is shown in Figure 11.14 for two operands, $A = +36.5$ and $B = +5.75$. Since the *implied 1* is part of the fractions, it must be considered when subtracting two normalized floating-point numbers — the implied 1 is shown as the high-order bit in Figure 11.14.

Before alignment												
$A =$	0	.	1	0	0	1	0	0	1	0	$\times 2^6$	+36.5
$B =$	0	.	1	0	1	1	1	0	0	0	$\times 2^3$	+5.75
After alignment												
$A =$	0	.	1	0	0	1	0	0	1	0	$\times 2^6$	+36.5
$B =$	0	.	0	0	0	1	0	1	1	1	$\times 2^6$	+5.75
Subtract fractions (Add 2s compl B)												
$A =$	0	.	1	0	0	1	0	0	1	0	$\times 2^6$	
+) $B' + 1 =$	0	.	1	1	1	0	1	0	0	1	$\times 2^6$	
$1 \leftarrow$	0	.	0	1	1	1	1	0	1	1	$\times 2^6$	+30.75
Postnormalize (SL1)	0	.	1	1	1	1	0	1	1	0	$\times 2^5$	+30.75

Figure 11.14 Example of floating-point subtraction.

11.5.1 Numerical Examples

Subtraction can yield a result that is either true addition or true subtraction. *True addition* produces a result that is the sum of the two operands disregarding the signs; *true subtraction* produces a result that is the difference of the two operands disregarding the signs. There are four cases that yield a true addition, as shown in Figure 11.15, and eight cases that yield a true subtraction, as shown in Figure 11.16.

	– Small number		– Large number
–)	+ Large number	–)	+ Small number
	True addition		True addition
	+ Large number		+ Small number
–)	– Small number	–)	– Large number
	True addition		True addition

Figure 11.15 Examples of true addition.

	+ Large number		+ Small number
–)	+ Small number	–)	+ Large number
	True subtraction		True subtraction
	– Small number		– Large number
–)	– Large number	–)	– Small number
	True subtraction		True subtraction
	+ Small number		– Small number
+)	– Large number	+)	+ Large number
	True subtraction		True subtraction
	+ Large number		– Large number
+)	– Small number	+)	+ Small number
	True subtraction		True subtraction

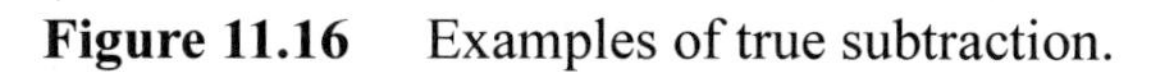

Figure 11.16 Examples of true subtraction.

An example of true addition is shown in Figure 11.17 in which +24 is subtracted from –40 to yield a result of –64.

Before alignment						
	$A =$	1 .	1 0 1 0	0 0 0 0	$\times 2^6$	–40
	$B =$	0 .	1 1 0 0	0 0 0 0	$\times 2^5$	+24
After alignment						
	$A =$	1 .	1 0 1 0	0 0 0 0	$\times 2^6$	–40
	$B =$	0 .	0 1 1 0	0 0 0 0	$\times 2^6$	+24
Add fractions						
	$A =$	1 .	1 0 1 0	0 0 0 0	$\times 2^6$	–40
	+) $B =$	1 .	0 1 1 0	0 0 0 0	$\times 2^6$	–24
	1 ←	.	0 0 0 0	0 0 0 0	$\times 2^6$	
Postnormalize		1 .	1 0 0 0	0 0 0 0	$\times 2^7$	–64

Figure 11.17 An example of true addition.

An example of true subtraction is shown in Figure 11.18 in which –13 is added to +45 to yield a result of +32.

Before alignment

$A =$	0	.	1	0	1	1	0	1	0	0	$\times 2^6$	+45
$B =$	1	.	1	1	0	1	0	0	0	0	$\times 2^4$	–13

After alignment

$A =$	0	.	1	0	1	1	0	1	0	0	$\times 2^6$	+45
$B =$	1	.	0	0	1	1	0	1	0	0	$\times 2^6$	–13

Add fractions

$A =$	0	.	1	0	1	1	0	1	0	0	$\times 2^6$	
+) $B' + 1 =$	1	.	1	1	0	0	1	1	0	0	$\times 2^6$	
1 ←		.	1	0	0	0	0	0	0	0	$\times 2^6$	
	0	.	1	0	0	0	0	0	0	0	$\times 2^6$	+32

Figure 11.18 An example of true subtraction.

11.5.2 Subtract Instructions

There are different versions of the *subtract* instruction. One version, FSUB, subtracts the single-operand floating-point 32-bit or a 64-bit source operand in memory from ST(0) of the register stack and stores the difference in ST(0). For some subtract instructions, the source operand can be a single-precision floating-point operand, a double-precision floating-point operand, an integer word operand, or an integer doubleword operand. The syntax for the FSUB instruction is shown below.

```
FSUB   m32fp (memory, 32 bits, floating-point)
FSUB   m64fp (memory, 64 bits, floating-point)
```

Another version of the subtract instruction subtracts the operand in register ST(*i*) from the operand in register ST(0) and stores the difference in ST(0). A similar version subtracts ST(0) from ST(*i*) and stores the difference in ST(*i*). The syntax for the two-operand FSUB instruction is shown below.

```
FSUB   ST(0), ST(i) (stores difference in ST(0))
FSUB   ST(i), ST(0) (stores difference in ST(i))
```

Another version of the subtract instruction, FSUBP, is similar to the double-operand version shown above, where the difference is stored in ST(*i*). The operand in ST(0) is subtracted from the operand in ST(*i*) and the difference is stored in ST(*i*). However, in this version the register stack is popped after the difference is stored. The syntax is shown below.

```
FSUBP   ST(i), ST(0)
```

Another version of the subtract instruction, FSUBP, is the no-operand version, which subtracts the operand in ST(0) from the operand in ST(1) and stores the difference in ST(1), then pops the register stack. The syntax is shown below.

```
FSUBP
```

Another version of the subtract instruction, FISUB, subtracts the 16-bit or a 32-bit single-operand integer source operand in memory from ST(0) and stores the difference in ST(0). The FISUB instruction converts the integer source operand to a double extended-precision floating-point number before subtracting it from ST(0). The syntax is shown below.

```
FISUB   m16int (memory, 16 bits, integer)
FISUB   m32int (memory, 32 bits, integer)
```

There are also a variety of *reverse subtract* instructions. These instructions are similar to those listed above, except that the subtract operation is reversed. For example, the FSUBR instruction subtracts ST(0) from the single-operand floating-point 32-bit or a 64-bit source operand in memory stores the difference in ST(0). These instructions include the following:

```
FSUBR   m32fp (memory, 32 bits, floating-point,
              subtracts ST(0) from memory and
              stores the difference in ST(0))
FSUBR   m64fp (memory, 64 bits, floating-point,
              subtracts ST(0) from memory and
              stores the difference in ST(0))

FSUBR   ST(0), ST(i)  (subtracts ST(0) from ST(i) and
                      stores the difference in ST(0))
FSUBR   ST(i), ST(0)  (subtracts ST(i) from ST(0) and
                      stores the difference in ST(i))

FSUBRP   ST(i), ST(0)  (subtracts ST(i) from ST(0),
                       stores the difference in ST(i),
                       then pops stack)
FSUBRP                (subtracts ST(1) from ST(0),
                       stores the difference in ST(1),
                       then pops stack)
```

```
FISUBR   m16int (memory, 16 bits, integer,
                subtracts ST(0) from the integer
                in memory and stores the difference
                in ST(0)
FISUBR   m32int (memory, 32 bits, integer,
                subtracts ST(0) from the integer
                in memory and stores the difference
                in ST(0))
```

The advantage of using the reverse subtraction instructions is that it is not necessary to exchange the operand in ST(0) with the operand in another register in the stack in order to perform a subtraction.

Figure 11.19 shows an assembly language module embedded in a C program that illustrates utilizing different versions of the FSUB instruction. The FSUB single-operand instruction, the FSUB double-operand instruction, and the FSUBP instruction are used in the program. Four floating-point numbers are entered from the keyboard for use in the program: two negative numbers and two positive numbers, as shown below. The results of the five subtract instructions are also shown below.

```
flp1_num = -296.125
flp2_num = -77.625
flp3_num = +156.750
flp4_num = +127.500
```

The *initialize floating-point unit* (FINIT) can be used to initialize the register stack. This instruction does not change the contents of the stack; however, each register is tagged as being empty — the tag register is set to 11_2.

FSUB flp2_num

	+127.500	ST(0)
–)	– 77.625	flp2_num
	+205.125	→ ST(0)

FSUB flp2_num

	+205.125	ST(0)
–)	– 77.625	flp2_num
	+282.750	→ ST(0)

FSUBP ST(1), ST(0)

	+156.750	ST(1)
–)	+282.750	ST(0)
	–126.000	→ ST(1) then pop

FSUB ST(0), ST(2)

	–126.000	ST(0)
–)	–296.125	ST(2)
	+170.125	→ ST(0)

FSUB ST(0), ST(2)

	+170.125	ST(0)
–)	–296.125	ST(2)
	+466.250	→ ST(0)

```
//fsub_versions.cpp
//show the use of versions of the FSUB instruction

#include "stdafx.h"

int main (void)
{
//define variables
   float flp1_num, flp2_num, flp3_num, flp4_num,
         flp_rslt1, flp_rslt2, flp_rslt3, flp_rslt4,
         flp_rslt5;

   printf ("Enter 4 flp numbers, 2 negative & 2 positive:
            \n");
   scanf ("%f %f %f %f", &flp1_num, &flp2_num, &flp3_num,
                         &flp4_num);

//switch to assembly
      _asm
      {
         FLD    flp1_num     //flp1_num (-296.125) -> ST(0)
         FLD    flp2_num     //flp2_num (-77.625) -> ST(0)
                             //(-296.125) -> ST(1)
         FLD    flp3_num     //flp3_num (+156.750) -> ST(0)
                             //(-77.625) -> ST(1)
                             //(-296.125) -> ST(2)
         FLD    flp4_num     //flp4_num (+127.500) -> ST(0)
                             //(+156.750) -> ST(1)
                             //(-77.625) -> ST(2)
                             //(-296.125) -> ST)3)
//--------------------------------------------------------
         FSUB   flp2_num     //ST(0) - flp2_num -> ST(0)
                             //(+127.50) - (-77.625) = +205.125
         FST    flp_rslt1    //+205.125 -> flp_rslt1
//--------------------------------------------------------
         FSUB   flp2_num     //ST(0) - flp2_num -> ST(0)
                             //(+205.125) - (-77.625) =
                             //+282.750
         FST    flp_rslt2    //+282.75 -> flp_rslt2
//--------------------------------------------------------
```

(a) //continued on next page

Figure 11.19 Program to illustrate using versions of the FSUB instruction: (a) the program and (b) the outputs.

```
//-----------------------------------------------------------
        FSUBP ST(1), ST(0)//ST(1) - ST(0) -> ST(1), then pop
                          //(+156.75) - (+282.750) =
                          //-126.000
        FST   flp_rslt3   //-126.00 -> flp_rslt3
//-----------------------------------------------------------
        FSUB  ST(0), ST(2)//ST(0) - ST(2) -> ST(0)
                          //(-126.00) - (-296.125) =
                          //+170.125
        FST   flp_rslt4   //+170.125 -> flp_rslt4
//-----------------------------------------------------------
        FSUB  ST(0), ST(2)//ST(0) - ST(2) -> ST(0)
                          //(+170.125) - (-296.125) =
                          //+466.250
        FST   flp_rslt5   //+466.25 -> flp_rslt5
//-----------------------------------------------------------
}

//print result
   printf ("\nflp_rslt1 = %f\n", flp_rslt1);
   printf ("flp_rslt2 = %f\n", flp_rslt2);
   printf ("flp_rslt3 = %f\n", flp_rslt3);
   printf ("flp_rslt4 = %f\n", flp_rslt4);
   printf ("flp_rslt5 = %f\n\n", flp_rslt5);

   return 0;
}
```

```
Enter 4 flp numbers, 2 negative & 2 positive:
-296.125 -77.625 +156.750 +127.500

flp_rslt1 = 205.125000
flp_rslt2 = 282.750000
flp_rslt3 = -126.000000
flp_rslt4 = 170.125000
flp_rslt5 = 466.250000

Press any key to continue . . . _
```

(b)

Figure 11.19 (Continued)

Figure 11.20 shows the register stack contents for different stages of the program. Figure 11.20(a) shows the result of the four floating-point numbers having been pushed onto the register stack. The remaining figures in Figure 11.20 portray the results of the various instructions after they have been executed.

(a)

Register	Contents	Stack
R0	+127.500	ST(0)
R1	+156.750	ST(1)
R2	–77.625	ST(2)
R3	–296.125	ST(3)
R4		ST(4)
R5		ST(5)
R6		ST(6)
R7		ST(7)

(b) **FSUB** `flp2_num`

Contents	Stack	
+205.125	ST(0)	→`flp_rslt1`
+156.750	ST(1)	
–77.625	ST(2)	
–296.125	ST(3)	
	ST(4)	
	ST(5)	
	ST(6)	
	ST(7)	

(c) **FSUB** `flp2_num`

Register	Contents	Stack	
R0	**+282.750**	ST(0)	→`flp_rslt2`
R1	+156.750	ST(1)	
R2	–77.625	ST(2)	
R3	–296.125	ST(3)	
R4		ST(4)	
R5		ST(5)	
R6		ST(6)	
R7		ST(7)	

(d) **FSUBP** `ST(1), ST(0)`

Contents	Stack	
–126.000	ST(0)	→`flp_rslt3`
–77.625	ST(1)	
–296.125	ST(2)	
	ST(3)	
	ST(4)	
	ST(5)	
	ST(6)	
	ST(7)	

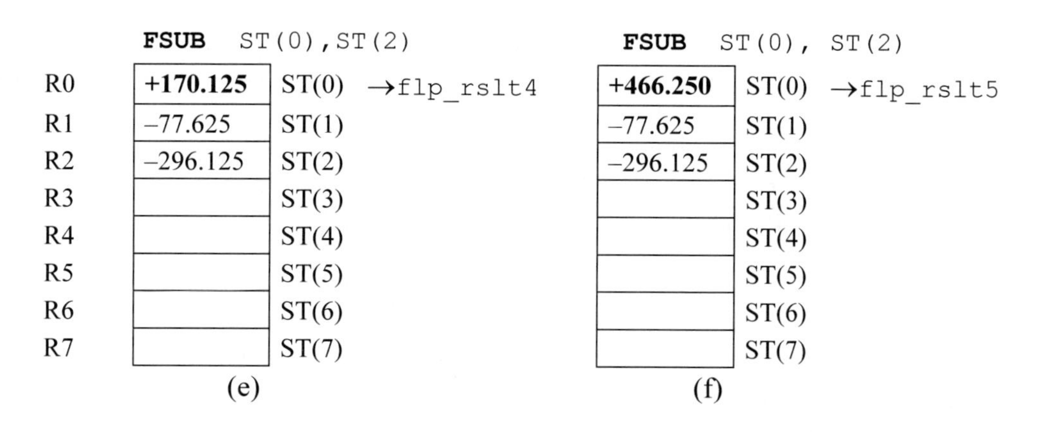

Figure 11.20 Register stack utilization for the program of Figure 11.19.

11.6 Multiplication Instructions

In floating-point multiplication, the fractions are multiplied and the exponents are added. Floating-point multiplication is simpler than floating-point addition or subtraction because there is no comparison of exponents and no alignment of fractions. Fraction multiplication and exponent addition are two independent operations and can be done in parallel. Floating-point multiplication is defined as shown in Equation 11.6.

$$
\begin{aligned}
A \times B &= (f_A \times r^{eA}) \times (f_B \times r^{eB}) \\
&= (f_A \times f_B) \times r^{(eA + eB)}
\end{aligned}
\tag{11.6}
$$

The sign of the product is determined by the signs of the operands as shown below.

$$A_{\text{sign}} \oplus B_{\text{sign}}$$

11.6.1 Double Bias

An n-bit multiplicand (A) and an n-bit multiplier (B) generate a $2n$-bit product (P), which, in conjunction with the exponent, should be of sufficient precision. Although it is not apparent in the numerical paper-and-pencil floating-point multiplication examples in the next section, there is a minor problem when adding two biased exponents. Since both exponents are biased, there will be a double bias in the resulting exponent, as shown below.

$$(e_A + \text{bias}) + (e_B + \text{bias}) = (e_A + e_B) + 2\ \text{bias}$$

The resulting exponent should be restored to a single bias before the multiplication operation begins. This is accomplished by subtracting the bias.

1. Check for zero operands. If $A = 0$ or $B = 0$, then the product $= 0$.
2. Determine the sign of the product.
3. Add exponents and subtract the bias.
4. Multiply fractions. Steps 3 and 4 can be done in parallel, but both must be completed before step 5.
5. Normalize the product.

An example will illustrate this concept. Let the exponents be $e_A = 0000\ 1010\ (10)$ and $e_B = 0000\ 0101\ (5)$. Each exponent will be biased, then added to produce a double bias. The bias will then be subtracted to produce a single bias, then subtracted again to produce the sum of the two unbiased exponents: $e_A = 0000\ 1010\ (10)$ and $e_B = 0000\ 0101\ (5) = (e_A + e_B)_{\text{unbiased}} = 0000\ 1111\ (15)$.

$$
\begin{array}{rcl}
e_{A(\text{unbiased})} & = & 0000\_1010 \\
+)\ \text{bias} & = & \underline{0111\_1111} \\
e_{A(\text{biased})} & = & 1000\_1001
\end{array}
$$

$$
\begin{array}{rcl}
e_{B(\text{unbiased})} & = & 0000\_0101 \\
+)\ \text{bias} & = & \underline{0111\_1111} \\
e_{B(\text{biased})} & = & 1000\_0100
\end{array}
$$

$$
\begin{array}{rcr}
e_{A(\text{biased})} = & & 1000\_1001 \\
+)\ e_{B(\text{biased})} = & & \underline{1000\_0100} \\
\text{Double bias} = & 1 \leftarrow & 0000\_1101
\end{array}
$$

Restore to single bias by subtracting the bias; that is, by adding the 2s complement of 0111 1111 (1000 0001).

$$
\begin{array}{rcr}
e_{A(\text{biased})} + e_{B(\text{biased})} = & & 0000\_1101 \\
+)\ \text{2s complement of bias} = & & \underline{1000\_0001} \\
(e_A + e_B)_{\text{single bias}} = & 1 \leftarrow & 1000\_1110
\end{array}
$$

$$
\begin{array}{rcr}
(e_A + e_B)_{\text{single bias}} = & & 1000\_1110 \\
+)\ \text{2s complement of bias} = & & \underline{1000\_0001} \\
(e_A + e_B)_{\text{no bias}} = & 1 \leftarrow & 0000\_1111
\end{array}
$$

11.6.2 Numerical Examples

Examples will now be presented that illustrate multiplication using the paper-and-pencil method for 4-bit operands in 2s complement notation. If the operands are in 2s complement notation, then the sign bit is treated in a manner identical to the other bits; however, the sign bit of the multiplicand is extended left in the partial product to accommodate the $2n$-bits of the product. The only requirement is that the multiplier must be positive — the multiplicand can be either positive or negative. This is not a problem when using the X86 assembly language — the assembler resolves this problem automatically. The assembler also resolves exponent biasing and significand alignment for addition and subtraction.

Example 11.1 The multiplicand and multiplier are two positive 4-bit operands, where *a[3:0]* = 0111 (+7) and *b[3:0]* = 0101 (+5) to yield a product *p[7:0]* = 0010 0011 (+35). A multiplier bit of 1 copies the multiplicand to the partial product; a multiplier bit of 0 enters 0s in the partial product.

Multiplicand *A*						0	1	1	1	+7
Multiplier *B*					×)	0	1	0	1	+5
	0	0	0	0		0	1	1	1	
Partial	0	0	0	0		0	0	0		
products	0	0	0	1		1	1			
	0	0	0	0		0				
Product *P*	0	0	1	0		0	0	1	1	+35

Example 11.2 This example multiplies a positive multiplicand by a negative multiplier to demonstrate that the multiplier must be positive. The multiplicand is *a[3:0]* = 0101 (+5); the multiplier is *b[3:0]* = 1101 (–3). The product should be –15; however, since the multiplier is treated as an unsigned number (1101 = 13), the result is 0100 0001 (65).

Multiplicand *A*						0	1	0	1	+5
Multiplier *B*					×)	1	1	0	1	(–3) 13
	0	0	0	0		0	1	0	1	
Partial	0	0	0	0		0	0	0		
products	0	0	0	1		0	1			
	0	0	1	0		1				
Product *P*	0	1	0	0		0	0	0	1	65

The problem can be resolved by either 2s complementing both operands or by 2s complementing the multiplier, performing the multiplication, then 2s complementing the result. The method shown below 2s complements both operands.

Multiplicand *A*						1	0	1	1	–5
Multiplier *B*					×)	0	0	1	1	+3
	1	1	1	1		1	0	1	1	
Partial	1	1	1	1		0	1	1		
products	0	0	0	0		0	0			
	0	0	0	0		0				
Product *P*	1	1	1	1		0	0	0	1	–15

When both operands are negative, the correct result can be obtained by 2s complementing both operands before the operation begins, since a negative multiplicand multiplied by a negative multiplier yields a positive product.

11.6.3 Multiply Instructions

There are different versions of the *multiply* instruction. One version, FMUL, multiplies the multiplicand in ST(0) by the single-operand floating-point 32-bit or a 64-bit multiplier source operand in memory and stores the product in ST(0). For some multiply instructions, the source operand can be a single-precision floating-point operand, a double-precision floating-point operand, an integer word operand, or an integer doubleword operand. The syntax for the FMUL instruction is shown below.

```
FMUL    m32fp (memory, 32 bits, floating-point)
FMUL    m64fp (memory, 64 bits, floating-point)
```

Another version of the multiply instruction multiplies the operand in register ST(0) by the operand in register ST(*i*) and stores the product in ST(0). A similar version multiplies ST(*i*) by ST(0) and stores the product in ST(*i*). The syntax for the two-operand FMUL instruction is shown below.

```
FMUL    ST(0), ST(i) (stores product in ST(0))
FMUL    ST(i), ST(0) (stores product in ST(i))
```

Another version of the multiply instruction, FMULP, is similar to the double-operand version shown above, where the product is stored in ST(*i*). The operand in ST(*i*) is multiplied by the operand in ST(0) and the product is stored in ST(*i*). However, in this version, the register stack is popped after the product is stored. The syntax is shown below.

```
FMULP   ST(i), ST(0)
```

Another version of the multiply instruction, FMULP, is the no-operand version, which multiplies the operand in ST(1) by the operand in ST(0) and stores the product in ST(1), then pops the register stack. The syntax is shown below.

```
FMULP
```

Another version of the multiply instruction, FIMUL, multiplies ST(0) by the 16-bit or a 32-bit single-operand integer source operand in memory and stores the product in ST(0). The FIMUL instruction converts the integer source operand to a double extended-precision floating-point number before the multiplication operation. The syntax is shown below.

```
FIMUL   m16int (memory, 16 bits, integer)
FIMUL   m32int (memory, 32 bits, integer)
```

Figure 11.21 shows an assembly language module embedded in a C program that illustrates utilizing different versions of the FMUL instruction. The FMUL single-operand instruction, the FMUL double-operand instruction, and the FMULP instruction are used in the program. Four floating-point numbers are entered from the keyboard for use in the program: two negative numbers and two positive numbers, as shown below. The results of the five multiply instructions are also shown below.

```
flp1_num = -10.500
flp2_num = -5.000
flp3_num = +7.700
flp4_num = +12.500
```

FMUL flp2_num

	+ 12.500	ST(0)
×)	− 5.000	flp2_num
	− 62.500	→ ST(0)

FMUL flp4_num

	− 62.500	ST(0)
×)	+ 12.500	flp4_num
	−781.250	→ ST(0)

FMUL ST(0), ST(2)

	− 781.250	ST(0)
×)	− 5.000	ST(2)
	+3906.250	→ ST(0)

FMULP ST(1), ST(0)

	+ 7.700	ST(1)
×)	+ 3906.250	ST(0)
	+30078.125	→ ST(1) Then pop

```
//fmul_versions.cpp
//show the use of versions of the FMUL instruction

#include "stdafx.h"

int main (void)
{
//define variables
   float flp1_num, flp2_num, flp3_num, flp4_num,
         flp_rslt1, flp_rslt2, flp_rslt3, flp_rslt4;

   printf ("Enter 4 flp numbers, 2 negative & 2 positive:
               \n");
   scanf ("%f %f %f %f", &flp1_num, &flp2_num, &flp3_num,
                         &flp4_num);

                                    //continued on next page
```

(a)

Figure 11.21 Program to illustrate using versions of the FMUL instruction: (a) the program and (b) the outputs.

```
//switch to assembly
      _asm
      {
         FLD    flp1_num     //flp1_num (-50.5) -> ST(0)
         FLD    flp2_num     //flp2_num (-5.0) -> ST(0)
                             //(-50.5) -> ST(1)
         FLD    flp3_num     //flp3_num (+7.7) -> ST(0)
                             //(-5.0) -> ST(1)
                             //(-50.5) -> ST(2)
         FLD    flp4_num     //flp4_num (+25.7) -> ST(0)
                             //(+7.7) -> ST(1)
                             //(-5.0) -> ST(2)
                             //(-50.5) -> ST)3)
//---------------------------------------------------------
         FMUL   flp2_num     //ST(0) × flp2_num -> ST(0)
                             //(+12.5) x (-5.0) = -62.5
         FST    flp_rslt1    //-62.5 -> flp_rslt1
//---------------------------------------------------------
         FMUL   flp4_num     //ST(0) × flp4_num -> ST(0)
                             //(-62.5) x (+12.5) = -781.25
         FST    flp_rslt2    //-781.25 -> flp_rslt2
//---------------------------------------------------------
         FMUL   ST(0), ST(2)//ST(0) × ST(2) -> ST(0)
                             //(-781.25) x (-5.0) = +3906.25
         FST    flp_rslt3    //+16512.25 -> flp_rslt3
//---------------------------------------------------------
         FMULP ST(1), ST(0)//ST(1) × ST(0) -> ST(1)
                             //(+7.700) × (+3906.25) =
                             //+30078.125
         FST    flp_rslt4    //+30078.125 -> flp_rslt4
//---------------------------------------------------------
      }

//print result
   printf ("\nflp_rslt1 = %f\n", flp_rslt1);
   printf ("flp_rslt2 = %f\n", flp_rslt2);
   printf ("flp_rslt3 = %f\n", flp_rslt3);
   printf ("flp_rslt4 = %f\n\n", flp_rslt4);

   return 0;
}                                  //continued on next page
```

Figure 11.21 (Continued)

```
Enter 4 flp numbers, 2 negative & 2 positive:
-10.500 -5.000 +7.700 +15.500

flp_rslt1 = -62.500000
flp_rslt2 = -781.250000
flp_rslt3 = 3906.250000
flp_rslt4 = 30078.125000

Press any key to continue . . . _
```
(b)

Figure 11.21 (Continued)

Figure 11.22 shows the register stack contents for different stages of the program. Figure 11.22(a) shows the result of the four floating-point numbers having been pushed onto the register stack. The remaining figures in Figure 11.22 portray the results of the various instructions after they have been executed.

(a)

R0	+12.500	ST(0)
R1	+7.700	ST(1)
R2	–5.000	ST(2)
R3	–10.500	ST(3)
R4		ST(4)
R5		ST(5)
R6		ST(6)
R7		ST(7)

(b) **FMUL** `flp2_num`

–62.500	ST(0)	→`flp_rslt1`
+7.700	ST(1)	
–5.000	ST(2)	
–10.500	ST(3)	
	ST(4)	
	ST(5)	
	ST(6)	
	ST(7)	

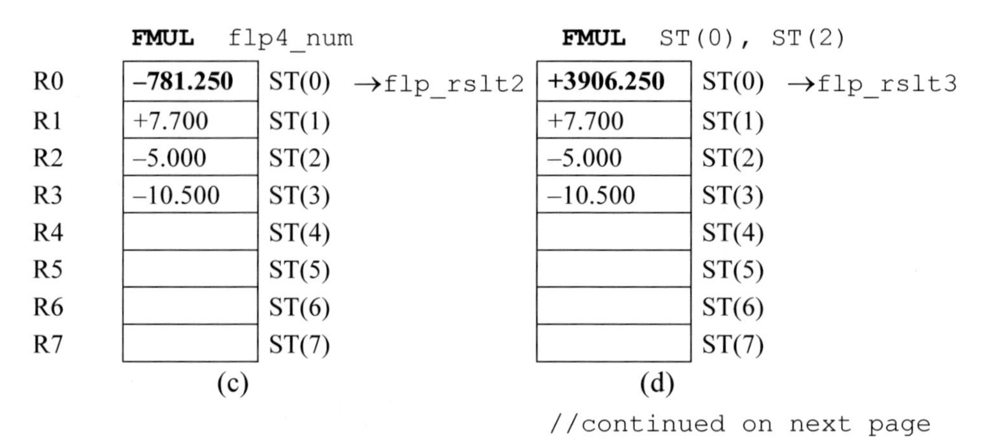

//continued on next page

Figure 11.22 Register stack utilization for the program of Figure 11.21.

FMULP ST(1), ST(0)

R0	**+30078.125**	ST(0) →flp_rslt4
R1	−5.000	ST(1)
R2	−10.500	ST(2)
R3		ST(3)
R4		ST(4)
R5		ST(5)
R6		ST(6)
R7		ST(7)

(e)

Figure 11.22 (Continued)

An example of floating-point multiplication is shown in Figure 11.23 to illustrate the concept of adding the exponents to obtain the correct resulting exponent. The example uses the sequential add-shift method with 8-bit operands. In this example, the multiplicand fraction *fract_a* = 0.1010 0000 × 2^3 (+5) is multiplied by a multiplier *fract_b* = 0.1100 0000 × 2^2 (+3) with partial product D = 0000 0000 to produce a product of *prod* = 0.1111 0000 0000 0000 × 2^4 (+15).

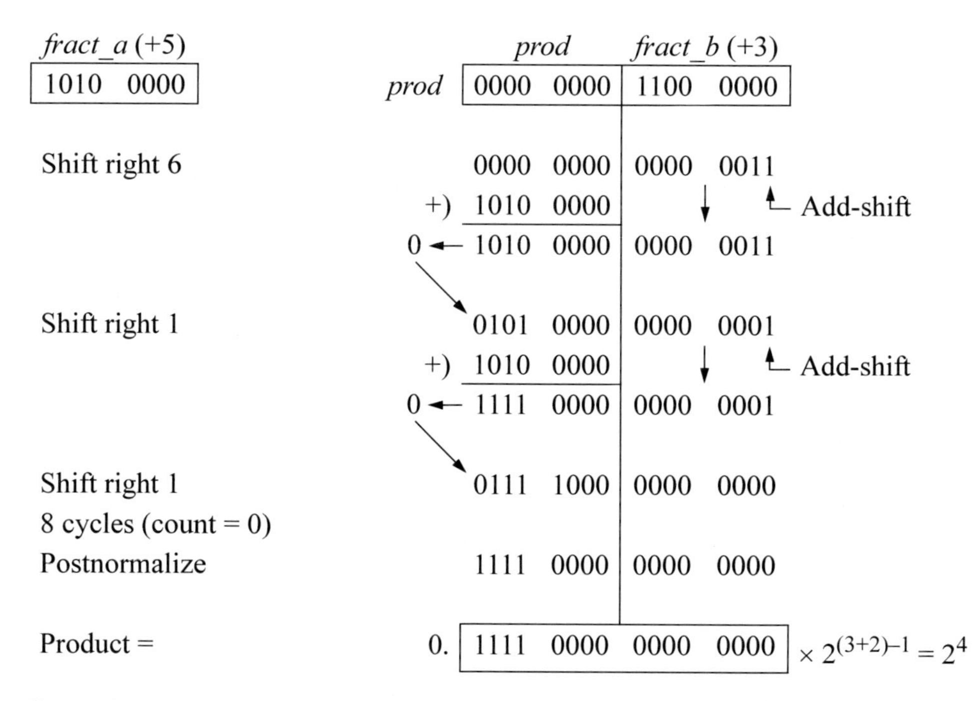

Figure 11.23 Example of floating-point multiplication using the sequential add-shift method.

Since the multiplication involves two n-bit operands, a count-down sequence counter, *count*, is set to a value that represents the number of bits in one of the operands. The counter is decremented by one for each step of the add-shift sequence. When the counter reaches a value of zero, the operation is finished and the product is normalized, if necessary.

If the low-order bit of register *fract_b* is equal to zero, then zeroes are added to the partial product and the sum is loaded into register *prod*. In this case, it is not necessary to perform an add operation — a right shift can accomplish the same result. The sequence counter is then decremented by one. If the low-order bit of register *fract_b* is equal to one, then the multiplicand is added to the partial product. The sum is loaded into register *prod* and the sequence counter is decremented.

11.7 Division Instructions

Floating-point division performs two operations in parallel: fraction division and exponent subtraction. The dividend is usually $2n$ bits and the divisor is n bits. Divide overflow is determined in the same way as in fixed-point division; that is, if the high-order half of the dividend is greater than or equal to the divisor, then divide overflow occurs. The problem is resolved by shifting the dividend right one bit position and incrementing the exponent by one. Since both operands were normalized, this assures that the entire dividend is smaller than the divisor, as shown below.

$$\begin{aligned} \text{High-order half of Dividend} &= 0.01xxx\ldots xx \\ \text{Divisor} &= 0.1xxx\ldots xx \end{aligned}$$

This is referred to as *dividend alignment*, providing the ranges for the two operands, as shown below.

$$\begin{aligned} 1/4 \leq \text{Dividend} &< 1/2 \\ 1/2 \leq \text{Divisor} &< 1 \end{aligned}$$

Both operands are checked for a value of zero. If the dividend is zero, then the exponent, quotient, and remainder are set to zero. If the divisor is zero, then the result is infinity and the operation is terminated. Division is performed on normalized floating-point operands A and B using biased exponents, such that

$$A = f_A \times r^{e_A}$$

$$B = f_B \times r^{e_B}$$

where f is the normalized fraction, e is the exponent, and r is the radix. Floating-point division is defined as shown in Equation 11.7, which shows fraction division and exponent subtraction performed simultaneously.

$$\begin{aligned} A / B &= (f_A \times r^{e_A}) / (f_B \times r^{e_B}) \\ &= (f_A / f_B) \times r^{e_A - e_B} \end{aligned} \tag{11.7}$$

The sign of the quotient is determined by the signs of the floating-point numbers. If the signs are the same, then the sign of the quotient is positive; if the signs are different, then the sign of the quotient is negative. This can be determined by the exclusive-OR of the two signs, as shown in Equation 11.8. The sign of the remainder is the same as the sign of the dividend.

$$\text{Quotient sign} = A_{\text{sign}} \oplus B_{\text{sign}} \tag{11.8}$$

11.7.1 Zero Bias

As was stated previously, the divisor exponent is subtracted from the dividend exponent in parallel with fraction division. The exponents are subtracted and the carry-out is examined. If the carry-out = 1, then the dividend exponent was greater than or equal to the divisor exponent ($e_A \geq e_B$). If the carry-out = 0, then the dividend exponent was less than the divisor exponent ($e_A < e_B$). Since both exponents were initially biased, the difference generates a result with no bias, as shown in Equation 11.9.

$$\begin{aligned} e_A - e_B &= (e_A + \text{bias}) - (e_B + \text{bias}) \\ &= e_A + \text{bias} - e_B - \text{bias} \\ &= (e_A - e_B)_{\text{unbiased}} \end{aligned} \tag{11.9}$$

Therefore, the bias must be added to the difference so that the resulting exponent is properly biased. Thus, for the single-precision format:

$$(e_A - e_B)_{\text{biased}} = (e_A - e_B)_{\text{unbiased}} + 0111\ 1111$$

Restoring the bias may result in an exponent overflow, in which case the division operation is terminated. Examples will now be presented that illustrate the previous statements and are chosen for $e_A > e_B$, $e_A = e_B$, and $e_A < e_B$.

Example 11.3 Let $e_A > e_B$, where $e_{A(\text{unbiased})}$ = 0001 0110 (22) and $e_{B(\text{unbiased})}$ = 0000 1010 (10). Therefore, $e_A - e_B = 22 - 10 = 12$.

$$\begin{array}{rr} e_{A(\text{unbiased})} = & 0001\ \ 0110 \\ \text{Add bias +)} & \underline{0111\ \ 1111} \\ e_{A(\text{biased})} = & 1001\ \ 0101 \end{array}$$

$$\begin{array}{rr} e_{B(\text{unbiased})} = & 0000\ \ 1010 \\ \text{Add bias +)} & \underline{0111\ \ 1111} \\ e_{B(\text{biased})} = & 1000\ \ 1001 \end{array}$$

$e_{A(\text{biased})} - e_{B(\text{biased})}$

$$\begin{array}{rrl} e_{A(\text{biased})} = & 1001\ \ 0101 & \\ \text{+) 2s complement of } e_{B(\text{biased})} = & \underline{0111\ \ 0111} & \\ 1 \leftarrow & 0000\ \ 1100 & \quad 12 \end{array}$$

Restore to single bias by adding the bias.

$$\begin{array}{rr} (e_A - e_B)_{\text{unbiased}} = & 0000\ \ 1100 \\ \text{Add bias} = & \underline{0111\ \ 0111} \\ (e_A - e_B)_{\text{biased}} = & 1001\ \ 1011 \end{array}$$

Example 11.4 Let $e_A = e_B$, where $e_{A(\text{unbiased})} = 0001\ 0101\ (21)$ and $e_{B(\text{unbiased})} = 0001\ 0101\ (21)$. Therefore, $e_A - e_B = 21 - 21 = 0$.

$$\begin{array}{rr} e_{A(\text{unbiased})} = & 0001\ \ 0101 \\ \text{Add bias +)} & \underline{0111\ \ 1111} \\ e_{A(\text{biased})} = & 1001\ \ 0100 \end{array}$$

$$\begin{array}{rr} e_{B(\text{unbiased})} = & 0001\ \ 0101 \\ \text{Add bias +)} & \underline{0111\ \ 1111} \\ e_{B(\text{biased})} = & 1001\ \ 0100 \end{array}$$

$e_{A(\text{biased})} - e_{B(\text{biased})}$

$$\begin{array}{rrl} e_{A(\text{biased})} = & 1001\ \ 0100 & \\ \text{+) 2s complement of } e_{B(\text{biased})} = & \underline{0110\ \ 1100} & \\ 1 \leftarrow & 0000\ \ 0000 & \quad 0 \end{array}$$

Restore to single bias by adding the bias.

$$\begin{aligned} (e_A - e_B)_{\text{unbiased}} &= 0000\ \ 0000 \\ \text{Add bias} &= \underline{0111\ \ 0111} \\ (e_A - e_B)_{\text{biased}} &= 0111\ \ 1111 \end{aligned}$$

Example 11.5 Let $e_A < e_B$, where $e_{A(\text{unbiased})} = 0000\ 1001\ (9)$ and $e_{B(\text{unbiased})} = 0001\ 0011\ (19)$. Therefore, $e_A - e_B = 9 - 19 = -10$.

$$\begin{aligned} e_{A(\text{unbiased})} &= 0000\ \ 1001 \\ \text{Add bias} +) &\ \underline{0111\ \ 1111} \\ e_{A(\text{biased})} &= 1000\ \ 1000 \end{aligned}$$

$$\begin{aligned} e_{B(\text{unbiased})} &= 0001\ \ 0011 \\ \text{Add bias} +) &\ \underline{0111\ \ 1111} \\ e_{B(\text{biased})} &= 1001\ \ 0010 \end{aligned}$$

$e_{A(\text{biased})} - e_{B(\text{biased})}$

$$\begin{aligned} e_{A(\text{biased})} &= 1000\ \ 1000 \\ +)\ \text{2s complement of } e_{B(\text{biased})} &= \underline{0110\ \ 1110} \\ 0 \leftarrow &\ 1111\ \ 0110 \quad -10 \end{aligned}$$

If carry-out = 0, then 2s complement to obtain the difference of 0000 1010 (10).

Restore to single bias by adding the bias.

$$\begin{aligned} (e_A - e_B)_{\text{unbiased}} &= 1111\ \ 0110 \\ \text{Add bias} &= \underline{0111\ \ 1111} \\ (e_A - e_B)_{\text{biased}} &= 0111\ \ 0101 \end{aligned}$$

11.7.2 Numerical Example

This section presents a numerical example using the sequential shift-subtract/add restoring division method with 4-bit divisors and 8-bit dividends. Register A contains the $2n$-bit normalized dividend fraction, *fract_a*, which will eventually contain the n-bit quotient and n-bit remainder. Register B contains the n-bit normalized divisor fraction, *fract_b*.

Since the division process involves one n-bit divisor and one $2n$-bit dividend, a count-down sequence counter, *count*, is set to a value that represents the number of bits in the divisor. The counter is decremented by one for each step of the shift-subtract/add sequence. When the counter reaches a value of zero, the operation is finished and the quotient resides in *fract_a[3:0]* and the remainder resides in *fract_a[7:4]*.

If the value of the high-order half of the dividend is greater than or equal to the value of the divisor, then an overflow condition exists. To resolve this problem, the dividend is shifted right one bit position and the dividend exponent is incremented by one. Each sequence in the division process consists of a shift left of one bit position followed by a subtraction of the divisor.

Example 11.6 A dividend fraction *fract_a* = 0.1010 0100 $\times 2^7$ (+82) is divided by a divisor fraction *fract_b* = 0.1001 $\times 2^4$ (+9) to yield a quotient of 1001 $\times 2^4$ (+9) and a remainder of 0001 $\times 2^4$ (+1), as shown in Figure 11.24.

fract_b (+9)		*fract_a* (+82)	
1001		1010 0100	
Align		0101 0010	$\times 2^{(7+1)} = 2^8$
Shift left 1		1010 010–	
Subtract *B*	+)	0111	
	1 ←	0001	
No restore		0001 0101	
Shift left 1		0010 101–	
Subtract *B*	+)	0111	
	0 ←	1001	
Restore	+)	1001	
		0010 1010	
Shift left 1		0101 010–	
Subtract *B*	+)	0111	
	0	1100	
Restore	+)	1001	
		0101 0100	

Continued on next page

Figure 11.24 Example of sequential shift-subtract/add restoring division.

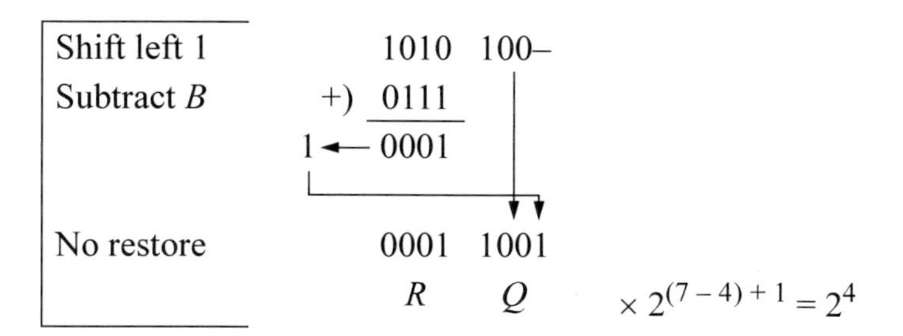

Figure 11.24 (Continued)

The example of Figure 11.24 was presented only to provide a review of the sequential shift-subtract/add restoring division algorithm and does *not* reflect the floating-point division procedure. Floating-point division yields a quotient only — there is no remainder. For example, the operands of Figure 11.24 will yield a floating-point result of 9.111111; that is, 82 / 9 = quotient of 9 and a remainder of 1 / 9 = 0.111111. This is shown in the program of Figure 11.25, using the FDIV instruction, which is explained in Section 11.7.3. The remainder can be obtained by using the *partial remainder* FPREM1 instruction described in Section 11.10.

```
//fdiv_versions2.cpp
//show the use of the FDIV instruction
#include "stdafx.h"
int main (void)
{
//define variables
   float flp_dvdnd, flp_dvsr, flp_rslt;
   printf ("Enter a dividend and a divisor: \n");
   scanf ("%f %f", &flp_dvdnd, &flp_dvsr);

//switch to assembly
      _asm
      {
         FLD    flp_dvdnd    //flp_dvdnd -> ST(0)
         FDIV   flp_dvsr     //ST(0) / flp_dvsr -> ST(0)
         FST    flp_rslt
}
   printf ("\nflp_rslt = %f\n\n", flp_rslt);
   return 0;
}                        (a)     //continued on next page
```

Figure 11.25 Program to show a divide operation of 82.00 / 9.00 to yield a quotient of 9.111111: (a) the program and (b) the outputs.

```
Enter a dividend and a divisor:
82.00 9.00

flp_rslt = 9.111111

Press any key to continue . . . _
-----------------------------------------------------------
Enter a dividend and a divisor:
12.25 7.45

flp_rslt = 1.644295

Press any key to continue . . . _
```

(b)

Figure 11.25 (Continued)

11.7.3 Divide Instructions

There are different versions of the *divide* instruction. One version, FDIV, divides the dividend in ST(0) by the single-operand floating-point 32-bit or a 64-bit divisor source operand in memory and stores the result in ST(0). For some divide instructions, the source operand can be a single-precision floating-point operand, a double-precision floating-point operand, an integer word operand, or an integer doubleword operand. The syntax for the FDIV instruction is shown below.

```
FDIV   m32fp (memory, 32 bits, floating-point)
FDIV   m64fp (memory, 64 bits, floating-point)
```

Another version of the divide instruction divides the operand in register ST(0) by the operand in register ST(*i*) and stores the result in ST(0). A similar version divides ST(*i*) by ST(0) and stores the result in ST(*i*). The syntax for the two-operand FDIV instruction is shown below

```
FDIV   ST(0), ST(i) (stores result in ST(0))
FDIV   ST(i), ST(0) (stores result in ST(i))
```

Another version of the divide instruction, FDIVP, is similar to the double-operand version shown above, where the result is stored in ST(*i*). The operand in ST(*i*) is divided by the operand in ST(0) and the result is stored in ST(*i*). However, in this version, the register stack is popped after the result is stored. The syntax is shown below.

```
FDIVP   ST(i), ST(0)
```

Another version of the divide instruction, FDIVP, is the no-operand version, which divides the operand in ST(1) by the operand in ST(0) and stores the result in ST(1), then pops the register stack. The syntax is shown below.

```
FDIVP
```

Another version of the divide instruction, FIDIV, divides ST(0) by the 16-bit or 32-bit integer source operand in memory and stores the result in ST(0). The FIDIV instruction converts the integer source operand to a double extended-precision floating-point number before the division operation. The syntax is shown below.

```
FIDIV   m16int (memory, 16 bits, integer)
FIDIV   m32int (memory, 32 bits, integer)
```

There are also a variety of *reverse divide* instructions. These instructions are similar to those listed above, except that the divide operation is reversed. For example, the FDIVR instruction divides the single-operand floating-point 32-bit or a 64-bit source operand in memory by ST(0) stores the result in ST(0). These instructions include the following:

```
FDIVR    m32fp (memory, 32 bits, floating-point,
               divides memory operand by ST(0) and
               stores the result in ST(0))
FDIVR    m64fp (memory, 64 bits, floating-point,
               divides memory operand by ST(0) and
               stores the result in ST(0))

FDIVR    ST(0), ST(i)  (divides ST(i) by ST(0) and
                       stores the result in ST(0))
FDIVR    ST(i), ST(0)  (divides ST(0) by ST(i) and
                       stores the result in ST(i))

FDIVRP    ST(i), ST(0)  (divides ST(0) by ST(i),
                        stores the result in ST(i),
                        then pops stack)

FDIVRP                 (divides ST(0) by ST(1),
                        stores the result in ST(1),
                        then pops stack)

FIDIVR    m16int (memory, 16 bits, integer,
                 divides memory operand by ST(0) and
                 stores the result in ST(0))
FIDIVR    m32int (memory, 32 bits, integer,
                 divides memory operand by ST(0) and
                 stores the result in ST(0))
```

Figure 11.26 shows an assembly language module embedded in a C program that illustrates utilizing different versions of the FDIV instruction. The FDIV single-operand instruction, the FDIV double-operand instruction, and the FDIVP instruction are used in the program. Four floating-point numbers are entered from the keyboard for use in the program: a positive and negative dividend and a positive and negative divisor, as shown below. The results of the four divide instructions are also shown below.

```
flp1_dvdnd = +547.125
flp2_dvdnd = -15.750
flp1_dvsr =  +65.175
flp2_dvsr =  -50.650
```

FDIV flp1_dvsr

	+547.125	ST(0)
÷)	+65.175	flp1_dvsr
	+8.394706	→ ST(0)

FDIV flp2_dvsr

	+8.394706	ST(0)
÷)	–50.650	flp2_dvsr
	–0.165740	→ ST(0)

FDIV ST(0), ST(1)

	–15.7250	flp2_dvdnd
÷)	–0.165740	ST(1)
	+95.028641	→ ST(0)

FDIVP ST(1), ST(0)

	–0.165740	ST(1)
÷)	+95.028641	ST(0)
	–0.001744	→ ST(1) Then pop

```
//fdiv_versions.cpp
//show the use of versions of the FDIV instruction
#include "stdafx.h"
int main (void)
{
//define variables
   float flp1_dvdnd, flp2_dvdnd, flp1_dvsr, flp2_dvsr,
         flp_rslt1, flp_rslt2, flp_rslt3, flp_rslt4;

   printf ("Enter a pos & neg dvdnd and a pos & neg dvsr:
            \n");
   scanf ("%f %f %f %f", &flp1_dvdnd, &flp2_dvdnd,
                        &flp1_dvsr, &flp2_dvsr);
```

(a) //continued on next page

Figure 11.26 Program to illustrate using versions of the FDIV instruction: (a) the program and (b) the outputs.

```
//switch to assembly
      _asm
      {
         FLD    flp1_dvdnd  //flp1_dvdnd (+547.125) -> ST(0)
//-------------------------------------------------------------
         FDIV   flp1_dvsr   //ST(0) / flp1_dvsr -> ST(0)
                            //(+547.125) / (+65.175) =
                            //+8.394706
         FST    flp_rslt1   //+8.394706 -> flp_rslt1
//-------------------------------------------------------------
         FDIV   flp2_dvsr   //ST(0) / flp2_dvsr -> ST(0)
                            //(+8.394706) / (-50.650) =
                            //-0.165740
         FST    flp_rslt2   //-0.165740 -> flp_rslt2
//-------------------------------------------------------------
         FLD    flp2_dvdnd  //flp2_dvdnd (-15.750) -> ST(0)
                            //ST(0) (-0.165740) -> ST(1)
         FDIV   ST(0), ST(1)//ST(0) / ST(1) -> ST(0)
                            //(-15.750) / (-0.165740) =
                            //+95.028641
         FST    flp_rslt3   //+95.028641 -> flp_rslt3
//-------------------------------------------------------------
         FDIVP  ST(1), ST(0)//ST(1) / ST(0) -> ST(1), pop
                            //(-0.165740) / (+95.028641) =
                            //-0.001744
         FST    flp_rslt4   //-0.001744 -> flp_rslt4
//-------------------------------------------------------------
      }

//print result
   printf ("\nflp_rslt1 = %f\n", flp_rslt1);

   printf ("flp_rslt2 = %f\n", flp_rslt2);

   printf ("flp_rslt3 = %f\n", flp_rslt3);

   printf ("flp_rslt4 = %f\n\n", flp_rslt4);

   return 0;
}                              //continued on next page
```

Figure 11.26 (Continued)

```
Enter a pos & neg dvdnd and a pos & neg dvsr:
+547.125 -15.750 +65.175 -50.650

flp_rslt1 = 8.394706
flp_rslt2 = -0.165740
flp_rslt3 = 95.028641
flp_rslt4 = -0.001744

Press any key to continue . . . _
```

(b)

Figure 11.26 (Continued)

Figure 11.27 shows the register stack contents for different stages of program execution for Figure 11.26. The floating-point number *flp1_dvdnd* is initially stored in ST(0) by the first load instruction. Figure 11.27(a) through Figure 11.27(d) portray the results of the various instructions after they have been executed.

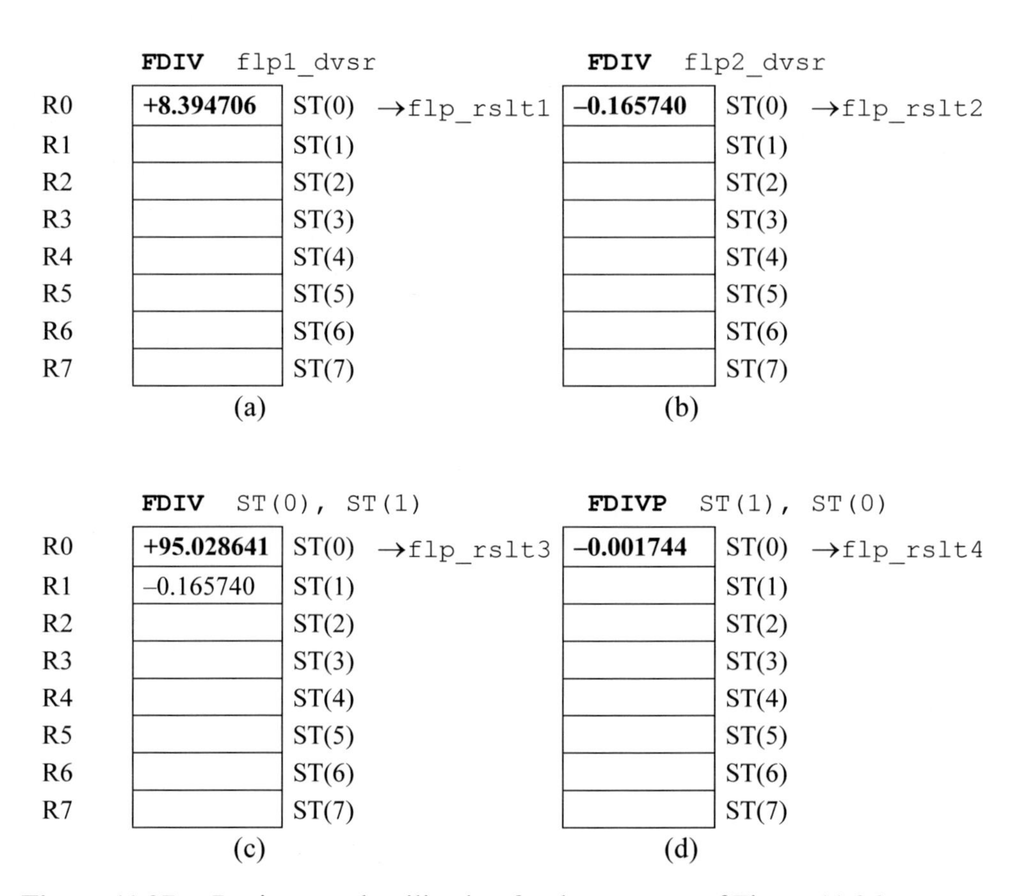

Figure 11.27 Register stack utilization for the program of Figure 11.26.

11.8 Compare Instructions

This section describes the floating-point instructions that compare different types of data. These include the *compare floating-point values* instructions: FCOM, FCOMP, and FCOMPP; the *compare floating-point values and set flags* instructions: FCOMI, FCOMIP, FUCOMI, and FUCOMIP; the *compare integer* instructions: FICOM and FICOMP; the *test* instruction: FTST; and the *unordered compare floating-point values* instructions: FUCOM, FUCOMP, and FUCOMPP. These instructions are explained in the sections that follow.

11.8.1 Compare Floating-Point Values

There are nine *compare floating-point values* instructions that compare the contents of stack register ST(0) with the source operand. The condition code flags are then set in the floating-point unit (FPU) status word or in the EFLAGS register, depending on the type of instruction and the results of the operation. The FPU status word is reproduced in Figure 11.28 and the EFLAGS register is reproduced in Figure 11.29.

15	14	13	12	11	10	9	8
B	C3	TOS			C2	C1	C0

7	6	5	4	3	2	1	0
ES	SF	PE	UE	OE	ZE	DE	IE

Figure 11.28 Floating-point unit status word format.

31	30	29	28	27	26	25	24
0	0	0	0	0	0	0	0

23	22	21	20	19	18	17	16
0	0	ID	VIP	VIF	AC	VM	RF

15	14	13	12	11	10	9	8
0	NT	IOPL		OF	DF	IF	TF

7	6	5	4	3	2	1	0
SF	ZF	0	AF	0	PF	1	CF

Figure 11.29 EFLAGS register.

The meaning of bits C3, C2, and C0 in the floating-point unit status word are defined in Table 11.1. Bits C3, C2, and C0 map into bits ZF, PF, and CF, respectively in the EFLAGS register. Unlike integer comparison instructions, floating-point comparison instructions have four — rather than three — results: ST(0) greater than source, ST(0) less than source, ST(0) equal to source, and *unordered.* An *unordered* condition is detected if an operand is *not-a-number* (NaN) or is in an undefined format. In this case, a floating-point invalid-operation exception (#IA) is produced. If the #IA exception is masked, then the condition code flags are set to the *unordered* state.

Table 11.1 X87 Condition Code Flags in the FPU Status Word for the Compare Floating-Point Values Instructions

Condition	C3	C2	C1	C0
ST(0) > source	0	0	–	0
ST(0) < source	0	0	–	1
ST(0) = source	1	0	–	0
Unordered	1	1	–	1

There are different versions of the *compare floating-point values* instruction. The source operand can be a register in the FPU stack or a memory location. However, if no source operand is given, then the operand in ST(0) is compared with the operand in ST(1).

One version, FCOM, compares the operand in ST(0) with the floating-point 32-bit or a 64-bit source operand in memory and sets the X87 FPU condition code flags. The syntax for the FCOM instruction is shown below.

```
FCOM    m32fp (memory, 32-bit floating-point)
FCOM    m64fp (memory, 64-bit floating-point)
```

Another version of the FCOM instruction compares the operand in register ST(0) with the operand in register ST(*i*) and sets the X87 FPU condition code flags. The syntax is shown below.

```
FCOM    ST(i) (compare ST(0) with ST(i))
```

Another version of the FCOM instruction compares the operand in register ST(0) with the operand in register ST(1) and sets the X87 FPU condition code flags. This version does not define a source operand. The syntax is shown below.

```
FCOM    (compare ST(0) with ST(1))
```

Another version, FCOMP, of the instruction compares the operand in register ST(0) with the floating-point 32-bit or a 64-bit source operand in memory, sets the X87 FPU condition code flags, then pops the register stack. The syntax is shown below.

```
FCOMP   m32fp (memory, 32-bit floating-point, pop stack)
FCOMP   m64fp (memory, 64-bit floating-point, pop stack)
```

Another version of the FCOMP instruction compares the operand in register ST(0) with the operand in ST(*i*), sets the X87 FPU condition code flags, then pops the register stack. The syntax is shown below.

```
FCOMP   ST(i) (compare ST(0) with ST(i), pop stack)
```

Another version of the FCOMP instruction compares the operand in register ST(0) with the operand in ST(1), sets the X87 FPU condition code flags, then pops the register stack. This version does not define a source operand. The syntax is shown below.

```
FCOMP   (compare ST(0) with ST(1), pop stack)
```

Another version of the instruction compares the operand in register ST(0) with the operand in ST(1), sets the X87 FPU condition code flags, then pops the register stack twice. This version does not define a source operand. The syntax is shown below.

```
FCOMPP   (compare ST(0) with ST(1), pop stack twice)
```

11.8.2 Compare Floating-Point Values and Set EFLAGS

These instructions perform an *unordered* comparison of the operands in stack registers ST(0) and ST(*i*). The result of the comparison sets the zero flag (ZF), the parity flag (PF), and the carry flag (CF) in the EFLAGS register, as shown in Table 11.2.

Table 11.2 Status Flag Bits for the Compare Floating-Point Values and Set EFLAGS Instructions

Condition	ZF	PF	CF
ST(0) > source	0	0	0
ST(0) < source	0	0	1
ST(0) = source	1	0	0
Unordered	1	1	1

An *unordered* comparison checks the type of numbers being compared; for example, unsupported, NaN, normal finite, infinity, zero, empty, or denormal. Denormalized numbers are very small numbers, where the biased exponent is zero and there are leading zeroes in the significand (fraction). There are four different versions of this type of instruction, which are described below. Each version has two operands and there is no destination.

One version, FCOMI, compares the operand in register stack ST(0) with the operand in register stack ST(*i*), then sets the three status flags in the EFLAGS register, as shown in Table 11.2. This instruction operates identically to the FCOM instruction, but sets the status flags in the EFLAGS register instead of the condition code flags in the X87 FPU status word register. The syntax is shown below.

```
FCOMI   ST(0), ST(i) (compare ST(0) with ST(i), set flags)
```

Another version, FCOMIP, compares the operand in register stack ST(0) with the operand in register stack ST(*i*), sets the three flags in the EFLAGS register, then pops the register stack. This instruction operates identically to the FCOM instruction, but sets the status flags in the EFLAGS register instead of the condition code flags in the X87 FPU status word register. The syntax is shown below.

```
FCOMIP   ST(0), ST(i) (compare ST(0) with ST(i),
                       set status flags, then pop stack)
```

Another version, FUCOMI, compares the operand in register stack ST(0) with the operand in register stack ST(*i*) for ordered operands, then sets the status flags in the EFLAGS register instead of the condition code flags in the X87 FPU status word register. This instruction operates identically to the FCOMI instruction, but does not yield a floating-point invalid-operation exception. The syntax is shown below.

```
FUCOMI   ST(0), ST(i) (compare ST(0) with ST(i)
                        for ordered operands,
                        then set status flags)
```

Another version, FUCOMIP, compares the operand in register stack ST(0) with the operand in register stack ST(*i*) for ordered operands, sets the status flags in the EFLAGS register instead of the condition code flags in the X87 FPU status word register, then pops the register stack. This instruction operates identically to the FCOMIP instruction, but does not yield a floating-point invalid-operation exception, except for NaNs or unsupported formats. The syntax is shown below.

```
FUCOMIP   ST(0), ST(i) (compare ST(0) with ST(i)
                         for ordered operands,
                         set status flags,
                         then pop stack)
```

11.8.3 Compare Integer

There are two different versions of the *compare integer* instruction, both of which are described below. The operation of both versions, FICOM and FICOMP, is identical to the operation of the FCOM and FCOMP instructions; however, the source operand is an integer in a memory location. The integer operand is changed to a double extended-precision floating-point value before the operands are compared.

One version, FICOM, compares ST(0) with a 16-bit or 32-bit integer source operand in memory, then sets the condition code flags in the X87 floating-point unit status word. Refer to Table 11.1 for the meaning of bits C3, C2, and C0. The syntax is shown below.

```
FICOM   m16int (compare ST(0) with a 16-bit integer
               in memory, then set flags)

FICOM   m32int (compare ST(0) with a 32-bit integer
               in memory, then set flags)
```

Another version, FICOMP, compares ST(0) with a 16-bit or 32-bit integer source operand in memory, sets the condition code flags in the X87 floating-point unit status word, then pops the register stack. The syntax is shown below.

```
FICOMP  m16int (compare ST(0) with a 16-bit integer
               in memory, set flags,
               then pop stack)

FICOMP  m32int (compare ST(0) with a 32-bit integer
               in memory, set flags,
               then pop stack)
```

11.8.4 Test

This instruction, FTST, performs an operation identical to the FCOM instruction, but compares the operand in ST(0) with a value of 0.0, then sets the condition code flags — C3, C2, C0 — in the X87 floating-point unit status word. The syntax is shown below.

```
FTST (compare ST(0) with 0.0)
```

11.8.5 Unordered Compare Floating-Point Values

There are different versions of the *unordered compare floating-point values* instruction, all of which are described below. The operation of FUCOM, FUCOMP, and

FUCOMPP are identical to the operation of the FCOM, FCOMP, and FCOMPP instructions, respectively. However, the floating-point invalid-operation exception is set only when one or both operands are an SNaN (defined below) or are in an unsupported format. When one or both operands are a QNaN (defined below), the condition code flags are set to *unordered* and do not set the floating-point invalid-operation exception.

There are two types of NaNs that are used in the architecture. An SNaN is defined as a *signaling* NaN, in which the high-order significand bit is reset. A QNaN is defined as a *quiet* NaN, in which the high-order significand bit is set. These instructions execute an unordered comparison of ST(0) with ST(*i*) or ST(1) and set the condition code flags — C3, C2, C0 — in the X87 floating-point unit status word.

One version, FUCOM, compares ST(0) with ST(*i*), then sets the condition code flags in the X87 floating-point unit status word. Refer to Table 11.1 for the meaning of bits C3, C2, and C0. The syntax is shown below.

```
FUCOM   ST(i) (compare ST(0) with ST(i), set flags)
```

Another version, FUCOM with no operand, compares ST(0) with ST(1), then sets the condition code flags in the X87 floating-point unit status word. Refer to Table 11.1 for the meaning of bits C3, C2, and C0. The syntax is shown below.

```
FUCOM   (compare ST(0) with ST(1), set flags)
```

Another version, FUCOMP, compares the operand in register stack ST(0) with the operand in register stack ST(*i*), sets the condition code flags in the X87 floating-point unit status word, then pops the register stack. Refer to Table 11.1 for the meaning of bits C3, C2, and C0. The syntax is shown below.

```
FUCOMP   ST(i) (compare ST(0) with ST(i),
               set flags, then pop stack)
```

Another version, FUCOMP with no operand, compares the operand in register stack ST(0) with the operand in register stack ST(1), sets the condition code flags in the X87 floating-point unit status word, then pops the register stack. Refer to Table 11.1 for the meaning of bits C3, C2, and C0. The syntax is shown below.

```
FUCOMP   (compare ST(0) with ST(1),
         set flags, then pop stack)
```

Another version, FUCOMPP with no operand, compares the operand in register stack ST(0) with the operand in register stack ST(1), sets the condition code flags in the X87 floating-point unit status word, then pops the register stack twice. Refer to Table 11.1 for the meaning of bits C3, C2, and C0. The syntax is shown below.

```
FUCOMPP   (compare ST(0) with ST(1),
          set flags, then pop stack twice)
```

A pop operation on the register stack is accomplished by setting the stack top tag register to a value of 11_2, indicating empty. Then the stack pointer is incremented by 1 — bits 13 through 11 (TOS) of the X87 floating-point status word, reproduced in Figure 11.30.

15	14	13	12	11	10	9	8
B	C3	TOS			C2	C1	C0

7	6	5	4	3	2	1	0
ES	SF	PE	UE	OE	ZE	DE	IE

Figure 11.30 X87 floating-point status word.

11.9 Trigonometric Instructions

This section describes the floating-point instructions that calculate the *cosine* FCOS, *partial tangent* FPTAN, *sine* FSIN, *sine and cosine* FSINCOS, and *partial arctangent* FPATAN of source operands that are expressed in radians. These instructions are explained in the sections that follow.

A radian is defined as an angular measurement that is equal to the angle at the center of a circle subtended by an arc that is equal to the radius of the circle. One radian is approximately equal to 57.296 degrees. Figure 11.31 shows a drawing that illustrates one radian. Since one radian ≈ 57.296 degrees, therefore, one degree ≈ 0.01745 radians.

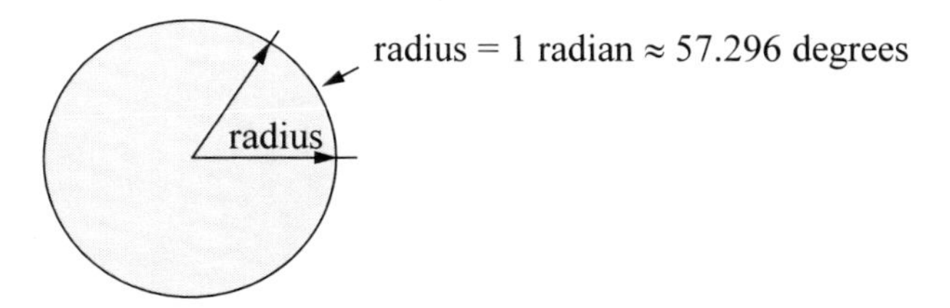

Figure 11.31 Angular measurement of one radian.

11.9.1 Cosine

The *cosine*, FCOS, instruction calculates the cosine of the source operand — given in radians — in the stack top register ST(0) and stores the result in ST(0). If the operand is not within a specified range (-2^{63} to $+2^{63}$), then bit C2 is set in the floating-point unit status word. This however, does alter the operand in ST(0) and does not generate an exception. The syntax is shown below and has no operands specified, because the source operand was previously loaded into ST(0).

```
FCOS (cosine -> ST(0))
```

11.9.2 Partial Arctangent

The *partial arctangent*, FPATAN, instruction is the inverse tangent function specified by $\tan^{-1}$ or arctan. The arctangent can be defined as follows:

$$(\tan^{-1} x = y) \equiv (\tan y = x)$$

The domain of the arctangent function is normally in the interval $-\pi/2$ to $+\pi/2$, as shown in Figure 11.32. The tangent function yields the ratio of the opposite / adjacent sides of a right triangle; the arctangent yields the angle of the ratio.

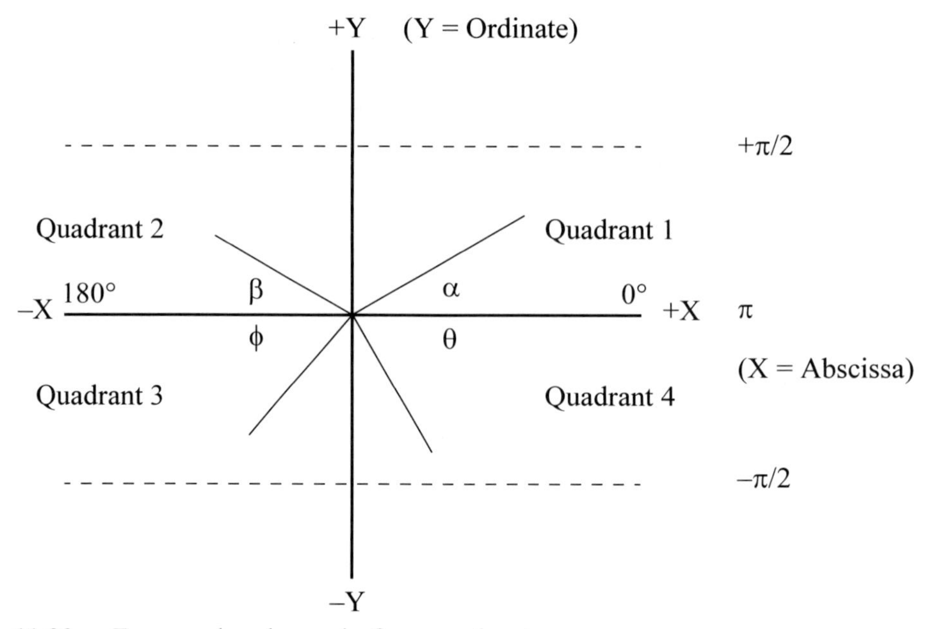

Figure 11.32 Four angles shown in four quadrants.

The FPATAN instruction has no operands specified in the instruction. It calculates the arctangent of the source operand in ST(1) divided by the second source operand in ST(0), then pops the register stack, which places the result in ST(0). The abscissa (X) is in ST(0) and the ordinate Y is in ST(1). The FPATAN instruction yields the angle between the X axis and the line drawn from the origin — center of the circle — to a point (X,Y) in a particular quadrant, as shown in Figure 11.32.

Since there are four quadrants, the angles in the quadrants have the following X and Y coordinates: (+X, +Y) in quadrant 1, (–X, +Y) in quadrant 2, (–X, –Y) in quadrant 3, (+X, –Y) in quadrant 4. The angle is a function of the sign of both X (the abscissa) and Y (the ordinate). An X, Y coordinate in quadrant 1 yields a positive angle; an X, Y coordinate in quadrant 2 yields an angle between $\pi/2$ and π; an X, Y coordinate in quadrant 3 yields an angle between $-\pi/2$ and $-\pi$, and an X, Y coordinate in quadrant 4 yields an angle between 0 and $-\pi/2$.

Figure 11.33 shows a short assembly language module embedded in a C program that illustrates the application of the FPATAN instruction. The inputs represent the opposite (ordinate *y*) and the adjacent (abscissa *x*) sides of a right triangle. The first set of inputs (+1.0, +1.0) represent a 45 degree angle in quadrant 1 whose arctangent is 0.785398 radians. The second set of inputs (+1.0, +1.75) represent a 30 degree angle, also in quadrant 1, whose arctangent is 0.519146 radians. The third set of inputs (+1.0, –1.75) represents a 30 degree angle in quadrant 2, whose arctangent is 2.622447 radians — an angle between $\pi/2$ (1.570796) and π (3.141592).

The fourth set of inputs (–1.0, –1.75) represent a 30 degree angle in quadrant 3 whose arctangent is –2.622447 radians — an angle between $-\pi/2$ (–1.570796) and $-\pi$ (–3.141592). The fifth, and final, set of inputs (–1.0, +1.75) represent a 30 degree angle in quadrant 4 whose arctangent is –0.519146 radians — an angle between 0 and $-\pi/2$ (–1.570796).

```
//arctan.cpp
//shows the use of the FPATAN instruction
#include "stdafx.h"
int main (void)
{
   float ordinate_y, abscissa_x, arctan_rslt;

   printf ("Enter ordinate (y), then abscissa (x): \n");
   scanf ("%f %f", &ordinate_y, &abscissa_x);
//switch to assembly
      _asm
      {
         FLD    ordinate_y      //ordinate -> ST(0)
         FLD    abscissa_x      //abscissa -> ST(0))
                                //ordinate -> ST(1)
         FPATAN                 //generate angle
         FST    arctan_rslt     //store angle
         }
   printf ("\nArctangent = %f radians\n\n", arctan_rslt);
   return 0;
}                        (a)          //continued on next page
```

Figure 11.33 Program to illustrate the use of the partial arctangent FPATAN: (a) the program and (b) the outputs.

```
Enter ordinate (y), then abscissa (x):
1.0 1.0                      //quadrant 1

Arctangent = 0.785398 radians

Press any key to continue . . . _
------------------------------------------------------------
Enter ordinate (y), then abscissa (x):
1.0 1.75                     //quadrant 1

Arctangent = 0.519146 radians

Press any key to continue . . . _
------------------------------------------------------------
Enter ordinate (y), then abscissa (x):
1.0 -1.75                    //quadrant 2

Arctangent = 2.622447 radians

Press any key to continue . . . _
------------------------------------------------------------
Enter ordinate (y), then abscissa (x):
-1.0 -1.75                   //quadrant 3

Arctangent = -2.622447 radians

Press any key to continue . . . _
------------------------------------------------------------
Enter ordinate (y), then abscissa (x):
-1.0 1.75                    //quadrant 4

Arctangent = -0.519146 radians

Press any key to continue . . . _
```

(b)

Figure 11.33 (Continued)

11.9.3 Partial Tangent

The *partial tangent*, FPTAN, instruction calculates the tangent of the source operand — expressed in radians — in ST(0) of the register stack, stores the result in ST(0), then pushes a value of +1.0 onto the stack, which maintains compatibility with X87 processors. The tangent for the angle θ of a right triangle is defined as follows: $\tan \theta =$ opposite / adjacent.

11.9.4 Sine

The *sine*, FSIN, instruction calculates the sign of the source operand — expressed in radians — in ST(0) of the register stack, and stores the result in ST(0). If the operand is not within a specified range (-2^{63} to $+2^{63}$), then bit C2 is set in the floating-point unit status word. This, however, does alter the operand in ST(0) and does not generate an exception. The syntax is shown below and has no operands specified, because the source operand was previously loaded into ST(0).

```
FSIN (sine -> ST(0)
```

Figure 11.34 shows an assembly language module embedded in a C program that illustrates the usage of the FSIN and FCOS instructions. The four sets of inputs are entered as radians.

```
//radian.cpp
//compute sine and cosine from radians
#include "stdafx.h"
int main (void)

{
   float radian, sine_result, cosine_result;

   printf ("Enter number of radians: \n");
   scanf ("%f", &radian);

//switch to assembly
      _asm
      {
         FLD    radian          //radians -> ST(0)
         FSIN                   //compute sine
         FST    sine_result     //store sine in result area

         FLD    radian          //radians -> ST(0)
         FCOS                   //compute cosine
         FST    cosine_result   //store cosine in result area
      }

   printf ("\nSine result = %f\n", sine_result);
   printf ("Cosine result = %f\n\n", cosine_result);
   return 0;
}
```

(a) //continued on next page

Figure 11.34 Program to illustrate the use of the FSIN and FCOS instructions: (a) the program and (b) the outputs.

```
Enter number of radians:
0.523598775               //30 degrees

Sine result = 0.500000
Cosine result = 0.866025

Press any key to continue . . . _
------------------------------------------------------------
Enter number of radians:
4.712388898               //270 degrees

Sine result = -1.000000
Cosine result = 0.000000

Press any key to continue . . . _
------------------------------------------------------------
Enter number of radians:
0.5                       //28.648 degrees

Sine result = 0.479426
Cosine result = 0.877583

Press any key to continue . . . _
------------------------------------------------------------
Enter number of radians:
1                         //57.296 degrees

Sine result = 0.841471
Cosine result = 0.540302

Press any key to continue . . . _
------------------------------------------------------------
Enter number of radians:
1.2                       //68.7552 degrees

Sine result = 0.932039
Cosine result = 0.362358

Press any key to continue . . . _
------------------------------------------------------------
Enter number of radians:
1.5                       //85.944 degrees

Sine result = 0.997495
Cosine result = 0.070737

Press any key to continue . . . _(b)
```

Figure 11.34 (Continued)

11.9.5 Sine and Cosine

The *sine and cosine*, FSINCOS, instruction calculates the sine and cosine of the source operand that was previously stored in ST(0). The FSINCOS instruction stores the sine of the operand in ST(0) of the register stack, then pushes the cosine onto the stack, so that ST(0) contains the cosine and ST(1) contains the sine. The source operand is expressed in radians. If the operand is not within a specified range (-2^{63} to $+2^{63}$), then bit C2 is set in the floating-point unit status word. This, however, does alter the operand in ST(0) and does not generate an exception.

Figure 11.35 contains an assembly language module embedded in a C program that uses the trigonometric instructions FSINCOS and FPTAN to obtain the sine, cosine, and tangent of radians that are entered from the keyboard. The program also uses a new instruction *exchange register contents* FXCH, which exchanges the contents of register ST(0) and register ST(i). The *load constant* FLD1 and the *add* FADD instructions are also utilized in the program.

```
//sine_cos_tan2.cpp
//calculate sine, cosine, and tangent

#include "stdafx.h"
int main (void)
{
   float radian, sine_rslt, cos_rslt, rad_rslt,
         cos_rslt2, tan_rslt;

   printf ("Enter number of radians: \n");
   scanf ("%f", &radian);

//switch to assembly
      _asm
      {
         FLD    radian          //radian -> ST(0)
         FSINCOS                //compute sine and cosine
                                //cosine -> ST(0)
                                //sine -> ST(1)
         FST    cos_rslt        //store cosine in result area
         FXCH   ST(1)           //exchange ST(0) with ST(1)
                                //sine -> ST(0), cos -> ST(1)
         FST    sine_rslt       //store sine in result area
                                   //continued on next page
```

(a)

Figure 11.35 Program to illustrate utilization of the trigonometric instructions FSINCOS and FPTAN: (a) the program and (b) the outputs.

```
//-----------------------------------------------------------
        FLD    radian          //radian -> ST(0)
        FLD1                   //+1.0 -> ST(0)
                               //radian -> ST(1)
        FADD   ST(0), ST(1)    //ST(0) + ST(1) -> ST(0)
                               //radian + 1.0 -> ST(0)
        FST    rad_rslt        //radian + 1 -> result area
//-----------------------------------------------------------
        FADD   ST(0), ST(3)    //ST(0) + cos -> ST(0)
                               //(1 + rad) + cos -> ST(0)
        FST    cos_rslt2       //1 + rad + cos -> result area
//-----------------------------------------------------------
        FLD    radian          //radian -> ST(0)
        FPTAN                  //1.0 -> ST(0)
                               //tangent -> ST(1)
        FXCH   ST(1)           //exchange ST(0) with ST(1)
                               //exchange 1.0 and tangent
        FST    tan_rslt        //store tan in result area
}

//print result
   printf ("\nSine result = %f\n", sine_rslt);
   printf ("Cosine result = %f\n", cos_rslt);
   printf ("Radian result = %f\n", rad_rslt);
   printf ("Cosine result2 = %f\n", cos_rslt2);
   printf ("Tangent result = %f\n\n", tan_rslt);

   return 0;
}
```

```
Enter number of radians:
1                              //57.296 degrees; quadrant 1

Sine result = 0.841471         //sine = opposite / hypotenuse
Cosine result = 0.540302       //cosine = adjacent / hypotenuse
Radian result = 2.000000       //1 + radian
Cosine result2 = 2.540302      //1 + radian + cosine
Tangent result = 1.557408      //tangent = opposite / adjacent

Press any key to continue . . . _
-----------------------------------------------------------
```

//continued on next page

(b)

Figure 11.35 (Continued)

```
Enter number of radians:
2                           //114.592 degrees; quadrant 2

Sine result = 0.909297      //sine = opposite / hypotenuse
Cosine result = -0.416147   //cosine = adjacent / hypotenuse
Radian result = 3.000000    //1 + radian
Cosine result2 = 2.583853   //1 + radian + cosine
Tangent result = -2.185040 //tangent = opposite / adjacent

Press any key to continue . . . _
-------------------------------------------------------------
Enter number of radians:
1.5                         //85.944 degrees; quadrant 1

Sine result = 0.997495      //sine = opposite / hypotenuse
Cosine result = 0.070737    //cosine = adjacent / hypotenuse
Radian result = 2.500000    //1 + radian
Cosine result2 = 2.570737   //1 + radian + cosine
Tangent result = 14.101420 //tangent = opposite / adjacent

Press any key to continue . . . _
-------------------------------------------------------------
Enter number of radians:
3                           //171.888 degrees; quadrant 2

Sine result = 0.141120      //sine = opposite / hypotenuse
Cosine result = -0.989992   //cosine = adjacent / hypotenuse
Radian result = 4.000000    //1 + radian
Cosine result2 = 3.010008   //1 + radian + cosine
Tangent result = -0.142547 //tangent = opposite / adjacent

Press any key to continue . . . _
-------------------------------------------------------------
Enter number of radians:
4                           //229.184 degrees; quadrant 3

Sine result = -0.756802     //sine = opposite / hypotenuse
Cosine result = -0.653644   //cosine = adjacent / hypotenuse
Radian result = 5.000000    //1 + radian
Cosine result2 = 4.346356   //1 + radian + cosine
Tangent result = 1.157821   //tangent = opposite / adjacent

Press any key to continue . . . _
-------------------------------------------------------------


                                  //continued on next page
```

Figure 11.35 (Continued)

```
Enter number of radians:
5                          //286.480 degrees; quadrant 4

Sine result = -0.958924    //sine = opposite / hypotenuse
Cosine result = 0.283662   //cosine = adjacent / hypotenuse
Radian result = 6.000000   //1 + radian
Cosine result2 = 6.283662  //1 + radian + cosine
Tangent result = -3.380515 //tangent = opposite / adjacent

Press any key to continue . . . _
```

Figure 11.35 (Continued)

The third set of outputs in Figure 11.35(b) has an angle of 85.944 degrees, placing the angle in quadrant 1, as shown below. Since the angle is close to 90 degrees and the sine is opposite / hypotenuse, the result will be close to a value of 1. The cosine has an adjacent side that is relatively small; therefore, since the cosine is adjacent / hypotenuse, the value of the cosine is very small. In a similar manner, since the tangent is opposite / adjacent, the tangent value has a relatively large value.

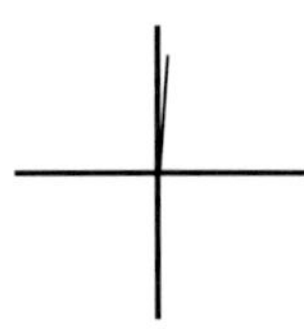

The fourth set of outputs has an angle of 171.888 degrees, placing the angle in quadrant 2, as shown below. Since the angle is close to 180 degrees and the sine is opposite / hypotenuse, the result will be a small value. The cosine has an adjacent side that is negative and relatively large; therefore, since the cosine is adjacent / hypotenuse, the value of the cosine is negative and relatively large. In a similar manner, since the tangent is opposite / adjacent, the tangent value has a relatively small negative value.

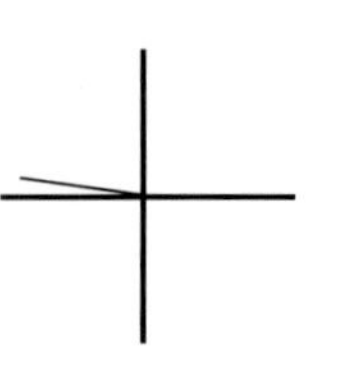

The fifth set of outputs has an angle of 229.184 degrees, placing the angle in quadrant 3, as shown below. The sine has a negative opposite side; therefore, since the sine is opposite / hypotenuse, the result will be a negative value. The cosine has an adjacent side that is negative; therefore, since the cosine is adjacent / hypotenuse, the value of the cosine is also negative. Since the tangent is opposite / adjacent, the tangent value has a positive value.

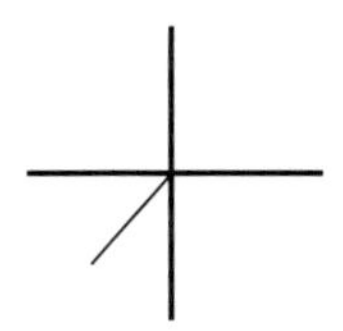

The sixth set of outputs has an angle of 286.480 degrees, placing the angle in quadrant 4, as shown below. The sine has a negative opposite side; therefore, since the sine is opposite / hypotenuse, the result will be a relatively large negative value. The cosine has an adjacent side that is positive; therefore, since the cosine is adjacent / hypotenuse, the cosine has a relatively small positive value. Since the tangent is opposite / adjacent, the tangent value has a negative value.

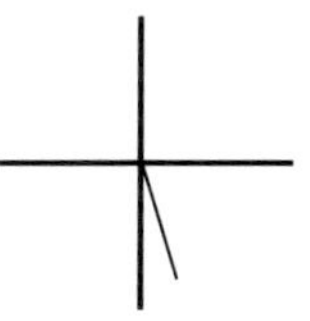

Figure 11.36 shows the register stack contents for different stages of program execution for Figure 11.35.

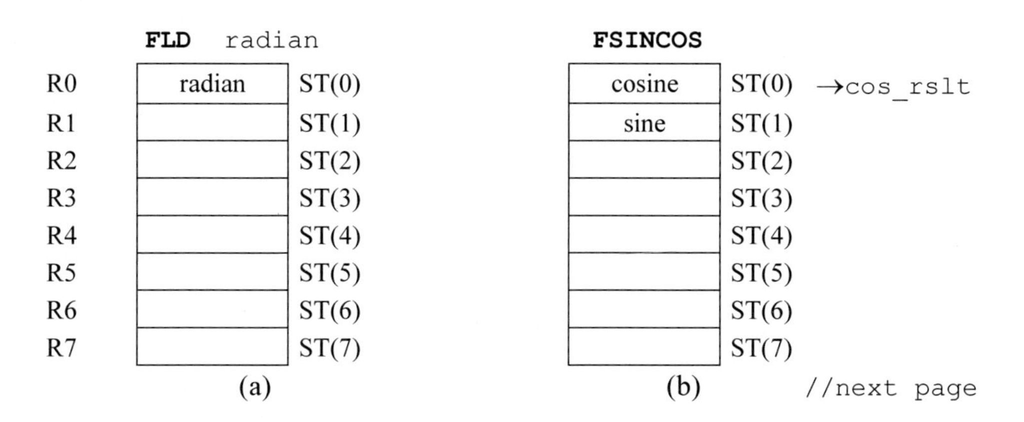

Figure 11.36 Register stack utilization for the program of Figure 11.35.

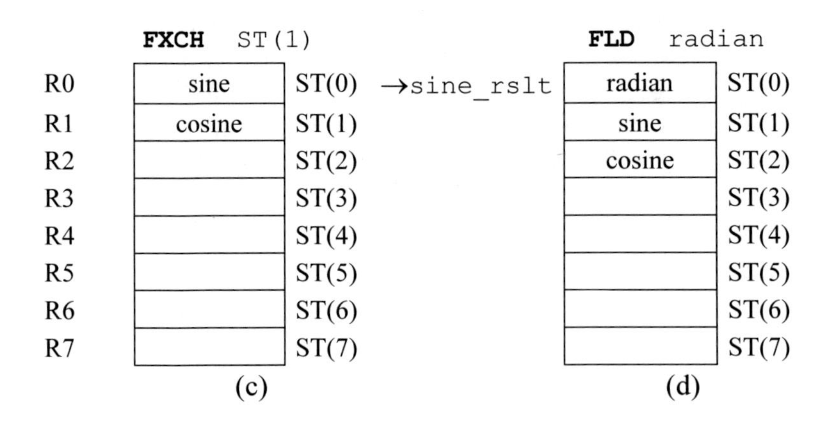

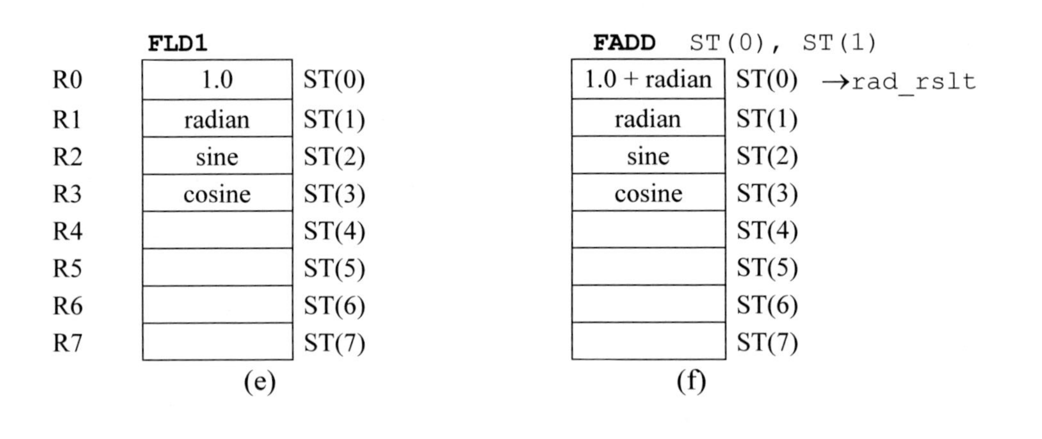

FADD ST(0), ST(3)
FLD radian

R0 1+rad+cos ST(0) →cos_rslt2 radian ST(0)
R1 radian ST(1) radian ST(1)
R2 sine ST(2) sine ST(2)
R3 cosine ST(3) cosine ST(3)
R4 ST(4) ST(4)
R5 ST(5) ST(5)
R6 ST(6) ST(6)
R7 ST(7) ST(7)

(g) (h)

//continued on next page

Figure 11.36 (Continued)

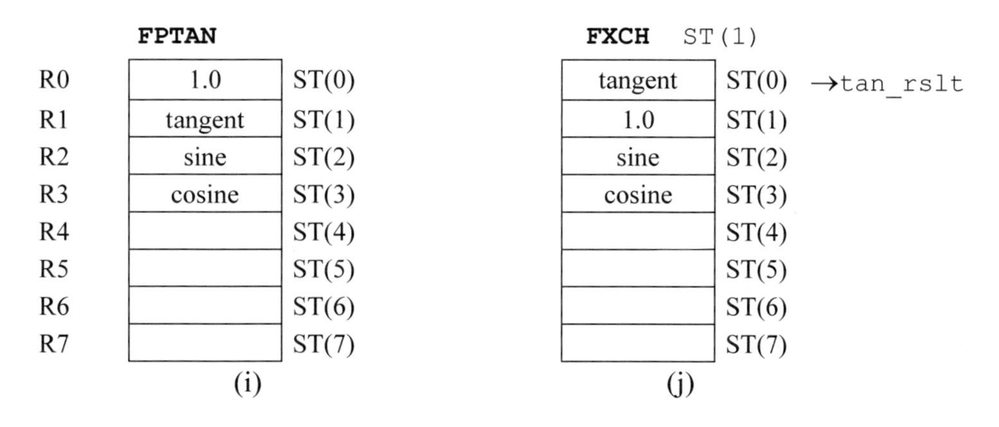

Figure 11.36 (Continued)

11.10 Additional Instructions

This section describes some additional floating-point instructions that perform basic arithmetic operations and have only one syntax. Most of the previous instructions in this chapter had more than one syntax. These additional instructions include the *absolute value* instruction: FABS; the *change sign* instruction: FCHS; the *partial remainder* instruction: FPREM1; the *round to integer* instruction: FRNDINT; and the *square root* instruction: FSQRT.

Some additional nonarithmetic instructions are also included. These include the *decrement stack-top pointer* instruction: FDECSTP; the *free floating-point register* instruction: FFREE; the *increment stack-top pointer* instruction: FINCSTP; and the *exchange register contents* instruction: FXCH. All of the above instructions are explained in the sections that follow and are listed alphabetically.

11.10.1 Absolute Value

The *absolute value*, FABS, instruction resets the sign bit of the operand in ST(0) of the register stack, thereby generating the absolute value of the operand. The absolute value of a number is a nonnegative number with the same numerical value without regard to the sign. Therefore, the absolute value of –12 or +12 is 12, and is written as $|12|$ — this does not change a positive number; however, a negative number is changed to a positive number.

The absolute value of a number can also be regarded as its unit distance from the number zero. The numbers –12 and +12 are both twelve units distance from the number zero. The condition code flags that are affected in the floating-point unit status word are as follows: C1 (bit 9) is reset; C0 (bit 8), C2 (bit 10), and C3 (bit 14) are undefined. The syntax for the FABS instruction is shown below — there is no operand specified.

```
FABS
```

11.10.2 Change Sign

The *change sign*, FCHS, instruction complements the sign bit of the operand in ST(0) of the register stack. The instruction changes a negative value to positive value or positive value to a negative value — the absolute value of the operand does not change. The condition code flags that are affected in the floating-point unit status word are as follows: C1 (bit 9) is reset; C0 (bit 8), C2 (bit 10), and C3 (bit 14) are undefined. The syntax for the FCHS instruction is shown below — there is no operand specified.

```
FCHS
```

11.10.3 Decrement Stack-Top Pointer

The *decrement stack-top pointer*, FDECSTP, instruction decrements the *top-of-stack* (TOS) field — bits 13 through 11 — in the floating-point unit status word. The instruction subtracts one from the TOS field. For example, if the top-of-stack register was ST(0), then the FDECSTP rotates the stack by one register causing ST(7) to become the new the top-of-stack register. The contents of the stack registers and the associated tag registers are not affected.

The condition code flags that are affected in the floating-point unit status word are as follows: C1 (bit 9) is reset; C0 (bit 8), C2 (bit 10), and C3 (bit 14) are undefined. The syntax for the FDECSTP instruction is shown below — there is no operand specified.

```
FDECSTP
```

11.10.4 Free Floating-Point Register

The *free floating-point register*, FFREE, instruction sets the tag field associated with stack register ST(*i*) to indicate empty; that is, the tag field is set to 11_2. The contents of stack register ST(*i*) and the stack-top pointer, however, are not affected. The condition code flags in the floating-point unit status word are undefined. The syntax for the FFREE instruction is shown below, where ST(*i*) indicates a register in the register stack.

```
FFREE    ST(i)
```

11.10.5 Increment Stack-Top Pointer

The *increment stack-top pointer*, FINCSTP, instruction increments the *top-of-stack* (TOS) field — bits 13 through 11 — in the floating-point unit status word. The instruction adds one to the TOS field. For example, if the top-of-stack register was ST(7), then the FINCSTP rotates the stack by one register causing ST(0) to become the new top-of-stack register. The contents of the stack registers and the associated tag registers are not affected; therefore, this instruction is not analogous to a pop operation.

The condition code flags that are affected in the floating-point unit status word are as follows: C1 (bit 9) is reset; C0 (bit 8), C2 (bit 10), and C3 (bit 14) are undefined. The syntax for the FINCSTP instruction is shown below — there is no operand specified.

```
FINCSTP
```

11.10.6 Partial Remainder

The *partial remainder*, FPREM1, instruction calculates the remainder that is obtained from dividing the dividend in register ST(0) by the divisor in register ST(1) and stores the result in register ST(0). The remainder is as specified in the Institute of Electrical and Electronics Engineers (IEEE) floating-point Standard 754. The remainder is also referred to as the *modulus*. The result that is obtained from the division process is rounded to the nearest integer.

The partial remainder is obtained by a process of repeated subtraction of no more than 63 iterations of one instruction execution. If the operation yields a result that is less than half the modulus, then the condition code flag C2 in the floating-point unit status word is reset, otherwise C2 is set. If necessary, the software can reexecute the FPREM1 instruction until the condition code flag C2 contains a value of zero. In this case, the result previously obtained and stored in ST(0) is used as the dividend. It should be noted, however, that a higher-priority interrupt can override the second iteration process. The syntax for the FPREM1 instruction is shown below — there is no operand specified.

```
FPREM1
```

11.10.7 Round to Integer

The *round to integer*, FRNDINT, instruction rounds the operand in ST(0) of the register stack to the nearest integer. The operand is rounded to an integer using the rounding method specified by bits 11 and 10 of the rounding control (RC) field in the floating-point control word register, reproduced in Figure 11.37. The RC field is defined as shown in Table 11.3. The rounded operand is stored in ST(0) of the register stack. A floating-point exception, indicating an indeterminate result, is produced if the source operand is not an integer. The condition code flags that are affected in the floating-point unit status word are as follows: C1 (bit 9) is reset if a stack underflow has occurred — C1 is set if the operand was rounded up; otherwise, C1 is reset; C0 (bit 8), C2 (bit 10), and C3 (bit 14) are undefined. The syntax for the FINCSTP instruction is shown below — there is no operand specified.

```
FRNDINT
```

15	14	13	12	11	10	9	8
			IC	RC		PC	

7	6	5	4	3	2	1	0
		PM	UM	OM	ZM	DM	IM

Figure 11.37 Floating-point unit control word register.

Table 11.3 Rounding Control Field

Rounding Method	11	10
Round to nearest (default mode)	0	0
Round down toward minus infinity	0	1
Round up toward positive infinity	1	0
Round toward zero (truncate)	1	1

11.10.8 Square Root

The *square root*, FSQRT, instruction calculates the square root of the source operand in ST(0) of the register stack and stores the solution in ST(0). The square root of a number a is written using the symbol $\sqrt{a}$, which specifies the square root of a. The square root symbol is also referred to as a *radical* sign. The square root of a can also be written as $a^{0.5}$, where $a \geq 0$.

There are two rules when using the square root operation. The *product rule* states that the square root of the product of two operands is equal to the product of the square roots of the operands, as shown below. The *quotient rule* states that the square root of the division of two operands is equal to the division of the square roots of the operands, also shown below.

$$\sqrt{a \times b} = \sqrt{a} \times \sqrt{b}$$

$$\sqrt{a/b} = (\sqrt{a})/(\sqrt{b})$$

Examples of the product rule and the quotient rule are shown below.

$$\sqrt{144 \times 16} = \sqrt{144} \times \sqrt{16} = 12 \times 4 = 48$$

$$\sqrt{(144)/(16)} = (\sqrt{144})/(\sqrt{16}) = 12 \div 4 = 3$$

The condition code flags that are affected in the floating-point unit status word are as follows: C1 (bit 9) is reset if a stack underflow has occurred — C1 is set if the result was rounded up; otherwise, C1 is reset; C0 (bit 8), C2 (bit 10), and C3 (bit 14) are undefined. The syntax for the FSQRT instruction is shown below — there is no operand specified.

```
FSQRT
```

11.10.9 Exchange Register Contents

The *exchange register contents*, FXCH, instruction was introduced in Section 11.9.5. The instruction exchanges the contents of register ST(0) and the source register ST(*i*) if a source operand is specified. If there is no source operand stipulated, then the FXCH instruction exchanges the contents of ST(0) and ST(1).

Some floating-point instructions operate only on the ST(0) register. The FXCH instruction provides a convenient method of exchanging the contents of the top-of-stack register ST(0) with another register in the stack.

The condition code flags that are affected in the floating-point unit status word are as follows: C1 (bit 9) is reset if a stack underflow has occurred; otherwise, C1 is set; C0 (bit 8), C2 (bit 10), and C3 (bit 14) are undefined. The syntax for the two versions of the FXCH instruction are shown below.

```
FXCH   ST(i)  (exchange the contents of ST(0) and ST(i))

FXCH          (exchange the contents of ST(0) and ST(1))
```

Figure 11.38 illustrates an assembly language module embedded in a C program that demonstrates the use of some of the additional instructions described in this section. The program uses the *change sign* FCHS instruction, the *round to integer* FRNDINT instruction, and the *square root* FSQRT instruction. A positive or negative floating-point number is entered from the keyboard and used with all three instructions.

The first number entered is +30. The sign is changed, the number is rounded to 30.000000 using the default mode of *round to nearest*, and a square root of 5.477226 is generated. The second number entered is +625.789, which is rounded up to 626.000000 and yields a square root of 25.015776. The third number entered is

+75.498, which is rounded down to 75.000000 and yields a square root of 8.688958. The fourth number entered is –25.473, which is rounded down to –25.000000. The square root of –25.473 is not a real number, since there is no real number with a square of –25.473. Therefore, the FSQRT instruction specifies a floating-point invalid-arithmetic-operand exception, which is indicated in the outputs as –1.#IND00 in this version of the Visual C++ software.

```
//additional_instr.cpp
//uses the FCHS, FRNDINT, and FSQRT instructions

#include "stdafx.h"

int main (void)
{
   float flp1, fchs_rslt, round_rslt, sq_root_rslt;

   printf ("Enter a floating-point number: \n");
   scanf ("%f", &flp1);

//switch to assembly
      _asm
      {
         FLD    flp1            //flp1 -> ST(0)
         FCHS                   //change sign of flp1
         FST    fchs_rslt       //ST(0) -> fchs_rslt
//------------------------------------------------------------
         FLD    flp1            //flp1 -> ST(0)
         FRNDINT                //integer -> ST(0)
         FST    round_rslt      //ST(0) -> round_rslt
//------------------------------------------------------------
         FLD    flp1            //flp1 -> ST(0)
         FSQRT                  //square root -> ST(0)
         FST    sq_root_rslt    //ST(0) -> sq_root_rslt
      }

//print result
   printf ("\nChange sign result = %f\n", fchs_rslt);
   printf ("Rounded result = %f\n", round_rslt);
   printf ("Square root result = %f\n\n", sq_root_rslt);
   return 0;
}                                    //continued on next page
```

(a)

Figure 11.38 Program to illustrate using the instructions FCHS, FRNDINT, and FSQRT: (a) the program and (b) the outputs.

```
Enter a floating-point number:
30.0

Change sign result = -30.000000
Rounded result = 30.000000
Square root result = 5.477226

Press any key to continue . . . _
------------------------------------------------------------
Enter a floating-point number:
625.789

Change sign result = -625.789001
Rounded result = 626.000000
Square root result = 25.015776

Press any key to continue . . . _
------------------------------------------------------------
Enter a floating-point number:
75.498

Change sign result = -75.498001
Rounded result = 75.000000
Square root result = 8.688958

Press any key to continue . . . _
------------------------------------------------------------
Enter a floating-point number:
-25.473

Change sign result = 25.473000
Rounded result = -25.000000
Square root result = -1.#IND00   //invalid operand exception

Press any key to continue . . . _
```

(b)

Figure 11.38 (Continued)

11.11 Problems

11.1 Convert $+19.5_{10}$ into a 32-bit single-precision floating-point number with a biased exponent and an implied 1.

11.2 Convert $+38.125_{10}$ into a 32-bit single-precision floating-point number with a biased exponent and an implied 1.

11.3 Obtain the unbiased exponent for the floating-point number shown below in which the exponent is biased.

s	exponent	fraction
0	1 1 0 1 0 1 1 0	1 0 0 0 0 0 0 0 ... 0 0 0 0

11.4 The floating-point number shown below has an unbiased exponent and an unnormalized fraction. Show the same floating-point number with a biased exponent and a normalized fraction in the single-precision floating-point format.

s	exponent	fraction
1	1 0 1 1 0 0 1 1	0 0 0 1 1 1 0 1 0 ... 0 0 0

11.5 For a 23-bit fraction with an 8-bit exponent and a sign bit, determine the most negative number with the most negative unbiased exponent.

11.6 Indicate whether the following statements are true or false:

(a) Adder-based rounding requires no more time than truncation.
(b) von Neumann rounding as also referred to as chopping.

11.7 Convert the decimal number 0.080078125_{10} into the 64-bit double-precision floating-point number.

11.8 Round the following floating-point number to eight bits using the rounding methods shown below:

.0001 0111 1100

(a) Chopping
(b) Adder-based
(c) von Neumann

11.9 Convert the single-precision floating-point number shown below to an equivalent decimal number. The exponent is biased.

s	exponent	significand
1	1 0 0 0 0 0 1 1	1 0 0 1 0 0 1 0 ... 0 0 0

↑ Implied 1

11.10 Write an assembly language module embedded in a C program that uses the *add* (FADDP) instruction and the *add* (FIADD) instruction. For the FADDP instruction, use the FADDP ST(*i*), ST(0) version. Enter three floating-point numbers and one integer number from the keyboard. Display the results of the program and show the register stack for each sequence of the program.

11.11 Perform an addition operation for the following floating-point numbers:

$$\begin{array}{rl} & 1.10100000 \times 2^4 \\ +) & 1.10001100 \times 2^6 \\ \hline \end{array}$$

11.12 Perform an addition operation for the following floating-point numbers:

$$\begin{array}{rl} & 0.00101100 \times 2^7 \\ +) & 0.00111000 \times 2^4 \\ \hline \end{array}$$

11.13 Perform an addition operation for the following floating-point numbers:

$$\begin{array}{rl} & 0.10010000 \times 2^6 \\ +) & 0.11110000 \times 2^2 \\ \hline \end{array}$$

11.14 Add the two floating-point numbers shown below.

31	exponent 24	fraction 0
0	0 0 0 0 0 1 1 0	1 1 0 0 1 0 0 0 0 ... 0 0 0
0	0 0 0 0 0 0 1 1	1 0 0 0 1 0 0 0 0 ... 0 0 0

11.15 Add the two floating-point numbers shown below.

$$
\begin{array}{rl}
 & A = 0\ .\ 1\ 0\ 0\ 1\ 1\ 0\ 0\ 0 \quad \times\ 2^{6} \\
+) & B = \underline{1\ .\ 1\ 0\ 1\ 0\ 0\ 0\ 0\ 0} \quad \times\ 2^{5}
\end{array}
$$

11.16 Convert the most negative 8-bit unbiased exponent to a biased exponent.

11.17 Write an assembly language module embedded in a C program that uses the *sub* (FSUBP) instruction and the *sub* (FISUB) instruction. For the FSUBP instruction, use the FSUBP ST(*i*), ST(0) version. Enter three floating-point numbers and one integer number from the keyboard. Enter the three separate sequences shown below. Display the results of the program and show the register stack for the third sequence of the program.

```
+1.2        +2.3        +3.4        +5
+32.456     +45.789     +16.123     +10
-237.658    +128.125    -279.463    -75
```

11.18 Perform the following operation on the two operands: (+127) – (–77).

11.19 Perform the following operation on the two operands: (–13) – (+54).

11.20 Perform the following operation on the two operands: (–47.25) – (–18.75).

11.21 Perform the following operation on the two operands: (+36.50) – (+5.75).

11.22 Write an assembly language module embedded in a C program that uses the *mul* (FMULP) instruction and the *mul* (FIMUL) instruction. For the FMULP instruction, use the FMULP ST(*i*), ST(0) version. Enter three floating-point numbers and one integer number from the keyboard. Enter the three separate sequences shown below. Display the results of the program and show the register stack for the third sequence of the program.

```
+1.200      +2.300      +3.400      +5
+32.400     +45.700     +16.100     +10
-23.600     +28.500     -27.400     -12
```

11.23 Write an assembly language module embedded in a C program to obtain the sum of cubes of *n* positive floating-point numbers. The sum of cubes can be represented by the following expression:

$$1^3 + 2^3 + 3^3 + 4^3 + \ldots + n^3$$

11.24 Comment on the biasing problem when the exponents are operated on during floating-point multiplication.

11.25 Write a C program to calculate the area of a flat surface using floating-point numbers. Enter three sets of floating-point numbers for the width and length.

11.26 Write an assembly language module embedded in a C program that calculates the area of a circle from a radius that is entered from the keyboard. Enter both integer radii and noninteger radii. Display the resulting areas.

11.27 How is quotient overflow determined in floating-point division? How is overflow resolved? After overflow is resolved, what is the numerical range of the dividend and divisor?

11.28 A resistor is one of four passive circuit elements: resistor, capacitor, inductor, and the recently discovered memristor. The equivalent resistance R_{eq} — specified in ohms — of resistors connected in series is the sum of the resistor values. However, the equivalent resistance of resistors connected in parallel is shown in the equation below.

$$\frac{1}{R_{eq}} = \frac{1}{R_1} + \frac{1}{R_2} + \frac{1}{R_3} + \cdots + \frac{1}{R_n}$$

The value of R_{eq} is smaller than the resistance of the smallest resistor in the parallel circuit. The circuit shown below contains three parallel resistors.

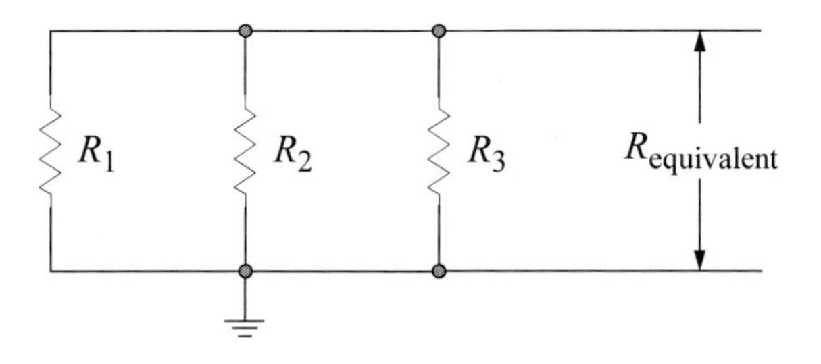

Write an assembly language module embedded in a C program that calculates the equivalent resistance of a three-resistor parallel network. Enter four sets of values for the resistors and display the equivalent resistance.

11.29 Using an assembly language module embedded in a C program, find the equivalent resistance of the circuit shown below for three sets of resistance values.

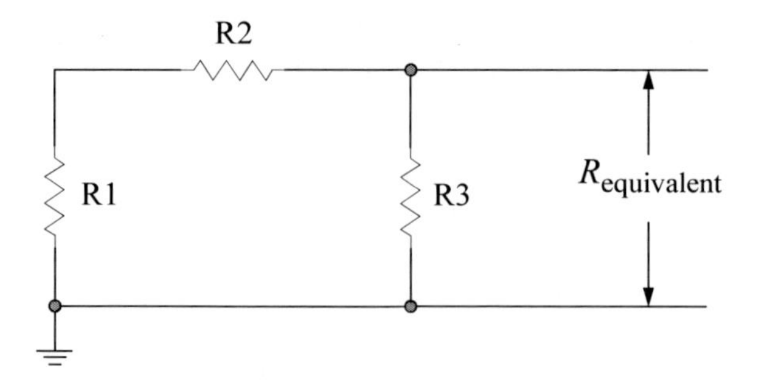

11.30 Write an assembly language module embedded in a C program to find the average of five floating-point numbers that are entered from the keyboard. Enter two sets of numbers. Display the sum and the average.

11.31 Given the program shown below, obtain the result of the program execution.

```
//fdivp_ex.cpp
//calculate result of program

#include "stdafx.h"
int main (void)
{
   float rslt;

//switch to assembly
      _asm
      {
         FLD1
         FADD  ST(0), ST(0)
         FLDPI
         FMUL  ST(0), ST(0)
         FDIVP ST(1), ST(0)
         FST   rslt
      }

//print result
   printf ("\nResult = %f\n\n", rslt);

   return 0;
}
```

11.32 Write an assembly language module embedded in a C program that calculates the sine, cosine, and tangent of a radian value that is entered from the keyboard. Enter radian values that correspond to the following degrees: 28.648°, 45°, 85.944°, 180°, 225°, 315°.

11.33 Write an assembly language module embedded in a C program that calculates the result of the expression shown below for different values of the floating-point variables *flp*1 and *flp*2.

$$\frac{1}{2\pi \sqrt{flp1 \times flp2}}$$

12

Procedures

A *procedure* is a set of instructions that perform a specific task. They are invoked from another procedure — the calling procedure (or program) — and provide results to the calling program at the end of execution. Procedures (also called subroutines) are utilized primarily for routines that are called frequently by other procedures. The procedure routine is written only once, but used repeatedly; thereby, saving storage space. Procedures permit a program to be coded in modules; thus, making the program easier to code and test.

A program can contain many procedures. Each procedure is delimited by the PROC and ENDP directives and contains identical names for each directive, as shown below. Each procedure must have a unique name.

```
Calc  PROC     NEAR/FAR
        .
        .
        .
Calc  ENDP
```

Comments are normally placed at the beginning of a procedure to indicate the purpose of the procedure. A procedure is invoked from a program by means of the CALL instruction, which can occur anywhere in the calling program, and terminated by the RET instruction, which returns control to the calling program. Control is returned to the instruction that immediately follows the CALL instruction in the invoking program. A calling procedure can pass parameters or the addresses of the parameters on the stack to be used by the invoked procedure or pass parameters by placing them in general-purpose registers that can be accessed by the invoked procedure.

12.1 Call a Procedure

The CALL instruction is used to call a procedure, either in the current code segment (near call — also referred to as an intrasegment call) or in a different code segment (far call — also referred to as an intersegment call). The syntax for a CALL instruction is shown below.

```
CALL    procedure_name_label
```

A *near call* pushes the value of the updated (E)IP register onto the stack; thus, it points to the instruction that immediately follows the CALL instruction — this instruction is executed upon returning from the called procedure to the calling program. Then a branch occurs to the destination (target) address. The destination address for a near call can be either an absolute offset or a relative offset.

An *absolute offset* points to an address that is an offset from the base of the current code segment. An absolute offset is obtained indirectly through a general-purpose register or from a memory location, as shown below using the EBX register for an indirect call. The contents of the EBX register contains an offset in memory.

```
CALL    [EBX]
```

The absolute offset is stored in the (E)IP register; if the operand-size attribute is 16, then the high-order half of the EIP register is reset. A *relative offset* is a signed displacement that is added to the (E)IP register and is usually specified as a label that is encoded as an immediate value. The CS register is not changed for near calls, because the target address is in the current code segment.

A *far call* pushes the CS register onto the stack and the updated value of the (E)IP register onto the stack. Then the far call loads the CS register with the segment selector of the invoked procedure and loads the offset of the invoked procedure into the (E)IP register. The target address for the called procedure is a far address obtained directly from a pointer in the instruction or indirectly through a memory location.

With the *direct method*, the target address of the invoked procedure — containing the segment and offset — is encoded in the instruction using either a 4-byte or a 6-byte immediate value. For a 4-byte immediate value, the low-order two bytes are loaded into the IP register and the high-order two bytes are loaded into the CS register. For a 6-byte immediate value, the low-order four bytes are loaded into the EIP register and the high-order two bytes are loaded into the CS register. A branch is then executed to the subroutine.

With the *indirect method*, the target address is located in memory and contains either a 4-byte or a 6-byte address. For a 4-byte value, the low-order two bytes are loaded into the IP register and the high-order two bytes are loaded into the CS register. For a 6-byte value, the low-order four bytes are loaded into the EIP register and the high-order two bytes are loaded into the CS register. A branch is then executed to the subroutine.

12.2 Return from a Procedure

The RET instruction transfers control from an invoked procedure to the instruction immediately following the CALL instruction in the calling procedure. When executing an intrasegment *near return*, the processor pops the value at the stack top to the (E)IP register, adds an optional immediate value to the stack pointer, then continues execution of the invoking procedure.

When executing an intersegment *far return*, the processor pops the value at the stack top to the (E)IP register, then pops the value at the new stack top to the CS register and adds an optional immediate value to the stack pointer, then continues execution of the invoking procedure. The immediate optional value is used to remove parameters from the stack that were placed on the stack by the invoking procedure. The syntax for a RET instruction is shown below.

```
RET    optional_immediate_value
```

12.3 Passing Parameters to a Procedure

Parameters (arguments) can be passed to a subroutine (procedure) by means of the stack, general-purpose registers, or by placing the addresses of the parameters in general-purpose registers. Since the contents of the GPRs are not saved prior to executing a CALL instruction, all six GPRs — EAX, EBX, ECX, EDX, ESI, and EDI — can be used to pass parameters to the invoked procedure. Registers ESP and EBP, however, cannot be used to pass parameters. The calling program can save the GPRs on the stack, in memory, or in a data segment before calling the procedure.

Passing parameters using the stack The code segment in Figure 12.1 illustrates using the stack to pass parameters to the called procedure. The calling program passes the augend and addend to a procedure by pushing them onto the stack. The invoked procedure then performs the addition operation and returns the sum to the calling program in the EAX register.

```
PUSH  AUGEND          ADD_PROC    PROC  NEAR
PUSH  ADDEND                      PUSH  EBP
CALL  ADD_PROC                    MOV   EBP, ESP
      .                           MOV   EAX, [EBP + 12]
      .                           ADD   EAX, [EBP + 8]
      .                           POP   EBP
                                  RET   8
                      ADD_PROC    ENDP
                              (a)   //continued on next page
```

Figure 12.1 Program segment to illustrate passing parameters to an invoked procedure using the stack; the procedure adds two operands and returns the sum: (a) the program segment and (b) the stack.

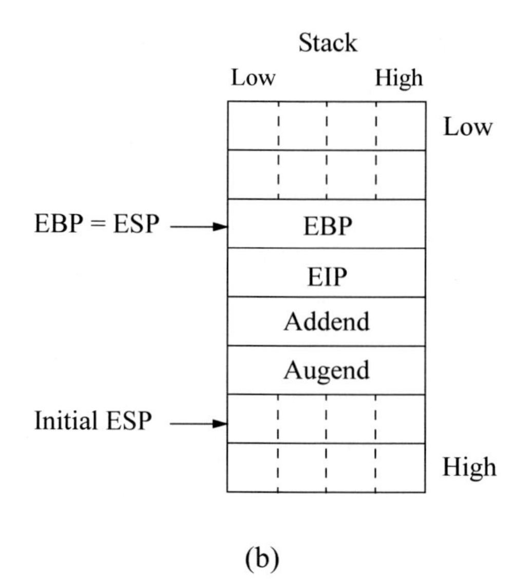

Figure 12.1 (Continued)

The calling program pushes the augend and the addend onto the stack, then calls the ADD_PROC procedure. The procedure is a near procedure; therefore, only the EIP register is placed on the stack. The invoked procedure pushes the EBP register onto the stack — the EBP register will be used to access the augend and the addend. Then the value of the ESP register — which points to the location of EBP on the stack — is moved to EBP. The EBP register can now be used to access data on the stack.

The augend is then moved to register EAX by the instruction shown below. The augend is located 12 bytes above the address of EBP (ESP); therefore, a value of 12 is added to the contents of the EBP register.

```
MOV    EAX, [EBP + 12]
```

An ADD instruction then adds the contents of register EAX (augend) to the contents of the stack at location EBP + 8 (addend) and places the sum in register EAX. Then the initial contents of register EBP are popped off the stack and stored in EBP. A near return instruction is then executed, which pops EIP off the stack; register ESP now points to the addend. In order to restore the contents of the ESP register to its initial value, an immediate value of eight is added to the ESP register upon returning to the calling procedure. The invoked procedure is delimited by the PROC and ENDP directives, both of which contain the name of the invoked procedure.

Passing parameters using general-purpose registers All six general-purpose registers — EAX, EBX, ECX, EDX, ESI, and EDI — can be used to transfer parameters to an invoked procedure. The operands are moved to the applicable registers prior to calling the procedure; the procedure can then access the registers directly to perform the specified operation. Since the stack is not used to store the operands before calling the procedure, there is no immediate value added to a near return. Figure 12.2 shows a program segment to add two operands using a near procedure. Although this program segment is relatively simple, it illustrates how general-purpose registers can be used to pass parameters to an invoked procedure.

```
            MOV   EAX, AUGEND
            MOV   EBX, ADDEND
            CALL  ADD_PROC
                  .
                  .
                  .
ADD_PROC    PROC  NEAR
            ADD   EAX, EBX
            RET
ADD_PROC    ENDP
```

Figure 12.2 A near procedure to add two operands by passing the operands to the procedure using general-purpose registers.

Passing parameters by indirect addressing In this method, the effective addresses of the parameters are loaded into general-purpose registers prior to calling the procedure. The procedure then uses the indirect addressing mode to access the parameters. Figure 12.3 uses this technique to add two operands.

```
            LEA   ECX, AUGEND
            LEA   EDX, ADDEND
            CALL  ADD_PROC
                  .
                  .
                  .
ADD_PROC    NEAR
            MOV   EAX, [ECX]
            ADD   EAX, [EDX]
            RET
ADD_PROC    ENDP
```

Figure 12.3 A near procedure to add two operands by passing the addresses of the operands to the procedure using general-purpose registers.

Example 12.1 This example provides a sequence of calls and returns to and from near and far procedures. There will be a total of four calls and four returns, as shown below. There is a main program, a far procedure called PROC_A, a near procedure called PROC_B, and a near procedure called PROC_C.

1. The first operation is a **call** from the main program to far PROC_A.

2. The second operation is a **call** from PROC_A to near PROC_B.

3. The third operation is a **call** from PROC_B to near PROC_C.

4. The fourth operation is a **return** from PROC_C to PROC_B.

5. The fifth operation is a **return** from PROC_B to PROC_A.

6. The sixth operation is a **call** from PROC_A to near PROC_C.

7. The seventh operation is a **return** from PROC_C to PROC_A.

8. The eighth operation is a **return** from PROC_A to the main program.

The above eight sequence of operations is shown pictorially in Figure 12.4.

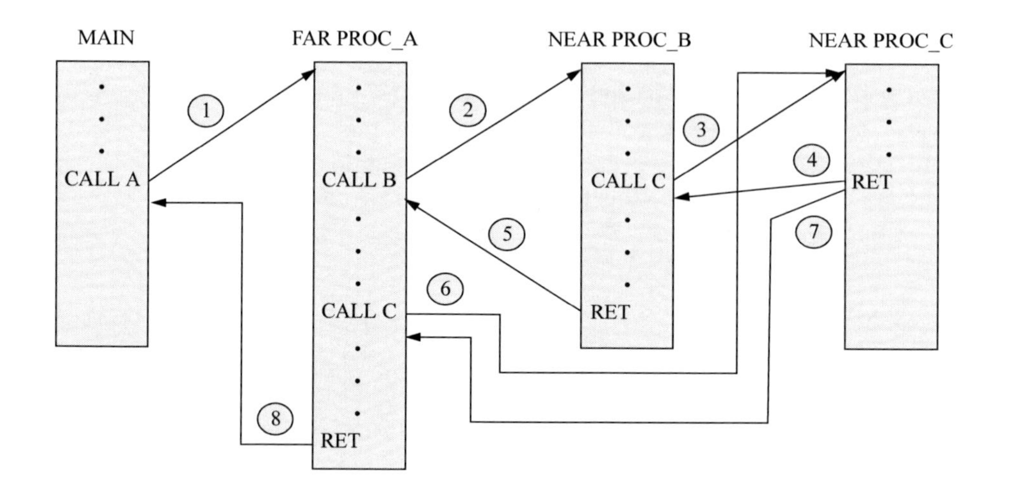

Figure 12.4 Sequence of calls and returns to and from near and far procedures.

The eight sequences for the various operations together with applicable addresses for the CS register and the EIP register are shown below. Figure 12.5 shows the sequence of operations and the contents of the stack for each operation.

1. MAIN program calls far procedure PROC_A.
   MAIN program return address from PROC_A call: CS = 0700, EIP = 2500.

2. PROC_A calls near PROC_B.
   PROC_A return address from PROC_B call: EIP = 4000.

3. PROC_B calls near PROC_C.
   PROC_B return address from PROC_C call: EIP = 5500.

4. Return to PROC_B.

5. Return to PROC_A.

6. PROC_A calls near PROC_C.
   PROC_A return address from PROC_C call: EIP = 6250.

7. Return to PROC_A.

8. Return to main program.

Before call to PROC_A:

ESP ⟶ TOS | (empty) |

---

1. After **call** to far PROC_A from MAIN:

	Stack	
ESP ⟶ TOS	2500	Push EIP
	0700	Push CS

---

2. After **call** to near PROC_B from PROC_A:

	Stack	
ESP ⟶ TOS	4000	Push EIP (PROC_A return addr from PROC_B)
	2500	
	0700	

---

3. After **call** to near PROC_C from PROC_B:

	Stack	
ESP ⟶ TOS	5500	Push EIP (PROC_B return addr from PROC_C)
	4000	
	2500	
	0700	

//Continued on next page

Figure 12.5 Sequence of operations and the contents of the stack for each operation.

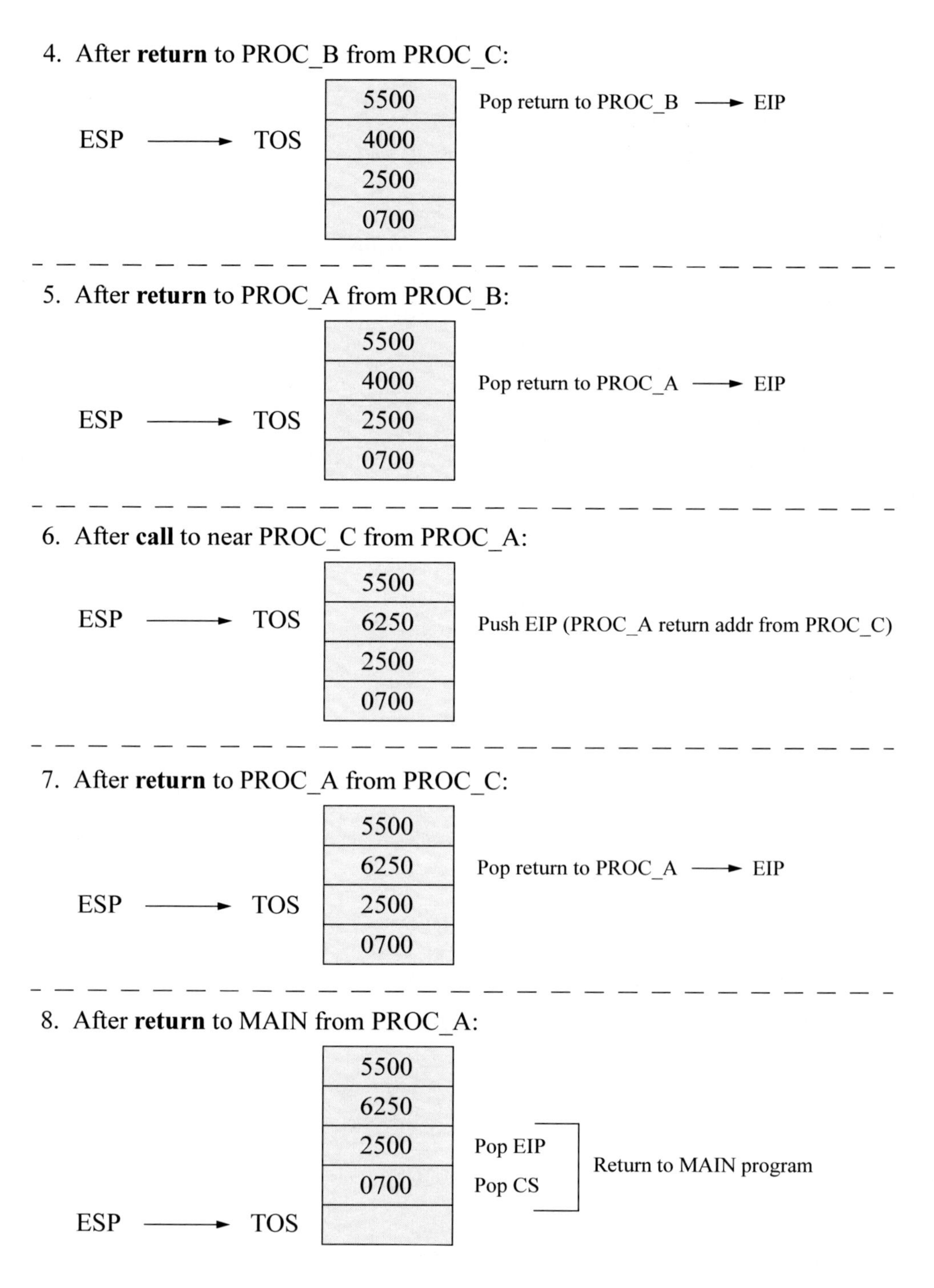

Figure 12.5 (Continued)

The parameter list is reproduced in Figure 12.6 for convenience. The name PARLST (parameter list) in the data segment of a program is the name of a one-dimensional array that is labelled as a byte array and accepts input data from the keyboard.

The first array element (MAXLEN) specifies the maximum number of characters that can be entered; the second array element (ACTLEN) indicates the actual number of characters that were entered from the keyboard; the third element of the array (OPFLD) contains the beginning of the operand field where the operands from the keyboard are stored — the last byte in the operand field is the *Enter* (carriage return) character (↵).

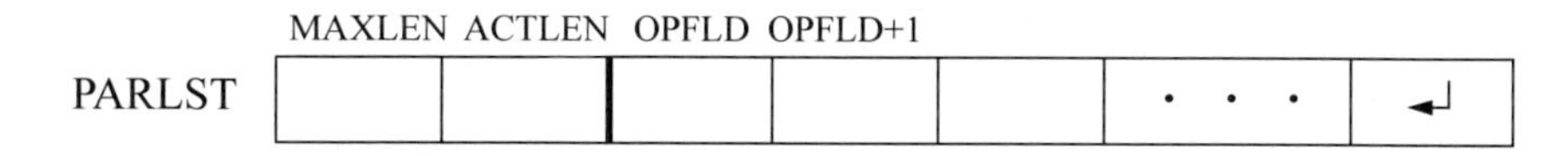

Figure 12.6 Parameter list one-dimensional array in which the keyboard input data are stored.

Figure 12.7 shows an assembly language program — not embedded in a C program — that illustrates the application of a procedure to add two single-digit numbers that are entered from the keyboard. The procedure utilizes the *ASCII adjust for addition* (AAA) instruction. The two numbers are prepared for the AAA instruction, then pushed onto the stack to be added by the CALC procedure.

Recall that the AAA instruction adjusts the result of an addition operation of two ASCII operands. ASCII numerical digits have a high-order four bits of 3H when entered from the keyboard. The AAA instruction produces an unpacked BCD number which contains zeroes in the high-order four bits of the byte; the numerical value of the number is contained in the low-order four bits of the byte. General-purpose register AL is the implied source and destination for the AAA instruction.

The sum obtained from adding the two ASCII numbers may not be a valid BCD value. If the low-order four bits of the sum in register AL are greater than nine or if the AF flag is set, then six is added to AL and register AH is incremented by 1. The AAA instruction converts the sum in register AL into a valid BCD number and stores the result in bits 3 through 0 of register AL and resets bits 7 through 4 of register AL.

Shown below is an example in which the low-order four bits of the result are not a valid BCD number; that is, a digit (C) that is greater than nine. When used in a program, the AAA instruction is preceded by an ADD instruction or an ADC instruction.

AX =	0 0	3 8	ASCII 8
+) BX =	0 0	3 4	ASCII 4
AX =	0 0	6C	Low half of AL > 9
AAA	1	6	
	0 1	0 2	After AAA

```
;add_proc2.asm

;----------------------------------------------------------------
.STACK

;----------------------------------------------------------------
.DATA
PARLST  LABEL BYTE
MAXLEN  DB 5
ACTLEN  DB ?
OPFLD   DB 5 DUP(?)
PRMPT   DB 0DH, 0AH, 'Enter two single-digit integers: $'
RSLT    DB 0DH, 0AH, 'Sum =        $'

;----------------------------------------------------------------
.CODE
BEGIN   PROC FAR

;set up pgm ds
        MOV    AX, @DATA     ;get addr of data seg
        MOV    DS, AX

;read prompt
        MOV    AH, 09H       ;display string
        LEA    DX, PRMPT
        INT    21H           ;dos interrupt

;kybd rtn to enter chars
        MOV    AH, 0AH       ;buffered kybd input
        LEA    DX, PARLST
        INT    21H

;get the two integers
        MOV    AL, OPFLD     ;get 1st digit from opfld
        MOV    AH, 00H       ;clear ah
        PUSH   AX

        MOV    BL, OPFLD+1 ;get 2nd digit from opfld
        MOV    BH, 00H       ;clear bh
        PUSH   BX

        CALL   CALC          ;push ip.  call calc rtn

                                    //continued on next page
```

(a)

Figure 12.7 Program to illustrate the application of a procedure: (a) the program and (b) the outputs.

```
;return to here after calc procedure
;move the sum to result
      MOV   RSLT+8, AH
      MOV   RSLT+9, AL

;display the resulting sum
      MOV   AH, 09H      ;display string
      LEA   DX, RSLT
      INT   21H

BEGIN ENDP

;------------------------------------------------------------
CALC  PROC NEAR
      PUSH  BP           ;bp will be used
      MOV   BP, SP       ;bp points to tos
      MOV   AX, [BP+4]   ;get opnd a (addend)
      ADD   AX, [BP+6]   ;add opnd b.  sum to ax
      AAA                ;ascii adjust for addition
      OR    AX, 3030H    ;convert to ascii
      POP   BP           ;restore bp
      RET   4            ;pop ip. add 4 to sp so that
                         ;sp points to initial location
CALC  ENDP

;------------------------------------------------------------
      END   BEGIN
```

```
Enter two single-digit integers: 23
Sum = 05
------------------------------------------------------------
Enter two single-digit integers: 54
Sum = 09
------------------------------------------------------------
Enter two single-digit integers: 56
Sum = 11
------------------------------------------------------------
Enter two single-digit integers: 78
Sum = 15
------------------------------------------------------------
Enter two single-digit integers: 99
Sum = 18
```

(b)

Figure 12.7 (Continued

The register stack for the program of Figure 12.7 is shown in Figure 12.8. The calling program first pushes the operand in AX onto the stack and then pushes the operand in BX onto the stack. The CALL to the near CALC procedure pushes IP onto the stack. The procedure then pushes the base pointer, BP, onto the stack to be used as an offset in the stack segment. In order to accomplish this, BP is made equal to the stack pointer, SP, so that BP points to the top of stack.

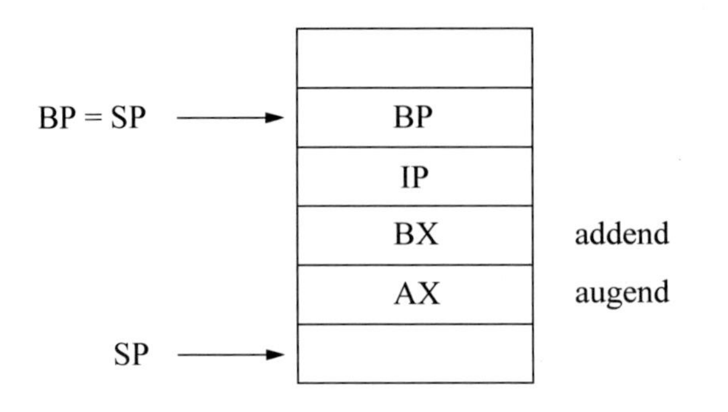

Figure 12.8 Stack usage for the program of Figure 12.7.

12.4 Problems

12.1 Write an assembly language program — not embedded in a C program — that uses a procedure to multiply two single-digit operands. Enter several operands for the multiplicand and multiplier and display the products.

12.2 Write an assembly language program — not embedded in a C program — that uses a procedure to obtain the area of a triangle from two integers that are entered from the keyboard. Enter several sets of single-digit numbers for the base and height and display the areas.

12.3 Calculate the area of a triangle using an assembly language module embedded in a C program. Floating-point numbers for the base and height are entered from the keyboard. In this problem, a procedure is not necessary. Unlike Problem 12.2, odd-valued bases will not be truncated. Enter several floating-point numbers for the base and height and display the areas.

12.4 Write an assembly language program — not embedded in a C program — that uses a procedure to exclusive-OR six hexadecimal characters that are entered from the keyboard. The following characters are exclusive-ORed: the first and third; the second and fourth; the third and fifth; and the fourth and sixth. The keyboard data can be any hexadecimal characters; for example, 2T/}b*.

12.5 Write an assembly language program — not embedded in a C program — that uses a procedure to convert an 8-bit binary code number to the corresponding Gray code number. The binary number is entered from the keyboard.

13

String Instructions

A string is a sequence of bytes, words, or doublewords that are stored in contiguous locations in memory as a one-dimensional array. Strings can be processed from low addresses to high addresses or from high addresses to low addresses, depending on the state of the *direction flag* (DF). If the direction flag is set (DF = 1), then the direction of processing is from high addresses to low addresses — also referred to as *auto-decrement*. If the direction flag is reset (DF = 0), then the direction of processing is from low addresses to high addresses — also referred to as *auto-increment*. The state of the direction flag can be set by the *set direction flag* (STD) instruction and can reset by the *clear direction flag* (CLD) instruction. The direction flag is located in bit 10 of the 32-bit EFLAGS register, which is reproduced below in Figure 13.1.

31	30	29	28	27	26	25	24
0	0	0	0	0	0	0	0

23	22	21	20	19	18	17	16
0	0	ID	VIP	VIF	AC	VM	RF

15	14	13	12	11	**10**	9	8
0	NT	IOPL		OF	**DF**	IF	TF

// continued on next page

Figure 13.1 EFLAGS register.

7	6	5	4	3	2	1	0
SF	ZF	0	AF	0	PF	1	CF

Figure 13.1 (Continued)

There are several instructions that operate specifically on strings. These include the *compare string operands* instructions: CMPS, CMPSB, CMPSW, CMPSD, and CMPSQ; the *load string* instructions: LODS, LODSB, LODSW, LODSD, and LODSQ; the *move data from string to string* instructions: MOVS, MOVSB, MOVSW, MOVSD, and MOVSQ; the *scan string* instructions: SCAS, SCASB, SCASW, and SCASD; and the *store string* instructions: STOS, STOSB, STOSW, STOSD, and STOSQ. All of the above instructions will be described in later sections.

There are additional instructions, flags, and registers that are associated with the string instructions. These include the *repeat string operation prefix* instructions: REP, REPE, REPZ, REPNE, and REPNZ; the *clear direction flag* (CLD) instruction and the *set direction flag* (STD) instruction, both of which were previously described; the general-purpose registers (E)SI and (E)DI, which are the source pointer and the destination pointers for string operations, respectively; the (E)CX register, which is the counter for certain string operations; and the general-purpose registers AL, AX, and EAX.

13.1 Repeat Prefixes

The *repeat* prefixes are placed before the string instruction and specify the condition for which the instruction is to be executed. The general-purpose registers (E)SI and (E)DI are automatically incremented or decremented after each execution of the string instruction to point to the next byte, word, or doubleword in the string.

The direction flag (DF) determines whether the registers are incremented (DF = 0) or decremented (DF = 1). As mentioned previously, the state of the direction flag is determined by the *set direction flag* (STD) instruction and the *clear direction flag* (CLD) instruction. When the string operation is completed, the (E)SI and (E)DI registers point to the first data element after or before the string. The repeat prefix causes successive iterations of the string instruction until the condition stipulated by the prefix is fulfilled. The general-purpose register (E)CX is also used to determine the cessation of the string instruction by specifying a count to indicate the number of iterations of the string instruction. The repeat prefixes apply only to the instruction that they immediately precede.

If a block of instructions is to be executed, then a LOOP instruction can be utilized. A string operation can be delayed by an exception or an interrupt, in which case, the registers are saved so that the operation can continue upon the completion of the exception or interrupt. Thus, the (E)SI and (E)DI registers point to the next string elements; the (E)IP register points to the string instruction; and the (E)CX register contains the count that it held prior to the exception or interrupt.

When operating in 64-bit mode, the source operand address and the destination operand address are stipulated by RSI (or ESI) and by RDI (or EDI), respectively, using an REX prefix. The count is contained in RCX or ECX depending on the address size attribute.

13.1.1 REP Prefix

The REP prefix allows a string instruction to be repeated a specified number of times as indicated by the count in register (E)CX. The REP prefix can be utilized with different versions of the following string instructions: MOVS, LODS, and STOS. The REP prefix can also be used with different versions of the input/output string instructions: *input from port to string* INS and *output string to port* OUTS.

The syntax for the REP prefix is shown below for the MOVS instruction, which moves string elements from the data segment source to the extra segment destination. Recall that the brackets stipulate indirect memory addressing. Thus, DS:[(E)SI] indicates a memory address in the data segment with an offset specified by the contents of register ESI or register SI. The memory address ES:[(E)DI] is similarly defined.

```
REP    MOVS m8, m8   (move (E)CX bytes from DS:[(E)SI]
                     to ES:[(E)DI])

REP    MOVS m8, m8   (move RCX bytes from [RSI]
                     to [RDI], 64-bit mode)

REP    MOVS m16, m16 (move (E)CX words from DS:[(E)SI]
                      to ES:[(E)DI])

REP    MOVS m32, m32 (move (E)CX doublewords from
                      DS:[(E)SI] to ES:[(E)DI])

REP    MOVS m64, m64 (move RCX quadwords from
                      [RSI] to [RDI], 64-bit mode)
```

Another version of the REP instruction is shown below for the LODS instruction, which loads string elements from the data segment to general-purpose registers. The REP prefix is normally not used. The syntax is shown below.

```
REP    LODS AL    (load (E)CX bytes from DS:[(E)SI]
                  to register AL)

REP    LODS AL    (load RCX bytes from [RSI]
                  to register AL, 64-bit mode)
```

```
REP    LODS AX   (load (E)CX words from
                 DS:[(E)SI] to register AX)

REP    LODS EAX  (load (E)CX doublewords from
                 DS:[(E)SI] to register EAX)

REP    LODS RAX  (load RCX quadwords from
                 [RSI] to register RAX, 64-bit mode)
```

Another version of the REP instruction is shown below for the STOS instruction, which stores string elements from a general-purpose register to the extra segment destination at ES:[(E)DI]. The syntax is shown below.

```
REP    STOS m8   (store (E)CX bytes from
                 register AL to ES:[(E)DI])

REP    STOS m8   (store RCX bytes from
                 register AL to [RDI], 64-bit mode)

REP    STOS m16  (store (E)CX words from
                 register AX to ES:[(E)DI])

REP    STOS m32  (store (E)CX doublewords from
                 register EAX to ES:[(E)DI])

REP    STOS m64  (store RCX quadwords from
                 register RAX to [RDI], 64-bit mode)
```

13.1.2 REPE / REPZ Prefix

Another version of the REP prefix is the *repeat while equal* (REPE) prefix. The REPE prefix can be used to find nonmatching string elements in memory location DS:[(E)SI] by comparing them with string elements in memory location ES:[(E)DI]. The operation continues as long as the count in register (E)CX is nonzero and the string elements are equal — zero flag (ZF) equals 1. The REPZ prefix is synonymous with the REPE prefix. The REPE / REPZ prefixes and the REPNE / REPNZ prefixes (covered in the next section) are used only with the *compare string operands* (CMPS) instruction and the *scan string* (SCAS) instruction. The syntax for the REPE prefix is shown below for the CMPS instruction.

```
REPE    CMPS m8, m8   (compare bytes in DS:[(E)SI]
                      with bytes in ES:[(E)DI]
                      to find nonmatching bytes)

REPE    CMPS m8, m8   (compare bytes in [RSI]
                      with bytes in [RDI]
                      to find nonmatching bytes)
```

```
REPE    CMPS m16, m16   (compare words in DS:[(E)SI]
                        with words in ES:[(E)DI]
                        to find nonmatching words)

REPE    CMPS m32, m32   (compare doublewords in DS:[(E)SI]
                        with doublewords in ES:[(E)DI]
                        to find nonmatching doublewords)

REPE    CMPS m64, m64   (compare quadwords in [RSI]
                        with quadwords in [RDI]
                        to find nonmatching quadwords,
                        64-bit mode)
```

Another application of the REPE instruction is shown below for the SCAS instruction, which compares the destination string element in ES:[(E)DI] with the contents of a general-purpose register and sets the status flags in the EFLAGS register based on the result. This REPE prefix, together with the SCAS instruction, is used to find string elements that do not match the contents of a general-purpose register. The syntax is shown below.

```
REPE    SCAS m8   (compare a byte in AL with
                  a byte in ES:[(E)DI]
                  to find nonmatching bytes)

REPE    SCAS m8   (compare a byte in AL with
                  a byte in [RDI]
                  to find nonmatching bytes,
                  64-bit mode)

REPE    SCAS m16 (compare a word in AX with
                  a word in ES:[(E)DI]
                  to find nonmatching words)

REPE    SCAS m32 (compare a doubleword in EAX with
                  a doubleword in ES:[(E)DI]
                  to find nonmatching doublewords)

REPE    SCAS m64 (compare a quadword in RAX with
                  a quadword in [RDI]
                  to find nonmatching quadwords,
                  64-bit mode)
```

13.1.3 REPNE / REPNZ Prefix

Another version of the REP prefix is the *repeat while not equal* (REPNE) prefix. The REPNE prefix can be used to find matching string elements in memory location

DS:[(E)SI] by comparing them with string elements in memory location ES:[(E)DI]. The operation continues as long as the count in register (E)CX is nonzero and the string elements are not equal — zero flag (ZF) equals 0. The REPNZ prefix is synonymous with the REPNE prefix. The REPE / REPZ prefixes and the REPNE / REPNZ prefixes are used only with the *compare string operands* (CMPS) instruction and the *scan string* (SCAS) instruction. The syntax for the REPNE prefix is shown below for the CMPS instruction.

```
REPNE   CMPS m8, m8    (compare bytes in DS:[(E)SI]
                       with bytes in ES:[(E)DI]
                       to find matching bytes)

REPNE   CMPS m8, m8    (compare bytes in [RSI]
                       with bytes in [RDI]
                       to find matching bytes,
                       64-bit mode)

REPNE   CMPS m16, m16  (compare words in DS:[(E)SI]
                       with words in ES:[(E)DI]
                       to find matching words)

REPNE   CMPS m32, m32  (compare doublewords in DS:[(E)SI]
                       with doublewords in ES:[(E)DI]
                       to find matching doublewords)

REPNE   CMPS m64, m64  (compare quadwords in [RSI]
                       with quadwords in [RDI]
                       to find matching quadwords,
                       64-bit mode)
```

Another application of the REPNE instruction is shown below for the SCAS instruction, which compares the destination string element in ES:[(E)DI] with the contents of a general-purpose register and sets the status flags in the EFLAGS register based on the result. This REPNE prefix, together with the SCAS instruction, is used to find string elements that match the contents of a general-purpose register. The syntax is shown below.

```
REPNE   SCAS m8   (compare a byte in AL with
                  a byte in ES:[(E)DI]
                  to find matching bytes)

REPNE   SCAS m8   (compare a byte in AL with
                  a byte in [RDI]
                  to find matching bytes, 64-bit mode)

REPNE   SCAS m16  (compare a word in AX with
                  a word in ES:[(E)DI]
                  to find matching words)
```

```
REPNE    SCAS m32 (compare a doubleword in EAX with
                  a doubleword in ES:[(E)DI]
                  to find matching doublewords)

REPNE    SCAS m64 (compare a quadword in RAX with
                  a quadword in [RDI]
                  to find matching quadwords,
                  64-bit mode)
```

13.2 Move String Instructions

The *move string* (MOVS) instructions transfer a string element — byte, word, or doubleword — from memory location DS:(E)SI to memory location ES:(E)DI. There are abbreviated mnemonics for the three different string element sizes: *move string byte* (MOVSB), *move string word* (MOVSW), and *move string doubleword* (MOVSD). In 64-bit mode, a quadword can be moved using the instruction *move string quadword* (MOVSQ).

After the transfer is completed, the (E)SI and (E)DI registers are automatically incremented or decremented depending on the state of the direction flag (DF) in the EFLAGS register. Figure 13.2 pictorially illustrates the MOVS instructions in transferring data from the source segment (DS) with offset (E)SI to the destination segment (ES) with offset (E)DI.

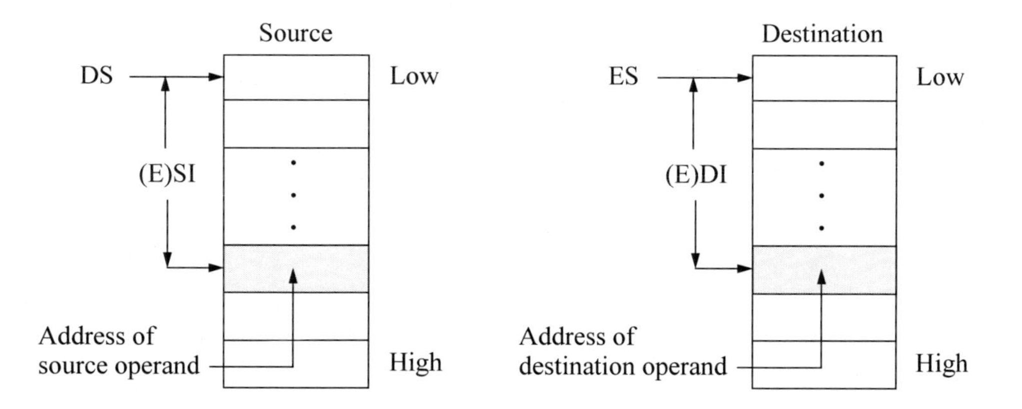

Figure 13.2 Illustration of transferring data from DS:(E)SI to ES:(E)DI.

The source and destination operand addresses should be initialized prior to the execution of the string instructions. This can be accomplished by using the *load far pointers* LDS and LES instructions as shown in Figure 13.3.

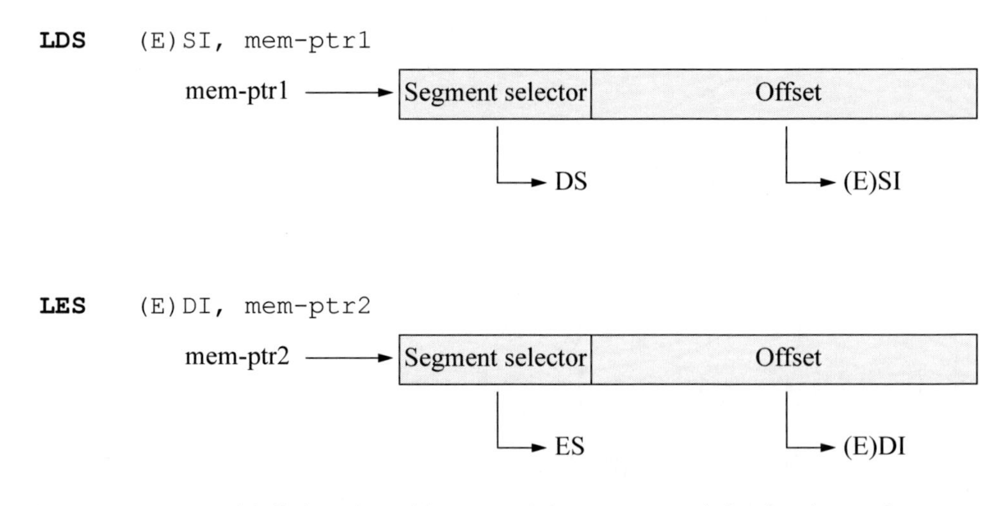

Figure 13.3 Initializing the addresses of the source and destination strings.

The data segment (DS) can be overridden with a *segment override prefix*. The segment override operator is specified by a colon (:). An example of a segment override prefix to move a word from general-purpose register BX to segment ES with an offset contained in general-purpose register AX is shown below. The extra segment (ES), however, cannot be overridden. Different versions of the move string instructions are described in the next sections.

```
MOV    ES:[AX], BX
```

13.2.1 Move Data from String to String (Explicit Operands) Instructions

This section describes the *explicit operands* form of the *move strings* instructions. The explicit operands form *explicitly* specifies the size of the source and destination operands as part of the instruction. The location of the source and destination operands are determined by contents of the DS:(E)SI and ES:(E)DI registers, respectively. There are four *move strings* instructions that have explicit operands, as shown in the syntax below.

```
MOVS   m8, m8   (transfer a byte from source in DS:(E)SI
                to destination in ES:E(DI);
                (R/E)SI and (R/E)DI for 64-bit mode)

MOVS   m16, m16 (transfer a word from source in DS:(E)SI
                 to destination in ES:E(DI);
                 (R/E)SI and (R/E)DI for 64-bit mode)
```

```
MOVS    m32, m32 (transfer a doubleword from source
                 in DS:(E)SI to destination in ES:E(DI);
                 (R/E)SI and (R/E)DI for 64-bit mode)

MOVS    m64, m64 (transfer a quadword from source in (R/E)SI
                 to destination in (R/E)DI, 64-bit mode)
```

13.2.2 Move Data from String to String (No Operands) Instructions

The *no-operands* form specifies the size of the operands by the mnemonic; for example, the MOVSW instruction specifies a word transfer. Like the explicit operand form, the no-operand form assumes that the location of the source and destination operands are determined by contents of the DS:(E)SI and ES:(E)DI registers, respectively. The syntax is shown below.

```
MOVSB    (transfer a byte from source in DS:(E)SI
         to destination in ES:E(DI);
         (R/E)SI and (R/E)DI for 64-bit mode)

MOVSW    (transfer a word from source in DS:(E)SI
         to destination in ES:E(DI);
         (R/E)SI and (R/E)DI for 64-bit mode)

MOVSD    (transfer a doubleword from source in DS:(E)SI
         to destination in ES:E(DI);
         (R/E)SI and (R/E)DI for 64-bit mode)

MOVSQ    (transfer a quadword from source in (R/E)SI
         to destination in (R/E)DI, 64-bit mode)
```

One application of the MOVS instruction is shown in Figure 13.4, which moves string operands from an input buffer to a working area in memory. This allows the system to operate on the relocated string data while the input buffer is being filled with additional string data.

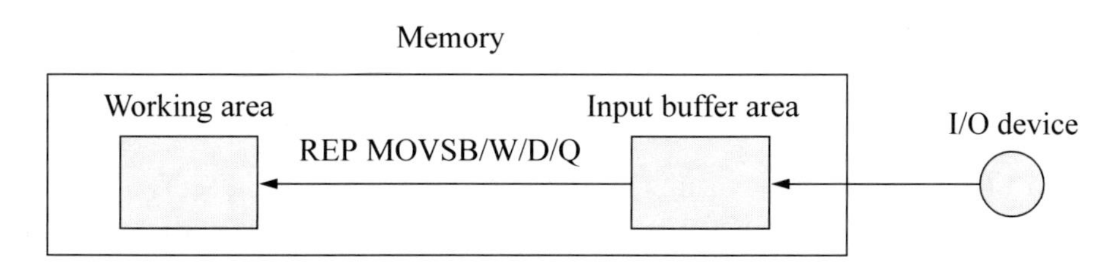

Figure 13.4 Application of a repeat move strings operation.

Examples of moving source strings to destination locations in which the strings are overlapping, or are in identical memory locations, or are nonoverlapping, are shown in Figure 13.5. Figure 13.5(a) illustrates overlapping strings, where the higher addresses of the source string overlap the lower addresses of the destination locations. In this case, the direction flag must be set (DF = 1) — *auto-decrement*; otherwise, the source string would overwrite some of the higher addresses of the source string.

In Figure 13.5(b), the strings are also overlapping, but in a reverse orientation to that of Figure 13.5(a). In this case, the direction flag must be reset (DF = 0) — *auto-increment*; otherwise, the source string would overwrite some of the lower addresses of the source string. In Figure 13.5(c), the source string and the destination locations are identical; therefore, there is no data transfer. In Figure 13.5(d) and Figure 13.5(e), the strings are nonoverlapping; the direction flag can be set or reset (DF = 1 or DF = 0).

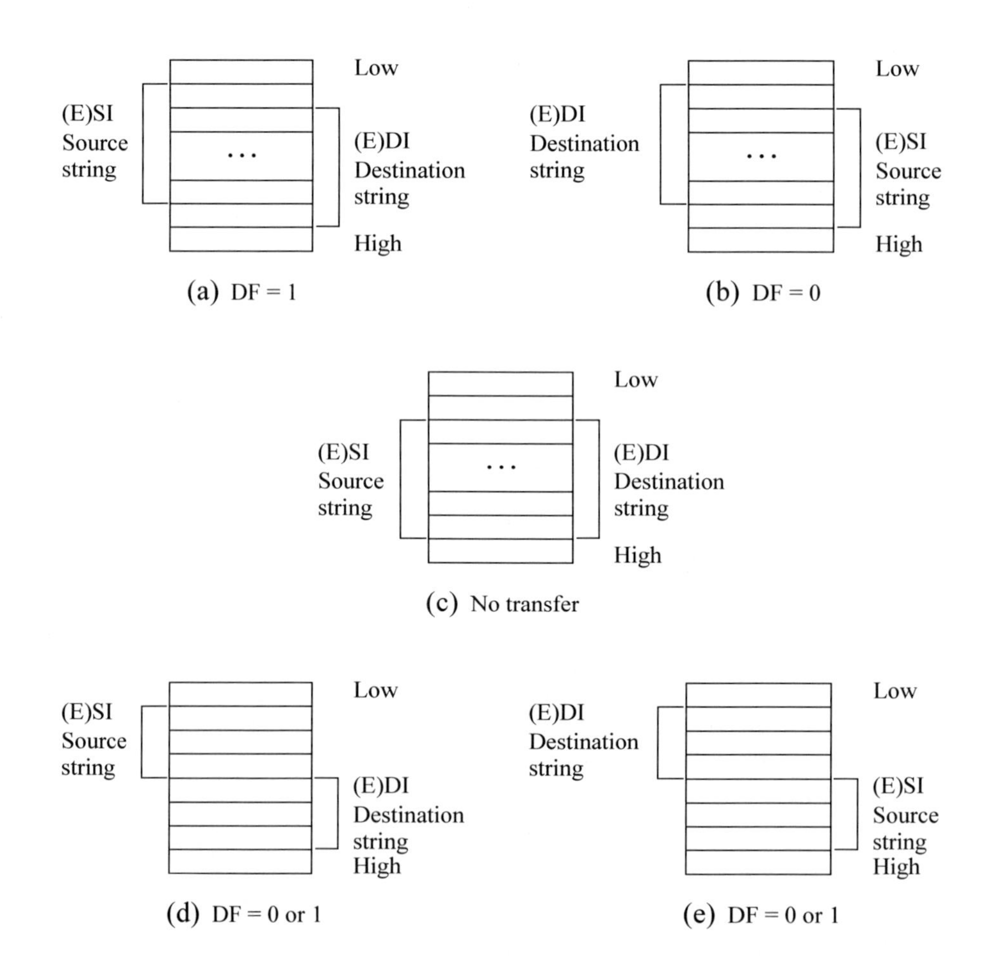

Figure 13.5 Different string orientations when executing a MOVS operation: (a) and (b) overlapping strings; (c) identical memory locations; and (d) and (e) nonoverlapping strings.

Figure 13.6 shows an assembly language program — not embedded in a C program — that illustrates using the MOVSB instruction with the REP prefix. The data segment (DS as source) and the extra segment (ES as destination) are made identical by the following instructions:

```
MOV    AX, @DATA     ;get addr of data seg
MOV    DS, AX        ;move addr to ds
MOV    ES, AX        ;move addr to es
```

The source index, SI, is assigned the address of OPFLD; the destination index, DI, is assigned the address of RSLT+15, which contains the string ABCDEFGHI. Nine hexadecimal characters are entered from the keyboard and stored in the OPFLD area. The program then moves the first five characters of OPFLD to the last five locations of the result area, effectively overwriting the last five locations. The result area is then displayed.

```
;movs_byte4.asm
;move the first 5 string elements in opfld
;to the result area last 5 string positions
;------------------------------------------------------------
.STACK

;------------------------------------------------------------
.DATA
PARLST    LABEL    BYTE
MAXLEN    DB   10
ACTLEN    DB   ?
OPFLD     DB   10   DUP(?)
PRMPT     DB   0DH, 0AH, 'Enter nine hexadecimal characters: $'
RSLT      DB   0DH, 0AH, 'Result = ABCDEFGHI     $'

;------------------------------------------------------------
.CODE
BEGIN PROC   FAR

;set up pgm ds
      MOV    AX, @DATA        ;get addr of data seg
      MOV    DS, AX           ;move addr to ds
      MOV    ES, AX           ;move addr to es

;read prompt
      MOV    AH, 09H          ;display string
      LEA    DX, PRMPT        ;put addr of prompt in dx
      INT    21H              ;dos interrupt
                              (a)     //continued on next page
```

Figure 13.6 Program to illustrate an application of using the MOVSB byte instruction with the REP prefix: (a) the program and (b) the outputs.

```
;keyboard request rtn to enter characters
       MOV    AH, 0AH          ;buffered keyboard input
       LEA    DX, PARLST       ;put addr of parlst in dx
       INT    21H              ;dos interrupt

;-------------------------------------------------------------
;set up source and destination addresses and
;move first 5 string elements of opfld to
;last 5 positions of rslt
       LEA    SI, OPFLD        ;addr of opfld -> si as src
       LEA    DI, RSLT+15      ;addr of rslt+15 -> di as dst
       MOV    CX, 5            ;count in cx

REP    MOVSB                   ;move bytes to dst

;-------------------------------------------------------------
;display result
       MOV    AH, 09H          ;display string
       LEA    DX, RSLT         ;put addr of rslt in dx
       INT    21H              ;dos interrupt

BEGIN ENDP
       END    BEGIN
```

```
Enter nine hexadecimal characters: 123456789
Result = ABCD12345
-------------------------------------------------------------
Enter nine hexadecimal characters: ABCDE6789
Result = ABCDABCDE
-------------------------------------------------------------
Enter nine hexadecimal characters: 111111111
Result = ABCD11111
```

(b)

Figure 13.6 (Continued)

13.3 Load String Instructions

The *load string* (LODS) instructions transfer a string element — byte, word, or doubleword — from memory location DS:(E)SI to registers AL, AX, or EAX, respectively. There are abbreviated mnemonics for the three different string element sizes: *load string byte* (LODSB), *load string word* (LODSW), and *load string doubleword* (LODSD). In 64-bit mode, a quadword can be loaded using the instruction *load string quadword* (LODSQ).

After the operand is loaded into register AL, AX, or EAX, the (E)SI register is automatically incremented or decremented depending on the state of the direction flag (DF) in the EFLAGS register. The source operand address should be initialized prior to the execution of the load string instruction. This can be accomplished by using the *load effective address* (LEA) instruction.

As mentioned in Section 13.2, the data segment (DS) can be overridden with a *segment override prefix*. The segment override operator is specified by a colon (:). The extra segment (ES), however, cannot be overridden. There is no need to use the REP prefix, because the previous data would be overwritten. The flags are not affected. Different versions of the load string instructions are described in the next sections.

13.3.1 Load String (Explicit Operands) Instructions

This section describes the *explicit operands* form of the *load string* instructions. The explicit operands form *explicitly* specifies the size of the source operand as part of the instruction. The location of the source operand is determined by contents of the DS:(E)SI register. There are four *load string* instructions that have explicit operands, as shown in the syntax below.

```
LODS   m8   (load a byte from source in DS:(E)SI
            into AL, (R)SI for 64-bit mode)

LODS   m16  (load a word from source in DS:(E)SI
            into AX, (R)SI for 64-bit mode)

LODS   m32  (load a doubleword from source in DS:(E)SI
            into EAX, (R)SI for 64-bit mode)

LODS   m64  (load a quadword from source in (R)SI
            into RAX, 64-bit mode)
```

13.3.2 Load String (No Operands) Instructions

The *no-operands* form specifies the size of the operands by the mnemonic; for example, the LODSB instruction specifies a *load byte into register AL* instruction. Like the explicit operand form, the no-operand form assumes that the location of the source operand is determined by contents of the DS:(E)SI register. The syntax is shown below.

```
LODSB   (load a byte from source in DS:(E)SI
        into AL, (R)SI for 64-bit mode)
```

LODSW (load a word from source in DS:(E)SI into AX, (R)SI for 64-bit mode)

LODSD (load a doubleword from source in DS:(E)SI into EAX, (R)SI for 64-bit mode)

LODSQ (load a quadword from source in (R)SI into RAX, 64-bit mode)

Figure 13.7 shows an assembly language program — not embedded in a C program — that illustrates an application using the LODSB (no operands) instruction. The program reverses the order of nine characters that are entered from the keyboard. The source index, SI, is assigned the address of OPFLD, which stores the nine characters; the destination index, DI, is assigned the address of RSLT+19, which contains the resulting reversed string. A count of nine is stored in register CX. Nine hexadecimal characters are entered from the keyboard and stored in the OPFLD area. A loop is used to read the nine characters and reverse their order. The result area is then displayed.

```
;lods_reverse.asm
;move the string elements in opfld to
;the result area in reverse order
;--------------------------------------------------------------
.STACK

;--------------------------------------------------------------
.DATA
PARLST   LABEL BYTE
MAXLEN   DB  10
ACTLEN   DB  ?
OPFLD    DB  10  DUP(?)
PRMPT    DB  0DH, 0AH, 'Enter nine hexadecimal characters: $'
RSLT     DB  0DH, 0AH, 'Result =                          $'

;--------------------------------------------------------------
.CODE
BEGIN PROC  FAR

;set up pgm ds
      MOV   AX, @DATA        ;get addr of data seg
      MOV   DS, AX           ;move addr to ds
                          (a)      //continued on next page
```

Figure 13.7 Program to illustrate an application of the LODSB instruction: (a) the program and (b) the outputs.

```
;read prompt
      MOV    AH, 09H          ;display string
      LEA    DX, PRMPT        ;put addr of prompt in dx
      INT    21H              ;dos interrupt

;keyboard request rtn to enter characters
      MOV    AH, 0AH          ;buffered keyboard input
      LEA    DX, PARLST       ;put addr of parlst in dx
      INT    21H              ;dos interrupt

;-------------------------------------------------------------
;set up source and destination addresses and
;move 9 string elements of opfld to
;rslt area in reverse order
      CLD                     ;left-to-right
      MOV    CX, 9            ;count in cx
      LEA    SI, OPFLD        ;addr of opfld -> si as src
      LEA    DI, RSLT+19      ;addr of rslt+19 -> di as dst

LP1:  LODSB                   ;load byte into al
      MOV    [DI], AL         ;move al to where di points
      DEC    DI               ;decrement di for reverse order
      LOOP   LP1              ;loop if cx != 0

;-------------------------------------------------------------
;display result
      MOV    AH, 09H          ;display string
      LEA    DX, RSLT         ;put addr of rslt in dx
      INT    21H              ;dos interrupt

BEGIN ENDP
      END    BEGIN
```

```
Enter nine hexadecimal characters: 123456789
Result = 987654321
-------------------------------------------------------------
Enter nine hexadecimal characters: IHGFEDCBA
Result = ABCDEFGHI
-------------------------------------------------------------
Enter nine hexadecimal characters: 987654321
Result = 123456789
-------------------------------------------------------------
Enter nine hexadecimal characters: abc123def
Result = fed321cba
```

(b)

Figure 13.7 (Continued)

13.4 Store String Instructions

The *store string* (STOS) instructions transfer a string element — byte, word, or doubleword — from registers AL, AX, or EAX to a destination memory location specified by ES:(E)DI. There are abbreviated mnemonics for the three different string element sizes: *store string byte* (STOSB), *store string word* (STOSW), and *store string doubleword* (STOSD). In 64-bit mode, a quadword can be loaded using the instruction *store string quadword* (STOSQ).

After the operand from register AL, AX, or EAX is stored in memory, the ES:(E)DI register is automatically incremented or decremented depending on the state of the direction flag (DF) in the EFLAGS register. The source operand address should be initialized prior to the execution of the load string instruction. This can be accomplished by using the *load effective address* (LEA) instruction.

The data segment (DS) can be overridden with a *segment override prefix*. The segment override operator is specified by a colon (:). The extra segment (ES), however, cannot be overridden. The REP prefix can be used to store a specific value from a general-purpose register into several contiguous areas in memory by setting the number of operands to be transferred into register (E)CX. The flags are not affected. Different versions of the store string instructions are described in the next sections.

13.4.1 Store String (Explicit Operands) Instructions

This section describes the *explicit operands* form of the *store string* instructions. The explicit operands form *explicitly* specifies the size of the destination operand as part of the instruction. The location of the destination operand is determined by contents of the ES:(E)DI register. There are four *store string* instructions that have explicit operands, as shown in the syntax below.

```
STOS    m8   (store a byte from AL to destination
              in ES:(E)DI, RDI or EDI for 64-bit mode)

STOS    m16  (store a word from AX to destination
              in ES:(E)DI, RDI or EDI for 64-bit mode)

STOS    m32  (store a doubleword from EAX to destination
              in ES:(E)DI, RDI or EDI for 64-bit mode)

STOS    m64  (store a quadword from RAX to destination
              in RDI or EDI for 64-bit mode)
```

13.4.2 Store String (No Operands) Instructions

The *no-operands* form specifies the size of the operands by the mnemonic; for example, the STOSW instruction specifies a *store word from register AX* instruction. Like the explicit operand form, the no-operand form assumes that the location of the destination operand is determined by contents of the ES:(E)DI register. The syntax is shown below.

```
STOSB    (store a byte from AL to destination
         in ES:(E)DI, RDI or EDI for 64-bit mode)

STOSW    (store a word from AX to destination
         in ES:(E)DI, RDI or EDI for 64-bit mode)

STOSD    (store a doubleword from EAX to destination
         in ES:(E)DI, RDI or EDI for 64-bit mode)

STOSQ    (store a quadword from RAX to destination
         in RDI or EDI for 64-bit mode)
```

One application of the store string instructions is to replace a string in contiguous locations in memory with a different string. Figure 13.8 shows an assembly language program — not embedded in a C program — that replaces a string of nine characters with asterisks. The program uses the store string (no operands) instruction with the REP prefix. The initial string is displayed before it is changed, then the new string is displayed.

```
;stos_byte_asterisk.asm
;uses store string no operand to
;replace a given string with asterisks

;-----------------------------------------------------------
.STACK

;-----------------------------------------------------------
.DATA
RSLT  DB  0DH, 0AH, '123456789    $'

;-----------------------------------------------------------
.CODE
BEGIN PROC  FAR

;set up pgm ds
      MOV   AX, @DATA          ;get addr of data seg
      MOV   DS, AX             ;move addr to ds
      MOV   ES, AX             ;move addr to es
;-----------------------------------------------------------
                                  //continued on next page
```

Figure 13.8 Using the STOSB string instruction to replace a string in memory.

```
;display result before changing
        MOV     AH, 09H             ;display string
        LEA     DX, RSLT            ;put addr of rslt in dx
        INT     21H                 ;dos interrupt

;------------------------------------------------------------
;move string  elements
        CLD                         ;left-to-right
        MOV     CX, 9               ;count in cx
        LEA     DI, RSLT+2          ;addr of rslt+2 -> di as dst
        MOV     AL, 2AH             ;* -> al
REP     STOSB                       ;move bytes to dst

;------------------------------------------------------------
;display result after changing
        MOV     AH, 09H             ;display string
        LEA     DX, RSLT            ;put addr of rslt in dx
        INT     21H                 ;dos interrupt

BEGIN ENDP
        END     BEGIN
;------------------------------------------------------------
123456789
*********
```

Figure 13.8 (Continued)

13.5 Compare Strings Instructions

The *compare strings* (CMPS) instructions contain two source operands — there is no destination operand. The instructions compare a string element — byte, word, doubleword, or quadword — in the first source operand with the byte, word, doubleword, or quadword in the second source operand. The comparison is accomplished by subtracting the first source operand from the second source operand. The status flags in the EFLAGS register reflect the result of the comparison. Both source operands are unaffected by the comparison; that is, both operands are unaltered. The DS:(E)SI and ES:(E)DI registers are automatically incremented or decremented depending on the state of the direction flag (DF) in the EFLAGS register.

Both operands reside in memory locations. The memory address of the first source operand is obtained from the contents of registers DS:(E)SI or RSI; the memory address of the second source operand is obtained from the contents of registers ES:(E)DI or RDI. There are abbreviated mnemonics for the different string element sizes: *compare strings byte* (CMPSB), *compare strings word* (CMPSW), and *compare strings doubleword* (CMPSD). In 64-bit mode, quadwords can be compared using the *compare strings quadword* (CMPSQ) instruction.

Variations of the REP prefix can be utilized with the compare strings instructions. These include: the *repeat while equal/zero* (REPE/REPZ) prefixes and the *repeat while not equal/not zero* (REPNE/REPNZ) prefixes. If a CMPS instruction is preceded by REPE or REPZ, then the operation is depicted as *compare while strings are equal* (ZF = 1) and *not end of string* [E(CX) ≠ 0]. If a CMPS instruction is preceded by REPNE or REPNZ, then the operation is depicted as *compare while strings are not equal* (ZF = 0) and *not end of string* [E(CX) ≠ 0]. If these prefixes are used, then the compare operation will terminate as soon as the specified condition becomes untrue or E(CX) = 0.

The data segment (DS) can be overridden with a *segment override prefix*. The segment override operator is specified by a colon (:). The extra segment (ES), however, cannot be overridden. Different versions of the compare strings instructions are described in the next sections.

13.5.1 Compare Strings (Explicit Operands) Instructions

This section describes the *explicit operands* form of the *compare strings* instructions. The explicit operands form *explicitly* specifies the size of the first and second source operands as part of the instruction. There are four *compare string* instructions that have explicit operands, as shown in the syntax below.

```
CMPS    m8, m8     (compare byte at DS:(E)SI with
                   byte at ES:(E)DI,
                   (R/E)SI, (R/E)DI for 64-bit mode)

CMPS    m16, m16   (compare word at DS:(E)SI with
                   word at ES:(E)DI,
                   (R/E)SI, (R/E)DI for 64-bit mode)

CMPS    m32, m32   (compare doubleword at DS:(E)SI with
                   doubleword at ES:(E)DI,
                   (R/E)SI, (R/E)DI for 64-bit mode)

CMPS    m64, m64   (compare quadword at (R/E)SI with
                   quadword at (R/E)DI for 64-bit mode)
```

13.5.2 Compare Strings (No Operands) Instructions

The *no-operands* form specifies the size of the operands by the mnemonic; for example, the CMPSW instruction specifies a *compare strings word* instruction. Like the explicit operand form, the no-operand form assumes that the locations of the operands are determined by the DS:(E)SI register for the first source operand and by the ES:(E)DI register for the second source operand. The syntax is shown below.

CMPSB (compare byte at DS:(E)SI with byte at ES:(E)DI, (R/E)SI, (R/E)DI for 64-bit mode)

CMPSW (compare word at DS:(E)SI with word at ES:(E)DI, (R/E)SI, (R/E)DI for 64-bit mode)

CMPSD (compare doubleword at DS:(E)SI with doubleword at ES:(E)DI, (R/E)SI, (R/E)DI for 64-bit mode)

CMPSQ (compare quadword at (R/E)SI with quadword at (R/E)DI, for 64-bit mode)

Figure 13.9 shows an assembly language program — not embedded in a C program — that illustrates an application using the CMPSB (no operands) instruction with the REPE prefix. Two strings are entered from the keyboard and stored in the OPFLD area of the parameter list (PARLST). Then the strings are compared and the resulting zero flag (ZF) is displayed. If ZF = 1, then the strings are equal; if ZF = 0, then the strings are not equal.

```
;cmps_byte2.asm
;compare two strings in opfld

;--------------------------------------------------------------
.STACK

;--------------------------------------------------------------
.DATA
PARLST  LABEL   BYTE
MAXLEN  DB  15
ACTLEN  DB  ?
OPFLD   DB  15  DUP(?)
PRMPT   DB  0DH, 0AH, 'Enter two 6-digit hex strings: $'
FLAGS   DB  0DH, 0AH, 'ZF flags =     $'

;--------------------------------------------------------------
.CODE
BEGIN PROC    FAR

;set up pgm ds
      MOV   AX, @DATA         ;get addr of data seg
      MOV   DS, AX            ;move addr to ds
      MOV   ES, AX            ;move addr to es
                              (a)    //continued on next page
```

Figure 13.9 Program to illustrate using the CMPSB instruction with the REPE prefix: (a) the program and (b) the outputs.

```
;read prompt
       MOV    AH, 09H          ;display string
       LEA    DX, PRMPT        ;put addr of prompt in dx
       INT    21H              ;dos interrupt
;-----------------------------------------------------------
;keyboard request rtn to enter characters
       MOV    AH, 0AH          ;buffered keyboard input
       LEA    DX, PARLST       ;put addr of parlst in dx
       INT    21H              ;dos interrupt
;-----------------------------------------------------------
;compare strings
       LEA    SI, OPFLD        ;addr of first string -> si
       LEA    DI, OPFLD+6      ;addr of second string -> di
       CLD                     ;left-to-right
       MOV    CX, 6            ;count in cx

REPE   CMPSB                   ;compare strings while equal
;-----------------------------------------------------------
;move flags to flag area
       PUSHF                   ;push flags
       POP    AX               ;flag word -> ax
       AND    AL, 40H          ;isolate zf
       SHR    AL, 6            ;shift right logical 6
       OR     AL, 30H          ;add ascii bias
       MOV    FLAGS+10, AL     ;move zf to flag area
;-----------------------------------------------------------
;display zf flag
       MOV    AH, 09H          ;display string
       LEA    DX, FLAGS        ;put addr of rslt in dx
       INT    21H              ;dos interrupt

BEGIN ENDP
       END    BEGIN
```

```
Enter two 6-digit hex strings: 123456123456
ZF flag = 1 //strings are equal
-----------------------------------------------------------
Enter two 6-digit hex strings: 123456123457
ZF flag = 0 //strings are not equal
-----------------------------------------------------------
Enter two 6-digit hex strings: ABCDEFABCDEF
ZF flag = 1 //strings are equal
-----------------------------------------------------------
Enter two 6-digit hex strings: ABCDEFABCDE2
ZF flag = 0 //strings are not equal
```

(b)

Figure 13.9 (Continued)

13.6 Scan String Instructions

The *scan string* (SCAS) instructions contain only one operand, which is in a general-purpose register. The instructions compare a string element — byte, word, doubleword, or quadword — in register AL, AX, EAX, or RAX, respectively — with the byte, word, doubleword, or quadword in a memory location addressed by ES:E(DI) or RDI. The comparison is accomplished by subtracting the memory operand from the general-purpose register. The status flags in the EFLAGS register reflect the result of the comparison. Both operands are unchanged by the comparison. The (E)DI register is automatically incremented or decremented depending on the state of the direction flag (DF) in the EFLAGS register.

There are abbreviated mnemonics for the different string element sizes: *scan string byte* (SCASB), *scan string word* (SCASW), and *scan string doubleword* (SCASD). In 64-bit mode, quadwords can be compared using the *scan string quadword* (SCASQ) instruction.

Variations of the REP prefix can be utilized with the scan string instructions. These include the *repeat while equal / zero* (REPE / REPZ) prefixes and the *repeat while not equal / not zero* (REPNE / REPNZ) prefixes. These prefixes can be utilized for block comparisons as specified by the count in the (E)CX register. If these prefixes are used, then the comparison operation (scanning) will terminate as soon as the specified condition becomes untrue or E(CX) = 0.

The data segment (DS) can be overridden with a *segment override prefix*. The segment override operator is specified by a colon (:). The extra segment (ES), however, cannot be overridden. Different versions of the compare strings instructions are described in the next sections.

13.6.1 Scan String (Explicit Operands) Instructions

This section describes the *explicit operands* form of the *scan string* instructions. The explicit operands form *explicitly* specifies the size of the operands as part of the instruction. There are four *scan string* instructions that have explicit operands, as shown in the syntax below.

```
SCAS  m8  (compare byte in AL with
          byte at ES:(E)DI, RDI for 64-bit mode)

SCAS  m16 (compare word in AX with
          word at ES:(E)DI, RDI for 64-bit mode)

SCAS  m32 (compare doubleword in EAX with
          doubleword at ES:(E)DI, RDI for 64-bit mode)

SCAS  m64 (compare quadword in RAX with
          quadword at EDI or RDI for 64-bit mode)
```

13.6.2 Scan String (No Operands) Instructions

The *no-operands* form specifies the size of the operands by the mnemonic; for example, the SCASB instruction specifies a *scan string byte* instruction. Like the explicit operand form, the no-operand form assumes that the location of the memory operand is determined by the ES:(E)DI register. The register operands are contained in the general-purpose registers. The syntax is shown below.

```
SCASB    (compare byte in AL with
         byte at ES:(E)DI, RDI for 64-bit mode)

SCASW    (compare word in AX with
         word at ES:(E)DI, RDI for 64-bit mode)

SCASD    (compare doubleword in EAX with
         doubleword at ES:(E)DI, RDI for 64-bit mode)

SCASQ    (compare quadword in RAX with
         quadword at EDI or RDI for 64-bit mode)
```

Figure 13.10 shows an assembly language program — not embedded in a C program — that illustrates an application using the SCASB (no operands) instruction with the REPNE prefix. If the scan character (A) in register AL matches a string character, then ZF = 1, otherwise ZF = 0. The count is displayed together with the flags.

```
;scans_byte2.asm
;use scans with repne prefix
;-----------------------------------------------------------
.STACK
;-----------------------------------------------------------
.DATA
STR      DB 'BCDERASW  $'
CL_RSLT  DB 0DH, 'CL =     $'
FLAGS    DB 0DH, 0AH, 'ZF flag =    $'
;-----------------------------------------------------------
.CODE
BEGIN PROC  FAR

;set up pgm ds and es
      MOV   AX, @DATA        ;get addr of data seg
      MOV   DS, AX           ;move addr to ds
      MOV   ES, AX           ;move addr to es
                             (a)    //continued on next page
```

Figure 13.10 Program to illustrate an application of the SCASB with the REPNE prefix: (a) the program and (b) the outputs.

```
;--------------------------------------------------------------
        MOV     AL, 'A'          ;ascii character A -> al
        LEA     DI, STR          ;addr of str -> di
        CLD                      ;left-to-right
        MOV     CL, 8            ;count in cl

REPNE
        SCASB                    ;compare char while not equal

        PUSHF                    ;push flags onto stack
        POP     AX               ;pop flags to ax
        AND     AL, 40H          ;isolate zf
        SHR     AL, 6            ;shift right logical 6
        OR      AL, 30H          ;add ascii bias
        MOV     FLAGS+12, AL     ;move zf to flag area

;--------------------------------------------------------------
;display residual count
        OR      CL, 30H          ;add ascii bias
        MOV     CL_RSLT+6, CL    ;cl -> result area

        MOV     AH, 09H          ;display string
        LEA     DX, CL_RSLT      ;addr of cl_rslt -> dx
        INT     21H              ;dos interrupt

;--------------------------------------------------------------
;display flags
        MOV     AH, 09H          ;display string
        LEA     DX, FLAGS        ;addr of flags -> dx
        INT     21H              ;dos interrupt

BEGIN ENDP
        END     BEGIN
```

```
CL = 2          if AL = W, then CL = 0
ZF flag = 1     and ZF flag = 1     (b)       //characters match
```

Figure 13.10 (Continued)

13.7 Problems

13.1 Write an assembly language program — not embedded in a C program — that receives six hexadecimal characters entered from the keyboard and stores them in the OPFLD area. The program then moves the characters to the result area to be utilized as a second string. Then the first three characters from OPFLD are moved to replace the last three characters in the result area. Display the resulting contents of the result area.

13.2 Determine the contents of RSLT after execution of the following program:

```
;movs_byte5.asm
;-----------------------------------------------------------
.STACK

;-----------------------------------------------------------
.DATA
RSLT     DB   0DH, 0AH, '123456789    $'

;-----------------------------------------------------------
.CODE
BEGIN PROC   FAR

;set up pgm ds
      MOV    AX, @DATA          ;get addr of data seg
      MOV    DS, AX             ;move addr to ds
      MOV    ES, AX             ;move addr to es

;-----------------------------------------------------------
;move string  elements
      CLD
      MOV    CX, 4              ;count in cx
      LEA    SI, RSLT+2         ;addr of rslt+2 -> si as src
      LEA    DI, RSLT+4         ;addr of rslt+4 -> di as dst

REP   MOVSB                     ;move bytes to dst

;-----------------------------------------------------------
;display result
      MOV    AH, 09H            ;display string
      LEA    DX, RSLT           ;put addr of rslt in dx
      INT    21H                ;dos interrupt

BEGIN ENDP
      END    BEGIN
```

13.3 Determine the contents of RSLT after execution of the following program:

```
;movs_byte_rev.asm
;--------------------------------------------------------
.STACK

;--------------------------------------------------------
.DATA
RSLT    DB  0DH, 0AH, '123456789    $'

;--------------------------------------------------------
.CODE
BEGIN PROC  FAR

;set up pgm ds
      MOV   AX, @DATA       ;get addr of data seg
      MOV   DS, AX          ;move addr to ds
      MOV   ES, AX          ;move addr to es

;--------------------------------------------------------
;move string  elements
      STD                   ;right-to-left
      MOV   CX, 4           ;count in cx
      LEA   SI, RSLT+10     ;addr of rslt+10 -> si as src
      LEA   DI, RSLT+8      ;addr of rslt+8 -> di as dst

REP   MOVSB                 ;move bytes to dst

;--------------------------------------------------------
;display result
      MOV   AH, 09H         ;display string
      LEA   DX, RSLT        ;put addr of rslt in dx
      INT   21H             ;dos interrupt

BEGIN ENDP
      END   BEGIN
```

13.4 Write an assembly language program — not embedded in a C program — that receives six hexadecimal characters entered from the keyboard and stores them in the OPFLD area. The result area contains a second string of ABC-DEF. Then the first three characters from OPFLD are moved to replace the last three characters in the result area. Display the resulting contents of the result area. This program is similar to Problem 13.1, except that the second string is given; therefore, only one LOOP instruction is required. The REP prefix is not used.

13.5 Although this problem does not use the MOVS instruction, it does move modified characters from a source to a destination. Assume that the following characters are entered from the keyboard:

```
0FF, 00, 0EE, 22, 0C6, 0F5
```

Then show the result after the program shown below has been executed.

```
//array_xor2.cpp
#include "stdafx.h"
int main (void)
{
//define variables
   unsigned char hex1, hex2, hex3, hex4, hex5, hex6;

   printf ("Enter six 2-digit hexadecimal characters:
            \n");
   scanf ("%X %X %X %X %X %X", &hex1, &hex2, &hex3,
            &hex4, &hex5, &hex6);

//switch to assembly
      _asm
      {
         MOV    AL, hex1
         XOR    hex3, AL

         MOV    AL, hex2
         XOR    hex4, AL

         MOV    AL, hex3
         XOR    hex5, AL

         MOV    AL, hex4
         XOR    hex6, AL
      }

   printf ("\nhex1 = %X", hex1);
   printf ("\nhex2 = %X", hex2);
   printf ("\nhex3 = %X", hex3);
   printf ("\nhex4 = %X", hex4);
   printf ("\nhex5 = %X", hex5);
   printf ("\nhex6 = %X\n\n", hex6);
   return 0;
}
```

13.6 Write an assembly language module embedded in a C program using the LODS instruction with explicit operands that receive byte, word, and doubleword operands that are entered from the keyboard. Then display the operands.

13.7 Given the program shown below, obtain the result when the following four-digit hexadecimal characters are entered from the keyboard separately:

```
1233  9999  2E+F  )2<=
```

```
;lods_stos.asm
;illustrates using the load string
;and store string no operand instructions
;-----------------------------------------------------
.STACK
;-----------------------------------------------------
.DATA
PARLST   LABEL BYTE
MAXLEN   DB 10
ACTLEN   DB ?
OPFLD    DB  10 (?)
PRMPT    DB 0DH, 0AH, 'Enter four 1-digit hex chars: $'
RSLT     DB 0DH, 0AH, 'Result =                 $'
;-----------------------------------------------------
.CODE
BEGIN PROC FAR

;set up pgm ds
      MOV   AX, @DATA        ;get addr of data seg
      MOV   DS, AX           ;move addr to ds

;read prompt
      MOV   AH, 09H          ;display string
      LEA   DX, PRMPT        ;put addr of prompt in dx
      INT   21H              ;dos interrupt
;-----------------------------------------------------
;keyboard request rtn to enter characters
      MOV   AH, 0AH          ;buffered keyboard input
      LEA   DX, PARLST       ;put addr of parlst in dx
      INT   21H              ;dos interrupt
;-----------------------------------------------------
      MOV   CX, 3            ;# of iterations for loop
      LEA   SI, OPFLD        ;addr of opfld -> si
      LEA   DI, OPFLD+1      ;addr of opfld+1 -> di
      LEA   BX, RSLT+11      ;addr of rslt+11 -> bx
      CLD                    ;left-to-right transfer

                             //continued on next page
```

```
LP1:    LODSB
        ADC    AL, [DI]         ;add with carry
        STOSB
        MOV    [BX], AL         ;al -> rslt area
        INC    BX
        LOOP   LP1              ;if cx != 0, then loop

;-----------------------------------------------------------
;display result
        MOV    AH, 09H
        LEA    DX, RSLT
        INT    21H

BEGIN ENDP
        END    BEGIN
```

13.8 Given the program segment shown below, determine the contents of the count in register CL and the ZF flag after the program has been executed. Then write an assembly language program — not embedded in a C program — to verify the results.

```
                                        . . .
.DATA
STR1     DB 'ABCDEF  $'
STR2     DB 'AB1234  $'
CL_RSLT  DB 0DH, 0AH, 'CL =        $'
FLAGS    DB 0DH, 0AH, 'ZF flag =     $'

;-----------------------------------------------------------
.CODE
BEGIN    PROC FAR

;set up pgm ds and es
    MOV      AX, @DATA         ;get addr of data seg
    MOV      DS, AX            ;move addr to ds
    MOV      ES, AX            ;move addr to es

;-----------------------------------------------------------
    LEA      SI, STR1          ;addr of str1 -> si
    LEA      DI, STR2          ;addr of str2 -> di
    CLD                        ;left-to-right
    MOV      CL, 6             ;count in cl

REPNE
   CMPSB                   ;compare strings while not equal
                                        . . .
```

13.9 Given the program segment shown below, obtain the count in register CL after the program executes.

```
                                        . . .
.DATA
STR       DB 'ABCDEFGHI  $'
CL_RSLT DB 0DH, 'CL =    $'
FLAGS     DB 0DH, 0AH, 'ZF flag =    $'

;------------------------------------------------------
.CODE
BEGIN     PROC    FAR

;set up pgm ds and es
      MOV    AX, @DATA          ;get addr of data seg
      MOV    DS, AX             ;move addr to ds
      MOV    ES, AX             ;move addr to es

;------------------------------------------------------
      MOV    AL, 'G'
      LEA    DI, STR            ;addr of str2 -> di
      CLD                       ;left-to-right
      MOV    CL, 9              ;count in cl

REPNE
      SCASB                     ;compare char while ≠
                                        . . .
```

13.10 Given the program segment shown below, obtain the count in register CL after the program executes.

```
                                        . . .
.DATA
STR       DB 'ABCDEFGHI  $'
CL_RSLT DB 0DH, 'CL =    $'
FLAGS     DB 0DH, 0AH, 'ZF flag =    $'

;------------------------------------------------------
.CODE
BEGIN     PROC    FAR

;set up pgm ds and es
      MOV    AX, @DATA          ;get addr of data seg
      MOV    DS, AX             ;move addr to ds
      MOV    ES, AX             ;move addr to es

                                //continued on next page
```

```
;---------------------------------------------------------------
      MOV   AL, 'R'
      LEA   DI, STR           ;addr of str2 -> di
      CLD                     ;left-to-right
      MOV   CL, 9             ;count in cl

REPE  SCASB                   ;compare char while ≠
                             . . .
```

14

Arrays

An array is a data structure that contains a list of elements of the same data type (*homogeneous*) with a common name whose elements can be accessed individually. Arrays are also synonymous with tables and can be utilized to implement other data structures, such as strings and lists. An array can consist of characters, integers, floating-point numbers, or any other data types; however, the data types cannot be mixed. Array elements are usually stored in contiguous locations in memory, allowing easier access to the array elements. There are two main types of arrays: one-dimensional arrays and multi-dimensional arrays.

14.1 One-Dimensional Arrays

A *one-dimensional array* — also called a *linear array* — is an array that is accessed by a single index. A one-dimensional array is shown in Figure 14.1, which contains a list of ten characters: *a* through *j*. The array elements are accessed by an index — also called an offset or a subscript — beginning with the first array element. The base address of the array is the address of the first element — in the location whose offset is zero. Thus, the first element (*a*), in Figure 14.1, is addressed by *array_name [0]*, the fourth array element (*d*) is addressed by *array_name [3]*, and the last array element (*j*) is addressed by *array_name [9]*. In X86 assembly language, arrays are defined by directives, such as define byte (DB) and define word (DW), followed by the list of elements.

Index	0	1	2	3	4	5	6	7	8	9
array_name	a	b	c	d	e	f	g	h	i	j

Figure 14.1 A one-dimensional linear array.

The name of the array is the *symbolic* name of the memory address that specifies the first element of the array — in Figure 14.1 this would be *array_name*. This type of addressing is referred to as *register indirect addressing* and uses a register that contains the address of the data, rather than the actual data to be accessed.

Register indirect addressing uses the general-purpose registers within brackets; for example, [BX], [BP], [EAX], [EBX], etc., where the brackets specify that the indicated register contains an offset that points to the data within a specific segment; that is, the value in the register is added to the base address of the array. For example, the following instruction moves the contents of register AL into the array labelled *array_name* at a location specified by the contents of register BX, which are added to the base address of *array_name*:

```
MOV    array_name [BX], AL
```

The registers contain values for indexing — or subscripting — into an array, such as (E)BX, (E)SI, and (E)DI, which contain offsets for processing operands in the data segment; for example, DS:BX, DS:ESI, and DS:DI. Register (E)BP is used to reference data in the stack segment, such as SS:EBP.

Data stored in an array is organized as bytes, words, doublewords, or quadwords. A byte array and a word array can be defined as shown below for arrays of 20 elements in which the array elements are undefined.

```
array_byte    DB    20 DUP(?)
array_word    DW    20 DUP(?)
```

Assumed that register BX is used as the index into both arrays. Then the byte array elements can be accessed by incrementing or decrementing register BX by one; the word array elements can be accessed by incrementing or decrementing register BX by two.

A stack can be considered as a one-dimensional array. The base register can point to the stack top; the contents of the base register plus the displacement points to the beginning of the array. The index register then selects elements in the array. The effective address (EA) is calculated as shown below.

```
EA = EBP + displacement + ESI
```

One-dimensional arrays can be combined to generate a one-dimensional array that consists of contiguous one-dimensional arrays, as shown in Figure 14.2. Accessing elements in this type of array can be achieved by the *base plus index times scale factor plus displacement* addressing mode, as shown below, where the *scale* is 1, 2, 4, or 8.

$$\text{Base register} + (\text{index register} \times 2^{\text{scale}}) + \text{Displacement}$$

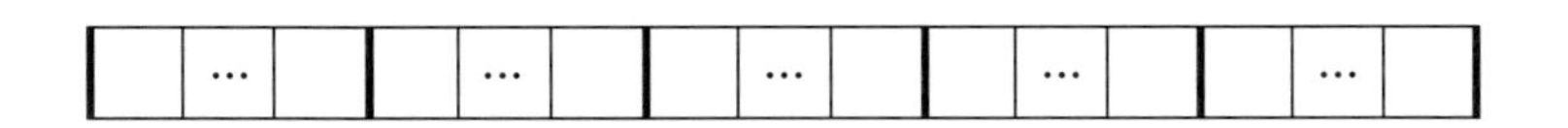

Figure 14.2 A one-dimensional array of one-dimensional arrays.

A typical example of a one-dimensional (linear) array composed of contiguous linear arrays is a list of elements of the same type with the same length. An array of numbers is representative of this type of array. Since the numbers are of different lengths, spaces are added to certain numbers in order to make them the same length as the longest number. This can be considered as an array of arrays.

The array of numbers can be declared either in one of two ways:

```
array_number    DB    'TWELVE'
            . . .
                DB    'ONE   '
```

or

```
array_number    DB    'TWELVE',
            . . .
                      'ONE   '
```

To select a particular number, a 2-digit number (01 — 12) is entered from the keyboard, then the corresponding number is displayed (ONE — TWELVE). The number is stored as equivalent hexadecimal numbers in the OPERAND field (OPFLD and OPFLD + 1) locations of the parameter list (PARLST), as shown in Figure 14.3 for the number SIX.

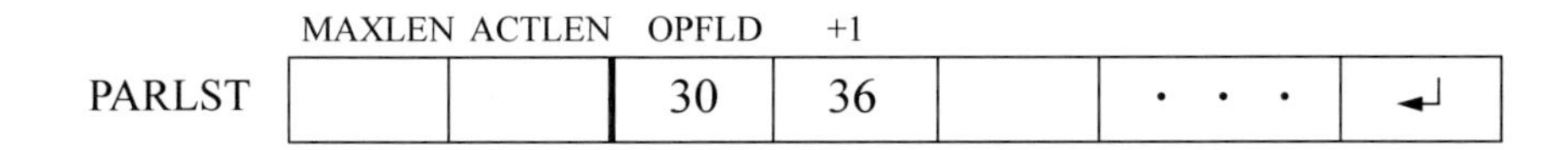

Figure 14.3 Parameter list one-dimensional array in which the keyboard input data are stored.

Figure 14.4 lists an assembly language program — not embedded in a C program — that illustrates this concept. The first digit (30H), for the number SIX, is moved to register AH to remove the ASCII bias of 3H; then the second digit is moved to register AL to remove the ASCII bias. The *ASCII adjust AX before division* (AAD) instruction is then executed. Recall that the AAD instruction converts two unpacked digits in register AX — high-order digit in register AH and low-order digit in register AL — to an equivalent binary value, stores the result in register AL, and resets register AH.

Then a value of 1 is subtracted from register AL to adjust for the offset in the array labelled *TABLE*. A value of 6 is moved to register CL, because there are six characters per number. Register AL is then multiplied by 6, allowing the first letter in SIX to be accessed. A loop instruction then moves the addressed number one letter at a time to the result (RSLT) area to be displayed.

```
;number_table.asm
;enter a number (01 -- 12) from the keyboard
;and display the spelling of the corresponding number

;-------------------------------------------------------------
.STACK

;-------------------------------------------------------------
.DATA
PARLST    LABEL    BYTE
MAXLEN    DB       40
ACTLEN    DB       ?
OPFLD     DB       40         DUP(?)
TABLE     DB       'ONE   '
          DB       'TWO   '
          DB       'THREE '
          DB       'FOUR  '
          DB       'FIVE  '
          DB       'SIX   '
          DB       'SEVEN '
          DB       'EIGHT '
          DB       'NINE  '
          DB       'TEN   '
          DB       'ELEVEN'
          DB       'TWELVE'
PRMPT     DB       0DH, 0AH, 'Enter 2-digit number: $'
RSLT      DB       0DH, 0AH, 'Number =           $'

;-------------------------------------------------------------
```

(a) //continued on next page

Figure 14.4 Program to illustrate selecting a number from the keyboard to be spelled and displayed: (a) the program and (b) the outputs.

```
;---------------------------------------------------------------
.CODE
BEGIN PROC   FAR

;set up pgm ds
      MOV    AX, @DATA        ;get addr of data seg
      MOV    DS, AX           ;move addr to ds

;read prompt
      MOV    AH, 09H          ;display string
      LEA    DX, PRMPT        ;put addr of prompt in dx
      INT    21H              ;dos interrupt

;keyboard request rtn to enter characters
      MOV    AH, 0AH          ;buffered keyboard input
      LEA    DX, PARLST       ;put addr of parlst in dx
      INT    21H              ;dos interrupt

;---------------------------------------------------------------
;set up source and destination
      LEA    SI, OPFLD        ;point to first number
      LEA    DI, RSLT + 10    ;point to destination

;get number
      MOV    AH, [SI]         ;get 1st number
      SUB    AH, 30H          ;unpack it
      INC    SI               ;point to opfld+1
      MOV    AL, [SI]         ;get 2nd number
      SUB    AL, 30H          ;unpack it

;convert unpacked number in ax
;to the binary equivalent in al
      AAD
      SUB    AL, 01H          ;adjust for offset in table
      MOV    CL, 06H          ;6 characters per number
      MUL    CL               ;mul al by 6 to find number

;---------------------------------------------------------------
;get number from table
      LEA    SI, TABLE        ;put addr of table in si
      MOV    AH, 00H          ;clear ah
      ADD    SI, AX           ;si points to number
      MOV    CX, 06H          ;for 6 characters

                              //continued on next page
```

Figure 14.4 (Continued)

```
NUM:   MOV   AL, [SI]        ;get number letter
       MOV   [DI], AL        ;put letter in rslt
       INC   SI              ;point to next letter
       INC   DI              ;point to next destination
       LOOP  NUM             ;loop until cx = 0

;--------------------------------------------------------------
;display the result
       MOV   AH, 09H         ;display string
       LEA   DX, RSLT        ;put addr of rslt in dx
       INT   21H             ;dos interrupt

BEGIN ENDP
       END   BEGIN
```

```
Enter 2-digit number: 01
Number = ONE
--------------------------------------------------------------
Enter 2-digit number: 06
Number = SIX
--------------------------------------------------------------
Enter 2-digit number: 09
Number = NINE
--------------------------------------------------------------
Enter 2-digit number: 10
Number = TEN
--------------------------------------------------------------
Enter 2-digit number: 11
Number = ELEVEN
--------------------------------------------------------------
Enter 2-digit number: 12
Number = TWELVE
```

(b)

Figure 14.4 (Continued)

Figure 14.5 illustrates an assembly language module embedded in a C program that adds select integers in an array and displays the resulting sums. The array is declared as a *char* data type with five elements containing both positive and negative integers. A *char* type array specifies elements of one byte with a range from –128 to +127.

```
//array.cpp
//add two numbers in an array
#include "stdafx.h"
int main (void)

{
//define variables
   char array1 [5] = {2, -4, 56, -87, 35};
   char sum1, sum2, sum3, sum4, sum5;

//switch to assembly
      _asm
      {
         MOV   AL, array1[0]      ;+2
         ADD   AL, array1[1]      ;-4
         MOV   sum1, AL           ;-2
;------------------------------------------------------------
         MOV   AL, array1[1]      ;-4
         ADD   AL, array1[3]      ;-87
         MOV   sum2, AL           ;-91
;------------------------------------------------------------
         MOV   AL, array1[2]      ;+56
         ADD   AL, array1[4]      ;+35
         MOV   sum3, AL           ;+91
;------------------------------------------------------------
         MOV   AL, array1[3]      ;-87
         ADD   AL, array1[2]      ;+56
         MOV   sum4, AL           ;-31
;------------------------------------------------------------
         MOV   AL, array1[0]      ;+2
         ADD   AL, array1[4]      ;+35
         MOV   sum5, AL           ;+37
      }
//display sums
   printf ("Sum1 = %d\n", sum1);
   printf ("Sum2 = %d\n", sum2);
   printf ("Sum3 = %d\n", sum3);
   printf ("Sum4 = %d\n", sum4);
   printf ("Sum5 = %d\n\n", sum5);
   return 0;
}
```

(a) //continued on next page

Figure 14.5 Program to add two integers in an array: (a) the program and (b) the outputs.

```
Sum1 = -2
Sum2 = -91
Sum3 = 91
Sum4 = -31
Sum5 = 37

Press any key to continue . . . _
```

(b)

Figure 14.5 (Continued)

14.1.1 One-Dimensional Arrays in C

An array can be initialized by simply listing the array elements, as shown below. The *type* can be **char**, **int**, **float**, **double**, or any valid C type. An array is a sequence of constants of the same data type enclosed in braces. The size of the array does not have to be specified, also shown below.

```
int array[5] = {1, 4, 9, 16, 25};
char array[3] = {'A', 'B', 'C'};

int array_odd[] = {1, 3, 5, 7, 9, 11);
```

The length of the array is one element longer than the size of the array — the compiler supplies the null terminator, also called the null zero (\0). The null character is a special character indicating the end of the array. An *automatic* array is defined inside the *main ()* function; an *external* array is defined outside a function, usually before the *main ()* function.

Figure 14.6 illustrates program to obtain the sum of six positive and negative floating-point numbers in an array labelled *array_flp*. The keyword *double* is a double-precision floating-point data type containing eight bytes for very large values. The operator += is an *arithmetic assignment operator* defined as follows:

```
a += b;
```

is equivalent to

```
a = a + b;
```

The **for** loop in the program repeats a statement or block of statements a specific number of times. When the **for** loop has completed the final loop, the program exits the loop and transfers control to the first statement following the block of statements. The syntax for the **for** loop is shown below.

```
for (initialization; conditional test; increment)
   {one or more statements}
```

```
//array_flp.cpp
//add six floating-point numbers in an array

#include "stdafx.h"
int main (void)

{
//define variables
   double array_flp[6] = {72.5, -104.8, -56.7, 110.5,
                                 -255.3, 30.6};

   double flp_sum = 0.0;

   int ctr;

   for (ctr = 0; ctr < 7; ctr++)
      {flp_sum += array_flp[ctr];}

//display result
   printf ("Floating-point sum = %f\n\n", flp_sum);

   return 0;
}
```

(a)

```
Floating-point sum = -203.200000

Press any key to continue . . . _
```

(b)

Figure 14.6 Program to illustrate adding consecutive floating-point numbers in an array: (a) the program and (b) the outputs.

Another example of initializing and operating on arrays in C is shown in Figure 14.7. Two 5-element arrays are initialized using two different methods: method one uses the *assignment operator* (=) to establish floating-point values to the individual elements of *array1[0]* through *array1[4]*; method two assigns floating-point values to the elements of *array2[0]* through *array2[4]* using a **for** loop.

The sum of each array is then printed together with the average of the values in each array. The *arithmetic assignment operator* (+=) is utilized to obtain the sum of the array elements in both arrays. The average of the array elements is obtained by dividing the sum by the number of elements.

```
//array_init2.cpp
//two ways to initialize arrays.
//print the sum and average of the array elements

#include "stdafx.h"

int main (void)
{
//define variables
   int i;

   double array1 [5];
   float array2 [5];

   double average1;
   double average2;

   double sum1 = 0.0;
   float sum2 = 0.0;

//-----------------------------------------------------------
//define array1 by assigning values to individual elements
   array1[0] = 111.1;
   array1[1] = 222.2;
   array1[2] = 333.3;
   array1[3] = 444.4;
   array1[4] = 555.5;

//-----------------------------------------------------------
//print array1 elements, sum, and average
   for (i = 0; i < 5; i++)
   {
      printf ("Array1 [%d] = %f\n", i, array1 [i]);
      sum1 += array1[i];
   }

   printf ("\nSum1 = %f\n\n", sum1);

   average1 = sum1/5;
   printf ("Average1 = %f\n\n", average1);
```

(a) //continued on next page

Figure 14.7 Program to initialize two arrays and obtain the sum of the array elements and the average: (a) the program and (b) the outputs.

```
//define array2 using a for loop
   for (i = 0; i < 5; i++)
   {
      printf ("Enter a floating-point number: ");
      scanf ("%f", &array2[i]);
      sum2 += array2[i];
   }

//print array2 sum and average
   printf ("\nSum2 = %f\n\n", sum2);

   average2 = sum2/5;
   printf ("Average2 = %f\n\n", average2);

   return 0;
}
```

```
Array1 [0] = 111.1
Array1 [1] = 222.2
Array1 [2] = 333.3
Array1 [3] = 444.4
Array1 [4] = 555.5

Sum1 = 1666.500000

Average1 = 333.300000

---------------------------------------------------------------
Enter a floating-point number: 111.1
Enter a floating-point number: 222.2
Enter a floating-point number: 333.3
Enter a floating-point number: 444.4
Enter a floating-point number: 555.5

Sum2 = 1666.500000

Average2 = 333.300000

Press any key to continue . . . _
```

(b)

Figure 14.7 (Continued)

14.2 Multi-Dimensional Arrays

A *two-dimensional array* — also called a *multi-dimensional array* — is an array that is accessed by two subscripts, as shown below in C for an array of integers consisting of three rows and three columns (3 × 3) labelled *array_int1*.

```
int array_int1 [3] [3];
```

Arrays can be initialized in C by a variety of ways, as shown below for a 3 × 3 array. The initialized array is shown in Figure 14.8.

```
int array_int2 [3][3] = {1, 2, 3, 4, 5, 6, 7, 8, 9};

int array_int2 [3][3] =
{
   1, 2, 3,
   4, 5, 6,
   7, 8, 9
};
```

```
int array_int2 [3][3] ={{1, 2, 3}, {4, 5, 6}, {7, 8, 9}};
```

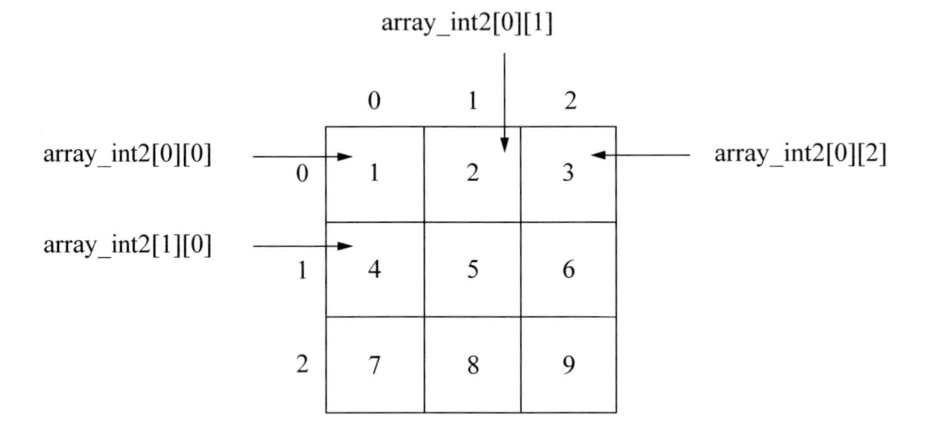

Figure 14.8 A 3 × 3 array initialized to {1, 2, 3, 4, 5, 6, 7, 8, 9};

Figure 14.9 illustrates a C program that initializes a 5 × 8 array — labelled *array_int3* — to the products of the indices, then prints the resulting array. Two **for** loops are utilized: one to address the row index and one to address the column index. Note that in the first **printf()** instruction, the notation `"%4d"` represents a format specifier. All format specifiers begin with the percent sign (%). In this case, the notation `"%4d"` is a *minimum-field-width* specifier, where the 4 represents the minimum distance from the end of the previous column of digits to the end of the next column of digits, allowing for right-aligned columns. The minimum field width of 4 is inserted after the percent sign and before the format specifier (d), which indiicates a signed decimal integer.

```
//array_int3.cpp
//load a 5 x 8 array with the products of the indices,
//then display the array in a row-column format

#include "stdafx.h"

int main (void)
{
//define variables
   int array_int3 [5][8];          //declare a 5 x 8 array

   int i, j;

//-----------------------------------------------------------
   for (i = 0; i < 5; i++)         //i is the row index
      for (j = 0; j < 8; j++)      //j is the column index
         array_int3 [i][j] = i*j;//row index x column index

//-----------------------------------------------------------
   for (i = 0; i < 5; i++)
   {
      for (j = 0; j < 8; j++)
//"%4d" in the printf ( ) is 4-digit spacing right aligned
         printf ("%4d", array_int3 [i][j]);
         printf ("\n");
   }
   printf ("\n");

   return 0;
}
```

(a)

```
   0   0   0   0   0   0   0   0
   0   1   2   3   4   5   6   7
   0   2   4   6   8  10  12  14
   0   3   6   9  12  15  18  21
   0   4   8  12  16  20  24  28

Press any key to continue . . . _
```

(b)

Figure 14.9 Program to initialize a 5×8 array to the product of the indices then print the array: (a) the program and (b) the outputs.

A C program will now be presented that demonstrates the four arithmetic operations of addition, subtraction, multiplication, and division on floating-point array elements. A 2×3 array is defined as shown below. Figure 14.10 illustrates the C program for the four arithmetic floating-point operations.

```
float   array_flp_arith [2][3] = {
                                   110.2, 75.5, 50.8,
                                   45.6, 80.7, 120.0
                                  };
```

```
//array_flp_arith.cpp
//perform add, sub, mul, and div
//on floating-point array elements

#include "stdafx.h"

int main (void)
{
   float array_flp [2][3] = {
                               110.2, 75.5, 50.8,
                               45.6, 80.7, 120.0
                              };

   int i, j;        //i is row index; j is column index

//------------------------------------------------------
//perform floating-point addition
   array_flp [0][2] = array_flp [1][0] + array_flp [1][2];
   //45.6 + 120.0 = 165.6 in [0][2]; replaces 50.8

//------------------------------------------------------
//perform floating-point subtraction
   array_flp [1][0] = array_flp [1][0] - array_flp [1][2];
   //45.6 - 120.0 = -74.4 in [1][0]; replaces 45.6

//------------------------------------------------------
//perform floating-point multiplication
   array_flp [0][1] = array_flp [0][2] * array_flp [1][0];
   //165.6 x -74.4 = 240.0 in [0][1]; replaces 75.5

                                  //continued on next page
```

(a)

Figure 14.10 Program to illustrate floating-point arithmetic operations on a 2×3 array: (a) the program and (b) the outputs.

```
//-----------------------------------------------------------
//perform floating-point division
   array_flp [1][2] = array_flp [1][2] / array_flp [1][1];
   //120.0 / 80.7 = 1.486988848 in [1][2]; replaces 80.7

//-----------------------------------------------------------
//print the array
   for (i = 0; i < 2; i++)
   {
      for (j = 0; j < 3; j++)
         printf ("%15f", array_flp [i][j]);
         printf ("\n\n");
   }
   printf ("\n");

   return 0;
}
```

```
After addition

     110.199997      75.500000     165.600006

      45.599998      80.699997     120.000000


Press any key to continue . . . _
-----------------------------------------------------------
After subtraction

     110.199997      75.500000     165.600006

     -74.400002      80.699997     120.000000


Press any key to continue . . . _
-----------------------------------------------------------
After multiplication

     110.199997  -12320.640625     165.600006

     -74.400002      80.699997     120.000000


Press any key to continue . . . _
-----------------------------------------------------------
```

(b) //continued on next page

Figure 14.10 (Continued)

```
--------------------------------------------------------------
After division - all operations finished

     110.199997  -12320.640625     165.600006

     -74.400002       80.699997       1.486989

Press any key to continue . . . _
```

Figure 14.10 (Continued)

14.3 Problems

14.1 Write an assembly language module embedded in a C program that defines an array of five positive and negative integers. Then calculate the cumulative sum of the five integers and display the resulting sum.

14.2 A table can be considered as a restructured one-dimensional array. Define an array of this type using six floating-point positive and negative numbers. Then write an assembly language module embedded in a C program to calculate the cumulative sum of the six floating-point numbers and display the resulting sum.

14.3 Write an assembly language module embedded in a C program that adds, subtracts, and multiplies two select floating-point numbers. Three positive and three negative floating-point numbers are to be defined, then used in the calculations. Perform two calculations on each arithmetic operation and display the corresponding results.

14.4 Write an assembly language module embedded in a C program that performs floating-point division on positive and negative floating-point numbers. Provide two examples for each of the following divide instructions, then display the corresponding results:

```
FDIV   memory
FDIV   ST(0), ST(i)
```

14.5 Write a C program that calculates the cubes of the integers 1 through 10 using an array, where array [0] contains a value of 1.

14.6 Write a C program that multiplies one 4-element array of integers by another 4-element array of integers one element at a time, as shown below. Then display the resulting products.

```
multiplier[0] x multiplicand[0] – [3], then
multiplier[1] x multiplicand[0] – [3], then
multiplier[2] x multiplicand[0] – [3], then
multiplier[3] x multiplicand[0] – [3]
```

14.7 Write a C program that multiplies a 5-element array of integers by a single integer, then display resulting array.

14.8 Write a C program that loads a 3×4 array with the products of the indices, then display the array in a row-column format.

14.9 Define a 3×5 integer array as shown below. Then write a C program to perform addition, subtraction, multiplication, and division on select elements of the array and display the modified array.

```
long array_int = {
                  10, 20, 30, 40, 50,
                  60, 70, 80, 90, 100,
                  200, 300, 400, 500, 600
                 };
```

15

Macros

Macros and procedures are similar because they both call a sequence of instructions to be executed in the main program. Procedures save memory and programming time, but require linkage to invoke the procedure, which requires a **CALL** and a **RET** instruction.

A *procedure* is a set of instructions that perform a specific task. They are invoked from another procedure — the calling procedure (or program) — and provide results to the calling program at the end of execution. Procedures (also called subroutines) are utilized primarily for routines that are called frequently by other procedures. The procedure routine is written only once, but used repeatedly; thereby, saving storage space. Procedures permit a program to be coded in modules; thus, making the program easier to code and test.

A *macro* is a segment of code that is written only once, but can be executed many times in the main program. When the macro is invoked (or called), the assembler replaces the *macro call* with the macro code, which is essentially an *instruction substitution* procedure. The macro code is then placed in-line with the main program, which usually requires more memory than a similar procedure. Macros generally make the program more readable and increase program execution, because there is no **CALL** or **RET** instruction.

15.1 Macro Definitions

A macro is specified by a unique name followed by the macro directive (MACRO). The macro may include a *parameter list* with a comma separating each parameter;

otherwise, there is no parameter list. This is followed by the body of the macro, which contains the instructions that define the macro. The macro code is terminated by the *end macro* directive (ENDM). Thus, the macro definition is bracketed by the MACRO and the ENDM directives.

A macro is a user-defined instruction that normally occurs at the beginning of a program or before the macro is called. The code between the MACRO and ENDM directives is a sequence of consecutive instructions that replace the macro being invoked. A general macro definition is shown below.

```
macro_name      MACRO   optional parameter list
                assembly language instructions
                ENDM
```

The macro is then called from the main program, as follows:

```
macro_name      optional parameter list
```

An example of a simple macro definition, called CALC, is shown in Figure 15.1, which adds two single-digit integers. The macro is called in the main program and includes the list of parameters that are to be utilized by the macro. Macro arguments A and B correspond to integers 3 and 4, respectively. Figure 15.2 shows the resulting code after the CALC macro has been invoked.

```
CALC    MACRO A, B          ;parameter list
        MOV   AL, A         ;operand A -> al
        AND   AL, 0FH       ;remove ascii bias

        MOV   BL, B         ;operand B -> bl
        AND   BL, 0FH       ;remove ascii bias

        ADD   AL, BL        ;add operand A and operand B
        OR    AL, 30H       ;add ascii bias to sum
        ENDM
;------------------------------------------------------------
.STACK
;------------------------------------------------------------
.DATA
;------------------------------------------------------------
.CODE
                            . . .

        CALC  3, 4          ;invoke the macro
        MOV   RSLT, AL      ;store result for display
                            . . .
```

Figure 15.1 Illustrates the use of a macro to add two single-digit integers showing the macro definition and invoking the macro.

```
CALC  MACRO A, B        ;parameter list
      MOV   AL, A       ;operand A -> al
      AND   AL, 0FH     ;remove ascii bias

      MOV   BL, B       ;operand B -> bl
      AND   BL, 0FH     ;remove ascii bias

      ADD   AL, BL      ;add operand A and operand B
      OR    AL, 30H     ;add ascii bias to sum
      ENDM
;-----------------------------------------------------------
.STACK
;-----------------------------------------------------------
.DATA
;-----------------------------------------------------------
.CODE
                                    . . .
;-----------------------------------------------------------
;macro is in-line with the main program
      CALC  3, 4        ;invoke the macro and pass parameters
      MOV   AL, 3       ;operand 3 -> al
      AND   AL, 0FH     ;remove ascii bias

      MOV   BL, 4       ;operand 4 -> bl
      AND   BL, 0FH     ;remove ascii bias

      ADD   AL, BL      ;add operand 3 and operand 4
      OR    AL, 30H     ;add ascii bias to sum

;-----------------------------------------------------------
      MOV   RSLT, AL    ;store result for display
                                    . . .
```

Figure 15.2 Illustrates the program segment after the macro has been invoked.

A macro can be called many times in a program and the code in a macro is reproduced each time the macro is invoked; however, there is only one macro definition in a program. Figure 15.3 shows a complete assembly language program that illustrates using the CALC macro, as defined in Figure 15.1 and Figure 15.2. The CALC macro is invoked twice using different arguments (parameters) and the resulting sums are then displayed. The arguments cannot designate a CPU register. The macro is defined at the beginning of the program before any segment is declared so that it can replace the macro reference where it is invoked in the program. Since the macro is called twice, two different result areas are required to store the two sums that will be displayed.

```
;macro_add.asm
;calculate (A + B)

;define calc macro
;-----------------------------------------------------------------
CALC   MACRO A, B
       MOV   AL, A
       AND   AL, 0FH            ;remove ascii bias
       MOV   BL, B
       AND   BL, 0FH            ;remove ascii bias
       ADD   AL, BL
       OR    AL, 30H            ;add ascii bias to sum
       ENDM
;-----------------------------------------------------------------
.STACK
;-----------------------------------------------------------------
.DATA
RSLT1 DB 0DH, 0AH, 'Result1 =        $'
RSLT2 DB 0DH, 0AH, 'Result2 =        $'
;-----------------------------------------------------------------
.CODE
BEGIN PROC   FAR

;set up pgm ds
       MOV   AX, @DATA          ;put addr of data seg in ax
       MOV   DS, AX             ;put addr in ds
;-----------------------------------------------------------------
;call the macro
       CALC  3, 4
       MOV   RSLT1+12, AL
;-----------------------------------------------------------------
       MOV   AH, 09H            ;display string--to display sum
       LEA   DX, RSLT1          ;put addr of rslt in dx
       INT   21H
;-----------------------------------------------------------------
;call the macro
       CALC  6, 3
       MOV   RSLT2+12, AL
;-----------------------------------------------------------------
       MOV   AH, 09H            ;display string--to display sum
       LEA   DX, RSLT2          ;put addr of rslt in dx
       INT   21H

;-----------------------------------------------------------------
BEGIN ENDP
       END   BEGIN
```

(a) //continued on next page

Figure 15.3 Assembly language program invoking the CALC macro twice to add two single-digit integers: (a) the program and (b) the outputs.

```
Result1 = 7
Result2 = 9                    (b)
```

Figure 15.3 (Continued)

15.2 Macro Examples

This section presents examples using macros with and without parameters. The examples consist of arithmetic operations, a code conversion program, and a sorting program that sorts numbers in ascending numerical order, among others.

Example 15.1 This example uses a macro that contains no parameters. The program adds a 2-digit augend and a 2-digit addend that are entered from the keyboard. The assembly language program uses the *ASCII adjust after addition* (AAA) instruction in the execution of the program. Figure 15.4 shows the parameter list (PARLST) in the data segment that contains the augend in OPFLD and OPFLD + 1 and the addend in OPFLD + 2 and OPFLD + 3. Figure 15.5 illustrates the program to calculate the sum of the two operands.

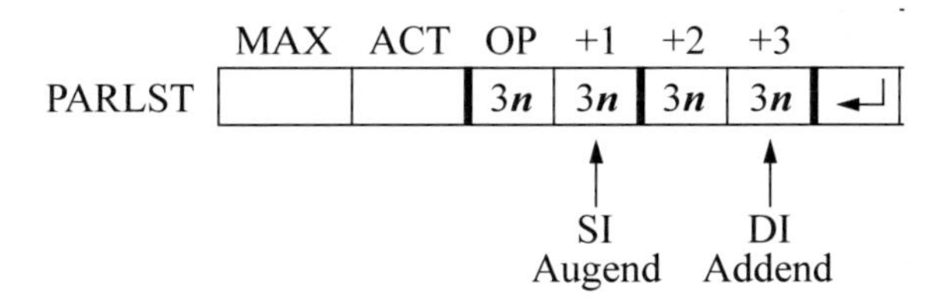

Figure 15.4 Parameter list containing the 2-digit augend and the 2-digit addend that are entered from the keyboard as hexadecimal numbers.

```
;macro_add2.asm
;calculate the sum of two 2-digit integers

;define calc macro
;--------------------------------------------------------------
CALC  MACRO
;set up addresses for ascii numbers
      CLC                        ;clear the carry flag (CF)
      LEA   SI, OPFLD+1          ;addr of augend low-order #
      LEA   DI, OPFLD+3          ;addr of addend low-order #
      LEA   BX, RSLT+10          ;addr of sum low-order #
      MOV   CX,2                 ;two loops for two digits
                            (a)        //continued on next page
```

Figure 15.5 Program that uses a macro to add two 2-digit operands: (a) the program and (b) the outputs.

```
;calculate sum
LP1:   MOV    AH, 00            ;clear ah; aaa adds 1 to ah
       MOV    AL, [SI]          ;get augend #
       ADC    AL, [DI]          ;add ascii # with carry
       AAA                      ;convert al sum to valid bcd #

       MOV    [BX], AL          ;move sum to next high-order loc
       DEC    SI                ;si points next high-order aug #
       DEC    DI                ;di points next high-order add #
       DEC    BX                ;bx points next high-order sum
       LOOP   LP1               ;if cx !=0, loop; get next #

;change sum to ascii for display
       LEA    BX, RSLT+10       ;addr of low-order sum
       MOV    CX, 2             ;two loops for two digits

LP2:   MOV    AL, [BX]          ;move next low-order sum # to al
       OR     AL, 30H           ;add ascii bias for display
       MOV    [BX], AL          ;move digit to sum location
       DEC    BX                ;bx points next high-order sum
       LOOP   LP2               ;if cx !=0, loop
                                ;change next # to ascii

       OR     AH, 30H           ;add ascii bias to display carry
       MOV    [BX], AH          ;when finished, store carry
       ENDM

;-------------------------------------------------------------
.STACK

;-------------------------------------------------------------
.DATA
PARLST   LABEL    BYTE
MAXLEN   DB 10
ACTLEN   DB ?
OPFLD    DB 10  DUP(?)
PRMPT    DB 0DH, 0AH, 'Enter 2 augend & 2 addend digits: $'
RSLT     DB 0DH, 0AH, 'Sum =            $'

;-------------------------------------------------------------
.CODE
BEGIN PROC  FAR

;set up pgm ds
       MOV    AX, @DATA         ;put addr of data seg in ax
       MOV    DS, AX            ;put addr in ds
                                      //continued on next page
```

Figure 15.5 (Continued)

```
;read prompt
      MOV    AH, 09H              ;display string
      LEA    DX, PRMPT            ;put addr of prompt in dx
      INT    21H                  ;dos interrupt

;keyboard request to enter characters
      MOV    AH, 0AH              ;buffered keyboard input
      LEA    DX, PARLST           ;put addr of parlst in dx
      INT    21H                  ;dos interrupt

;---------------------------------------------------------
;call the macro
      CALC

;---------------------------------------------------------
;display result
      MOV    AH, 09H              ;display string
      LEA    DX, RSLT             ;put addr of rslt in dx
      INT    21H                  ;dos interrupt

BEGIN ENDP
      END    BEGIN
```

```
Enter 2 augend & 2 addend digits: 1111
Sum = 022
----------------------------------------------------------
Enter 2 augend & 2 addend digits: 3478
Sum = 112
----------------------------------------------------------
Enter 2 augend & 2 addend digits: 7777
Sum = 154
----------------------------------------------------------
Enter 2 augend & 2 addend digits: 9999
Sum = 198
----------------------------------------------------------
Enter 2 augend & 2 addend digits: 2586
Sum = 111
----------------------------------------------------------
Enter 2 augend & 2 addend digits: 4444
Sum = 088
----------------------------------------------------------
Enter 2 augend & 2 addend digits: 9876
Sum = 174
```

(b)

Figure 15.5 (Continued)

Example 15.2 This example uses three calls to a macro that contains two parameters. The program obtains the product of two unpacked single-digit BCD operands. The assembly language program uses the *ASCII adjust AX after multiplication* (AAM) instruction in the execution of the program. Figure 15.6 illustrates the assembly language program to calculate the product of the two operands.

```
;macro_mul.asm
;use multiple macro calls to obtain the products
;of two unpacked bcd operands per macro execution

;define the MULT macro
;-----------------------------------------------------------
MULT   MACRO     A, B
;get the two numbers and unpack them
       MOV    AL, A                ;get multiplicand
       AND    AL, 0FH              ;unpack multiplicand

       MOV    AH, B                ;get multiplier
       AND    AH, 0FH              ;unpack multiplier

;multiply the two numbers,
;adjust them, and add ascii bias
       MUL    AH                   ;ax = al x ah
       AAM                         ;ax = 0d 0d
       OR     AX, 3030H            ;add ascii bias
       ENDM

;-----------------------------------------------------------
.STACK

;-----------------------------------------------------------
.DATA
RSLT1 DB   0DH, 0AH, 'Product1 =       $'
RSLT2 DB   0DH, 0AH, 'Product2 =       $'
RSLT3 DB   0DH, 0AH, 'Product3 =       $'

;-----------------------------------------------------------
.CODE
BEGIN PROC   FAR

;set up pgm ds
       MOV    AX, @DATA            ;put addr of data seg in ax
       MOV    DS, AX               ;move addr to ds
                                        //continued on next page
```

(a)

Figure 15.6 Program using three macro calls to obtain the product of two unpacked BCD operands: (a) the program and (b) the outputs.

```
;-----------------------------------------------------------
;call the macro
      MULT  9, 9

;display the result
      MOV   RSLT1+13, AH       ;move high product to result
      MOV   RSLT1+14, AL       ;move low product to result

      MOV   AH, 09H            ;display string
      LEA   DX, RSLT1          ;addr of rslt1 (product) -> dx
      INT   21H                ;dos interrupt

;-----------------------------------------------------------
;call the macro
      MULT  6, 5

;display the result
      MOV   RSLT2+13, AH       ;move high product to result
      MOV   RSLT2+14, AL       ;move low product to result

      MOV   AH, 09H            ;display string
      LEA   DX, RSLT2          ;addr of rslt2 (product) -> dx
      INT   21H                ;dos interrupt

;-----------------------------------------------------------
;call the macro
      MULT  3, 8

;display the result
      MOV   RSLT3+13, AH       ;move high product to result
      MOV   RSLT3+14, AL       ;move low product to result

      MOV   AH, 09H            ;display string
      LEA   DX, RSLT3          ;addr of rslt3 (product) -> dx
      INT   21H                ;dos interrupt

;-----------------------------------------------------------
BEGIN ENDP
      END   BEGIN
```

```
Product1 = 81
Product2 = 30
Product3 = 24
```

(b)

Figure 15.6 (Continued)

Example 15.3 This example presents an assembly language program that converts a binary code to the corresponding Gray code. The Gray code belongs to a class of cyclic codes called reflective codes and is an unweighted code where only one input changes between adjacent code words. The *i*th Gray code bit g_i can be obtained from the corresponding binary code word by the following algorithm:

$$g_{n-1} = b_{n-1}$$
$$g_i = b_i \oplus b_{i+1} \qquad (15.1)$$

for $0 \leq i \leq n - 2$, where the symbol $\oplus$ denotes modulo-2 addition defined as:

$$0 \oplus 0 = 0 \qquad 0 \oplus 1 = 1 \qquad 1 \oplus 0 = 1 \qquad 1 \oplus 1 = 0$$

Using the algorithm shown in Equation 15.1, an 8-bit binary code can be converted to the corresponding 8-bit Gray code, as follows:

Gray		Binary
g_7	=	b_7
g_6	=	$b_6 \oplus b_7$
g_5	=	$b_5 \oplus b_6$
g_4	=	$b_4 \oplus b_5$
g_3	=	$b_3 \oplus b_4$
g_2	=	$b_2 \oplus b_3$
g_1	=	$b_1 \oplus b_2$
g_0	=	$b_0 \oplus b_1$

For example, a binary code word of 10001101 converts to the corresponding Gray code word of 11001011, as shown in Figure 15.7.

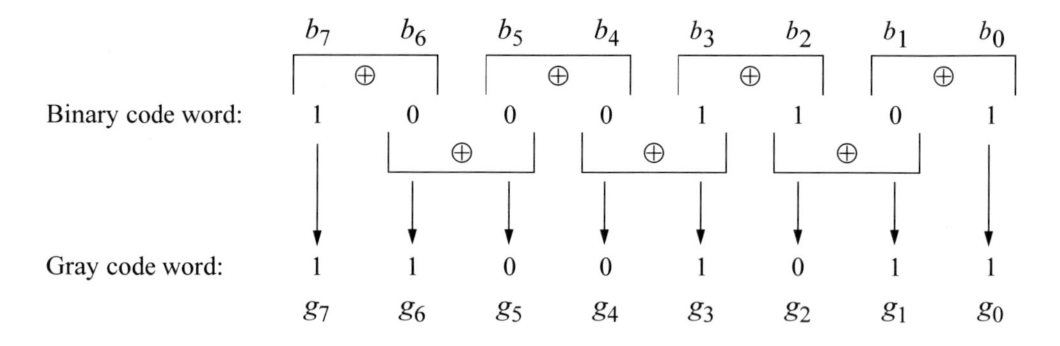

Figure 15.7 Example of converting a binary code word of 10001101 to the Gray code word of 11001011.

Figure 15.8 shows the parameter list (PARLST) in the data segment that contains representative binary data that was entered from the keyboard and stored in the OPFLD section. This data will be used by an assembly language program presented in this example. The first bit entered is the high-order bit and is stored in location OP (OPFLD + 0). The eight bits of the OPFLD depict a binary number of 10001101 in hexadecimal notation, which includes the ASCII bias of 3 for numerical digits. The appropriate bits are then exclusive-ORed, the ASCII bias is added to the result, and the resulting Gray code bit is stored in the result area (RSLT). When two ASCII digits are exclusive-ORed, the result is a digit with no ASCII bias.

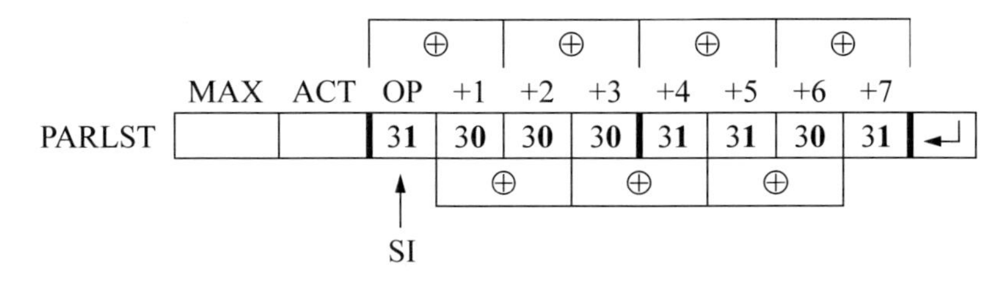

Figure 15.8 Example of binary data entered from the keyboard and stored in the OPFLD location of the parameter list.

Figure 15.9 illustrates an assembly language program — using a macro (BIN_GRAY) with no parameters — to translate an 8-bit binary code word to the corresponding 8-bit Gray code word. The binary data is entered from the keyboard. Since no operation is required to translate the high-order binary bit to the high-order Gray code bit, a count of seven is entered in the counter general-purpose register CX — Gray code bit g_7 is obtained directly from the binary bit b_7.

```
;macro_bin_gray.asm
;use a macro with no parameters
;to convert from binary to gray code

;define the binary-to-gray macro (BIN_GRAY)
;------------------------------------------------------------
BIN_GRAY MACRO
;move first bit unchanged to rslt
      MOV    AL, [SI]            ;si points to 1st bit in opfld
      MOV    [DI], AL            ;di points to the rslt + 14

                                      //continued on next page
```

(a)

Figure 15.9 An assembly language program to translate an 8-bit binary code word to the corresponding Gray code word: (a) the program and (b) the outputs.

```
;set up loop to process characters
NEXT: MOV    AL, [SI]              ;move bit to al
      INC    SI                    ;inc si to point to next bit
      MOV    AH, [SI]              ;move next bit to ah
      XOR    AL, AH                ;perform the xor operation
      ADD    AL, 30H               ;add ascii bias
      INC    DI                    ;inc di to point to
                                   ;... the next location in rslt
      MOV    [DI], AL              ;move the gray code bit to
                                   ;... the rslt area
      LOOP   NEXT                  ;decrement cx by 1.
                                   ;If cx != 0, then loop
      ENDM

;-------------------------------------------------------------
.STACK

;-------------------------------------------------------------
.DATA
PARLST   LABEL BYTE
MAXLEN   DB 12
ACTLEN   DB ?
OPFLD    DB 12 DUP(?)
PRMPT    DB 0DH, 0AH, 'Enter 8 binary bits: $'
RSLT     DB 0DH, 0AH, 'Gray code =           $'

;-------------------------------------------------------------
.CODE
BEGIN PROC   FAR

;set up pgm ds
      MOV    AX, @DATA             ;put addr of data seg in ax
      MOV    DS, AX                ;put addr in dx

;read prompt
      MOV    AH, 9                 ;display string
      LEA    DX, PRMPT             ;put addr of prompt in dx
      INT    21H                   ;dos interrupt

;keyboard request rtn to enter numbers
      MOV    AH, 0AH               ;buffered keyboard input
      LEA    DX, PARLST            ;load addr of parlst
      INT    21H                   ;dos interrupt

                                      //continued on next page
```

Figure 15.9 (Continued)

```
;set up count for number of bits
       MOV    CX, 7               ;set up count
       LEA    SI, OPFLD           ;set up addr of opfld
       LEA    DI, RSLT + 14       ;set up addr of rslt

;------------------------------------------------------------
;call the macro
         BIN_GRAY

;------------------------------------------------------------
;display the result
       MOV    AH, 9               ;display string
       LEA    DX, RSLT            ;put addr of rslt in dx
       INT    21H                 ;dos interrupt

BEGIN  ENDP
       END    BEGIN
```

```
Enter 8 binary bits: 10101110
Gray code = 11111001
-------------------------------------------------------------
Enter 8 binary bits: 01111010
Gray code = 01000111
-------------------------------------------------------------
Enter 8 binary bits: 10001101
Gray code = 11001011
-------------------------------------------------------------
Enter 8 binary bits: 11111111
Gray code = 10000000
-------------------------------------------------------------
Enter 8 binary bits: 01010110
Gray code = 01111101
```

(b)

Figure 15.9 (Continued)

Example 15.4 Figure 15.10 illustrates an assembly language program to calculate the area of a triangle from two single-digit integers using a macro with two parameters. The parameters represent the base and the height of a triangle. The ASCII bias is removed from the parameters and the base is divided by two using a *shift logical right* (SHR) instruction. The result is then multiplied by the height and converted to two unpacked decimal numbers in register AX by the *ASCII adjust AX after multiply* (AAM) instruction. The product is then converted to ASCII for display. The macro labelled AREA is invoked three times.

```
;macro_area.asm
;calculate the area of triangle
;(1/2 base x height) using
;a macro with parameters

;define area macro
AREA   MACRO A, B
;get the two numbers and unpack them
       MOV    AL, A              ;get base
       AND    AL, 0FH            ;remove ascii bias
       SHR    AL, 1              ;divide base by 2

       MOV    AH, B              ;get height
       AND    AH, 0FH            ;remove ascii bias

       MUL    AH                 ;multiply 1/2 base x height
       AAM                       ;convert to 2 unpacked
                                 ;decimal numbers in ax
       OR     AX, 3030H          ;convert to ascii
       ENDM

;------------------------------------------------------------
.STACK

;------------------------------------------------------------
.DATA

AREA1   DB 0DH, 0AH, 'Area of triangle1 (4, 8) =    $'
AREA2   DB 0DH, 0AH, 'Area of triangle2 (8, 8) =    $'
AREA3   DB 0DH, 0AH, 'Area of triangle3 (7, 9) =    $'
AREA4   DB 0DH, 0AH, 'Area of triangle4 (8, 5) =    $'

;------------------------------------------------------------
.CODE
BEGIN PROC FAR

;set up pgm ds
       MOV    AX, @DATA          ;put addr of data seg in ax
       MOV    DS, AX             ;move addr to ds

;------------------------------------------------------------
                                   //continued on next page
```

(a)

Figure 15.10 Program to calculate the area of a triangle using a macro with two parameters: (a) the program and (b) the outputs.

```
;----------------------------------------------------------
;call the area macro for area1
      AREA  4, 8
;display the area of triangle 1
      MOV   AREA1+29, AH       ;move high number
      MOV   AREA1+30, AL       ;move low number

      MOV   AH, 09H            ;display string
      LEA   DX, AREA1          ;addr of area1 to dx
      INT   21H                ;dos interrupt

;----------------------------------------------------------
;call the area macro for area2
      AREA  8, 8
;display the area of triangle 2
      MOV   AREA2+29, AH       ;move high number
      MOV   AREA2+30, AL       ;move low number

      MOV   AH, 09H            ;display string
      LEA   DX, AREA2          ;addr of area2 to dx
      INT   21H                ;dos interrupt

;----------------------------------------------------------
;call the area macro for area3
      AREA  7, 9
;display the area of triangle 3
      MOV   AREA3+29, AH       ;move high number
      MOV   AREA3+30, AL       ;move low number

      MOV   AH, 09H            ;display string
      LEA   DX, AREA3          ;addr of area3 to dx
      INT   21H                ;dos interrupt

;----------------------------------------------------------
;call the area macro for area4
      AREA  8, 5
;display the area of triangle 4
      MOV   AREA4+29, AH       ;move high number
      MOV   AREA4+30, AL       ;move low number

      MOV   AH, 09H            ;display string
      LEA   DX, AREA4          ;addr of area4 to dx
      INT   21H                ;dos interrupt

;----------------------------------------------------------
BEGIN ENDP
      END   BEGIN                    //continued on next page
```

Figure 15.10 (Continued)

```
Area of triangle1 (4, 8) = 16
Area of triangle2 (8, 8) = 32
Area of triangle3 (7, 9) = 27 //SHR rounds down if odd base
Area of triangle4 (8, 5) = 20
```

(b)

Figure 15.10 (Continued)

Example 15.5 Figure 15.11 presents an assembly language program that utilizes a macro to sort ten single-digit integers — that are entered from the keyboard — in ascending numerical order. Since there are ten digits, only nine comparisons are required. Also, it is not necessary to remove the ASCII bias from the digits, because all the numbers have the same bias — the only difference is the low-order four bits.

Each number is compared to all the other numbers in the OPFLD to determine the smallest number, which is then stored in the next lower-order location of the OPFLD area of the parameter list. The program enters five separate 10-digit sequences, then displays the sorted numbers in ascending numerical order.

```
;macro_sort.asm
;sort 10 integers in ascending numerical order

;-----------------------------------------------------------
;define the sort macro
SORT   MACRO
       LEA    SI, OPFLD          ;addr of first byte -> si
       MOV    DX, 9              ;# of bytes -> dx as ctr

LP1:   MOV    CX, DX             ;cx is loop2 ctr

LP2:   MOV    AL, [SI]           ;number -> al
       MOV    BX, CX             ;bx is base addr
       CMP    AL, [BX+SI]        ;is the number <= to al?
       JBE    NEXT               ;yes, get next number
       XCHG   AL, [BX+SI]        ;number is !<=; put # in al
       MOV    [SI], AL           ;put small number in opfld

NEXT:  LOOP   LP2                ;loop until all #s are compared
       INC    SI                 ;si points to next number
       DEC    DX                 ;decrement loop ctr
       JNZ    LP1                ;compare remaining numbers
       ENDM
```

(a) //continued on next page

Figure 15.11 Program to sort ten single-digit integers in ascending numerical order using a macro: (a) the program and (b) the outputs.

```
;---------------------------------------------------------------
.STACK

;---------------------------------------------------------------
.DATA
PARLST    LABEL   BYTE
MAXLEN    DB 15
ACTLEN    DB ?
OPFLD     DB 15  DUP(?)
PRMPT     DB 0DH, 0AH, 'Enter numbers: $'
RSLT      DB 0DH, 0AH, 'Sorted numbers =             $'

;---------------------------------------------------------------
.CODE
BEGIN PROC   FAR

;set up pgm ds
      MOV    AX, @DATA           ;put addr of data seg in ax
      MOV    DS, AX              ;move addr to ds

;read prompt
      MOV    AH, 09H             ;display string
      LEA    DX, PRMPT           ;put addr of prompt in dx
      INT    21H                 ;dos interrupt

;keyboard request rtn to enter characters
      MOV    AH, 0AH             ;buffered keyboard input
      LEA    DX, PARLST          ;put addr of parlst in dx
      INT    21H                 ;dos interrupt

;---------------------------------------------------------------
;and call sort macro
      SORT

;---------------------------------------------------------------
;set up src (opfld) and dst (rslt)
      LEA    SI, OPFLD
      LEA    DI, RSLT+20
      MOV    CL, 10

;move sorted numbers to result area
RSLT_AREA:
      MOV    BL, [SI]            ;opfld number to bl
      MOV    [DI], BL            ;number to result area
      INC    SI                  ;set up next source
      INC    DI                  ;set up next destination
      LOOP   RSLT_AREA           ;loop if cl != 0    //next page
```

Figure 15.11 (Continued)

```
;display sorted numbers
        MOV    AH, 09H              ;display string
        LEA    DX, RSLT             ;put addr of rslt in dx
        INT    21H                  ;dos interrupt

BEGIN ENDP
        END    BEGIN
```

```
Enter numbers: 9873453241
Sorted numbers = 1233445789
------------------------------------------------------------
Enter numbers: 7652348971
Sorted numbers = 1234567789
------------------------------------------------------------
Enter numbers: 5406798373
Sorted numbers = 0334567789
------------------------------------------------------------
Enter numbers: 9876543210
Sorted numbers = 0123456789
------------------------------------------------------------
Enter numbers: 5151515151
Sorted numbers = 1111155555
```

(b)

Figure 15.11 (Continued)

Example 15.6 This example uses a macro in a program to detect odd parity in an 8-bit code word that is entered from the keyboard. The parity bit is an extra bit that is added to a code word to make the overall parity — the code word bits plus the parity bit — contain either an odd or an even number of 1s, as shown in Table 15.1 for both odd and even parity using a 4-bit code word.

The parity bit can be generated by modulo-2 addition, as previously defined in Example 15.3. Equation 15.2 depicts a method to maintain even parity for a 4-bit message $m_3\, m_2\, m_1\, m_0$ using modulo-2 addition. The parity bit for odd parity generation is the complement of Equation 15.2.

$$\text{Parity bit (even)} = m_3 \oplus m_2 \oplus m_1 \oplus m_0 \tag{15.2}$$

Parity implementation can detect an odd number of errors, but cannot correct the errors, because the bits in error cannot be determined. If a single error occurred, then an incorrect code word would be generated and the error would be detected. If two errors occurred, then the parity would be unchanged and still be correct; however, the code word would be incorrect.

Table 15.1 Parity Bit Generation

Code Word	Parity Bit (odd)	Code Word	Parity Bit (even)
0000	1	0000	0
0001	0	0001	1
0010	0	0010	1
0011	1	0011	0
0100	0	0100	1
0101	1	0101	0
0110	1	0110	0
0111	0	0111	1
1000	0	1000	1
1001	1	1001	0
1010	1	1010	0
1011	0	1011	1
1100	1	1100	0
1101	0	1101	1
1110	0	1110	1
1111	1	1111	0

Figure 15.12 shows the parameter list containing one of the 8-bit code words — 10001101 — that will be used in the program. Bits (30H) and (31H) are hexadecimal notations for binary bits 0 and 1, respectively. Since the exclusive-OR function will be used in the macro, the ASCII bias does not have to be removed. Register SI points to the first bit in the OPFLD area of the parameter list.

Figure 15.13 shows the program utilizing a macro called PARITY that contains no parameters. General-purpose register BL is used to contain the generated odd parity bit for the 8-bit code word. The register is initialized to a value of 01H and will be toggled when each code word bit is examined, depending on the bits in the code word. Register CX contains a value of eight, which is the number of bits in the code word, and will be used with the LOOP instruction in the macro.

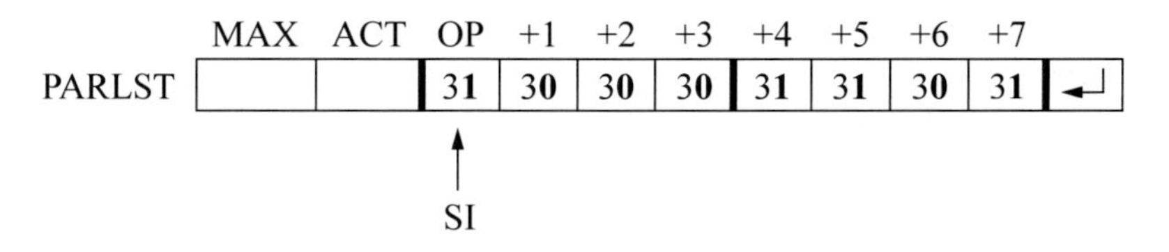

Figure 15.12 Example of the parameter list for a program to detect the parity in an 8-bit code word.

```
;macro_parity.asm
;use a macro to indicate the parity of 8 bits

;------------------------------------------------------------
;define the parity macro
PARITY   MACRO
NEXT: MOV    AL, [SI]             ;bit -> al
      CMP    AL, 31H              ;is bit = 1?
      JNE    ZERO                 ;if bit = 0, jump
      XOR    BL, 01H              ;if bit = 1, toggle parity flag
ZERO: INC    SI                   ;si points to next bit
      LOOP   NEXT                 ;if cx != 0, get next bit
      ENDM

;------------------------------------------------------------
.STACK

;------------------------------------------------------------
.DATA
PARLST  LABEL   BYTE
MAXLEN  DB      15
ACTLEN  DB      ?
OPFLD   DB      20  DUP(?)
PRMPT   DB      0DH, 0AH, 'Enter code word: $'
RSLT    DB      0DH, 0AH, 'Parity bit =   $'

;------------------------------------------------------------
.CODE
BEGIN PROC   FAR

;set up pgm ds
      MOV    AX, @DATA            ;put addr of data seg in ax
      MOV    DS, AX               ;move addr to ds

;read prompt
      MOV    AH, 09H              ;display string
      LEA    DX, PRMPT            ;addr of prompt -> dx
      INT    21H                  ;dos interrupt

;keyboard request to enter characters
      MOV    AH, 0AH              ;buffered keyboard input
      LEA    DX, PARLST           ;addr of parlst -> dx
      INT    21H                  ;dos interrupt
;------------------------------------------------------------
                                        //continued on next page
```

(a)

Figure 15.13 Program to generate the odd parity bit for an 8-bit code word: (a) the program and (b) the outputs.

```
;-----------------------------------------------------------
;initialize parity flag to 1 and count in cx to 8
;initialize source (si) and destination (di) addr
      MOV   BL, 01H           ;set parity flag in bl to 1
      MOV   CX, 08H           ;number of bits is 8
      LEA   SI, OPFLD         ;addr of opfld -> si
      LEA   DI, RSLT+15       ;addr of result -> di

;-----------------------------------------------------------
;call parity macro
        PARITY

;-----------------------------------------------------------
      OR    BL, 30H           ;add ascii bias for parity
      MOV   [DI], BL          ;parity bit -> result area

;-----------------------------------------------------------
;display result
      MOV   AH, 09H           ;display string
      LEA   DX, RSLT          ;addr of result -> dx
      INT   21H               ;dos interrupt

BEGIN ENDP
      END   BEGIN
```

```
Enter code word: 10001101
Parity bit = 1
-----------------------------------------------------------
Enter code word: 10110000
Parity bit = 0
-----------------------------------------------------------
Enter code word: 10101010
Parity bit = 1
-----------------------------------------------------------
Enter code word: 11111111
Parity bit = 1
-----------------------------------------------------------
Enter code word: 01111111
Parity bit = 0
-----------------------------------------------------------
Enter code word: 11000011
Parity bit = 1
-----------------------------------------------------------
Enter code word: 11001110
Parity bit = 0
```

(b)

Figure 15.13 (Continued)

15.3 Problems

All of the programs in this section are to be written in assembly language, not embedded in a C program.

15.1 Design a program using a macro that adds two single-digit packed BCD operands that are entered from the keyboard and display the results. Use hexadecimal characters 0 through 9 and higher-valued hexadecimal numbers as valid packed BCD numbers, such as the ASCII characters @ (40H) through *Y* (59H) and ' (60H) through *y* (79H). For example, 2 (32H) + R(52H) = 084 and s(73H) + 9(39H) = 112.

15.2 Write a program using a macro that sorts from two to ten single-digit integers in ascending numerical order that are entered from the keyboard. Enter several different sets of integers ranging from two integers to ten integers and display the results.

15.3 Write a program using a macro that is invoked twice to exchange two single-digit integers and to exchange two alphabetic characters and display the results. The macro is defined using two parameters.

15.4 Write a program using a macro to reverse the order and change to uppercase — if necessary — all alphabetic characters that are entered from the keyboard. Enter several sets of characters of different lengths, both lowercase and uppercase, and display the results.

15.5 Write a program that uses two macros. One macro generates an odd parity bit for an 8-bit code word that is entered from the keyboard. If the parity is odd, then display a 0; if the parity is even, then display a 1. The other macro removes all nonletters from data that are entered from the keyboard.

15.6 Write a program that uses two macros. One macro is called twice to switch the locations of two characters that are listed as parameters. The other macro has no parameters and makes all characters uppercase that are entered from the keyboard. It also reverses the order of the characters.

15.7 Write a program using a macro with no parameters that adds two single-digit integers that are entered from the keyboard. Enter several integers and display the results.

15.8 Write a program using a macro with no parameters that multiplies two single-digit integers that are entered from the keyboard. Enter several integers and display the results.

15.9 Write an assembly language program — not using a macro — that performs subtraction on two 2-digit integers. The program can easily be changed to a macro-oriented program by simply placing the arithmetic part of the program within the macro segment. Enter several different sets of integer pairs in which the entire minuend is greater than, equal to, or less than the entire subtrahend. Also enter integers in which the individual minuend digits are greater than, equal to, or less than the individual subtrahend digits. Display the results of all operations. If the subtraction operation results in a negative difference, then display a negative sign for the result. The parameter list is shown below for this problem.

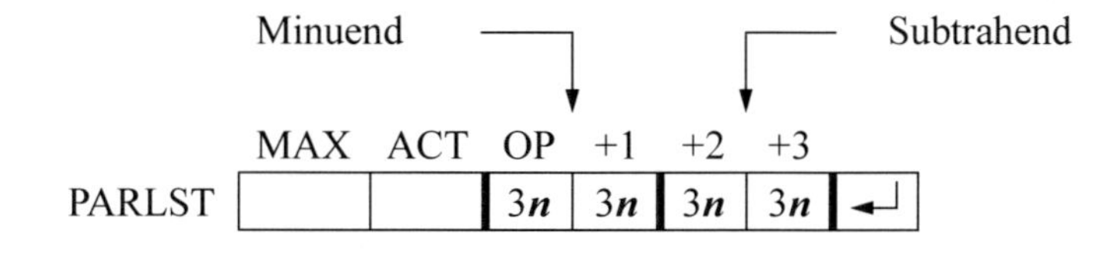

16

Interrupts and Input/Output Operations

An *interrupt* is an asynchronous occurrence that is generally initiated by an input/output device; for example, the device is ready to transfer data or to receive data. When an interrupt occurs, the processor stops execution of the current operation and transfers control to a procedure that is specifically designed to handle the interrupt. Application programs can access these procedures — whose addresses are located in an *interrupt descriptor table* (IDT) of the operating system — by means of instructions that are similar to the procedure CALL instruction. The IDT provides access to predefined and used-defined interrupts and each entry in the IDT is identified by an interrupt vector number.

An *exception* is similar to an interrupt, but is a synchronous event caused by a predefined condition that occurs during the execution of an instruction. This chapter also provides an introduction to input/output operations, such as *direct memory access* (DMA), *memory-mapped I/O*, and the *input/output instructions* (IN/OUT).

16.1 Interrupts

This section provides a brief introduction to calling interrupts and returning from interrupts. When an interrupt occurs, the processor suspends operation of the current program and pushes the contents of the following registers onto the stack: first the EFLAGS register, then the code segment (CS) register, and then the instruction pointer (EIP) register. The processor also resets the interrupt enable flag (IF) so that any additional interrupts will not affect the execution of the interrupt program. An error code may also be pushed onto the stack.

A software-generated interrupt is initiated by the *call to interrupt procedure* (INT *n*) instruction, which is a typical mnemonic for a call to an interrupt handler. The syntax for a typical interrupt call is shown below.

```
INT    immed8        //immed8 is an immediate
                     //byte that specifies an
                     //interrupt vector number
```

The destination operand of an interrupt or exception specifies an address range of 0 to 255 in the IDT, where each address contains an offset of 8-byte descriptors (protected mode) or 4-byte interrupt vectors (real-address mode: IP:CS). The IDT is referred to as the interrupt vector table (IVT) in real-address mode. The contents of the IDT and IVT addresses point to an operating system procedure for a particular type of interrupt. There are 18 predefined interrupts and exceptions, 224 user-defined interrupts, and 13 addresses in the IDT that are reserved.

Return from an interrupt is generated by the *interrupt return* (IRET) instruction, which is similar to the procedure far return (RET) instruction. The IRET instruction pops the previous contents of the CS register, the EIP register, and the EFLAGS register from the stack that were pushed onto the stack by the interrupt. The stack pointer (ESP) register is incremented to the correct setting and execution of the interrupted program continues at the instruction immediately following the INT instruction. Four typical interrupt vectors are shown below.

- The INT 00H instruction is used to generate an interrupt for divide by zero.

```
INT    00H           //generates an interrupt for
                     //DIV and IDIV instructions if an
                     //attempt is made to
                     //divide by zero,
                     //located at address 000H
```

- The INT 04H instruction is used to generate an interrupt for divide overflow.

```
INT    04H           //generates an interrupt for a
                     //an arithmetic operation if
                     //the overflow flag is set,
                     //located at address 010H
```

- The INT 21H instruction is used to generate an interrupt for screen display using a function code of 09H. This interrupt is used to display a string that is entered elsewhere in the program and is terminated by a $ sign (24H). The string is bracketed by apostrophes. All the INT 21H instructions require a function code to be stored in general-purpose register AH prior to executing the INT 21H instruction.

```
INT    21H           //A function code of 09H in AH
```

- The INT 21H instruction is used to generate an interrupt for buffered keyboard input using a function code of 0AH. This interrupt is used to store a character string that is entered from the keyboard into a parameter list. All the INT 21H instructions require a function code to be stored in general-purpose register AH prior to executing the INT 21H instruction.

```
INT   21H          //A function code of 0AH in AH
```

The interrupt flag (IF) in the EFLAGS register can be set by the *set interrupt flag* (STI) instruction. When the IF flag is set by the STI instruction, the processor commences responding to maskable interrupts after the next instruction has been executed, permitting interrupts to be enabled just prior to returning from a procedure. Interrupts can be masked by an external device so that they are ignored by the processor. Nonmaskable interrupts are those that must be acknowledged by the processor. The IF flag can be reset by the *clear interrupt flag* (CLI), permitting the processor to disregard maskable interrupts.

Input/output devices have highest priority over central processing units when transferring data to or from main memory, because they are inherently slower and cannot have the data transfer stopped temporarily. Disk drives and tape drives are examples of I/O subsystems in this category. This may be resolved by including in the I/O device data bus hardware a first-in, first-out (FIFO) buffer to temporarily store the data during a data transfer operation. Another technique is to grant the I/O device bus control when requested — a process called *cycle stealing* from the CPU, although this decreases the speed of the CPU.

In summary, interrupts transfer control from the current program to an *interrupt service routine* (ISR) when an internally or externally generated interrupt request has been generated. Internally generated interrupt requests originate from the CPU — for example, when a divide by zero, a divide overflow, or an INT 21H instruction occurs. Externally generated interrupt requests are asynchronous events that originate from an I/O device and suspend the currently running CPU program. When the interrupt operation is completed, the CPU restores the appropriate registers and returns to the interrupted program.

16.2 Direct Memory Access

Direct memory access (DMA) allows an I/O device control unit to transfer data directly to or from main memory without CPU intervention. This is a much faster data transfer operation allowing both the processor and the I/O device to operate concurrently in most cases. The DMA technique is used primarily to transfer large amounts of data between the I/O device and main memory without continual intervention by the CPU. DMA transfers involve a device control unit specifically designed for DMA operations and replaces software with hardware.

When the CPU receives a DMA request from an I/O device, it suspends its operation at certain breakpoints and releases control of the memory bus, address bus, and control signals, then asserts an acknowledgement to the I/O device. Possible breakpoints are shown in Figure 16.1 for a typical instruction cycle. These breakpoints allow the processor to suspend execution of the current program without loss of program continuity. The breakpoints occur just prior to when the processor requires use of the buses. Interrupt breakpoints are allowed only at the end of an instruction cycle.

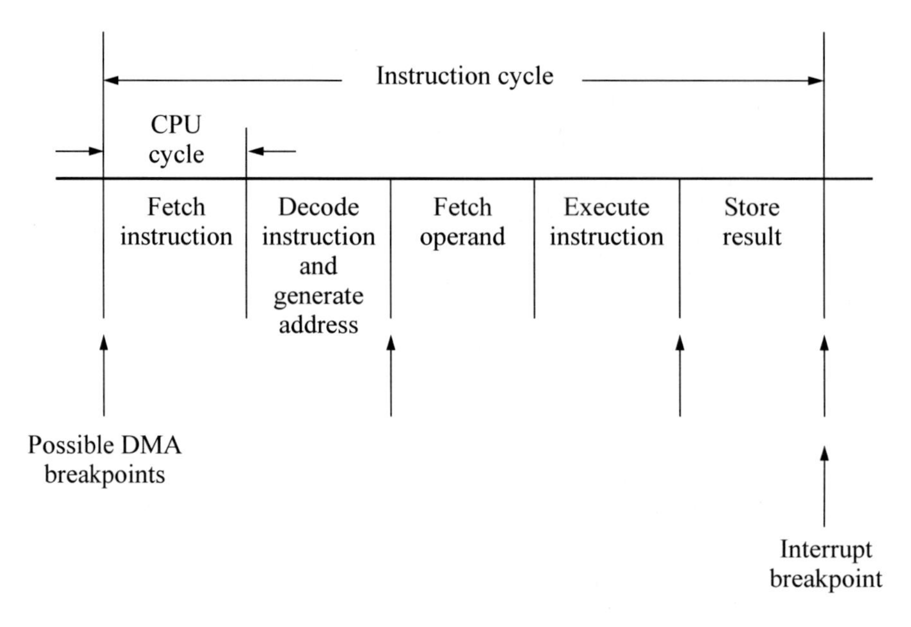

Figure 16.1 Typical instruction cycle showing possible DMA breakpoints.

Figure 16.2 shows a typical I/O device subsystem containing a dedicated DMA control unit and I/O device for a single-bus structure. The DMA device control unit contains the following registers:

- A memory data register (MDR), which contains the data to be transferred to memory or received from memory.
- A memory address register (MAR), which contains the starting address in memory where the data is to be stored or from where the data is to be retrieved. When a word of data is transferred to or from memory, the contents of MAR are incremented to the next contiguous address.
- A word count register (WCR), which is used to determine the number of words to be transferred. When a word of data is transferred to or from memory, the contents of WCR are decremented by a value of one.
- A command register, which is used to contain the type of DMA operation to be

performed — either a read or a write operation. The register may contain several different commands depending on the I/O device. This information may be contained in a separate control and status register.

- A control and status register (CSR), which may contain information on the status of the device — whether the device is ready, the type of data transfer operation — read or write, any errors that occured during the data transfer operation, an enable interrupt bit, and any other bits that are specific to the type of I/O device.

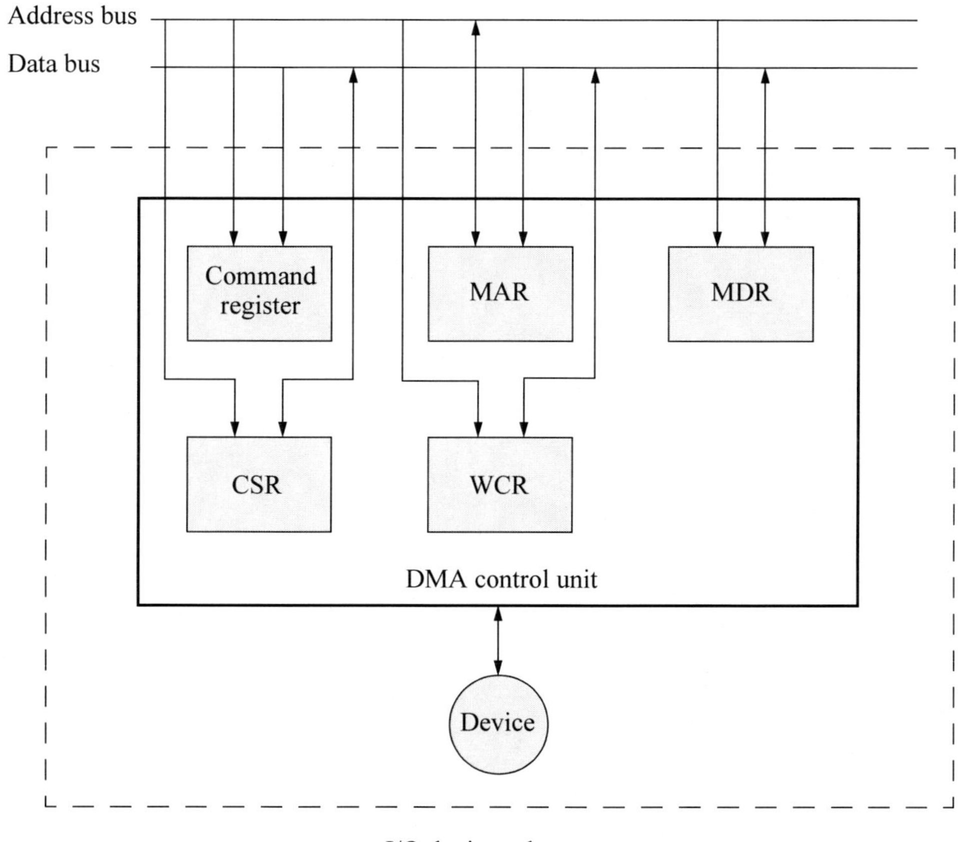

Figure 16.2 Typical single-bus hardware for a DMA input/output device.

A DMA transfer is initiated by the CPU by sending the device control unit the starting memory address, the number of words to be transferred, and a read or write command. When the data transfer is complete, the control unit signals the CPU by asserting an interrupt signal.

The address bus addresses all five blocks shown in Figure 16.2. The data bus contains the appropriate information depending on the specific register that is addressed. The CPU loads the starting address of memory into MAR, then loads the number of words to be transferred into WCR, then sends the command (read or write) to the command register — alternatively, the command may be placed in the CSR. The WCR may send data to the CPU that represents a residual count; that is, a count indicating that not all data was transferred to or from memory.

The command register in Figure 16.2 may contain several different commands depending on the type of I/O device. For example, a tape drive can be issued any of the following commands, among others:

- A *read* command to read a record on tape.
- A *write* command to write a record on tape.
- A *rewind* command to rewind the tape to the beginning of tape.
- A *back space* command to back space over a record or to back space over a specific number of records.
- A *forward space* command to forward space over a record or to forward over a specific number of records.
- A *write end of file* command.

16.3 Memory-Mapped I/O

For single-bus machines, the same bus can be utilized for both memory and I/O devices. Therefore, I/O devices may be assigned a unique address within main memory, which is partitioned into separate areas for memory and I/O devices. Using the memory-mapped technique, I/O devices are accessed in the same way as memory locations, providing significant flexibility in managing I/O operations. Thus, there are no separate I/O instructions, and the I/O devices can be accessed utilizing any of the memory read or write instructions and their addressing modes.

The memory space of an I/O device is associated with a specific set of address values used as registers. Therefore, when the CPU generates a memory address, it may refer to a main memory address or to a register in an I/O device, depending on the address. Using an I/O-mapped technique, I/O devices are accessed using the IN and OUT type of instructions described in Section 16.4.

Figure 16.3 illustrates the organization for a typical memory-mapped I/O structure. A set of four interface paths is required to communicate with memory and I/O devices: a data bus, an address bus, a read signal, and a write signal. The CPU decodes a memory reference instruction, then activates the read or write control signal. This initiates either a memory access cycle or an I/O transfer, depending on the address.

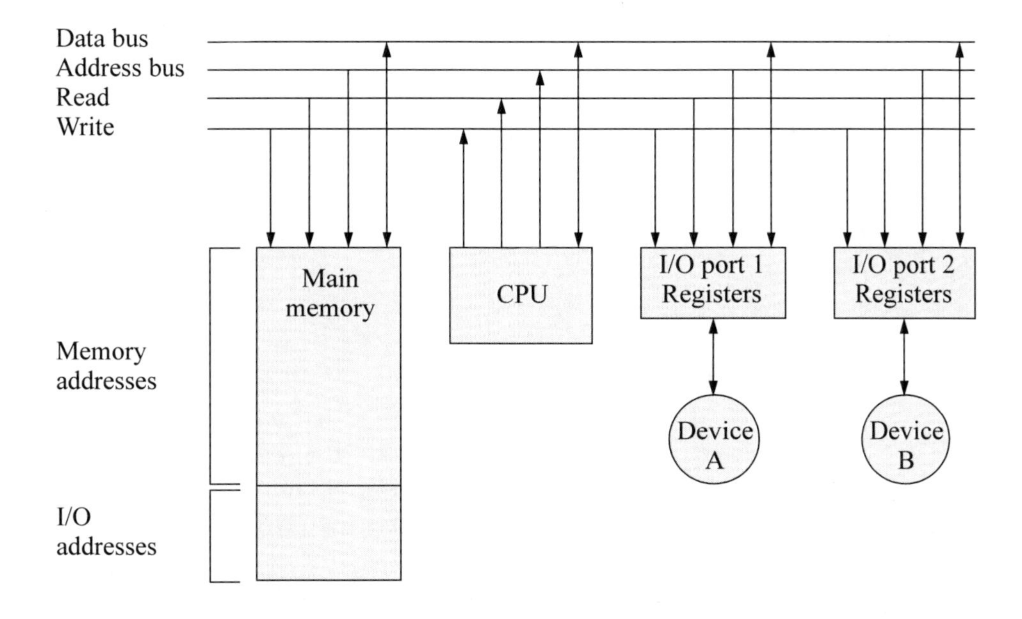

Figure 16.3 Typical memory-mapped I/O organization for a single-bus structure.

Communicating to an I/O device, therefore, is the same as reading and writing to memory addresses that are dedicated to the I/O device. The same protocol is used to communicate with the CPU and an I/O device as with the CPU and main memory. An I/O port is handled the same as a memory location; however, the memory-mapped I/O technique reduces the number of addresses available for main memory.

16.4 In/Out Instructions

There are two types of instructions that are specifically designed to communicate with I/O subsystems: IN and OUT. The IN instruction transfers data from an I/O port to register AL, AX, or EAX, depending on the size of the input data. The OUT instruction transfers data from register AL, AX, or EAX to an I/O port, depending on the size of the output data. These are referred to as *register I/O* instructions. The input or output port is specified by either an immediate byte or the contents of register DX.

There are also two types of instructions that are specifically designed to transfer string data between memory and an I/O port: INS and OUTS. The INS instructions transfer string data from an I/O port to memory. These are INSD, INSW, and INSB, which transfer doubleword, word, or byte data, respectively. The OUTS instructions transfer string data from memory to an I/O port. These are OUTD, OUTW, and OUTB, which transfer doubleword, word, or byte data, respectively. These are referred to as *string (block) I/O* instructions. The input or output port is specified by either an immediate byte or the contents of register DX.

The operation of the *string I/O* string instructions is similar to the string instructions presented in Chapter 13. The (E)SI and (E)DI registers are used for string operations as source and destination pointers, respectively — they point to string and destination memory elements. The *repeat* (REP) prefixes specify the condition for which the instruction is to be executed. The general-purpose registers (E)SI and (E)DI are automatically incremented or decremented after each repetition of the string instruction to point to the next byte, word, or doubleword to implement block moves for string operations.

16.4.1 Register I/O IN Instructions

There are six categories of the IN instruction referred to as *register I/O* instructions: three using immediate operands to indicate the port address for byte, word, or doubleword operands; and three using general-purpose register DX to indicate the port address for byte, word, or doubleword operands.

The syntax for the three IN instructions using immediate port addresses is shown below, where the source operand is the I/O port and the destination operand is a general-purpose register.

```
IN   register, immediate (input from port specified by
                          immediate to AL, AX, or EAX)
```

The syntax for the three IN instructions using DX to contain the port addresses is shown below, where the source operand is the I/O port address in DX and the destination operand is a general-purpose register.

```
IN   register, DX        (input from port specified by
                          DX to AL, AX, or EAX)
```

If an immediate byte operand is used to specify the port address, then the address has a range from 0 to 255 ($2^8 - 1$); if register DX is used to specify the port address, then the address has a range from 0 to 65,535 ($2^{16} - 1$). No flags are affected by the IN instructions.

16.4.2 Register I/O OUT Instructions

There are six categories of the OUT instruction referred to as *register I/O* instructions: three instructions using immediate operands to indicate the port address for byte, word, or doubleword operands; and three instructions using general-purpose register DX to indicate the port address for byte, word, or doubleword operands.

The syntax for the three OUT instructions using immediate port addresses is shown below, where the source operand is a general-purpose register and the I/O port address is the destination immediate operand.

```
OUT   immediate, register (output from AL, AX, or EAX
                           to port specified by
                           immediate value)
```

The syntax for the three OUT instructions using DX to contain the port addresses is shown below, where the source operand is a general-purpose register and the destination operand in DX contains the I/O port address.

```
OUT   DX, register         (output from AL, AX, or EAX
                            to port specified by DX)
```

If an immediate byte operand is used to specify the port address, then the address has a range from 0 to 255 ($2^8 - 1$); if register DX is used to specify the port address, then the address has a range from 0 to 65,535 ($2^{16} - 1$). No flags are affected by the OUT instructions.

16.4.3 String I/O IN Instructions

A string is a sequence of bytes, words, or doublewords that are stored in contiguous locations in memory as a one-dimensional array. Strings can be processed from low addresses to high addresses or from high addresses to low addresses, depending on the state of the *direction flag* (DF). If the direction flag is set (DF = 1), then the direction of processing is from high addresses to low addresses — also referred to as *auto-decrement*. If the direction flag is reset (DF = 0), then the direction of processing is from low addresses to high addresses — also referred to as *auto-increment*. The state of the direction flag can be set by the *set direction flag* (STD) instruction and can reset by the *clear direction flag* (CLD) instruction. The direction flag is located in bit 10 of the 32-bit EFLAGS register.

The *input from port to string* (INS) instructions transfer a string element — byte, word, or doubleword — from the I/O port designated by the source operand to the destination operand located in memory. The source (second) operand is an I/O port address contained in general-purpose register DX. The destination (first) operand is a memory location specified by the contents of registers ES:DI, ES:EDI, or RDI for 16, 32, or 64 addresses, respectively.

After the transfer is completed, the destination register DI, EDI, or RDI is automatically incremented or decremented depending on the state of the direction flag (DF) in the EFLAGS register. Thus, after each repetition of the string instruction the registers point to the next byte, word, or doubleword to implement block moves for string operations. There are two types of *string I/O* IN instructions: the *explicit-operands* form and the *no-operands* form, as described below.

Explicit-operands form instructions The explicit-operands form *explicitly* specifies the source and destination operands as part of the INS instruction. The address of the I/O source operand port is in register DX and the destination operand is a memory location whose size is explicitly defined as 8, 16, or 32 bits. There are three *input from port to string* INS instructions that have explicit operands, as shown in the syntax below.

```
INS    memory, DX (transfer byte, word, or doubleword
                  from I/O port to a memory location
                  specified by the contents of
                  ES:(E)DI or RDI)
```

No-operands form instructions The *no-operands* form specifies the size of the operands by the mnemonic; for example, the INSW instruction specifies a word transfer from an I/O port to memory. Like the explicit-operands form, the no-operands form assumes that the address of the source operand port is contained in register DX and the destination memory operand is determined by contents of registers ES:(E)DI or RDI. The syntax for the three no-operands INS instructions is shown below.

```
INSB    (transfer a byte from port addressed by
        DX to memory addressed by ES:(E)DI or RDI)

INSW    (transfer a word from port addressed by
        DX to memory addressed by ES:(E)DI or RDI)

INSD    (transfer a doubleword from port addressed by
        DX to memory addressed by ES:(E)DI or RDI)
```

After the transfer is completed, the destination register DI, EDI, or RDI is automatically incremented or decremented depending on the state of the direction flag (DF) in the EFLAGS register, as stated previously. If the direction flag is set, then the direction of processing is from high addresses to low addresses; if the direction flag is reset, then the direction of processing is from low addresses to high addresses.

The *input from port to string* instructions can utilize the *repeat* (REP) prefixes to specify the condition for which the instruction is to be executed. A count in register (E)CX specifies the number of bytes, words, or doublewords that are transferred from the source I/O port to the memory destination for block transfers.

16.4.4 String I/O OUT Instructions

The *output string to port* (OUTS) instructions transfer a string element — byte, word, or doubleword — from a source operand in a memory location to the I/O port destination operand. The source (second) operand is a memory location specified by the contents of registers DS:SI, DS:ESI, or RSI for 16, 32, or 64 addresses, respectively. The destination (first) operand is an I/O port address contained in general-purpose register DX.

After the transfer is completed, the destination register SI, ESI, or RSI is automatically incremented or decremented depending on the state of the direction flag (DF) in the EFLAGS register. Thus, after each repetition of the string instruction the registers point to the next byte, word, or doubleword to implement block moves for string operations. There are two types of *string I/O* OUTS instructions: the *explicit-operands* form and the *no-operands* form, as described below.

Explicit-operands form instructions The explicit-operands form *explicitly* specifies the source and destination operands as part of the OUTS instruction. The address of the source operand is a memory location whose size is explicitly defined as 8, 16, or 32 bits. The source operand memory location is specified by the contents of the DS:SI, DS:ESI, or RSI registers. The address of the I/O port destination operand is in register DX. There are three *output string to port* OUTS instructions that have explicit operands, as shown in the syntax below.

```
OUTS   DX, memory (transfer byte, word, or doubleword
                   from a memory location specified by
                   the contents of DS:(E)SI or RSI to
                   the I/O port addressed by DX)
```

No-operands form instructions The *no-operands* form specifies the size of the operands by the mnemonic; for example, the OUTSB instruction specifies a byte transfer from memory to an I/O port. Like the explicit-operands form, the no-operands form assumes that the address of the source operand is specified by the contents of registers DS:(E)SI or RSI and the contents of register DX specify the destination port address. The syntax for the three no-operands OUTS instructions is shown below.

```
OUTSB    (transfer a byte from memory addressed by
         DS:(E)SI or RSI to the port addressed by DX)

OUTSW    (transfer a word from memory addressed by
         DS:(E)SI or RSI to the port addressed by DX)

OUTSD   (transfer a doubleword from memory addressed by
         DS:(E)SI or RSI to the port addressed by DX)
```

After the transfer is completed, the destination register SI, ESI, or RSI is automatically incremented or decremented depending on the state of the direction flag (DF) in the EFLAGS register, as stated previously. If the direction flag is set, then the direction of processing is from high addresses to low addresses; if the direction flag is reset, then the direction of processing is from low addresses to high addresses.

The *output string to port* instructions can utilize the *repeat* (REP) prefixes to specify the condition for which the instruction is to be executed. A count in register (E)CX specifies the number of bytes, words, or doublewords that are transferred from the memory source to the destination I/O port for block transfers.

16.5 Problems

16.1 Define an interrupt and indicate the purpose of an interrupt.

16.2 Describe what happens when an interrupt occurs.

16.3 Explain why interrupt breakpoints occur only at the end of an instruction cycle.

16.4 Indicate which of the following instructions are valid or invalid.

```
IN  AX, EA60H
IN  AL, 297
```

16.5 Indicate the use of an interrupt with a vector of 00H.

16.6 Which interrupt type has a higher priority, maskable or nonmaskable interrupts?

16.7 Explain why DMA access to main memory has a higher priority than central processing units access to main memory.

17

Additional Programming Examples

This chapter presents programming examples from topics presented in previous chapters in the book. The programming designs are accomplished using assembly language only and also using assembly language modules embedded in C programs. Each example shows the outputs obtained from executing the program. There are numerous examples designed in the text and also numerous programs to be designed in the problem set at the end of the chapter.

17.1 Programming Examples

The various topics that will be covered in the examples include: logic instructions, bit test instructions, compare instructions, unconditional and conditional jump instructions, unconditional and conditional loop instructions, fixed-point instructions, floating-point instructions, string instructions, and arrays. The problem set includes the above instructions plus additional select instructions.

Example 17.1 This example uses the following instructions: the bit test instructions *bit test and set* (BTS), *bit test and reset* (BTR), and *bit test and complement* (BTC). These instructions are used in conjunction with the *load status flags into AH register* (LAHF) instruction and the *move* (MOV) instruction. The program is designed using an assembly language module embedded in a C program. Figure 17.1(a) shows the program. Figure 17.1(b) shows the outputs.

```
//bts_btr_btc.cpp
//apply the bit test instructions
//bit test and set (bts),
//bit test and reset (btr), and
//bit test and complement (btc)
//to data entered from the keyboard

#include "stdafx.h"
int main (void)
{
//define variables
   int hex_data, bts_rslt, btr_rslt, btc_rslt;
   unsigned char bts_flags, btr_flags, btc_flags;

   printf ("Enter eight hexadecimal characters: \n");
   scanf ("%X", &hex_data);

//switch to assembly
      _asm
      {
//BTS instruction
         MOV   EBX, hex_data  ;data entered -> ebx
         BTS   EBX, 2         ;bit 2 -> flags, set bit 2
         MOV   bts_rslt, EBX
         LAHF                 ;status flags -> ah
         MOV   bts_flags, AH

//BTR instruction
         MOV   EBX, bts_rslt  ;bts result -> ebx
         BTR   EBX, 7         ;bit 7 -> flags, reset bit 7
         MOV   btr_rslt, EBX
         LAHF                 ;status flags -> ah
         MOV   btr_flags, AH

//BTC instruction
         MOV   EBX, btr_rslt  ;btr result -> ebx
         BTC   EBX, 12        ;bit 12 -> flags, compl bit 12
         MOV   btc_rslt, EBX
         LAHF                 ;status flags -> ah
         MOV   btc_flags, AH
      }                    (a)      //continued on next page
```

Figure 17.1 Program illustrating the use of bit test instructions: (a) the program and (b) the outputs.

```
    printf ("\nBTS result = %X\nBTS flags = %X\n",
                bts_rslt, bts_flags);

    printf ("\nBTR result = %X\nBTR flags = %X\n",
                btr_rslt, btr_flags);

    printf ("\nBTC result = %X\nBTC flags = %X\n\n",
                btc_rslt, btc_flags);

    return 0;
    }
```

```
Enter eight hexadecimal characters:
FFFFFFFF

BTS result = FFFFFFFF
BTS flags = 47          //ZF, PF, CF

BTR result = FFFFFF7F
BTR flags = 47          //ZF, PF, CF

BTC result = FFFFEF7F
BTC flags = 47          //ZF, PF, CF

Press any key to continue . . . _
------------------------------------------------------------
Enter eight hexadecimal characters:
00000000

BTS result = 4
BTS flags = 46          //ZF, PF

BTR result = 4
BTR flags = 46          //ZF, PF

BTC result = 1004
BTC flags = 46          //ZF, PF

Press any key to continue . . . _
------------------------------------------------------------

                                  //continued on next page
```

(b)

Figure 17.1 (Continued)

```
Enter eight hexadecimal characters:
ABCD9A01

BTS result = ABCD9A05
BTS flags = 46          //ZF, PF

BTR result = ABCD9A05
BTR flags = 46          //ZF, PF

BTC result = ABCD8A05
BTC flags = 47          //ZF, PF, CF

Press any key to continue . . . _
```

(b)

Figure 17.1 (Continued)

Example 17.2 This example calculates the sum of cubes for numbers in the following range: 1 to n, where n is ≤ 22. The range number is entered from the keyboard. The sum of cubes can be obtained from any of the expressions shown in Equation 17.1. The three expressions will be used in Examples 17.2, Example 17.3, and Example 17.4, respectively. This example uses the first expression listed using an assembly language module embedded in a C program, as shown in Figure 17.2.

$$1^3 + 2^3 + 3^3 + 4^3 + \ldots + n^3$$

$$[n(n + 1)/2]^2$$

$$(1 + 2 + 3 + 4 + \ldots + n)^2 \quad (17.1)$$

```
//sum_of_cubes8.cpp
//obtain the sum of cubes for numbers with a range from
//1 through n <= 22, where n is entered from the keyboard

#include "stdafx.h"
int main (void)
{
//define variables
   unsigned short n, range, sum;
```

(a) //continued on next page

Figure 17.2 Program to calculate the sum of cubes in the range from 1 to ≤ 22: (a) the program and (b) the outputs.

```
   n = 0;
   sum = 0;

   printf ("Enter a number n indicating a range: \n");
   scanf ("%d", &range);

//switch to assembly
      _asm
      {
//-------------------------------------------------------------
LP1:     INC    n
         MOV    CX, 2          //count in cx
         MOV    AX, n          //n -> ax

LP2:     MUL    n              //ax x n -> dx:ax
         LOOP   LP2
         ADD    sum, AX        //ax + sum -> sum
         MOV    BX, range      //range -> bx
         CMP    n, BX          //compare current value to range
         JB     LP1            //jump if n < range
      }
//-------------------------------------------------------------
//print sum
   printf ("\nSum of cubes 1 through n = %d\n\n", sum);

   return 0;
}
```

```
Enter a number n indicating a range:
3

Sum of cubes 1 through n = 36

Press any key to continue . . . _
-------------------------------------------------------------
Enter a number n indicating a range:
10

Sum of cubes 1 through n = 3025

Press any key to continue . . . _
-------------------------------------------------------------
```

(b) //continued on next page

Figure 17.2 (Continued)

```
Enter a number n indicating a range:
3

Sum of cubes 1 through n = 36

Press any key to continue . . . _
-----------------------------------------------------------
Enter a number n indicating a range:
10

Sum of cubes 1 through n = 3025

Press any key to continue . . . _
-----------------------------------------------------------
Enter a number n indicating a range:
16

Sum of cubes 1 through n = 18496

Press any key to continue . . . _
-----------------------------------------------------------
Enter a number n indicating a range:
17

Sum of cubes 1 through n = 23409

Press any key to continue . . . _
-----------------------------------------------------------
Enter a number n indicating a range:
18

Sum of cubes 1 through n = 29241

Press any key to continue . . . _
-----------------------------------------------------------
Enter a number n indicating a range:
22

Sum of cubes 1 through n = 64009

Press any key to continue . . . _
```

Figure 17.2 (Continued)

Example 17.3 This example calculates the sum of cubes for numbers in the following range: 1 to n, where n is ≤ 22. The range number is entered from the keyboard. The sum of cubes for this example is obtained from second expression listed in Equation 17.1 and shown below. This example uses an assembly language module embedded in a C program, as shown in Figure 17.3.

$$[n(n+1)/2]^2$$

```
//sum_of_cubes9.cpp
//obtain the sum of cubes for numbers with a range from
//1 through n <= 22, where n is entered from the keyboard.
//the following expression is used: [n(n+1)/2]^2

#include "stdafx.h"
int main (void)
{
//define variables
   unsigned short two, n, sum;
   two = 2;

   printf ("Enter a number indicating a range: \n");
   scanf ("%d", &n);

//switch to assembly
      _asm
      {
         MOV   AX, n        //n -> ax
         INC   AX           //n+1 -> ax
         MUL   n            //n(n+1) -> dx:ax
         DIV   two          //n(n+1)/2 -> dx:ax
         MUL   AX           //[n(n+1)/2]^2
         MOV   sum, AX
      }

//print sum
   printf ("\nSum of cubes 1 through n = %d\n\n", sum);

   return 0;
}
                                    //continued on next page
```

(a)

Figure 17.3 Program to calculate the sum of cubes in the range from 1 to ≤ 22 using the expression shown above: (a) the program and (b) the outputs.

```
Enter a number indicating a range:
3

Sum of cubes 1 through n = 36

Press any key to continue . . . _
-------------------------------------------------------------
Enter a number indicating a range:
10

Sum of cubes 1 through n = 3025

Press any key to continue . . . _
-------------------------------------------------------------
Enter a number indicating a range:
16

Sum of cubes 1 through n = 18496

Press any key to continue . . . _
-------------------------------------------------------------
Enter a number indicating a range:
17

Sum of cubes 1 through n = 23409

Press any key to continue . . . _
-------------------------------------------------------------
Enter a number indicating a range:
18

Sum of cubes 1 through n = 29241

Press any key to continue . . . _
-------------------------------------------------------------
Enter a number indicating a range:
22

Sum of cubes 1 through n = 64009

Press any key to continue . . . _
```

(b)

Figure 17.3 (Continued)

Example 17.4 This example calculates the sum of cubes for numbers in the following range: 1 to n, where n is ≤ 22. The range number is entered from the keyboard. The sum of cubes for this example is obtained from third expression listed in Equation 17.1 and shown below. This example uses an assembly language module embedded in a C program, as shown in Figure 17.4.

$$(1 + 2 + 3 + 4 + \ldots + n)^2$$

```
//sum_of_cubes10.cpp
//obtain the sum of cubes for numbers with a range from
//1 through n <= 22, where n is entered from the keyboard.
//use the following expression: (1 + 2 + 3 + 4 + ... + n)^2

#include "stdafx.h"
int main (void)
{
//define variables
   unsigned short n, sum;
   sum = 0;

   printf ("Enter a number n indicating a range: \n");
   scanf ("%d", &n);

//switch to assembly
      _asm
      {
         MOV    CX, n        //range -> cx
         MOV    AX, 0        //clear ax
LP1:     ADD    AX, 1        //inc ax
         ADD    sum, AX      //sum + ax -> sum
         LOOP   LP1          //if cx != 0, loop

         MOV    AX, sum      //sum -> ax
         MUL    AX           //ax^2 -> dx:ax
         MOV    sum, AX      //sum of cubes -> sum
                             //(1+2+3+4+ ... +n)^2 -> sum
      }
//print sum
   printf ("\nSum of cubes 1 through n = %d\n\n", sum);
   return 0;
}
```

(a) //continued on next page

Figure 17.4 Program to calculate the sum of cubes in the range from 1 to ≤ 22 using the expression shown above: (a) the program and (b) the outputs.

```
Enter a number n indicating a range:
3

Sum of cubes 1 through n = 36

Press any key to continue . . . _
-------------------------------------------------------------
Enter a number n indicating a range:
5

Sum of cubes 1 through n = 225

Press any key to continue . . . _
-------------------------------------------------------------
Enter a number n indicating a range:
10

Sum of cubes 1 through n = 3025

Press any key to continue . . . _
-------------------------------------------------------------
Enter a number n indicating a range:
16

Sum of cubes 1 through n = 18496

Press any key to continue . . . _
-------------------------------------------------------------
Enter a number n indicating a range:
17

Sum of cubes 1 through n = 23409

Press any key to continue . . . _
-------------------------------------------------------------
Enter a number n indicating a range:
18

Sum of cubes 1 through n = 29241

Press any key to continue . . . _
-------------------------------------------------------------
                                   //continued on next page
```

(b)

Figure 17.4 (Continued)

```
Enter a number n indicating a range:
20

Sum of cubes 1 through n = 44100

Press any key to continue . . . _
-----------------------------------------------------------
Enter a number n indicating a range:
21

Sum of cubes 1 through n = 53361

Press any key to continue . . . _
-----------------------------------------------------------
Enter a number n indicating a range:
22

Sum of cubes 1 through n = 64009

Press any key to continue . . . _
```

Figure 17.4 (Continued)

Example 17.5 This example replaces all spaces with hyphens in a character string that is entered from the keyboard. The character string is stored in the operand field (OPFLD) of the parameter list (PARLST). The parameter list is reproduced in Figure 17.5 for convenience and illustrates an example for this program. The program in this example uses an assembly language program — not embedded in a C program — as shown in Figure 17.6. This program does not use the MOVSB string instruction — the same program using the MOVSB instruction is left as a problem at the end of the chapter. The comments at the end of each line of code in Figure 17.6 explain the operation of that individual line.

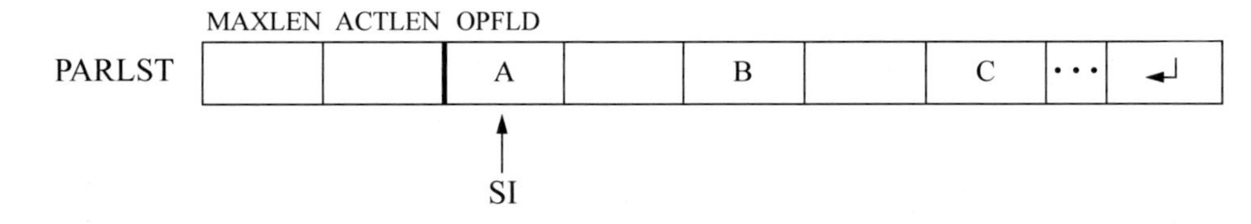

Figure 17.5 Parameter list one-dimensional array in which the keyboard input data are stored.

```
;hyphen.asm
;replace a space with a hyphen in a
;character string entered from the keyboard

;-------------------------------------------------------------
.STACK

;-------------------------------------------------------------
.DATA
PARLST  LABEL BYTE
MAXLEN  DB 20
ACTLEN  DB ?
OPFLD   DB 20 DUP(?)
PRMPT   DB 0DH, 0AH, 'Enter a character string: $'
RSLT    DB 0DH, 0AH, 'Hyphen string =                    $'

;-------------------------------------------------------------
.CODE
BEGIN PROC FAR

;set up pgm ds
      MOV   AX, @DATA       ;put addr of data seg in ax
      MOV   DS, AX          ;put addr in ds

;read prompt
      MOV   AH, 09H         ;display string
      LEA   DX, PRMPT       ;put addr of prompt in dx
      INT   21H             ;dos interrupt

;keyboard request rtn to enter characters
      MOV   AH, 0AH         ;buffered keyboard input
      LEA   DX, PARLST      ;put addr of parlst in dx
      INT   21H             ;dos interrupt

;-------------------------------------------------------------
;set up source and count
      LEA   SI, OPFLD       ;si points to input buffer

      MOV   CL, ACTLEN      ;actual length -> cl
      MOV   CH, 00H         ;clear ch

;-------------------------------------------------------------

                                   //continued on next page
```

(a)

Figure 17.6 Program to replace spaces with hyphens in a character string that is entered from the keyboard: (a) the program and (b) the outputs.

```
LP1:   MOV    AL, [SI]         ;byte in opfld -> al
       CMP    AL, 20H          ;20H is a space
       JE     ADD_HYPH         ;jump if char is a space
       INC    SI               ;point to next char
       LOOP   LP1              ;loop if cx != 0
       JMP    MOV_CHAR         ;jump if cx = 0

ADD_HYPH:
       MOV    [SI], 2DH        ;2DH is a hyphen
       INC    SI               ;point to next char
       LOOP   LP1              ;loop if cx != 0
       JMP    MOV_CHAR         ;jump if cx = 0

MOV_CHAR:
       MOV    CL, ACTLEN       ;actual length -> cl
       MOV    CH, 00H          ;clear ch

       LEA    SI, OPFLD        ;si points to input buffer
       LEA    DI, RSLT+18      ;di points to result area

LP2:   MOV    BL, [SI]         ;char in opfld -> bl
       MOV    [DI], BL         ;bl -> rslt area
       INC    SI               ;point to next char in opfld
       INC    DI               ;point to next loc in rslt area
       LOOP   LP2              ;loop if cx != 0

;------------------------------------------------------------
;display result
       MOV    AH, 09H          ;display string
       LEA    DX, RSLT         ;put addr of rslt in dx
       INT    21H              ;dos interrupt

BEGIN ENDP
       END    BEGIN
```

```
Enter a character string: A B C
Hyphen string = A-B-C
------------------------------------------------------------
Enter a character string: A B C D E F G H
Hyphen string = A-B-C-D-E-F-G-H
------------------------------------------------------------
Enter a character string: AB CDE FG H IJKL
Hyphen string = AB-CDE-FG-H-IJKL
------------------------------------------------------------
```

//continued on next page

(b)

Figure 17.6 (Continued)

```
Enter a character string: AB12CD34EF56GH
Hyphen string = AB12CD34EF56GH
-------------------------------------------------------------
Enter a character string:         //eight spaces
Hyphen string = --------
-------------------------------------------------------------
Enter a character string: 1234 5678 abc def
Hyphen string = 1234-5678-abc-def
-------------------------------------------------------------
Enter a character string: AB  CDE   FGHI    J
Hyphen string = AB--CDE---FGHI----J
```

Figure 17.6 (Continued)

Example 17.6 This example counts the number of times that the integer 6 occurs in an array of eight integers. The program in this example uses an assembly language program — not embedded in a C program — as shown in Figure 17.7. The array is defined in the data segment (.DATA) and is labelled LIST. The comments at the end of each line of code in Figure 17.7 explain the operation of that individual line. A similar program that uses data entered from the keyboard is left as a problem at the end of the chapter.

```
;count_occurrence.asm
;count the number of times that the number 6
;occurs in an array of eight numbers
;------------------------------------------------------------
.STACK
;------------------------------------------------------------
.DATA
LIST    DB 32H, 36H, 31H, 38H, 36H, 39H, 37H, 36H
RSLT    DB 0DH, 0AH, 'Occurs =   times  $'
;------------------------------------------------------------
.CODE
BEGIN PROC FAR

;set up pgm ds
      MOV   AX, @DATA       ;put addr of data seg in ax
      MOV   DS, AX          ;put addr in ds
      LEA   DI, RSLT + 11   ;put addr of rslt in di
;------------------------------------------------------------
                        (a)       //continued on next page
```

Figure 17.7 Program to count the number of times that the integer 6 occurs in an array of eight integers: (a) the program and (b) the outputs.

```
        MOV   DL, 36H          ;number to compare -> dl
        LEA   SI, LIST         ;put addr of list in si
        MOV   CX, 8            ;qty of numbers to compare -> cx
        MOV   AH, 30H          ;initialize number of occurrences

LP1:    MOV   AL, [SI]         ;move number to al
        CMP   DL, AL           ;compare dl to al
        JE    INC_COUNT        ;if equal, jump to inc count
        INC   SI               ;point to next number to compare
        LOOP  LP1              ;loop to compare next number
        JMP   MOVE_COUNT       ;comparison is finished

INC_COUNT:
        INC   AH               ;incr the number of occurrences
        INC   SI               ;point to next number
        LOOP  LP1              ;if cx != 0, compare next number

MOVE_COUNT:
        MOV   [DI], AH         ;# of occurrences to result area

;-----------------------------------------------------------
;display result
        MOV   AH, 09H          ;display string
        LEA   DX, RSLT         ;put addr of rslt in dx
        INT   21H              ;dos interrupt

BEGIN ENDP
        END   BEGIN
```

```
Occurs = 3 times
```

(b)

Figure 17.7 (Continued)

Example 17.7 This example determines the largest integer in an array of eight integers. The program in this example uses an assembly language program — not embedded in a C program — as shown in Figure 17.8. The array is defined in the data segment (.DATA) and is labelled LIST. The comments at the end of each line of code in Figure 17.8 explain the operation of that individual line. A similar program that uses data entered from the keyboard is left as a problem at the end of the chapter.

```
;largest_num.asm
;find the largest number in an array of numbers

;------------------------------------------------------------
.STACK
;------------------------------------------------------------
.DATA
LIST    DB 32H, 39H, 31H, 38H, 36H, 39H, 37H, 35H
RSLT    DB 0DH, 0AH, 'Largest =         $'
;------------------------------------------------------------
.CODE
BEGIN PROC FAR

;set up pgm ds
      MOV   AX, @DATA       ;put addr of data seg in ax
      MOV   DS, AX          ;put addr in ds
      LEA   DI, RSLT + 12   ;put addr of rslt in di

;------------------------------------------------------------
      LEA   SI, LIST        ;put addr of list in si
      MOV   CX, 8           ;count in cx
      MOV   AH, 30H         ;initialize maximum number

LP1:  MOV   AL, [SI]        ;move number to al
      CMP   AH, AL          ;compare hi in ah to al
      JAE   INC_SI          ;if ah >= al, jump
      MOV   AH, AL          ;if not, largest in al -> ah

INC_SI:
      INC   SI              ;point to next number
      LOOP  LP1             ;if cx != 0, compare next number

      MOV   [DI], AH        ;move largest to result area
;------------------------------------------------------------
;display result
      MOV   AH, 09H         ;display string
      LEA   DX, RSLT        ;put addr of rslt in dx
      INT   21H             ;dos interrupt

BEGIN ENDP
      END   BEGIN
```

(a)

```
Largest = 9
```

(b)

Figure 17.8 Program to determine the largest integer in an array of eight integers: (a) the program and (b) the outputs.

Example 17.8 This example calculates the factorial of single-digit integers in the range of 1 through 8 that are entered from the keyboard. The program in this example is designed using an assembly language module embedded in a C program, as shown in Figure 17.9. A similar program using assembly language only — that uses data entered from the keyboard — is left as a problem at the end of the chapter.

```
//factorial3.cpp
//calculate the factorial of an integer entered
//from the keyboard in the range of 1 -- 8

#include "stdafx.h"
int main (void)
{
//define variables
   unsigned short integer_num;
   unsigned long result;

   printf ("Enter a single-digit integer 1 -- 8: \n");
   scanf ("%d", &integer_num);

//switch to assembly
      _asm
      {
         MOV   AX, integer_num
         MOV   CX, integer_num

LP1:     DEC   CX
         JZ    PRNT
         MUL   CX             //ax x cx -> eax
         JMP   LP1

PRNT:    MOV   result, EAX
      }

//display result
   printf ("\nInteger = %d\nFactorial = %d\n\n",
            integer_num, result);

   return 0;
}
                                //continued on next page
```

(a)

Figure 17.9 Program to calculate the factorial of single-digit integers that are entered from the keyboard: (a) the program and (b) the outputs.

```
Enter a single-digit integer 1 -- 8:
1

Integer = 1
Factorial = 1

Press any key to continue . . . _
--------------------------------------------------------------
Enter a single-digit integer 1 -- 8:
2

Integer = 2
Factorial = 2

Press any key to continue . . . _
--------------------------------------------------------------
Enter a single-digit integer 1 -- 8:
3

Integer = 3
Factorial = 6

Press any key to continue . . . _
--------------------------------------------------------------
Enter a single-digit integer 1 -- 8:
4

Integer = 4
Factorial = 24

Press any key to continue . . . _
--------------------------------------------------------------
Enter a single-digit integer 1 -- 8:
5

Integer = 5
Factorial = 120

Press any key to continue . . . _
--------------------------------------------------------------
```

//continued on next page

(b)

Figure 17.9 (Continued)

```
Enter a single-digit integer 1 -- 8:
6

Integer = 6
Factorial = 720

Press any key to continue . . . _
-------------------------------------------------------------
Enter a single-digit integer 1 -- 8:
7

Integer = 7
Factorial = 5040

Press any key to continue . . . _
-------------------------------------------------------------
Enter a single-digit integer 1 -- 8:
8

Integer = 8
Factorial = 40320

Press any key to continue . . . _
```

Figure 17.9 (Continued)

Example 17.9 This example calculates the value of single-digit integer bases in the range of $1 \leq y \leq 6$ raised to the power of single-digit integer exponents in the range of $0 \leq x \leq 6$ — bases and exponents are entered from the keyboard. The program in this example is designed using an assembly language module embedded in a C program, as shown in Figure 17.10.

```
//exponent3.cpp
//obtain the value of base 1 <= y <= 6 raised
//to the power of exponent 0 <= x <= 6

#include "stdafx.h"
int main (void)
```

//continued on next page

(a)

Figure 17.10 Program to calculate single-digit bases raised to the power of single-digit exponents: (a) the program and (b) the outputs.

```
{
//define variables
   unsigned short base, exponent;
   unsigned long result;

   printf ("Enter a single-digit base: \n");
   scanf ("%d", &base);

   printf ("\nEnter a single-digit exponent: \n");
   scanf ("%d", &exponent);

//switch to assembly
      _asm
      {
//if exponent = 0, print 1
         MOV   AX, base
         MOV   CX, exponent
         CMP   CX, 0
         JE    PRNT_1

//------------------------------------------------------------
//if exponent = 1, print base
         CMP   CX, 1
         JE    PRNT_BASE

//------------------------------------------------------------
//calculate base ^ exponent
         MOV   AX, base
         MOV   BX, base

LP1:     MUL   BX
         MOV   result, EAX
         DEC   CX
         CMP   CX, 1
         JE    PRNT_RSLT
         JMP   LP1
      }

//------------------------------------------------------------
                                    //continued on next page
```

Figure 17.10 (Continued)

```
PRNT_RSLT:
   printf ("\nResult = %d\n\n", result);
   goto END_PGM;

PRNT_BASE:
   printf ("\nResult = %d\n\n", base);
   goto END_PGM;

PRNT_1:
   printf ("\nResult = 1\n\n");

END_PGM:
   return 0;
}
```

```
Enter a single-digit base:
5

Enter a single-digit exponent:
0

Result = 1

Press any key to continue . . . _
------------------------------------------------------------
Enter a single-digit base:
6

Enter a single-digit exponent:
1

Result = 6

Press any key to continue . . . _
------------------------------------------------------------
Enter a single-digit base:
2

Enter a single-digit exponent:
2

Result = 4

Press any key to continue . . . _
------------------------------------------------------------
```

(b) //continued on next page

Figure 17.10 (Continued)

```
Enter a single-digit base:
6

Enter a single-digit exponent:
2

Result = 36

Press any key to continue . . . _
-------------------------------------------------------------
Enter a single-digit base:
3

Enter a single-digit exponent:
3

Result = 27

Press any key to continue . . . _
-------------------------------------------------------------
Enter a single-digit base:
4

Enter a single-digit exponent:
3

Result = 64

Press any key to continue . . . _
-------------------------------------------------------------
Enter a single-digit base:
5

Enter a single-digit exponent:
4

Result = 625

Press any key to continue . . . _
-------------------------------------------------------------

                                  //continued on next page
```

Figure 17.10 (Continued)

```
------------------------------------------------------------
Enter a single-digit base:
6

Enter a single-digit exponent:
4

Result = 1296

Press any key to continue . . . _
------------------------------------------------------------
Enter a single-digit base:
5

Enter a single-digit exponent:
5

Result = 3125

Press any key to continue . . . _
------------------------------------------------------------
Enter a single-digit base:
5

Enter a single-digit exponent:
6

Result = 15625

Press any key to continue . . . _
------------------------------------------------------------
Enter a single-digit base:
6

Enter a single-digit exponent:
5

Result = 7776

Press any key to continue . . . _
------------------------------------------------------------
Enter a single-digit base:
6

Enter a single-digit exponent:
6

Result = 46656

Press any key to continue . . . _
```

Figure 17.10 (Continued)

Example 17.10 This example calculates the value of the following expression using integers for fixed-point arithmetic:

$$[(A \times B) + C + 500] \,/\, A$$

The program in this example is designed using an assembly language module embedded in a C program. Both the quotient and remainder will be obtained after the division operation is executed.

Recall that the DIV instruction divides the unsigned dividend in an implied general-purpose register by the unsigned divisor source operand in a register or memory location. The results are stored in the implied registers. The syntax for the DIV instruction is shown below.

```
DIV    register/memory
```

The locations of the DIV operands are as follows: the dividend registers are AX, DX:AX, EDX:EAX, and RDX:RAX and the size of the corresponding divisors are 8 bits, 16 bits, 32 bits, and 64 bits, respectively; the resulting quotients are stored in registers AL, AX, EAX, and RAX, respectively; the resulting remainders are stored in registers AH, DX, EDX, and RDX, respectively. Results that are not integers are truncated to integers. Figure 17.11 illustrates pictorially the division operation.

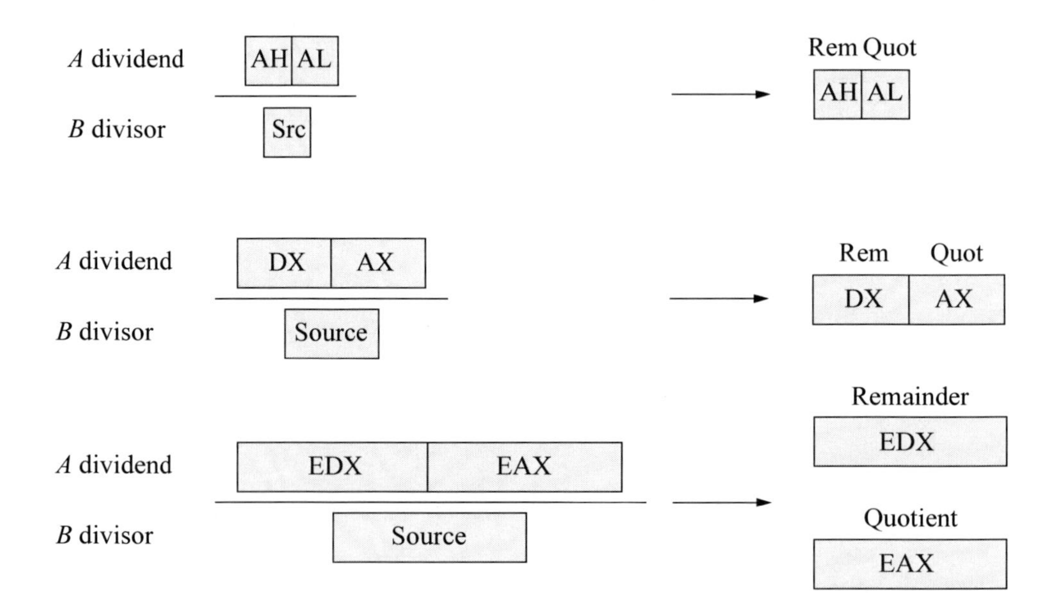

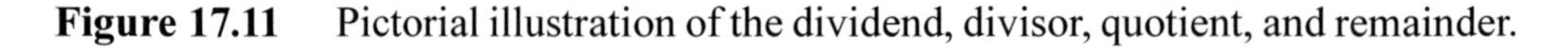

Figure 17.11 Pictorial illustration of the dividend, divisor, quotient, and remainder.

An example is shown below using the following values for variables A, B, and C: $A = 200$, $B = 200$, and $C = 1000$, yielding a quotient of 207 and a remainder of 100.

$$[(A \times B) + C + 500] \,/\, A$$
$$[(200 \times 200) + 1000 + 500] \,/\, 200$$
$$= (40000 + 1000 + 500) \,/\, 200$$
$$= 41500 \,/\, 200$$
$$= 207.5$$

A second example is shown below using the following values for variables A, B, and C: $A = 790$, $B = 650$, and $C = 1500$, yielding a quotient of 652 and a remainder of 420.

$$[(A \times B) + C + 500] \,/\, A$$
$$[(790 \times 650) + 1500 + 500] \,/\, 790$$
$$= (40000 + 1000 + 500) \,/\, 790$$
$$= 515500 \,/\, 790$$
$$= 652.531645569$$

Figure 17.12 illustrates a program to calculate the given expression using the fixed-point arithmetic instructions of MUL, ADD, and DIV.

```
//eval_expr5.cpp
//evaluate the following expression using integers
//and obtain the quotient and remainder:
//[(A x B) + C + 500)/A
#include "stdafx.h"
int main (void)
{
//define variables
   unsigned short int_a, int_b, int_c, rslt_quot, rslt_rem;
   printf ("Enter integers for x, y, and z: \n");
   scanf ("%d %d %d", &int_a, &int_b, &int_c);

//switch to assembly
      _asm
      {
         MOV   AX, int_a
         MUL   int_b    ;result in dx:ax
         ADD   AX, int_c
         ADD   ax, 500
         DIV   int_a
         MOV   rslt_quot, AX
         MOV   rslt_rem, DX
      }                    (a)       //continued on next page
```

Figure 17.12 Program to calculate the value of $[(A \times B) + C + 500] \,/\, A$: (a) the program and (b) the outputs.

```
//display result
   printf ("\nQuotient = %d\nRemainder = %d\n\n",
                  rslt_quot, rslt_rem);

   return 0;
}
```

```
Enter integers for a, b, and c:
200 200 1000

Quotient = 207
Remainder = 100

Press any key to continue . . . _
------------------------------------------------------------
Enter integers for a, b, and c:
400 400 2000

Quotient = 406
Remainder = 100

Press any key to continue . . . _
------------------------------------------------------------
Enter integers for a, b, and c:
350 175 1250

Quotient = 180
Remainder = 0

Press any key to continue . . . _
------------------------------------------------------------
Enter integers for a, b, and c:
790 650 1500

Quotient = 652
Remainder = 420

Press any key to continue . . . _
```

(b)

Figure 17.12 (Continued)

Example 17.11 This example calculates the value of the following expression using integers for fixed-point arithmetic:

$$[(A \times B) + C + 500] / A$$

This example is similar to Example 17.10, except that doublewords are used and only the quotient is obtained after the division operation is executed; therefore, all fractions are truncated. The program in this example is designed using an assembly language module embedded in a C program and is shown in Figure 17.13.

```
//eval_expr3.cpp
//evaluate the following expression using integers:
//[(A x B) + C + 500)/A.
//doublewords are used and only the quotient is obtained

#include "stdafx.h"
int main (void)
{
//define variables
   unsigned long int_a, int_b, int_c, rslt;

   printf ("Enter integers for a, b, and c: \n");
   scanf ("%d %d %d", &int_a, &int_b, &int_c);

//switch to assembly
      _asm
      {
         MOV   EAX, int_a       ;doubleword -> eax
         MUL   int_b            ;result -> edx:eax
         ADD   EAX, int_c
         ADD   EAX, 500
         DIV   int_a
         MOV   rslt, EAX
      }

//display result
   printf ("\nResult = %d\n\n", rslt);

   return 0;
}
```

//continued on next page

(a)

Figure 17.13 Program to calculate the value of [(A × B) + C + 500] / A using doubleword operands to obtain the quotient only: (a) the program and (b) the outputs.

```
Enter integers for a, b, and c:
60000 60000 10000

Result = 60000          //truncated from 60000.175
--------------------------------------------------------------
Enter integers for a, b, and c:
350000 175000 20000

Result = 175000         //truncated from 175000.05714
--------------------------------------------------------------
Enter integers for a, b, and c:
12345 67890 2468

Result = 67890          //truncated from 67890.19992
--------------------------------------------------------------
Enter integers for a, b, and c:
3500 175000 20000

Result = 175005         //truncated from 175005.8571
```

(b)

Figure 17.13 (Continued)

Example 17.12 This example calculates the value of the following expression using variables for floating-point arithmetic:

$$[(A \times B) + C + \text{constant}] \,/\, A$$

This example is similar to previous examples, except that floating-point operands are used and a floating-point quotient is obtained after the division operation is executed; therefore, all fractions are retained and are not truncated. The program in this example is designed using an assembly language module embedded in a C program.

Recall that floating-point numbers consist of the following three fields: a sign bit, s; an exponent, e; and a fraction, f. These parts represent a number that is obtained by multiplying the fraction, f, by a radix, r, raised to the power of the exponent, e, as shown in Equation 17.2 for the number A, where f and e are signed fixed-point numbers, and r is the radix (or base).

$$A = f \times r^e \tag{17.2}$$

Figure 17.14 shows the format for 32-bit single-precision floating-point numbers. The single-precision format consists of a sign bit that indicates the sign of the number, an 8-bit signed exponent, and a 23-bit unsigned fraction.

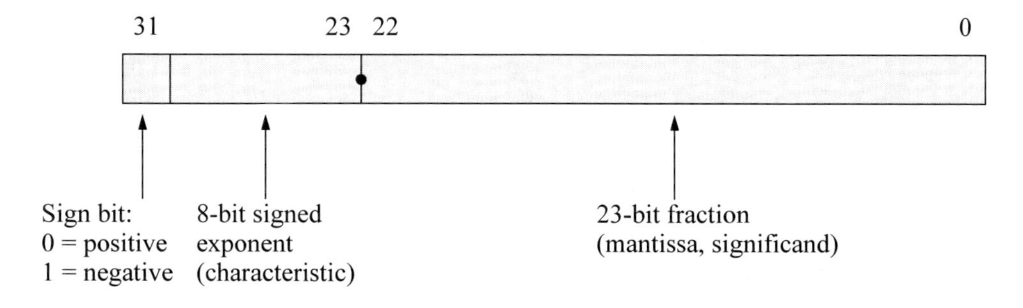

Figure 17.14 Format for single-precision floating-point numbers.

Figure 17.15 shows the eight data registers — called the register stack — used in the floating-point unit. The stack top, ST(0), is register R0 and, like a normal stack, builds towards lower-numbered registers. The register immediately below the stack top is referred to as ST(1); the register immediately below ST(1) is referred to as ST(2), etc.

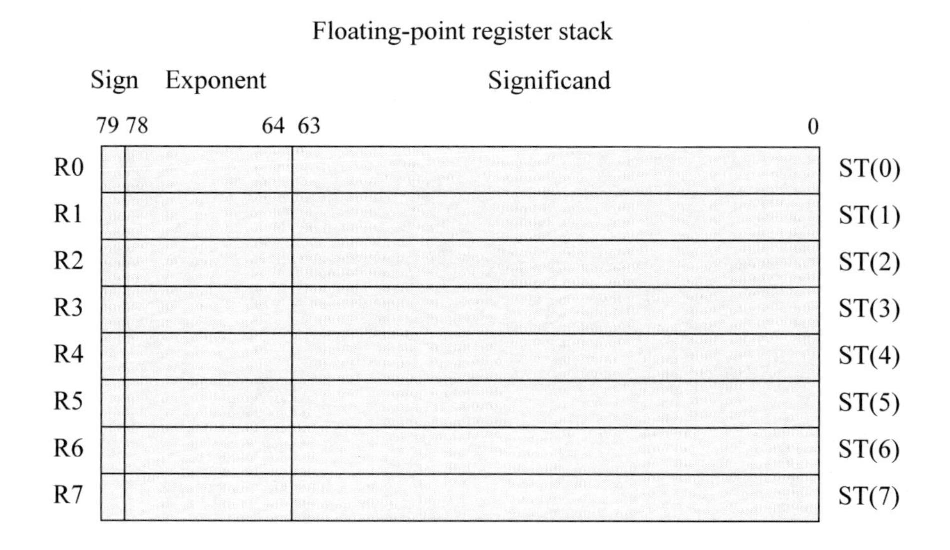

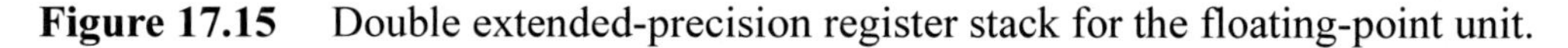

Figure 17.15 Double extended-precision register stack for the floating-point unit.

When the stack is full; that is, the registers at ST(0) through the register at ST(7) contain valid data, a stack wraparound occurs if an attempt is made to store additional data on the stack. This results in a stack overflow because the unsaved data is overwritten. The stack registers are specified by three bits — 000 through 111 — to reference ST(0) through ST(7), respectively. Therefore, ST(*i*) references the *i*th register from the current stack top. Chapter 11 can be reviewed for more details on floating-point operations.

Figure 17.16 shows the assembly language module embedded in a C program to perform the calculations for the expression shown in Figure 17.16. Figure 17.17 shows the contents of the register stack during execution of the program for variables *X*, *Y*, *Z*, and a constant that are equal to 4.0, 5.0, 6.0, and 7.0, respectively.

```
//eval_expr6.cpp
//evaluate the following expression using
//floating-point numbers for X, Y, Z, and constant:
//[(X x Y) + Z + constant]/X
#include "stdafx.h"
int main (void)
{
//define variables
   float flp_x, flp_y, flp_z, flp_num, flp_rslt;

   printf ("Enter floating-point numbers for x, y, z,
               and constant: \n");
   scanf ("%f %f %f %f", &flp_x, &flp_y, &flp_z, &flp_num);

//switch to assembly
      _asm
      {
         FLD    flp_x            //X -> ST(0)
         FLD    flp_y            //Y -> ST(0)
                                 //X -> ST(1)

         FMUL   ST(0), ST(1)     //X x Y -> ST(0)
                                 //ST(1) = X

         FLD    flp_z            //Z -> ST(0)
                                 //X x Y -> ST(1)
                                 //ST(2) = X
                                       //continued on next page
```

(a)

Figure 17.16 Program to perform calculations on [($X \times Y$) + Z + constant] / X: (a) the program and (b) the outputs.

```
        FADD  ST(0), ST(1)    //[(X x Y) + Z] -> ST(0)
                              //X x Y -> ST(1)
                              //ST(2) = X

        FLD   flp_num         //flp_num -> ST(0)
                              //[(X x Y) + Z] -> ST(1)
                              //X x Y -> ST(2)
                              //ST(3) = X

        FADD  ST(0), ST(1)    //[(X x Y) + Z + flp_num]
                              //-> ST(0)
                              //[(X x Y) + Z] -> ST(1)
                              //X x Y -> ST(2)
                              //ST(3) = X

        FDIV  ST(0), ST(3)    //[(X x Y) + Z + flp_num]/X
                              //-> ST(0)

        FST   flp_rslt        //ST(0) (result) -> flp_rslt
      }

//print result
   printf ("\nResult = %f\n\n", flp_rslt);

   return 0;
}
```

```
Enter floating-point numbers for x, y, z, and constant:
4.0 5.0 6.0 7.0

Result = 8.250000

Press any key to continue . . . _
-----------------------------------------------------------
Enter floating-point numbers for x, y, z, and constant:
60000.0 60000.0 10000.0 500.0

Result = 60000.175781

Press any key to continue . . . _
-----------------------------------------------------------
```

//continued on next page

(b)

Figure 17.16 (Continued)

```
Enter floating-point numbers for x, y, z, and constant:
400.0 400.0 2000.0 500.0

Result = 406.250000

Press any key to continue . . . _
------------------------------------------------------------
Enter floating-point numbers for x, y, z, and constant:
200.0 200.0 1000.0 5000.0

Result = 207.500000

Press any key to continue . . . _
------------------------------------------------------------
Enter floating-point numbers for x, y, z, and constant:
45000.0 62000.0 25000.0 8500.0

Result = 62000.746094

Press any key to continue . . . _
------------------------------------------------------------
Enter floating-point numbers for x, y, z, and constant:
125.125 750.500 875.125 917.375

Result = 764.825684

Press any key to continue . . . _
------------------------------------------------------------
Enter floating-point numbers for x, y, z, and constant:
8.123456 75.987654 350.2468 4666.375

Result = 693.535400

Press any key to continue . . . _
------------------------------------------------------------
Enter floating-point numbers for x, y, z, and constant:
1.111111 2.222222 3.333333 4.444444

Result = 9.222222

Press any key to continue . . . _
------------------------------------------------------------
Enter floating-point numbers for x, y, z, and constant:
123456.123456 234567.234567 345678.345678 999.999

Result = 234570.046875

Press any key to continue . . . _
```

Figure 17.16 (Continued)

Assume that instructions `FLD flp_x` and `FLD flp_y` have been executed.

FMUL `ST(0), ST(1)`

Register	Value	Stack
R0	20.0	ST(0)
R1	4.0	ST(1)
R2		ST(2)
R3		ST(3)
R4		ST(4)
R5		ST(5)
R6		ST(6)
R7		ST(7)

FLD `flp_z`

Register	Value	Stack
R0	6.0	ST(0)
R1	20.0	ST(1)
R2	4.0	ST(2)
R3		ST(3)
R4		ST(4)
R5		ST(5)
R6		ST(6)
R7		ST(7)

FADD `ST(0), ST(1)`

Register	Value	Stack
R0	26.0	ST(0)
R1	20.0	ST(1)
R2	4.0	ST(2)
R3		ST(3)
R4		ST(4)
R5		ST(5)
R6		ST(6)
R7		ST(7)

FLD `flp_num`

Register	Value	Stack
R0	7.0	ST(0)
R1	26.0	ST(1)
R2	20.0	ST(2)
R3	4.0	ST(3)
R4		ST(4)
R5		ST(5)
R6		ST(6)
R7		ST(7)

FADD `ST(0),ST(1)`

Register	Value	Stack
R0	33.0	ST(0)
R1	26.0	ST(1)
R2	20.0	ST(2)
R3	4.0	ST(3)
R4		ST(4)
R5		ST(5)
R6		ST(6)
R7		ST(7)

FDIV `ST(0), ST(3)`

Register	Value	Stack
R0	**8.25**	ST(0) →`flp_rslt`
R1	26.0	ST(1)
R2	20.0	ST(2)
R3	4.0	ST(3)
R4		ST(4)
R5		ST(5)
R6		ST(6)
R7		ST(7)

Figure 17.17 Register stack for Example 17.12.

Example 17.13 This example calculates the value of the following expression using x as a variable for floating-point arithmetic:

$$x^3 - 5x^2 - 5x + \sqrt{x}$$

Figure 17.18 shows the assembly language module embedded in a C program to perform the calculations for the given expression. Figure 17.19 shows the contents of the register stack during execution of the program for the variable x that is equal to 2.0.

```
//eval_expr9.cpp
//evaluate the following expression
//for different values of x:
//x^3 - 5x^2 - 5x + x^(1/2)

#include "stdafx.h"
int main (void)
{
//define variables
   float flp_x, five, rslt;
   five = 5.0;

   printf ("Enter a floating-point value for x: \n");
   scanf ("%f", &flp_x);

//switch to assembly
      _asm
      {
//calculate x^3
         FLD    flp_x           //x -> ST(0)
         FMUL   ST(0), ST(0)    //x^2 -> ST(0)
         FMUL   flp_x           //x^3 -> ST(0)

//calculate 5x^2
         FLD    flp_x           //x -> ST(0)
                                //x^3 ->ST(1)
         FMUL   ST(0), ST(0)    //x^2 -> ST(0)
                                //ST(1) = x^3
         FMUL   five            //5x^2 -> ST(0)

                                      //continued on next page
```

(a)

Figure 17.18 Program to calculate the given expression for different values of x: (a) the program and (b) the outputs.

```
//calculate 5x
        FLD    flp_x          //x -> ST(0)
        FMUL   five           //5x -> ST(0)
                              //ST(1) = 5x^2
                              //ST(2) = x^3
//calculate square root of x
        FLD    flp_x          //x -> ST(0)
        FSQRT                 //sq rt of x -> ST(0)

//calculate result
        FXCH   ST(3)          //x^3 -> ST(0)
        FSUB   ST(0), ST(2)   //x^3 - 5x^2 -> ST(0)
        FSUB   ST(0), ST(1)   //x^3 -5x^2 - 5x -> ST(0)
        FADD   ST(0), ST(3)   //x^3 -5x^2 - 5x + sq rt x
                              //-> ST(0)
        FST    rslt
}

//display result
   printf ("\nResult = %f\n\n", rslt);

   return 0;
}
```

```
Enter a floating-point value for x:
2.0

Result = -20.585787

Press any key to continue . . . _
-------------------------------------------------------------
Enter a floating-point value for x:
4.0

Result = -34.000000

Press any key to continue . . . _
-------------------------------------------------------------
Enter a floating-point value for x:
6.0

Result = 8.449490

Press any key to continue . . . _
-------------------------------------------------------------
```

//continued on next page

(b)

Figure 17.18 (Continued)

```
Enter a floating-point value for x:
7.250

Result = 84.708206

Press any key to continue . . . _
-------------------------------------------------------------
Enter a floating-point value for x:
15.456789

Result = 2424.903320

Press any key to continue . . . _
-------------------------------------------------------------
Enter a floating-point value for x:
125.375


Result = 1891545.750000

Press any key to continue . . . _
-------------------------------------------------------------
Enter a floating-point value for x:
4.450275

Result = -31.029083

Press any key to continue . . . _
```

Figure 17.18 (Continued)

Calculate x^3

R0	8.0	ST(0)
R1		ST(1)
R2		ST(2)
R3		ST(3)
R4		ST(4)
R5		ST(5)
R6		ST(6)
R7		ST(7)

Calculate $5x^2$

20.0	ST(0)
8.0	ST(1)
	ST(2)
	ST(3)
	ST(4)
	ST(5)
	ST(6)
	ST(7)

//continued on next page

Figure 17.19 Register stack for Example 17.13.

	Calculate 5x		Calculate $x^{1/2}$	
R0	10.0	ST(0)	1.414214	ST(0)
R1	20.0	ST(1)	10.0	ST(1)
R2	8.0	ST(2)	20.0	ST(2)
R3		ST(3)	8.0	ST(3)
R4		ST(4)		ST(4)
R5		ST(5)		ST(5)
R6		ST(6)		ST(6)
R7		ST(7)		ST(7)

	FXCH ST(3)		**FSUB** ST(0), ST(2)	
R0	8.0	ST(0)	-12.0	ST(0)
R1	10.0	ST(1)	10.0	ST(1)
R2	20.0	ST(2)	20.0	ST(2)
R3	1.414214	ST(3)	1.414214	ST(3)
R4		ST(4)		ST(4)
R5		ST(5)		ST(5)
R6		ST(6)		ST(6)
R7		ST(7)		ST(7)

	FSUB ST(0),ST(1)		**FADD** ST(0), ST(3)	
R0	-22.0	ST(0)	**-20.585787**	ST(0) → rslt
R1	10.0	ST(1)	10.0	ST(1)
R2	20.0	ST(2)	20.0	ST(2)
R3	1.414214	ST(3)	1.414214	ST(3)
R4		ST(4)		ST(4)
R5		ST(5)		ST(5)
R6		ST(6)		ST(6)
R7		ST(7)		ST(7)

Figure 17.19 (Continued)

Example 17.14 This example calculates the value of the following trigonometric expressions using floating-point arithmetic:

$$[(\text{sine } 30°) - (\text{cosine } 45°)]$$
$$[(\text{sine } 30°) \times (\text{cosine } 45°)]$$
$$[(\text{sine } 30°) \times (\text{cosine } 45°)]^2$$

Figure 17.20 shows the assembly language module embedded in a C program to perform the calculations for the given trigonometric expressions. Note that the degrees must be given in radians. Figure 17.21 shows the contents of the register stack during execution of the program for the [(sine 30°) – (cosine 45°)] operation.

```
//sine_cos.cpp
//calculate the following expressions:
//[(sine 30 deg) - (cos 45 deg)]
//[(sine 30 deg) x (cos 45 deg)]
//[(sine 30 deg) x (cos 45 deg)]^2

#include "stdafx.h"

int main (void)

{
   double angle1, angle2, rslt_sub, rslt_mul, rslt_mul_sq;

   angle1 = 0.523598775;        //30 deg must be in radians
   angle2 = 0.785398163;        //45 deg must be in radians

//switch to assembly
      _asm
      {
//-----------------------------------------------------------
//perform subtraction
         FLD    angle1          //angle1 -> ST(0)
         FSIN                   //sine of angle1 -> ST(0)

         FLD    angle2          //angle2 -> ST(0)
                                //sine angle1 -> ST(1)
         FCOS                   //cos of ST(0) -> ST(0)
                                //ST(1) = sine angle1

         FSUBP ST(1), ST(0)     //ST(1) - ST(0) -> ST(1), pop
                                //sine - cos -> ST(0)
         FST    rslt_sub        //ST(0) (diff) -> rslt_sub

//-----------------------------------------------------------
                                      //continued on next page
```

(a)

Figure 17.20 Program to calculate the values of the given trigonometric expressions: (a) the program and (b) the outputs.

```
//----------------------------------------------------------------
//perform multiplication
         FLD    angle1          //angle1 -> ST(0)
         FSIN                   //sine of angle1 -> ST(0)

         FLD    angle2          //angle2 -> ST(0)
                                //sine angle1 -> ST(1)
         FCOS                   //cos of ST(0) -> ST(0)
                                //ST(1) = sine angle1

         FMUL   ST(0), ST(1)    //sine x cos -> ST(0)
         FST    rslt_mul        //ST(0) (result) -> rslt_mul

//----------------------------------------------------------------
//perform multiplication squared
         FMUL   ST(0), ST(0)    //(sine x cos)^2 -> ST(0)
         FST    rslt_mul_sq     //ST(0) (result)
                                //-> rslt_mul_sq
      }

   printf ("Result subtracting = %f\n", rslt_sub);

   printf ("\nResult multiplication = %f\n", rslt_mul);

   printf ("\nResult multiplication squared = %f\n\n",
            rslt_mul_sq);

   return 0;
}
```

```
Result subtraction = -0.207107

Result multiplication = 0.353553

Result multiplication squared = 0.125000

Press any key to continue . . . _
```

(b)

Figure 17.20 (Continued)

FLD `angle1`

R0	0.523598775	ST(0)
R1		ST(1)
R2		ST(2)
R3		ST(3)
R4		ST(4)
R5		ST(5)
R6		ST(6)
R7		ST(7)

FSIN

0.5	ST(0)
	ST(1)
	ST(2)
	ST(3)
	ST(4)
	ST(5)
	ST(6)
	ST(7)

FLD `angle2`

R0	0.785398163	ST(0)
R1	0.5	ST(1)
R2		ST(2)
R3		ST(3)
R4		ST(4)
R5		ST(5)
R6		ST(6)
R7		ST(7)

FCOS

0.707106781	ST(0)
0.5	ST(1)
	ST(2)
	ST(3)
	ST(4)
	ST(5)
	ST(6)
	ST(7)

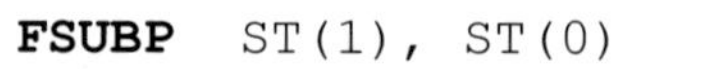

R0	**-0.207106781**	ST(0) → `rslt_sub`
R1		ST(1)
R2		ST(2)
R3		ST(3)
R4		ST(4)
R5		ST(5)
R6		ST(6)
R7		ST(7)

Figure 17.21 Register stack for Example 17.14.

Example 17.15 This example calculates the value of the following trigonometric expression using floating-point arithmetic:

$$(5 \times \cos 30°) - (2 \times \text{sine } 60°)$$

Figure 17.22 shows the assembly language module embedded in a C program to perform the calculation for the given trigonometric expression. Note that the degrees must be given in radians.

```
//cos_sine.cpp
//evaluate the following expression:
//(5 x cos 30 deg) - (2 x sin 60 deg)

#include "stdafx.h"
int main (void)

{
//define variables
   double angle1, angle2, rslt;

   angle1 = 0.523598775;       //30 deg must be in radians
   angle2 = 1.047197551;       //60 deg must be in radians

   float five, two;
   five = 5.0;
   two = 2.0;

//switch to assembly
      _asm
      {
         FLD    angle1         //angle1 (30) -> ST(0)
         FCOS                  //cos 30 deg -> ST(0
         FMUL   five           //ST(0) x 5 -> ST(0)

         FLD    angle2         //angle2 (60) -> ST(0)
                               //(cos 30) x 5 -> ST(1)
         FSIN                  //sine 60 deg -> ST(0)
                               //ST(1) = (cos 30) x 5
         FMUL   two            //ST(0) x 2 -> ST(0)
                               //ST(1) = (cos 30) x 5

         FSUBP ST(1), ST(0)    //ST(1) - ST(0) -> ST(1), pop
                               //(5 cos 30)-(2 sine 60)-> ST(0)

         FST    rslt
      }

//display result
   printf ("Result = %f\n\n", rslt);
   return 0;
}
```

(a)

//continued on next page

Figure 17.22 Program to calculate the value of the given trigonometric expression: (a) the program and (b) the outputs.

```
Result = 2.598076

Press any key to continue . . . _
```
(b)

Figure 17.22 (Continued)

Example 17.16 This example uses the *move data from string to string* for bytes (MOVSB) instruction to overwrite data in a given string. The repeat prefix (REP) is also used in the program to copy four bytes from the given string to a different location in the same string. Figure 17.23 shows the assembly language program — not embedded in a C program — to perform the operation.

```
;movs_byte6.asm
;replace the last 4 characters with the first 4 characters
.STACK
;----------------------------------------------------------------
.DATA
RSLT    DB  0DH, 0AH, 'ABCD123456781234    $'
;----------------------------------------------------------------
.CODE
BEGIN PROC    FAR

;set up pgm ds
      MOV   AX, @DATA         ;get addr of data seg
      MOV   DS, AX            ;move addr to ds
      MOV   ES, AX            ;move addr to es

;move string  elements
      CLD
      MOV   CX, 4             ;count in cx
      LEA   SI, RSLT + 2      ;addr of rslt+2 -> si as src
      LEA   DI, RSLT + 14     ;addr of rslt+4 -> di as dst

REP   MOVSB                   ;move bytes to dst

;display result
      MOV   AH, 09H           ;display string
      LEA   DX, RSLT          ;put addr of rslt in dx
      INT   21H               ;dos interrupt

BEGIN ENDP
      END   BEGIN        (a)       //continued on next page
```

Figure 17.23 Program to overwrite string locations: (a) the program and (b) the output.

```
ABCD12345678ABCD
                                (b)
```

Figure 17.23 (Continued)

Example 17.17 This example is similar to Example 17.16; however, the string is entered from the keyboard. The *move data from string to string* for bytes (MOVSB) instruction is also used together with the repeat prefix (REP) to replace the last four characters of a 10-character string with the first four characters. Figure 17.24 shows the assembly language program — not embedded in a C program — to perform the operation.

```
;movs_byte7.asm
;replace the last 4 characters of a
;10-character string entered from the
;keyboard with the first 4 characters

;------------------------------------------------------------
.STACK
;------------------------------------------------------------
.DATA
PARLST  LABEL BYTE
MAXLEN  DB 15
ACTLEN  DB ?
OPFLD   DB 20 DUP(?)
PRMPT   DB 0DH, 0AH, 'Enter a 10-character string: $'
RSLT    DB 0DH, 0AH, 'New string =                 $'
;------------------------------------------------------------
.CODE
BEGIN PROC FAR

;set up pgm ds
      MOV   AX, @DATA          ;addr of data seg -> ax
      MOV   DS, AX             ;addr of data seg -> ds
      MOV   ES, AX             ;addr of data seg -> es

;read prompt
      MOV   AH, 09H            ;display string
      LEA   DX, PRMPT          ;addr of prompt -> dx
      INT   21H                ;dos interrupt
                          (a)       //continued on next page
```

Figure 17.24 Program to exchange the last four characters of a 10-character string with the first four characters: (a) the program and (b) the outputs.

```
;keyboard rtn to enter characters
      MOV    AH, 0AH             ;buffered keyboard input
      LEA    DX, PARLST          ;addr of parlst -> dx
      INT    21H                 ;dos interrupt
;-------------------------------------------------------------
;move opfld to rslt area
      LEA    SI, OPFLD           ;addr of opfld -> si
      LEA    DI, RSLT+15         ;addr of rslt -> di

      MOV    CL, ACTLEN          ;actual length -> cl (or use 10)
      MOV    CH, 00H             ;clear ch

REP   MOVSB
;-------------------------------------------------------------
;overlap strings
      MOV    CX, 0004H
      LEA    SI, RSLT+15
      LEA    DI, RSLT+21

REP   MOVSB
;-------------------------------------------------------------
;display result
      MOV    AH, 09H             ;display string
      LEA    DX, RSLT            ;addr of rslt -> dx
      INT    21H                 ;dos interrupt
;-------------------------------------------------------------
BEGIN ENDP
      END    BEGIN
```

```
Enter a 10-character string: ABCDEFGHIJ
New string = ABCDEFABCD
----------------------------------------------------------
Enter a 10-character string: 1234567890
New string = 1234561234
----------------------------------------------------------
Enter a 10-character string: ^^^^&&&&&&
New string = ^^^^&&^^^^
----------------------------------------------------------
Enter a 10-character string: ----123456
New string = ----12----
----------------------------------------------------------
Enter a 10-character string: 12ab//////
New string = 12ab//12ab
```

(b)

Figure 17.24 (Continued)

Example 17.18 This example converts an integer (1 — 26) that is entered from the keyboard to the corresponding uppercase alphabetic character. An array is used to define the uppercase alphabetic characters. Figure 17.25 illustrates the program written in the C programming language.

```
//translate_2.cpp
//Convert an integer that represents an alphabetic
//character (displayed as uppercase); 1=A, 26=Z

#include "stdafx.h"
int main (void)
{
//define alphabetic character array
   int alpha[27] = {'0','A', 'B', 'C',
                    'D', 'E', 'F', 'G',
                    'H', 'I', 'J', 'K',
                    'L', 'M', 'N', 'O',
                    'P', 'Q', 'R', 'S',
                    'T', 'U', 'V', 'W',
                    'X', 'Y', 'Z'};

   int x;

   printf ("Enter an integer (1-26) that corresponds
               to an alphabetic character: \n");
   scanf ("%d", &x);

   printf ("\nInteger = %d, Character = %c\n\n",
               x, alpha[x]);

   return 0;
}
```

(a)

```
Enter an integer (1-26) that corresponds to an alphabetic char-
acter:
1

Integer = 1, Character = A

Press any key to continue . . . _
---------------------------------------------------------------
```

//continued on next page

(b)

Figure 17.25 Program to convert an integer to the corresponding uppercase alphabetic character: (a) the program and (b) the outputs.

```
Enter an integer (1-26) that corresponds to an alphabetic char-
acter:
2

Integer = 2, Character = B

Press any key to continue . . . _

-------------------------------------------------------------

Enter an integer (1-26) that corresponds to an alphabetic char-
acter:
8

Integer = 8, Character = H

Press any key to continue . . . _

-------------------------------------------------------------

Enter an integer (1-26) that corresponds to an alphabetic char-
acter:
20

Integer = 20, Character = T

Press any key to continue . . . _

-------------------------------------------------------------

Enter an integer (1-26) that corresponds to an alphabetic char-
acter:
26

Integer = 26, Character = Z

Press any key to continue . . . _
```

Figure 17.25 (Continued)

Example 17.19 This example utilizes an array to translate a 2-digit number that is entered from the keyboard to the correct spelling of the corresponding number. The 2-digit numbers are in the following range: 01 — 12. Figure 17.26 shows the assembly language program — not embedded in a C program — to perform the translation operation. Since the length of the longest number is six characters, each table entry in the array must contain six locations.

```
;number_table.asm
;enter a number (01-12) from the keyboard
;and display the spelling of the corresponding number

;--------------------------------------------------------------
.STACK
;--------------------------------------------------------------
.DATA
PARLST  LABEL   BYTE
MAXLEN  DB      40
ACTLEN  DB      ?
OPFLD   DB      40      DUP(?)
TABLE   DB      'ONE   '
        DB      'TWO   '
        DB      'THREE '
        DB      'FOUR  '
        DB      'FIVE  '
        DB      'SIX   '
        DB      'SEVEN '
        DB      'EIGHT '
        DB      'NINE  '
        DB      'TEN   '
        DB      'ELEVEN'
        DB      'TWELVE'
PRMPT   DB      0DH, 0AH, 'Enter 2-digit number (01-12): $'
RSLT    DB      0DH, 0AH, 'Number =         $'
;--------------------------------------------------------------
.CODE
BEGIN PROC    FAR

;set up pgm ds
      MOV   AX, @DATA           ;get addr of data seg
      MOV   DS, AX              ;move addr to ds

;read prompt
      MOV   AH, 09H             ;display string
      LEA   DX, PRMPT           ;put addr of prompt in dx
      INT   21H                 ;dos interrupt

;keyboard request rtn to enter characters
      MOV   AH, 0AH             ;buffered keyboard input
      LEA   DX, PARLST          ;put addr of parlst in dx
      INT   21H                 ;dos interrupt
;--------------------------------------------------------------
                                  //continued on next page
```

(a)

Figure 17.26 Program to translate a 2-digit number to the correct spelling of the number: (a) the program and (b) the outputs.

```
;----------------------------------------------------------
;set up source and destination
      LEA   SI, OPFLD           ;point to first number
      LEA   DI, RSLT + 11       ;point to destination

;get number
      MOV   AH, [SI]            ;get 1st number
      SUB   AH, 30H             ;unpack it
      INC   SI                  ;point to opfld+1
      MOV   AL, [SI]            ;get 2nd number
      SUB   AL, 30H             ;unpack it

;convert unpacked number in ax
;to the binary equivalent in al
      AAD
      SUB   AL, 01H             ;adjust for offset in table
      MOV   CL, 06H             ;6 characters per number
      MUL   CL                  ;mul al by 6 to find number

;----------------------------------------------------------

;get number from table
      LEA   SI, TABLE           ;put addr of table in si
      MOV   AH, 00H             ;clear ah
      ADD   SI, AX              ;si points to number
      MOV   CX, 06H             ;for 6 characters

NUM:  MOV   AL, [SI]            ;get number letter
      MOV   [DI], AL            ;put letter in rslt
      INC   SI                  ;point to next letter
      INC   DI                  ;point to next destination
      LOOP  NUM                 ;loop until cx = 0

;----------------------------------------------------------

;display the result
      MOV   AH, 09H             ;display string
      LEA   DX, RSLT            ;put addr of rslt in dx
      INT   21H                 ;dos interrupt

BEGIN ENDP
      END   BEGIN                  //continued on next page
```

Figure 17.26 (Continued)

```
Enter 2-digit number (01-12): 01
Number = ONE
------------------------------------------------------------
Enter 2-digit number (01-12): 06
Number = SIX
------------------------------------------------------------
Enter 2-digit number (01-12): 09
Number = NINE
------------------------------------------------------------
Enter 2-digit number (01-12): 10
Number = TEN
------------------------------------------------------------
Enter 2-digit number (01-12): 11
Number = ELEVEN
------------------------------------------------------------
Enter 2-digit number (01-12): 12
Number = TWELVE
```

(b)

Figure 17.26 (Continued)

17.2 Problems

17.1 Write a program in assembly language — not embedded in a C program — that determines the largest number from eight single-digit numbers that are entered from the keyboard, then display the number.

17.2 Write a program in assembly language — not embedded in a C program — that counts the number of times that a number occurs in an array of eight numbers. The first number entered from the keyboard is the number that is being compared to the remaining eight numbers. Display the result.

17.3 Write a program in assembly language — not embedded in a C program — that translates an ASCII character (0 through 9) that is entered from the keyboard to the letters *a* through *j*, respectively.

17.4 Write an assembly language module embedded in a C program that uses the exclusive-OR function to operate on a string of hexadecimal characters that are entered from the keyboard. Perform the operation on the following character positions in the specified sequence, then display the results:

Position 1 ⊕ position 3 → position 3
Position 2 ⊕ position 4 → position 4
Position 3 ⊕ position 5 → position 5
Position 4 ⊕ position 6 → position 6

17.5 Write a program in assembly language — not embedded in a C program — that calculates the factorial of numbers entered from the keyboard in the range of 1 through 6. Display the results.

17.6 Write an assembly language module embedded in a C program to obtain the sum of cubes for the numbers 1 through 5. Display the resulting sum of cubes.

17.7 Write an assembly language module embedded in a C program to calculate 2 to power of n (0 – 8), where the variable n is entered from the keyboard.

17.8 Write an assembly language module embedded in a C program to evaluate the following expression using fixed-point integers for bytes, words, and doublewords and obtain the quotients only:

$$[(X \times Y) + Z + \text{constant}]$$

17.9 Write an assembly language module embedded in a C program to evaluate the expression shown below for Y using a floating-point number for the variable X. The range for X is $-3.0 \leq X \leq +3.0$. Enter several numbers for X and display the corresponding results.

$$Y = X^3 - 10X^2 + 20X + 30$$

17.10 Write an assembly language module embedded in a C program to determine the result of the equation shown below, where L and C are floating-point numbers. The variables L and C are entered from the keyboard. Display the results.

$$\text{Result} = 1/[2 \times \pi \times (L \times C)^{1/2}]$$

17.11 Given the program segment shown below, obtain the results for the following floating-point numbers: 2.0, 10.0, and 15.0.

```
                                   . . .
FLD    flp_num
FMUL   ST(0), ST(0)
FLDPI
FMUL   ST(0), ST(1)
FST    rslt
                                   . . .
```

17.12 Given the program segment shown below, obtain the results for the following floating-point numbers:

```
flp_num1 = 125.0,      flp_num2 = 245.0

flp_num1 = 650.0,      flp_num2 = 375.0

flp_num1 = 755.125,    flp_num2 = 575.150
```

```
                    . . .
FLD    flp_num1
FSQRT
FLD    flp_num2
FSQRT
FSUBP  ST(1), ST(0)
FST    rslt
                    . . .
```

17.13 Write an assembly language module embedded in a C program to determine the result of the following equation:

(sine 90° – sine 30°) / (cosine 180° + cosine 60°)

17.14 Write an assembly language module embedded in a C program to calculate the tangent of the following angles using the sine and cosine of the angles:

30°, 45°, and 60°

17.15 Write an assembly language program — not embedded in a C program — that replaces all spaces with hyphens in a character string entered from the keyboard.

Appendix A

ASCII Character Codes

HEX	ASCII	HEX	ASCII	HEX	ASCII
20	SP	40	@	60	‘
21	!	41	A	61	a
22	“	42	B	62	b
23	#	43	C	63	c
24	$	44	D	64	d
25	%	45	E	65	e
26	&	46	F	66	f
27	’	47	G	67	g
28	(	48	H	68	h
29	)	49	I	69	i
2A	*	4A	J	6A	j
2B	+	4B	K	6B	k
2C	,	4C	L	6C	l
2D	-	4D	M	6D	m
2E	.	4E	N	6E	n
2F	/	4F	O	6F	o
30	0	50	P	70	p
31	1	51	Q	71	q
32	2	52	R	72	r
33	3	53	S	73	s
34	4	54	T	74	t
35	5	55	U	75	u
36	6	56	V	76	v
37	7	57	W	77	w
38	8	58	X	78	x
39	9	59	Y	79	y
3A	:	5A	Z	7A	z
3B	;	5B	[	7B	{
3C	<	5C	\	7C	|
3D	=	5D	]	7D	}
3E	>	5E	^	7E	~
3F	?	5F	_	7F	DEL

Appendix B

Answers to Select Problems

Chapter 1 Number Systems and Number Representations

1.5 Convert the octal number 173.25_8 to decimal.

$$
\begin{aligned}
473.26_8 &= (4 \times 8^2) + (7 \times 8^1) + (3 \times 8^0) + (2 \times 8^{-1}) + (6 \times 8^{-2}) \\
&= (256 + 56 + 3 + 0.25 + 0.09375)_{10} \\
&= 315.34375_{10}
\end{aligned}
$$

1.9 Add the following binary numbers to yield a 12-bit sum:

```
1111 1111 1111
1111 1111 1111
1111 1111 1111
1111 1111 1111
--------------
1111 1111 1100
```

1.16 Multiply the unsigned binary numbers 1111_2 and 0011_2.

```
                 1  1  1  1
             ×)  0  0  1  1
            ---------------
 0  0  0  0  1  1  1  1
 0  0  0  1  1  1  1
 ----------------------
 0  0  1  0  1  1  0  1
```

1.19 The numbers shown below are in sign-magnitude representation. Convert the numbers to 2s complement representation for radix 2 with the same numerical value using eight bits.

Sign magnitude		2s complement
1001 1001	–25	1110 0111
0001 0000	+16	0001 0000
1011 1111	–63	1100 0001

1.21 Add the following BCD numbers:

```
             1001       1000
        +)   1001       0111
                        ----
                1   ←   1111
             ----
     1 ←     0011       0110
                        ----
             0110       0101
             ----
             1001
     ↓        ↓          ↓
   0001      1001       0101
```

Chapter 2 X86 Processor Architecture

2.3 The 7-bit code words shown below are received using Hamming code with odd parity. Determine the syndrome word for each received code word.

(a)

Bit Position =	1	2	3	4	5	6	7
Received Code Word =	1	1	1	1	1	1	1
Syndrome Word =	1	1	1				

(b)

Bit Position =	1	2	3	4	5	6	7
Received Code Word =	0	0	0	0	0	0	0
Syndrome Word =	1	1	1				

2.7 Obtain the code word using the Hamming code with odd parity for the following message word: 1 1 0 1 0 1 0 1 1 1 1.

1	2	3	4	5	6	7	8	9	10	11	12	13	14	15
0	0	1	1	1	0	1	0	0	1	0	1	1	1	1

2.13 A fraction and the bits to be deleted (low-order three bits) are shown below. Determine the rounded results using truncation (round toward zero), adder-based rounding (round to nearest), and von Neumann rounding.

Fraction	
. 0 0 1 0 1 0 0 1	1 0 1

Truncation: .0010 1001
Adder-based: .0010 1010
von Neumann: .0010 1001

Chapter 3 Addressing Modes

3.3 Let DS = 1100H, DISPL = –126, and SI = 0500H. Using the real addressing mode with a 20-bit address, determine the physical memory address for the instruction shown below.

```
MOV   DISPL[SI], DX

        SI:  0500
 +) DISPL:  FF82
            -----
            0482H   Effective address

        DS:  11000
    +)  EA:  00482
            ------
            11482H  Real Memory address
```

3.6 Differentiate between the operation of the following two move instructions:

```
(a)    MOV    EBX, 1234ABCDH
(b)    MOV    [EBX], 1234H
```

(a) Moves the immediate value of 1234ABCDH to register EBX.
(b) Moves the immediate value of 1234H to the address pointed to by register EBX.

Chapter 4 C Programming Fundamentals

4.3 Write a program to change the following Fahrenheit temperatures to centigrade temperatures: 32.0, 100.0, and 85.0. Then print the results.

```
//fahr_to_cent.cpp
//change Fahrenheit to centigrade

#include "stdafx.h"

int main (void)
{
   float fahr1 = 32.0;
   float fahr2 = 100.0;
   float fahr3 = 85.0;

   float cent1, cent2, cent3;

   cent1 = (fahr1 - 32) * (5.0/9.0);
   cent2 = (fahr2 - 32) * (5.0/9.0);
   cent3 = (fahr3 - 32) * (5.0/9.0);

   printf ("fahrenheit 32.0 = centigrade %f\n",
            cent1);
   printf ("fahrenheit 100.0 = centigrade %f\n",
            cent2);
   printf ("fahrenheit 85.0 = centigrade %f\n\n",
            cent3);

   return 0;
}
```

```
Fahrenheit 32.0 = centigrade 0.000000
Fahrenheit 100.0 = centigrade 37.777779
Fahrenheit 85.0 = centigrade 29.444445

Press any key to continue . . . _
```

4.7 Write a program to illustrate the difference between integer division and floating-point division regarding the remainder, using the following dividends and divisors: 742 ÷ 16 and 1756 ÷ 24. For the floating-point result, let the integer be two digits and the fraction be four digits.

```
//div_int_flp.cpp
//show difference between integer division and
//floating-point division regarding the remainder
#include "stdafx.h"

int main (void)
{
      printf ("Integer division: %d\n", 742/16);
      printf ("Floating-point division:
                  %2.4f\n\n", 742.0/16.0);

      printf ("Integer division: %d\n", 1756/24);
      printf ("Floating-point division:
                  %2.4f\n\n", 1756.0/24.0);
      return 0;
}
```

```
Integer division: 46
Floating-point division: 46.3750

Integer division: 73
Floating-point division: 73.1667

Press any key to continue . . . _
```

4.10 Indicate the value to which each of the following expressions evaluate.

(a) (1 + * 3)	Evaluates to 7.
(b) 10 % 3 * 3 – (1 + 2)	Evaluates to 0 (* and % same precedence).
(c) ((1 + 2) * 3)	Evaluates to 9 (1 + 2) first.
(d) (5 == 5)	Evaluates to true (1)
(e) (5 = 5)	Evaluates to 5.

4.18 To what does the expression 5 + 3 * 8 / 2 + 2 evaluate?
Rewrite the expression, adding parentheses, so that it evaluates to 16.

```
//evaluate_expr.cpp
//evaluate expression without parentheses,
//then with parentheses
#include "stdafx.h"
int main (void)
{
	int result1, result2;
	result1 = 5 + 3 * 8 / 2 + 2;
	result2 = (5 + 3) * 8 / (2 + 2);

	printf ("Result1 = %d \nResult2 = %d\n\n",
			result1, result2);
	return 0;
}
```

```
Result1 = 19
Result2 = 16

Press any key to continue . . . _
```

4.22 Given the program shown below, obtain the outputs for variables *a* and *b* after the program has executed.

```
//pre_post_incr4.cpp
//example to illustrate pre- and post-increment
#include "stdafx.h"
int main (void)
{
	int a, b;
	int c = 0;
	a = ++c;
	b = c++;

	printf ("%d   %d   %d\n\n", a, b, ++c);
	return 0;
}
```

```
1   1   3

Press any key to continue . . . _
```

4.30 Use a **for ( )** loop to count from 10 to 100 in increments of 10. Display the numbers on a single line.

```
//for_loop4.cpp
//use a for loop to count from 10 to 100
//in increments of 10
#include "stdafx.h"
int main (void)
{
   int int1;

   for (int1 = 10; int1 < 110; int1 = int1 + 10)
      printf ("%d ", int1);

   printf ("\n\n");
   return 0;
}
```

```
10 20 30 40 50 60 70 80 90 100

Press any key to continue . . . _
```

4.33 Determine the number of *A*s that are printed by the program shown below.

```
//for_loop_nested.cpp
//nested for loop to yield multiple outputs
#include "stdafx.h"
int main (void)
{
   int x, y;

   for (x = 0; x < 5; x++)
      for (y = 5; y > 0; y--)
         printf ("A");

   printf ("\n\n");

   return 0;
}
```

```
AAAAAAAAAAAAAAAAAAAAAAAAA

Press any key to continue . . . _
```

4.36 Write a program that defines a string, then prints the string. Then reverse the order of the original string and print the new string.

```
//string_cpy_len.cpp
//print a string in reverse order

#include "stdafx.h"

#include "string.h"

int main (void)

{
   char string[15];
   int i;

   strcpy (string, "54321 edcba");

   for (i=0; string[i]; i++)
      printf ("%c", string[i]);

   printf ("\n\n");

   for (i=strlen(string)-1; i>=0; i--)
      printf ("%c", string[i]);

   printf ("\n\n");

   return 0;
}
```

```
54321 edcba

abcde 12345

Press any key to continue . . . _
```

4.38 Write a program to illustrate pointer postincrement. Assign an integer *int1* a value of 50 and assign a pointer variable the address of *int1*.

```
//ptr_incr.cpp
//program to illustrate pointer postincrement

#include "stdafx.h"

int main (void)
{
   int int1 = 50;
   int *pointer = &int1;

   int1++;
   (*pointer)++;

   printf ("int1 = %d\n\n", *pointer);

   return 0;
}
```

```
int1 = 52

Press any key to continue . . . _
```

Chapter 5 Data Transfer Instructions

5.3 The sign-magnitude notation for a positive number is represented by the equation shown below, where the sign bit of 0 indicates a positive number.

$$A = (0\ a_{n-2}a_{n-3}\ \cdots\ a_1a_0)_r$$

The sign-magnitude notation for a negative number is represented by the equation shown below, where the sign bit is the radix minus 1. In sign-magnitude notation, the positive version differs from the negative version only in the sign digit position. The magnitude portion $a_{n-2}a_{n-3}\ \cdots\ a_1a_0$ is identical for both positive and negative numbers of the same absolute value.

$$A' = [(r-1)\ a_{n-2}a_{n-3}\ \cdots\ a_1a_0]_r$$

The numbers shown below are in sign-magnitude representation for radix 2. Convert the numbers to 2s complement representation with the same numerical value using eight bits.

	Sign-magnitude	2s Complement
(a)	**0**111 1111	0111 1111
(b)	**1**111 1111	1000 0001
(c)	**0**000 1111	0000 1111
(d)	**1**000 1111	1111 0001
(e)	**1**001 0000	1111 0000

5.4 Convert the positive 2s complement numbers shown below to negative numbers in 2s complement representation.

$$0000\ 1100\ 0011_2\ (+195_{10}) = 1111\ 0011\ 1101_2\ (-195_{10})$$

$$0101\ 0101\ 0101_2\ (+1365_{10}) = 1010\ 1010\ 1011_2\ (-1365)$$

5.7 Write a program using C and assembly language that shows the application of the *conditional move greater than or equal* (CMOVGE) instruction for signed operands.

```
//mov_cond3.cpp
//uses cmovge (greater than or equal); sf xor of = 0.
//if signed user-entered x is greater than or equal to
//user-entered y, then print x; otherwise, print y.
//if integers are equal, then print x

#include "stdafx.h"

int main (void)
{
//define variables
   int   x, y, rslt;

   printf ("Enter two signed integers: \n");
   scanf ("%d %d", &x, &y);

//switch to assembly
      _asm
      {
         MOV       EAX, x
         MOV       EBX, y

         // move ebx to rslt if conditional move fails
         MOV       EDX, EBX
         CMP       EAX, EBX       //set flags

         //if eax >= ebx, move eax to rslt;
         //otherwise, move ebx to result
         CMOVGE    EDX, EAX
         MOV       rslt, EDX
      }

   printf ("Result = %d\n\n", rslt);

   return 0;
}
```

```
Enter two signed integers:
-256 -128
Result = -128

Press any key to continue . . . _
-----------------------------------------------------
Enter two signed integers:
555 444
Result = 555

Press any key to continue . . . _
-----------------------------------------------------
Enter two signed integers:
-400 400
Result = 400

Press any key to continue . . . _
-----------------------------------------------------
Enter two signed integers:
750 750
Result = 750

Press any key to continue . . . _
-----------------------------------------------------
Enter two signed integers:
-800 -800
Result = -800

Press any key to continue . . . _
-----------------------------------------------------
Enter two signed integers:
-100 -101
Result = -100

Press any key to continue . . . _
```

5.10 Write a program using C and assembly language that shows the application of the *conditional move sign* (CMOVS) instruction for signed operands. The compare instruction subtracts the second source operand from the first source operand and the state of the sign flag is set to either 0 or 1, depending on the sign of the result. If the sign of the difference is negative (SF = 1), then the move operation is executed.

```
//mov_cond7.cpp
//uses cmovs (sign is neg); sf = 1.
//signed user-entered x is compared to user-entered y.
//If sign flag is set, print x; otherwise, print y.

#include "stdafx.h"

int main (void)
{
//define variables
   int   x, y, rslt;
   printf ("Enter two signed integers: \n");
   scanf ("%d %d", &x, &y);

//switch to assembly
      _asm
      {
         MOV      EAX, x
         MOV      EBX, y
//move ebx to rslt if conditional move fails

         MOV      EDX, EBX
         CMP      EAX, EBX    //set flags
//if sf = 1 (sign is neg), move eax to rslt;
//otherwise, move ebx to result

         CMOVS    EDX, EAX
         MOV      rslt, EDX
      }
   printf ("Result = %d\n\n", rslt);

   return 0;
}
```

```
Enter two signed integers:
20000 2000          //4E20H 07D0H
Result = 2000       //Difference = 4650H
                    //Sign is positive

Press any key to continue . . . _
------------------------------------------------------
Enter two signed integers:
28000 30000         //6D60H 7530H
Result = 2800       //Difference = F830H
                    //Sign is negative

Press any key to continue . . . _
------------------------------------------------------
Enter two signed integers:
-4500 -6000         //EE6CH E890H
Result = -6000      //Difference = 05DCH
                    //Sign is positive

Press any key to continue . . . _
------------------------------------------------------
Enter two signed integers:
-6000 -4500         //E890H EE6CH
Result = -6000      //Difference = FA24H
                    //Sign is negative

Press any key to continue . . . _
------------------------------------------------------
Enter two signed integers:
750 -600            //02EEH FDA8H
Result = -600       //Difference = 0546H
                    //Sign is positive

Press any key to continue . . . _
------------------------------------------------------
Enter two signed integers:
-600 750            //FDA8H 02EEH
Result = -600       //Difference = FABAH
                    //Sign is negative

Press any key to continue . . . _
```

Chapter 6 Branching and Looping Instructions

6.3 Determine if an overflow occurs for the operation shown below. The operands are signed numbers in 2s complement representation.

$$\begin{array}{r} 1\ 1\ 1\ 1\quad 1\ 1\ 1\ 1 \\ +)\ \underline{1\ 1\ 1\ 1\quad 1\ 1\ 1\ 1} \\ 1\ 1\ 1\ 1\quad 1\ 1\ 1\ 0 \end{array}$$

No overflow, because $(a_{n-1} \bullet b_{n-1} \bullet s_{n-1}') = 1 \bullet 1 \bullet 1$.

6.7 Determine whether the conditional jump instructions shown below will cause a jump to DEST.

(a) 004FH + 200D **JS** DEST
004FH + 00C8H = 0117H
Sign is positive; therefore no jump.

(b) FF38H + 200D **JZ** DEST
FF38H + 00C8H = 1 ← 0000H
Result is zero; therefore, a jump will occur.

6.11 Let AX = –405D. Determine whether the conditional jump instruction shown below will cause a jump to BRANCH_ADDR.

```
CMP     AX, FE6CH
JGE     BRANCH_ADDR
```

For signed operands, a jump will not occur, because FE6CH (–404D) is greater than –405D (FE6BH).

6.13 Determine the number of times the following program segment executes the body of the loop:

```
        MOV     CX, -1
LP1:            .
                .
                .
        LOOP    LP1
```

65,535 times.

Chapter 7 Stack Operations

7.3 Determine the result of each instruction for the following program segment:

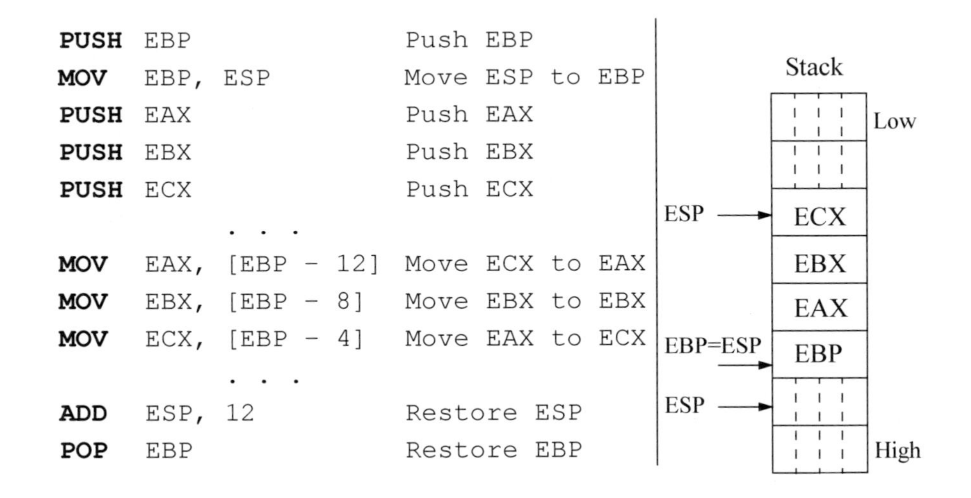

7.5 The partial contents of a stack are shown below before execution of the program segment listed below. Determine the contents of the stack after the program has been executed and indicate the new top of stack.

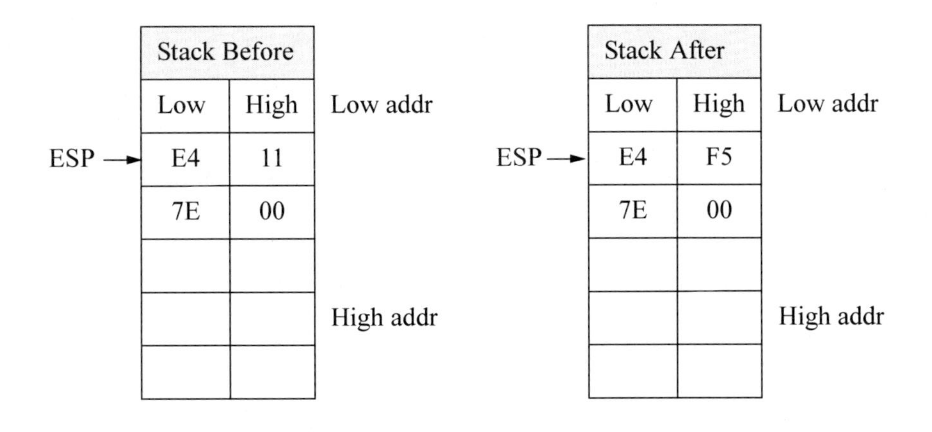


```
POP    BX      ;BX = 11 E4 H
MOV    AH, BH  ;AX = 11 __ H
ADD    AH, BL  ;AX = F5 __  (E4 + 11 = F5)
MOV    BH, AH  ;BX = F5 E4 H              BL = E4
PUSH   BX      ;Stack = E4 F5 H           Low addr = E4
```

7.8 Write an assembly language program — using the PUSH and POP instructions — that adds four decimal integers, then displays their sum. Embed the assembly module in a C program. The decimal integers are entered from the keyboard.

```
//push_pop_add_int.cpp
//add four decimal integers using push and pop

#include "stdafx.h"

int main (void)
{
//define variables
   int dec1, dec2, dec3, dec4, sum_1234;

   printf ("Enter four decimal integers: \n");
   scanf ("%d %d %d %d", &dec1, &dec2, &dec3, &dec4);

//switch to assembly
      _asm
      {
         MOV   EAX, dec1
         MOV   EBX, dec2
         ADD   EAX, EBX         ;dec1 + dec2 -> EAX
         PUSH  EAX              ;push dec1 + dec2

         MOV   EAX, dec3
         MOV   EBX, dec4
         ADD   EAX, EBX         ;dec3 + dec4 -> EAX

         POP   EBX              ;pop dec1 + dec2
         ADD   EAX, EBX         ;total sum -> EAX

         MOV   sum_1234, EAX    ;move sum to result area
      }

   printf ("\nsum of four integers = %d\n\n",
            sum_1234);

   return 0;
}
```

```
Enter four decimal integers:
1 2 3 4

sum of four integers = 10

Press any key to continue . . . _
-----------------------------------------------------
Enter four decimal integers:
25 50 75 100

sum of four integers = 250

Press any key to continue . . . _
-----------------------------------------------------
Enter four decimal integers:
1000 2000 3000 4000

sum of four integers = 10000

Press any key to continue . . . _
-----------------------------------------------------
Enter four decimal integers:
37 1200 750 875

sum of four integers = 2862

Press any key to continue . . . _
```

Chapter 8 Logical, Bit, Shift, and Rotate Instructions

8.3 Using only logic instructions, write one instruction for each of the following parts that will perform the indicated operations.

(a) Set the low-order four bits of register AX.

```
OR    AX, 000FH
```

(b) Reset the high-order three bits of register AX.

```
AND    AX, 1FFFH
```

(c) Invert bits 7, 8, and 9 of register AX.

```
XOR    AX, 0380H
```

8.8 Write an assembly language module embedded in a C program that uses the *bit scan forward* (BSF) instruction to detect whether the first bit detected is in the low-order half or the high-order half of a 32-bit operand. Do not determine the bit position.

```
//bsf_if_struc.cpp
//determine if first bit is in position <= 15 or > 15
//using an if structure

#include "stdafx.h"
int main (void)
{
//define variables
   int bsf_src_opnd;

   printf ("Enter an 8-digit hexadecimal number: \n");
   scanf ("%X", &bsf_src_opnd);

                           //continued on next page
```

```
//switch to assembly
        _asm
        {
            MOV     EBX, bsf_src_opnd
            CMP     EBX, 0
            JZ      NO_ONES
            BSF     EAX, EBX
            CMP     EAX, 15
            JBE     LESS_THAN_16
            JMP     GREATER_THAN_15
        }

NO_ONES:
    printf ("\nBSF opnd has no 1s\n\n");
    goto end;

LESS_THAN_16:
    printf ("\nBit is in position 15 -- 0\n\n");
    goto end;

GREATER_THAN_15:
    printf ("\nBit is in position 31 -- 16\n\n");

end:
    return 0;
}
```

```
Enter an 8-digit hexadecimal number:
00A07000

Bit is in position 15 -- 0

Press any key to continue . . . _
-----------------------------------------------------
Enter an 8-digit hexadecimal number:
7BC68000

Bit is in position 15 -- 0

Press any key to continue . . . _
-----------------------------------------------------
```

//continued on next page

```
Enter an 8-digit hexadecimal number:
FFFF0001

Bit is in position 15 -- 0

Press any key to continue . . . _
------------------------------------------------------
Enter an 8-digit hexadecimal number:
80D10000

Bit is in position 31 -- 16

Press any key to continue . . . _
------------------------------------------------------
Enter an 8-digit hexadecimal number:
0DC00000

Bit is in position 31 -- 16

Press any key to continue . . . _
------------------------------------------------------
Enter an 8-digit hexadecimal number:
80000000

Bit is in position 31 -- 16

Press any key to continue . . . _
------------------------------------------------------
Enter an 8-digit hexadecimal number:
00000000

BSF opnd has no 1s

Press any key to continue . . . _
```

8.12 Determine the contents of register EBX and the flags register after execution of the program segment shown below. Then write an assembly language module embedded in a C program to verify the results.

```
MOV   EBX, 0FFFFA85BH
SAL   EBX, 20
MOV   EBX, 0FFFFA85BH
SAR   EBX, 20
```

SAL result = 85B00000 SAR result = FFFFFFFF
SAL flags = 86 SAR flags = 87

```
//sal_sar.cpp
//program to illustrate shift arithmetic left
//and shift arithmetic right instructions
//for the same operand

#include "stdafx.h"
int main (void)
{
//define variables
   int dst_opnd, sal_rslt, sar_rslt;
   unsigned char sal_flags, sar_flags;

   printf ("Enter an 8-digit hexadecimal number: \n");
   scanf ("%X", &dst_opnd);

//switch to assembly
      _asm
      {
//SAL instruction
         MOV   EBX, dst_opnd
         SAL   EBX, 20
         MOV   sal_rslt, EBX
         LAHF
         MOV   sal_flags, AH


//SAR instruction
         MOV   EBX, dst_opnd
         SAR   EBX, 20
         MOV   sar_rslt, EBX
         LAHF
         MOV   sar_flags, AH
      }


   printf ("\nSAL result = %X\nSAL flags = %X\n\n",
               sal_rslt, sal_flags);

   printf ("\nSAR result = %X\nSAR flags = %X\n\n",
               sar_rslt, sar_flags);

   return 0;
}
```

```
Enter an 8-digit hexadecimal number:
FFFFA85B

SAL result = 85B00000
SAL flags = 86

SAR result = FFFFFFFF
SAR flags = 87

Press any key to continue . . . _
```

8.14 Write an assembly language module embedded in a C program that will multiply and divide a decimal number by eight using arithmetic shift instructions. When dividing, some numbers will have the fraction truncated.

```
//sal_sar_2.cpp
//shifting positive and negative numbers
//using SAL and SAR instructions

#include "stdafx.h"

int main (void)
{
//define variables
   int dst_opnd, sal_rslt, sar_rslt;

   printf ("Enter a decimal number: \n");
   scanf ("%d", &dst_opnd);

//switch to assembly
      _asm
      {
//SAL instruction
         MOV   EAX, dst_opnd
         SAL   EAX, 3
         MOV   sal_rslt, EAX

                                    //continued on next page
```

```
//SAR instruction
         MOV   EAX, dst_opnd
         SAR   EAX, 3
         MOV   sar_rslt, EAX
      }

   printf ("\nSAL result = %d\n", sal_rslt);

   printf ("SAR result = %d\n\n", sar_rslt);

   return 0;
}
```

```
Enter a decimal number:
8

SAL result = 64
SAR result = 1

Press any key to continue . . . _
------------------------------------------------------
Enter a decimal number:
-8

SAL result = -64
SAR result = -1

Press any key to continue . . . _
------------------------------------------------------
Enter a decimal number:
-10

SAL result = -80
SAR result = -2          //1.25; 1111 ... 1111 0110
                         //1111 ... 1111 1110
Press any key to continue . . . _
------------------------------------------------------
Enter a decimal number:
25

SAL result = 200
SAR result = 3           //3.125; 0000 ... 0001 1001
                         //0000 ... 0000 0011
Press any key to continue . . . _
------------------------------------------------------
                            //continued on next page
```

```
Enter a decimal number:
-25

SAL result = -200
SAR result = -4          //-3.125; 1111 ... 1110 0111
                         //1111 ... 1111 1100
Press any key to continue . . . _
-------------------------------------------------------
Enter a decimal number:
500

SAL result = 4000
SAR result = 62          //62.5; 0000 ... 0001 1111 0100
                         //0000 ... 0000 0011 1110
Press any key to continue . . . _
-------------------------------------------------------
Enter a decimal number:
1000

SAL result = 8000
SAR result = 125

Press any key to continue . . . _
```

8.16 Let register EAX contain AD3E14B5H and the carry flag be set. Determine the contents of register EAX after the following instruction is executed:

CF	**RCL** EAX, 6								
1	1010	1101	0011	1110	0001	0100	1011	0101	
1	0101	1010	0111	1100	0010	1001	0110	1011	After 1 shift
0	1011	0100	1111	1000	0101	0010	1101	0111	After 2 shifts
1	0110	1001	1111	0000	1010	0101	1010	1110	After 3 shifts
0	1101	0011	1110	0001	0100	1011	0101	1101	After 4 shifts
1	1010	0111	1100	0010	1001	0110	1011	1010	After 5 shifts
1	0100	1111	1000	0101	0010	1101	0111	0101	After 6 shifts

Chapter 9 Fixed-Point Arithmetic Instructions

9.3 Indicate whether an overflow occurs for the operation shown below. The numbers are in radix complementation for radix 3.

$$
\begin{array}{rrrrrrcl}
 & 2 & 1 & 0 & 1 & 2_3 & = & 2\times3^4 + 1\times3^3 + 0\times3^2 + 1\times3^1 + 2\times3^0 = 194 \\
+) & 0 & 2 & 1 & 2 & 2_3 & = & 0\times3^4 + 2\times3^3 + 1\times3^2 + 2\times3^1 + 2\times3^0 = 71 \\
\hline
1 & 0 & 0 & 2 & 1 & 1_3 & = & 1\times3^5 + 0\times3^4 + 0\times3^3 + 2\times3^2 + 1\times3^1 + 1\times3^0 = 265
\end{array}
$$

No overflow, because a positive and negative number are being added.
The sign for a positive number is 0.
The sign for a negative number is $r - 1 = 3 - 1 = 2$.

9.6 Determine the contents of registers AL and BL after the program segment shown below has executed. Then write an assembly language module embedded in a C program to verify the results.

			AL	BL	AL	BL	AL	BL	AL	BL	AL	BL
	MOV	AL, 00H	0									
	MOV	BL, -5	0	-5								
LOOP1:	**ADD**	BL, 2	0	-3	1	-2	2	-1	3	0	4	1
	INC	AL	1	-3	2	-2	3	-1	4	0	5	1
	ADD	BL, -1	1	-4	2	-3	3	-2	4	-1	**5**	**0**
	JNZ	LOOP1										

```
//inc_al_dec_bl.cpp
//program to determine the contents of registers
//AL and BL after a program has executed
#include "stdafx.h"
int main (void)
{
//define variables
   char al_sum, bl_sum;

                                  //continued on next page
```

```
//switch to assembly
      _asm
      {
         MOV   AL, 00H
         MOV   BL, -5
LOOP1:   ADD   BL, 2
         INC   AL
         ADD   BL, -1
         JNZ   LOOP1
         MOV   al_sum, AL
         MOV   bl_sum, BL
      }
   printf ("\nAL = %d\nBL = %d\n\n", al_sum, bl_sum);
   return 0;
}
```

```
AL = 5
BL = 0

Press any key to continue . . . _
```

9.9 Write an assembly language program — not embedded in a C program — that removes all nonletters from a string of characters that are entered from the keyboard.

```
PAGE 66, 80
;non letters.asm
;----------------------------------------------------------
.STACK

;----------------------------------------------------------
.DATA
PARLST   LABEL BYTE
MAXLEN   DB 40
ACTLEN   DB ?
OPFLD    DB 40 DUP(?)
PRMPT    DB  0DH, 0AH, 'Enter text $'
RSLT     DB  0DH, 0AH, 'Result =                $'

;----------------------------------------------------------
                                //continued on next page
```

```
.CODE
BEGIN PROC FAR

;set up pgm ds
      MOV   AX, @DATA
      MOV   DS, AX

;read prompt
      MOV   AH, 9
      LEA   DX, PRMPT
      INT   21H

;kybd rtn to enter characters
      MOV   AH, 0AH
      LEA   DX, PARLST
      INT   21H

      MOV   CL, ACTLEN      ;get # of chars entered
      MOV   CH, 0

      LEA   SI, OPFLD       ;source addr to si
      LEA   DI, RSLT + 11   ;destination addr to di

LP1:  MOV   AL, [SI]        ;get char in opfld
      CMP   AL, 41H         ;compare to uppercase A
      JL    LP2             ;if < A, jump
      CMP   AL, 5AH         ;compare to uppercase Z
      JLE   LP4             ;if between A and Z, jump

      CMP   AL, 7AH         ;compare to lowercase z
      JG    LP2             ;if > z, jump
      CMP   AL, 61H         ;compare to lowercase a
      JL    LP2             ;if < a, jump

LP4:  MOV   [DI], AL        ;move valid char to dst
      INC   DI              ;inc di to next dst

LP2:  INC   SI              ;addr of next character
      LOOP  LP1             ;jump, get char in opfld

;display result
      MOV   AH, 09H
      LEA   DX, RSLT
      INT   21H

BEGIN ENDP
      END   BEGIN
```

```
Enter text: the6h7h
Result = the
------------------------------------------------------
Enter text: 7HD3e23kl
Result = HDekl
------------------------------------------------------
Enter text: 1234567879A
Result = A
------------------------------------------------------
Enter text: ABcdEF7
Result = ABcdEF
------------------------------------------------------
Enter text: abcdefghij
Result = abcdefghij
------------------------------------------------------
Enter text: 123456789
Result =
```

9.17 Let register AL = 61H (97_{10}). Then execute the instruction shown below to obtain the difference and the state of the following flags: OF, SF, ZF, AF, PF, and CF.

```
SUB    AL, 65H        ;65H = 101₁₀
```

CF				AF						
1	1	0	0	1		1				
	0	1	1	0		0	0	0	1	+97
−)	0	1	1	0		0	1	0	1	+101
	1	1	1	1		1	1	0	0	−4

↓

	0	1	1	0		0	0	0	1	+97
+)	1	0	0	1		1	0_1	1_1	1	−101
	1	1	1	1		1	1	0	0	−4
0(1)					0(1)					
CF					AF					

OF = 0, SF = 1, ZF = 0, AF = 1, PF = 1, CF = 1

9.21 Write an assembly language program — not embedded in a C program — to reverse the order and change to uppercase all letters that are entered from the keyboard. The letters that are entered may be lowercase or uppercase.

```
;reverse uppercase
PAGE 66, 80
;-------------------------------------------------------
.STACK
;-------------------------------------------------------
.DATA
PARLST    LABEL BYTE
MAXLEN    DB    40
ACTLEN    DB    ?
OPFLD     DB    40    DUP(?)
PRMPT     DB    0DH, 0AH, 'Enter characters:  $'
RSLT      DB    0DH, 0AH, 'Result =            $'

.CODE
BEGIN     PROC  FAR

;setup pgm ds
          MOV   AX, @DATA       ;get addr of data seg
          MOV   DS, AX          ;move addr to ds

;read prompt
          MOV   AH, 09H         ;display string
          LEA   DX, PRMPT       ;put addr of prompt in dx
          INT   21H             ;dos interrupt

;keyboard request rtn to enter characters
          MOV   AH, 0AH         ;buffered keyboard input
          LEA   DX, PARLST      ;put addr of parlst in dx
          INT   21H             ;dos interrupt

;setup count and addresses
          MOV   CL, ACTLEN      ;number of characters
          MOV   CH, 0           ;... entered
          MOV   BX, CX          ;bx is displacement
          DEC   BX
          LEA   SI, OPFLD[BX]   ;addr of last character
          LEA   DI, RSLT + 11   ;setup destination addr

;change to uppercase and reverse order
SWAP:     MOV   AL, [SI]        ;move character to al
          CMP   AL, 61H         ;compare to lowercase
          JL    UPPER           ;jump if uppercase
          SUB   AL, 20H         ;if lowercase, change to
                                ;uppercase
                                //continued on next page
```

```
UPPER:    MOV   [DI], AL          ;move char to result area
          DEC   SI
          INC   DI
          LOOP  SWAP              ;loop if cx !=0

;display result
          MOV   AH, 09H           ;display string
          LEA   DX, RSLT          ;put addr of rslt in dx
          INT   21H               ;dos interrupt

BEGIN     ENDP
          END   BEGIN
```

```
Enter characters: aBCdeFGh
Result = HGFEDCBA
-----------------------------------------------------
Enter characters: ihgfedcba
Result = ABCDEFGHI
-----------------------------------------------------
Enter characters: ABCDEFGHIJ
Result = JIHGFEDCBA
```

9.23 Write an assembly language program — not embedded in a C program — to sort *n* single-digit numbers in ascending numerical order that are entered from the keyboard. A simple exchange sort is sufficient for this program. The technique is slow, but appropriate for sorting small lists. The method is as follows:

1. Search the list to find the smallest number.
2. After finding the smallest number, exchange this number with the first number.
3. Search the list again to find the next smallest number, starting with the second number.
4. Then exchange this (second smallest) number with the second number.
5. Repeat this process, starting with the third number, then the fourth number, etc.

```
;sort n numbers2.asm

;-------------------------------------------------------
.STACK

;-------------------------------------------------------
.DATA
PARLST    LABEL BYTE
MAXLEN    DB 15
ACTLEN    DB ?
OPFLD     DB 15 DUP(?)
PRMPT     DB 0DH, 0AH, 'Enter numbers: $'
RSLT      DB 0DH, 0AH, 'Sorted numbers =             $'

;-------------------------------------------------------
.CODE
BEGIN     PROC   FAR

;set up pgm ds
          MOV    AX, @DATA        ;get addr of data seg
          MOV    DS, AX           ;move addr to ds

;read prompt
          MOV    AH, 09H          ;display string
          LEA    DX, PRMPT        ;put addr of prompt in dx
          INT    21H              ;dos interrupt

;keyboard request rtn to enter characters
          MOV    AH, 0AH          ;buffered keyboard input
          LEA    DX, PARLST       ;put addr of parlst in dx
          INT    21H              ;dos interrupt

;sort numbers
          LEA    SI, OPFLD        ;addr of first byte to si
          MOV    DL, ACTLEN       ;# of bytes to dx
          MOV    DH, 00H          ;... as loop ctr
          DEC    DX               ;loop ctr - 1

LP1:      MOV    CX, DX           ;cx is loop2 ctr

LP2:      MOV    AL, [SI]         ;number to al
          MOV    BX, CX           ;bx is base addr
          CMP    AL, [BX+SI]      ;is number <= to al?
          JBE    NEXT             ;number is <=; get next #
          XCHG   AL, [BX+SI]      ;number is !<=; put in al
          MOV    [SI], AL         ;put # in input buffer

                                  //continued on next page
```

```
NEXT:   LOOP  LP2            ;loop until #s compared
        INC   SI             ;point to next number
        DEC   DX             ;dec internal loop ctr
        JNZ   LP1            ;compare remaining #s

;set up src (opfld) and dst (rslt)
        LEA   SI, OPFLD
        LEA   DI, RSLT+20
        MOV   CL, ACTLEN
        MOV   CH, 00H        ;cx is counter

;move sorted numbers to input buffer
INP_BUFF:
        MOV   BL, [SI]       ;opfld to bl
        MOV   [DI], BL       ;byte to result area
        INC   SI
        INC   DI
        LOOP  INP_BUFF

;display sorted numbers
        MOV   AH, 09H        ;display string
        LEA   DX, RSLT       ;put addr of rslt in dx
        INT   21H            ;dos interrupt

BEGIN   ENDP
        END   BEGIN
```

```
Enter numbers: 7658943034
Sorted numbers = 0334456789
------------------------------------------------------
Enter numbers: 7548391235
Sorted numbers = 1233455789
------------------------------------------------------
Enter numbers: 9876543210
Sorted numbers = 0123456789
```

9.27 Determine the hexadecimal contents of register AX after the following program segment has been executed:

```
        MOV   CX, 3
        MOV   AX, 2      ;AX = 2
LP1:    MUL   AX         ;AX = 4  16    256
        LOOP  LP1        ;CX = 3  2     1
```

AX = 0000 0001 0000 0000

9.30 Write an assembly language module embedded in a C program that uses the three-operand form for signed multiplication. Enter a signed decimal multiplicand from the keyboard and assign an immediate multiplier for the IMUL operation. Display the multiplicand, the multiplier, and the product.

```
//imul_3_opnds.cpp
//signed decimal multiplication using 3 operands

#include "stdafx.h"
int main (void)
{
//define variables
   long mpcnd, product;

   printf ("Enter a signed decimal multiplicand: \n");
   scanf ("%d", &mpcnd);

//switch to assembly
      _asm
      {
         MOV   EBX, mpcnd
         IMUL  EAX, EBX, 10      ;EAX = EBX x 10
         MOV   product, EAX
      }

   printf ("\nMultiplicand = %d", mpcnd);
   printf ("\nMultiplier = 10");
   printf ("\nProduct = %d\n\n", product);

   return 0;
}
```

```
Enter a signed decimal multiplicand:
450

Multiplicand = 450
Multiplier = 10
Product = 4500

Press any key to continue . . . _
-----------------------------------------------------
                          //continued on next page
```

```
Enter a signed decimal multiplicand:
-5

Multiplicand = -5
Multiplier = 10
Product = -50

Press any key to continue . . . _
-------------------------------------------------------
Enter a signed decimal multiplicand:
6000

Multiplicand = 6000
Multiplier = 10
Product = 60000

Press any key to continue . . . _
-------------------------------------------------------
Enter a signed decimal multiplicand:
-6000

Multiplicand = -6000
Multiplier = 10
Product = -60000

Press any key to continue . . . _
-------------------------------------------------------
Enter a signed decimal multiplicand:
65535

Multiplicand = 65535
Multiplier = 10
Product = 655350

Press any key to continue . . . _
-------------------------------------------------------
Enter a signed decimal multiplicand:
-65535

Multiplicand = -65535
Multiplier = 10
Product = -655350

Press any key to continue . . . _
```

9.34 Write an assembly language module embedded in a C program to perform unsigned division (DIV) on two decimal integers that are entered from the keyboard. Print the dividend, divisor, quotient, and remainder.

```
//div.cpp
//unsigned division of two integers
#include "stdafx.h"
int main (void)
{
//define variables
   unsigned short dvdnd, dvsr, quot, rem;

   printf ("Enter a decimal dividend and divisor: \n");
   scanf ("%d %d", &dvdnd, &dvsr);

//switch to assembly
      _asm
      {
         MOV   AX, dvdnd
         CWD
         DIV   dvsr

         MOV   quot, AX
         MOV   rem, DX
      }

   printf ("\nDividend = %d\nDivisor = %d\n",
            dvdnd, dvsr);
   printf ("\nQuotient = %d\nRemainder = %d\n\n",
            quot, rem);
   return 0;
}
```

```
Enter a decimal dividend and divisor:
450 20

Dividend = 450
Divisor = 20

Quotient = 22
Remainder = 10

Press any key to continue . . . _
-------------------------------------------------------
Enter a decimal dividend and divisor:
25000 250

Dividend = 25000
Divisor = 250

Quotient = 100
Remainder = 0

Press any key to continue . . . _
-------------------------------------------------------
Enter a decimal dividend and divisor:
25000 30

Dividend = 25000
Divisor = 30

Quotient = 833
Remainder = 10

Press any key to continue . . . _
-------------------------------------------------------
Enter a decimal dividend and divisor:
4660 86

Dividend = 4660
Divisor = 86

Quotient = 54
Remainder = 16

Press any key to continue . . . _
```

9.38 Write an assembly language module embedded in a C program that evaluates the expression shown below, where $x = 400_{10}$ (0190_{16}), $y = 5_{10}$ (0005_{16}), $z = 1000_{10}$ $(03E8_{16})$, and $v = 4000_{10}$ $(0FA0_{16})$.

$$[v - (x \times y + z - 500)] / x$$

$$\begin{aligned} & [v - (x \times y + z - 500)] / x \\ = \; & [4000 - (400 \times 5 + 1000 - 500)] / 400 \\ = \; & [4000 - 2500] / 400 \\ = \; & 1500 / 400 \\ = \; & 3.75 \\ = \; & \text{Quotient} = 3_{10}\ (3_{16}), \text{Remainder} = 300_{10}\ (12C_{16}) \end{aligned}$$

```
//solve_eqn.cpp
//use imul, add, sub, and idiv to solve the equation:
//[V - (X x Y + Z - 500)]/X

#include "stdafx.h"

int main (void)
{
//define variables
   short x, y, z, v, quot, rem;

   x = 0x0190;
   y = 0x0005;
   z = 0x03E8;
   v = 0x0FA0;

//switch to assembly
      _asm
      {
//X x Y
         MOV    AX, x
         IMUL   y
         MOV    CX, AX
         MOV    BX, DX

//+Z
         MOV    AX, z
         CWD
         ADD    CX, AX
         ADC    BX, DX
                                    //continued on next page
```

```
//-500
          SUB   CX, 01F4H       //500
          SBB   BX, 0

//V-( )
          MOV   AX, v
          CWD
          SUB   AX, CX
          SBB   DX, BX

///400
          IDIV  x

//move ax to quotient and dx to remainder
          MOV   quot, AX
          MOV   rem, DX
      }

   printf ("\nQuotient = %X, Remainder = %X\n\n",
               quot, rem);

   return 0;
}
```

```
Quotient = 3, Remainder = 12C

Press any key to continue . . . _
```

Chapter 10 Binary-Coded Decimal Arithmetic Instructions

10.5 In each part below, assume that the following instructions are executed:

```
ADD    AL, BL
DAA
```

(a) Give the values of AL after the ADD instruction has been executed, but before the DAA instruction has been executed for the indicated register values shown below.

(b) Give the values of AL, the carry flag (CF), and the auxiliary carry flag (AF) after the DAA instruction has been executed for the indicated register values shown below.

	After ADD	After DAA	CF	AF
(1) AL = 35H, BL = 48H	Sum = 7DH	Sum = 83	0	1
(2) AL = 47H, BL = 61H	Sum = A8H	Sum = 08	1	0
(3) AL = 75H, BL = 46H	Sum = BBH	Sum = 21	1	1

Then write an assembly language module embedded in a C program to verify the results.

```
//daa3.cpp
//illustrates the use of the daa instruction

#include "stdafx.h"

int main (void)
{
//define variables
   unsigned char augend, addend, add_sum, daa_sum,
                 flags;

   printf ("Enter two 2-digit hexadecimal numbers:
            \n");
   scanf ("%X %X", &augend, &addend);

                              //continued on next page
```

```
//switch to assembly
      _asm
      {
//obtain sum
         MOV   AL, augend
         ADD   AL, addend
         MOV   add_sum, AL
         DAA
         MOV   daa_sum, AL

//save flags
         PUSHF
         POP   AX
         MOV   flags, AL
      }

   printf ("\nAugend = %X, Addend = %X\n",
               augend, addend);

   printf ("\nADD sum = %X, DAA sum = %X\n",
               add_sum, daa_sum);

   printf ("Flags = %X\n\n", flags);

   return 0;
}
```

```
Enter two 2-digit hexadecimal numbers:
35 48

Augend = 35, Addend = 48

ADD sum = 7D, DAA sum = 83     //DAA sum = 083
Flags = 92                     //SF, AF

Press any key to continue . . . _
------------------------------------------------------
```

//continued on next page

```
---------------------------------------------------------
Enter two 2-digit hexadecimal numbers:
47 61

Augend = 47, Addend = 61

ADD sum = A8, DAA sum = 8       //DAA sum = 108
Flags = 3                       //CF

Press any key to continue . . . _
---------------------------------------------------------
Enter two 2-digit hexadecimal numbers:
75 46

Augend = 75, Addend = 46

ADD sum = BB, DAA sum = 21      //DAA sum = 121
Flags = 17                      //AF, PF, CF

Press any key to continue . . . _
```

10.8 Perform a subtraction and adjustment of the following two operands: 07 – 09. Show the details of the subtraction and adjustment.

```
                          AH     AL
                         0  0   0  7
                            +)  F  7     (–9)
                               ------
                                F  E     (–2)
AAS     Subtract 6          +)  F  A     (–6)
                               ------
                                0  8
               AH – 1:   F  F   0  8     AF = 1
                                ↑        CF = 1
                                └─────── Clear high half of AL
```

10.17 Write an assembly language program — not embedded in a C program — that multiplies a five-digit multiplicand by a one-digit multiplier using the MUL instruction in combination with the AAM instruction. The ADD instruction and the AAA instruction will also be used in implementing the program.

The five-digit multiplicand and the one-digit multiplier are entered from the keyboard and stored in the OPFLD, as shown below using a multiplicand of 12345 and a multiplier of 6. Enter several different multiplicands and multipliers, then display the operands and the resulting product.

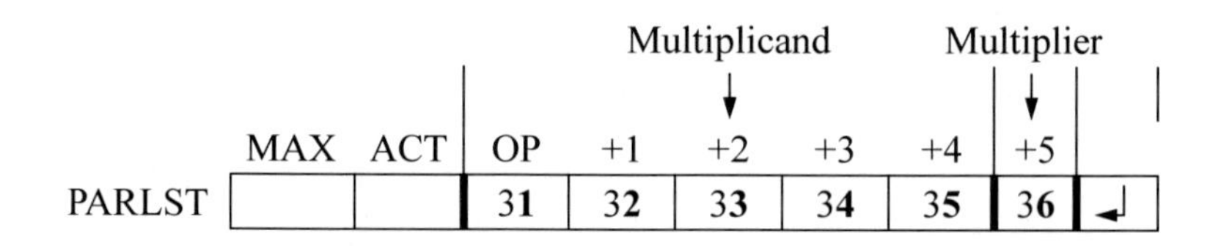

The paper-and-pencil example shown below may help to illustrate the algorithm used in the program.

```
            7   8   9
×)                  9
         ------------
                8   1     AAM, then store 1 in result area
+)          7   2
         --------
            8   0         AAM, AAA, then store 0 in result area
+)      6   3
     --------
        7   1             AAM, AAA, then store 7 and 1 in result area
                          Product = 7101
```

```
;aam4.asm
;obtain the product of a 5-digit multiplicand
;and a 1-digit multiplier

;-----------------------------------------------------------
.STACK

;-----------------------------------------------------------
.DATA
PARLST    LABEL BYTE
MAXLEN    DB 10
ACTLEN    DB ?
OPFLD     DB 10    DUP(?)
PRMPT     DB 0DH, 0AH, 'Enter 5-digit mpcnd and
                          1-digit mplyr: $'
RSLT      DB 0DH, 0AH, 'Product =          $'
                                //continued on next page
```

```
.CODE
BEGIN       PROC   FAR

;set up pgm ds
            MOV    AX, @DATA        ;get addr of data seg
            MOV    DS, AX           ;move addr to ds

;read prompt
            MOV    AH, 09H          ;display string
            LEA    DX, PRMPT        ;put addr of prompt in dx
            INT    21H              ;dos interrupt

;keyboard request rtn to enter characters
            MOV    AH, 0AH          ;buffered keyboard input
            LEA    DX, PARLST       ;put addr of parlst in dx
            INT    21H              ;dos interrupt

;------------------------------------------------------
;perform the multiplication
            MOV    CX, 05           ;five loops

            LEA    SI, OPFLD+4      ;si = addr of low-order
                                    ;mpcnd digit
            LEA    DI, RSLT+17      ;di = addr of low-order
                                    ;product digit
            AND    OPFLD+5, 0FH     ;unpack multiplier

LP1:        MOV    AL, [SI]         ;get next low-order
                                    ;mpcnd digit
            AND    AL, 0FH          ;unpack it
            MUL    OPFLD+5          ;multiply by the mplyr
            AAM                     ;adjust the product

            ADD    AL, [DI]         ;add low-order product
                                    ;digit to al
            AAA                     ;adjust the sum
            OR     AL, 30H          ;add ascii bias
            MOV    [DI], AL         ;move to low-order
                                    ;result area

            DEC    DI               ;get addr of next
                                    ;low-order result area
            OR     AH, 30H          ;add ascii bias
            MOV    [DI], AH         ;move to next low-order
                                    ;result area
            DEC    SI               ;address next high-order
                                    ;mpcnd digit
            LOOP   LP1              ;loop if cx != 0
                                //continued on next page
```

```
;display product
         MOV   AH, 09H          ;display string
         LEA   DX, RSLT         ;addr of product in dx
         INT   21H              ;dos interrupt

BEGIN    ENDP
         END   BEGIN
```

```
Enter 5-digit mpcnd and 1-digit mplyr: 123456
Product = 074070
------------------------------------------------------
Enter 5-digit mpcnd and 1-digit mplyr: 007899
Product = 007101
------------------------------------------------------
Enter 5-digit mpcnd and 1-digit mplyr: 999999
Product = 899991
------------------------------------------------------
Enter 5-digit mpcnd and 1-digit mplyr: 556667
Product = 389662
------------------------------------------------------
Enter 5-digit mpcnd and 1-digit mplyr: 111118
Product = 088888
------------------------------------------------------
Enter 5-digit mpcnd and 1-digit mplyr: 765432
Product = 153086
```

10.19 Repeat Problem 10.18, this time using an assembly language module embedded in a C program. Note the simplicity of this program. There is no requirement to remove the ASCII bias for the numbers that are entered from the keyboard, or to restore the ASCII bias before displaying the hours, minutes, and seconds. It is also not necessary to establish stack or data segments. The range is the same: 10,000 to 60,000 seconds.

```
//hrs_min_sec.cpp
//given a 5-digit number of seconds, calculate the
//equivalent number of hours, minutes, and seconds
#include "stdafx.h"
int main (void)
{
   unsigned short seconds, hrs, min, sec;

   printf ("Enter 5-digit #: range 10000 -- 60000:
            \n");
   scanf ("%d", &seconds);
                                   //continued on next page
```

```
//switch to assembly
      _asm
      {
         MOV   DX, 0
         MOV   AX, seconds
         MOV   BX, 3600
         DIV   BX
         MOV   hrs, AX

         MOV   AX, DX
         MOV   DX, 0
         MOV   BX, 60
         DIV   BX
         MOV   min, AX

         MOV   sec, DX
      }

   printf ("\nHours = %d", hrs);
   printf ("\nMinutes = %d", min);
   printf ("\nSeconds = %d\n\n", sec);

   return 0;
}
```

```
Enter 5-digit #: range 10000 -- 60000:
43210

Hours = 12
Minutes = 0
Seconds = 10

Press any key to continue . . . _
------------------------------------------------------
Enter 5-digit #: range 10000 -- 60000:
10000

Hours = 2
Minutes = 46
Seconds = 40

Press any key to continue . . . _
------------------------------------------------------
```

//continued on next page

```
Enter 5-digit #: range 10000 -- 60000:
00060

Hours = 0
Minutes = 1
Seconds = 0

Press any key to continue . . . _
-------------------------------------------------------
Enter 5-digit #: range 10000 -- 60000:
00059

Hours = 0
Minutes = 0
Seconds = 59

Press any key to continue . . . _
-------------------------------------------------------
Enter 5-digit #: range 10000 -- 60000:
20000

Hours = 5
Minutes = 33
Seconds = 20

Press any key to continue . . . _
-------------------------------------------------------
Enter 5-digit #: range 10000 -- 60000:
39760

Hours = 11
Minutes = 2
Seconds = 40

Press any key to continue . . . _
```

10.20 Write an assembly language program to repeat Problem 10.18 — not embedded in a C program — that calculates the number of hours, minutes, and seconds from a number of seconds that are entered from the keyboard. This time use a loop to calculate the sum that is represented by the digits that are entered from the keyboard. The number of seconds entered ranges from 10,000 to 60,000. Enter several different 5-digit numbers of seconds. This program will use the MUL, DIV, and AAM instructions.

```
;hrs_min_sec2.asm
;given a 5-digit number of seconds,
;calculate the equivalent number of
;hours, minutes, and seconds

;----------------------------------------------------
.STACK

;----------------------------------------------------
.DATA
PARLST   LABEL BYTE
MAXLEN   DB 10
ACTLEN   DB ?
OPFLD    DB 10 DUP(?)
PRMPT    DB 0DH, 0AH, 'Enter 5-digit #:
                        range 10000 -- 60000: $'
SUM      DW 8 DUP(0)
HOURS    DB 0DH, 0AH, 'Hours =        $'
MINUTES  DB 0DH, 0AH, 'Minutes =     $'
SECONDS  DB 0DH, 0AH, 'Seconds =     $'

;----------------------------------------------------
.CODE
BEGIN   PROC FAR

;----------------------------------------------------
;set up pgm ds
         MOV   AX, @DATA      ;get addr of data seg
         MOV   DS, AX         ;move addr to ds

;read prompt
         MOV   AH, 09H        ;display string
         LEA   DX, PRMPT      ;put addr of prompt in dx
         INT   21H            ;dos interrupt

;keyboard request rtn to enter characters
         MOV   AH, 0AH        ;buffered keyboard input
         LEA   DX, PARLST     ;put addr of parlst in dx
         INT   21H            ;dos interrupt

;----------------------------------------------------
;obtain the 5-digit number as a sum
         LEA   SI, OPFLD      ;addr of first digit
         MOV   BX, 10000      ;bx = 2710h
         MOV   CX, 5          ;five loops

                           //continued on next page
```

```
LP1:     MOV   AL, [SI]          ;get next digit
         AND   AL, 0FH           ;unpack it
         MOV   AH, 00H           ;ax = 00 al
         MUL   BX                ;dx ax = product
         ADD   SUM, AX           ;accumulate sum

         MOV   AX, BX            ;move multiply amount
                                 ;to ax for divide
         MOV   DX, 00H           ;dx ax = 00 ax
         MOV   DI, 10            ;10 is the divisor
         DIV   DI                ;divide bx (ax) by 10
         MOV   BX, AX            ;move quot in ax to bx
         INC   SI                ;si points to next digit
         LOOP  LP1               ;loop if cx != 0

;-----------------------------------------------------------
;calculate the hours
         MOV   DX, 0             ;clear dx
         MOV   AX, SUM           ;move sum to ax
         MOV   BX, 3600          ;to divide ax for hours
         DIV   BX                ;quot (hrs) in ax;
                                 ;rem (min) in dx
         AAM                     ;2 unpacked bcd #s in ax
         OR    AX, 3030H         ;add ascii bias
         MOV   HOURS+10, AH      ;move units hours
         MOV   HOURS+11, AL      ;move tens hours

;calculate the minutes
         MOV   AX, DX            ;move rem (min) to ax
         MOV   DX, 0             ;clear dx
         MOV   BX, 60            ;to divide ax for min
         DIV   BX                ;quot (min) in ax;
                                 ;rem (sec) in dx
         AAM                     ;2 unpacked bcd #s in ax
         OR    AX, 3030H         ;add ascii bias
         MOV   MINUTES+12, AH    ;move units minutes
         MOV   MINUTES+13, AL    ;move tens minutes

;calculate the seconds
         MOV   AX, DX            ;move rem (sec) to ax
         MOV   DX, 0             ;clear dx
         AAM                     ;2 unpacked bcd #s in ax
         OR    AX, 3030H         ;add ascii bias
         MOV   SECONDS+12, AH    ;move units seconds
         MOV   SECONDS+13, AL    ;move tens seconds

;-----------------------------------------------------------
                                 //continuedonnextpage
```

```
;display hours
        MOV    AH, 09H          ;display string
        LEA    DX, HOURS        ;put addr of hours in dx
        INT    21H              ;dos interrupt

;display minutes
        MOV    AH, 09H          ;display string
        LEA    DX, MINUTES      ;put addr of min in dx
        INT    21H              ;dos interrupt

;display seconds
        MOV    AH, 09H          ;display string
        LEA    DX, SECONDS      ;put addr of sec in dx
        INT    21H              ;dos interrupt

;------------------------------------------------------
BEGIN   ENDP
        END    BEGIN
```

```
Enter 5-digit #: range 10000 -- 60000: 43210
Hours = 12
Minutes = 00
Seconds = 10
------------------------------------------------------
Enter 5-digit #: range 10000 -- 60000: 10000
Hours = 02
Minutes = 46
Seconds = 40
------------------------------------------------------
Enter 5-digit #: range 10000 -- 60000: 12345
Hours = 03
Minutes = 25
Seconds = 45
------------------------------------------------------
Enter 5-digit #: range 10000 -- 60000: 00060
Hours = 00
Minutes = 01
Seconds = 00
------------------------------------------------------
Enter 5-digit #: range 10000 -- 60000: 00059
Hours = 00
Minutes = 00
Seconds = 59
------------------------------------------------------
```

//continued on next page

```
Enter 5-digit #: range 10000 -- 60000: 20000
Hours = 05
Minutes = 33
Seconds = 20
-------------------------------------------------------
Enter 5-digit #: range 10000 -- 60000: 39760
Hours = 11
Minutes = 02
Seconds = 40
-------------------------------------------------------
Enter 5-digit #: range 10000 -- 60000: 60000
Hours = 16
Minutes = 40
Seconds = 00
-------------------------------------------------------
Enter 5-digit #: range 10000 -- 60000: 50000
Hours = 13
Minutes = 53
Seconds = 20
```

Chapter 11 Floating-Point Arithmetic Instructions

11.4 The floating-point number shown below has an unbiased exponent and an unnormalized fraction. Show the same floating-point number with a biased exponent and a normalized fraction in the single-precision floating-point format.

s	exponent	fraction
1	1 0 1 1 0 0 1 1	0 0 0 1 1 1 0 1 0 ... 0 0 0

Add bias: 1011 0011 + 0111 1111 = 0011 0010 (for unnormalized fraction)
Normalize: shift fraction left 4 to yield an implied 1

s	exponent	fraction
1	0 0 1 0 1 1 1 0	1 1 0 1 0 0 0 0 0 ... 0 0 0

11.10 Write an assembly language module embedded in a C program that uses the *add* (FADDP) instruction and the *add* (FIADD) instruction. For the FADDP instruction, use the FADDP ST(*i*), ST(0) version. Enter three floating-point numbers and one integer number from the keyboard. Display the results of the program and show the register stack for each sequence of the program.

```
//fadd_versions2.cpp
//show the use of the FADDP and FIADD instructions

#include "stdafx.h"

int main (void)
{
//define variables
   float flp1_num, flp2_num, flp3_num, int1_num,
         flp_rslt1, flp_rslt2, flp_rslt3;

   int int1_num;
                              //continued on next page
```

```
   printf ("Enter 3 floating-point numbers
            and 1 integer: \n");

   scanf ("%f %f %f %d", &flp1_num, &flp2_num,
            &flp3_num, &int1_num);

//switch to assembly
      _asm
      {
         FLD   flp1_num     //flp1_num -> ST(0)
         FLD   flp2_num     //flp2_num -> ST(0)
                            //flp1_num -> ST(1)
         FLD   flp3_num     //flp3_num -> ST(0)
                            //fld2_num -> ST(1)
                            //fld1_num -> ST(2)

         FADDP ST(2), ST(0)//ST(0) + ST(2) -> ST(2),
                            //pop
         FST   flp_rslt1    //ST(0) -> flp_rslt1

         FADDP ST(1), ST(0)//ST(0) + ST(1) -> ST(1),
                            //pop
         FST   flp_rslt2    //ST(0) -> flp_rslt2

         FIADD int1_num     //ST(0) + int1_num -> ST(0)
         FST   flp_rslt3    //ST(0) -> flp_rslt3
      }

//print result
   printf ("\nflp_rslt1 = %f\n", flp_rslt1);

   printf ("flp_rslt2 = %f\n", flp_rslt2);

   printf ("flp_rslt3 = %f\n\n", flp_rslt3);

   return 0;
}
                                 //continued on next page
```

```
Enter 3 floating-point numbers and 1 integer:
1.2 2.3 3.4 5

flp_rslt1 = 2.300000
flp_rslt2 = 6.900000
flp_rslt3 = 11.900000

Press any key to continue . . . _
----------------------------------------------------
Enter 3 floating-point numbers and 1 integer:
32.456 45.789 16.123 10

flp_rslt1 = 45.789001
flp_rslt2 = 94.368004
flp_rslt3 = 104.368004

Press any key to continue . . . _
```

	FLD flp1_num		**FLD** flp2_num	
R0	32.456	ST(0)	45.789	ST(0)
R1		ST(1)	32.456	ST(1)
R2		ST(2)		ST(2)
R3		ST(3)		ST(3)
R4		ST(4)		ST(4)
R5		ST(5)		ST(5)
R6		ST(6)		ST(6)
R7		ST(7)		ST(7)

	FLD flp3_num		**FADDP** ST(2), ST(0)	
R0	16.123	ST(0)	16.123	ST(0)
R1	45.789	ST(1)	45.789	ST(1)
R2	32.456	ST(2)	48.579	ST(2)
R3		ST(3)		ST(3)
R4		ST(4)		ST(4)
R5		ST(5)		ST(5)
R6		ST(6)		ST(6)
R7		ST(7)		ST(7)

//continued on next page

	FADDP ST(1), ST(0)			**FIADD** int1_num		
R0	**45.789**	ST(0)	→ flp_rslt1	**94.368001**	ST(0)	→ flp_rslt2
R1	48.579	ST(1)			ST(1)	
R2		ST(2)			ST(2)	
R3		ST(3)			ST(3)	
R4		ST(4)			ST(4)	
R5		ST(5)			ST(5)	
R6		ST(6)			ST(6)	
R7		ST(7)			ST(7)	

11.20 Perform the following operation on the two operands: (–47.25) – (–18.75).

True subtraction

Before alignment

$A =$	1 . 1 0 1 1	1 1 0 1	$\times 2^6$	–47.25
$B =$	1 . 1 0 0 1	0 1 1 0	$\times 2^5$	–18.75

After alignment

$A =$	1 . 1 0 1 1	1 1 0 1	$\times 2^6$	–47.25
$B =$	1 . 0 1 0 0	1 0 1 1	$\times 2^6$	–18.75

Subtract fractions (Add 2s compl B)

$A =$	1 . 1 0 1 1	1 1 0 1	$\times 2^6$	
+) $B' + 1 =$	1 . 1 0 1 1	0 1 0 1	$\times 2^6$	
1 ←	. 0 1 1 1	0 0 1 0	$\times 2^6$	
Postnormalize	1 . 1 1 1 0	0 1 0 0	$\times 2^5$	–28.5

11.22 Write an assembly language module embedded in a C program that uses the *mul* (FMULP) and the *mul* (FIMUL) instructions. For the FMULP instruction, use the FMULP ST(*i*), ST(0) version. Use the three sequences shown below to execute the FMULP, FMULP, and FIMUL instructions in that order and display the results of the program. Show the register stack for the third sequence of the program.

```
+1.200      +2.300      +3.400      +5
+32.400     +45.700     +16.100     +10
-23.600     +28.500     -27.400     -12
```

```
//fmul_versions2.cpp
//show the use of the FMULP and FIMUL instructions
#include "stdafx.h"
int main (void)
{
//define variables
   float flp1_num, flp2_num, flp3_num,
         flp_rslt1, flp_rslt2, flp_rslt3;
   int int1_num;

   printf ("Enter 3 floating-point numbers and
               1 integer: \n");
   scanf ("%f %f %f %d", &flp1_num, &flp2_num,
                         &flp3_num, &int1_num);

//switch to assembly
      _asm
      {
         FLD    flp1_num     //flp1_num -> ST(0)
         FLD    flp2_num     //flp2_num -> ST(0)
                             //flp1_num -> ST(1)
         FLD    flp3_num     //flp3_num -> ST(0)
                             //fld2_num -> ST(1)
                             //fld1_num -> ST(2)
//-------------------------------------------------------
         FMULP ST(2), ST(0)//ST(2) × ST(0)->ST(2), pop
         FST    flp_rslt1    //ST(0) -> flp_rslt1
//-------------------------------------------------------
         FMULP ST(1), ST(0)//ST(1) × ST(0)->ST(1), pop
         FST    flp_rslt2    //ST(0) -> flp_rslt2
//-------------------------------------------------------
                             //continued on next page
```

```
//-----------------------------------------------------------
        FIMUL int1_num     //ST(0) × int1_num->ST(0)
        FST   flp_rslt3    //ST(0) -> flp_rslt3
//-----------------------------------------------------------
}

//print result
   printf ("\nflp_rslt1 = %f\n", flp_rslt1);
   printf ("flp_rslt2 = %f\n", flp_rslt2);
   printf ("flp_rslt3 = %f\n\n", flp_rslt3);

   return 0;
}
```

```
Enter 3 floating-point numbers and 1 integer:
1.2 2.3 3.4 5

flp_rslt1 = 2.300000
flp_rslt2 = 9.384001
flp_rslt3 = 46.920002

Press any key to continue . . . _
------------------------------------------------------
Enter 3 floating-point numbers and 1 integer:
32.4 45.7 16.1 10

flp_rslt1 = 45.700001
flp_rslt2 = 23838.949219
flp_rslt3 = 238389.500000

Press any key to continue . . . _
------------------------------------------------------
Enter 3 floating-point numbers and 1 integer:
-23.6 +28.5 -27.4 -12

flp_rslt1 = 28.500000
flp_rslt2 = 18429.240234
flp_rslt3 = -221150.875000

Press any key to continue . . . _
```

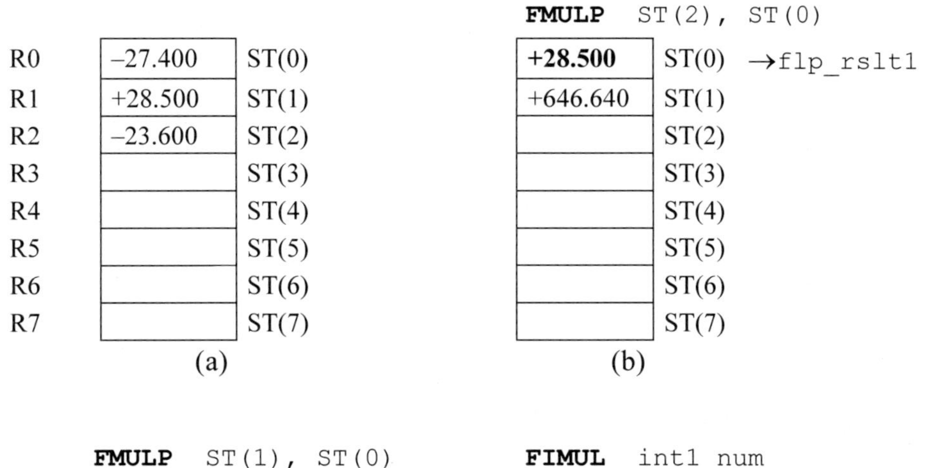

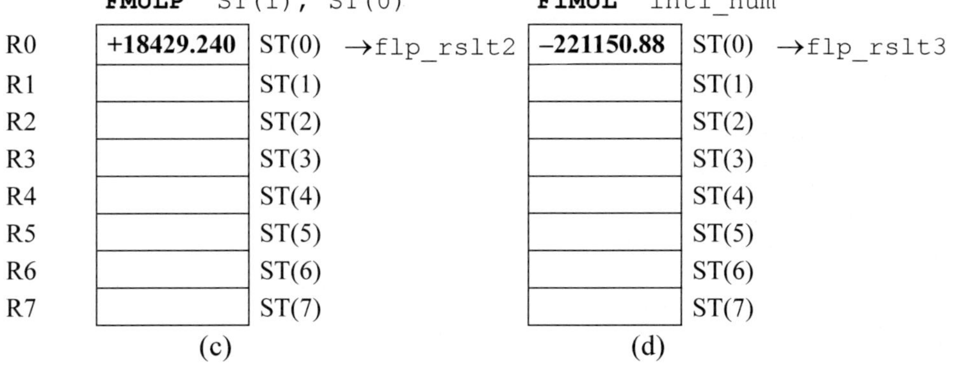

11.28 A resistor is one of four passive circuit elements: resistor, capacitor, inductor, and the recently discovered memristor. The equivalent resistance R_{eq} — specified in ohms — of resistors connected in series is the sum of the resistor values. However, the equivalent resistance of resistors connected in parallel is shown in the equation below.

$$\frac{1}{R_{eq}} = \frac{1}{R_1} + \frac{1}{R_2} + \frac{1}{R_3} + \cdots + \frac{1}{R_n}$$

The value of R_{eq} is smaller than the resistance of the smallest resistor in the parallel circuit. The circuit shown below contains three parallel resistors.

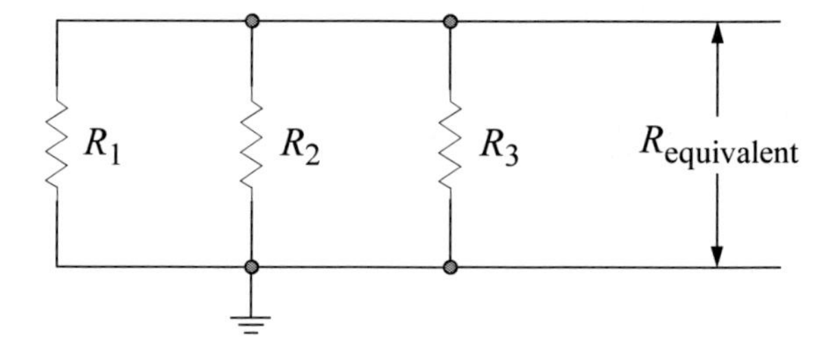

Write an assembly language module embedded in a C program that calculates the equivalent resistance of a three-resistor parallel network. Enter four sets of values for the resistors and display the equivalent resistance.

```
//parallel_resistors.cpp
//find the equivalent resistance of a
//three-resistor parallel network

#include "stdafx.h"

int main (void)
{
   float res1, res2, res3,
         res1_inv, res2_inv, res3_inv,
         resn_sum, res_equiv;

   printf ("Enter three resistor values: \n");

   scanf ("%f %f %f", &res1, &res2, &res3);

                           //continued on next page
```

```
//switch to assembly
      _asm
      {
//--------------------------------------------------------
//calculations for resistor 1
         FLD1               //+1.0 -> ST(0)
         FLD   res1         //res1 -> ST(0)
                            //+1.0 -> ST(1)
         FDIVP ST(1), ST(0)//+1.0 / res1 -> ST(1), pop
         FST   res1_inv     //ST(0) (+1.0 / res1) ->
                            //res1_inv
//--------------------------------------------------------
//calculations for resistor 2
         FLD1               //+1.0 -> ST(0)
                            //+1.0 / res1 -> ST(1)
         FLD   res2         //res2 -> ST(0)
                            //+1.0 -> ST(1)
                            //+1.0 / res1 -> ST(2)
         FDIVP ST(1), ST(0)//+1.0 / res2 -> ST(1), pop
         FST   res2_inv     //ST(0) (+1.0 / res2) ->
                            //res2_inv
//--------------------------------------------------------
//calculations for resistor 3
         FLD1               //+1.0 -> ST(0)
                            //+1.0 / res2 -> ST(1)
                            //+1.0 / res1 -> ST(2)
         FLD   res3         //res3 -> ST(0)
                            //+1.0 -> ST(1)
                            //+1.0 / res2 -> ST(2)
                            //+1.0 / res1 -> ST(3)
         FDIVP ST(1), ST(0)//+1.0 /res3 -> ST(1), pop
         FST   res3_inv     //ST(0) (+1.0 / res3) ->
                            //res3_inv
//--------------------------------------------------------
//calculate the sum of the inverted resistors
         FADD  ST(0), ST(1)//(+1.0 / res3) +
                            //(+1.0 / res2) -> ST(0)
         FADD  ST(0), ST(2)//ST(0) + (+1.0 / res1) ->
                            //ST(0). Sum of inverted
                            //resistors -> ST(0)
         FST   resn_sum     //sum of inverted resistors
                            //-> resn_sum
//--------------------------------------------------------
                            //continued on next page
```

```
//-------------------------------------------------------
//calculate the equivalent resistance
         FLD1                 //+1.0 -> ST(0)
         FDIV  resn_sum       //+1.0 / resn_sum -> ST(0)
         FST   res_equiv      //ST(0) -> equivalent
                              //resistance
//-------------------------------------------------------
}

//print result
   printf ("\nEquivalent resistance = %f ohms\n\n",
               res_equiv);

   return 0;
}
```

```
Enter three resistor values:
1000.0 1000.0 1000.0

Equivalent resistance = 333.333344 ohms

Press any key to continue . . . _
-------------------------------------------------------
Enter three resistor values:
72.25 1000 64.38

Equivalent resistance = 32.923325 ohms

Press any key to continue . . . _
-------------------------------------------------------
Enter three resistor values:
750.0 75000.0 3600.0

Equivalent resistance = 615.595032 ohms

Press any key to continue . . . _
-------------------------------------------------------
Enter three resistor values:
1.0 2.0 3.0

Equivalent resistance = 0.545455 ohms

Press any key to continue . . . _
```

11.33 Write an assembly language module embedded in a C program that calculates the result of the expression shown below for different values of the floating-point variables *flp*1 and *flp*2.

$$\frac{1}{2\pi\ \sqrt{flp1 \times flp2}}$$

```
//square_root.cpp
//uses FMUL, FSQRT, FLDPI, FLD1, FDIV instructions
#include "stdafx.h"
int main (void)
{
   float flp1, flp2, denom_rslt, flp_rslt;
   float two;
   two = 2.0;

   printf ("Enter two floating-point numbers: \n");
   scanf ("%f %f", &flp1, &flp2);

//switch to assembly
      _asm
      {
         FLD    flp1         //flp1 -> ST(0)
         FMUL   flp2         //flp1 x flp2 -> ST(0)
         FSQRT               //sq root of ST(0) -> ST(0)
         FMUL   two          //sq root of ST(0) x 2.0 ->
                             //ST(0)
         FLDPI               //pi -> ST(0)
                             //sq root of ST(0) x 2.0 ->
                             //ST(1)
         FMUL   ST(0), ST(1)//2.0 x sq root x pi ->
                             //ST(0)
         FSTP   denom_rslt   //denominator ->
                             //denom_rslt, pop stack
         FLD1                //1.0 -> ST(0)
         FDIV   denom_rslt   //1.0 / denom_rslt -> ST(0)
         FST    flp_rslt     //result -> flp_rslt area
      }
//print result
   printf ("\nResult = %f\n\n", flp_rslt);

   return 0;
}
```

```
Enter two floating-point numbers:
10.0 10.0

Result = 0.015915

Press any key to continue . . . _
---------------------------------------------------------
Enter two floating-point numbers:
20.0 30.0

Result = 0.006497

Press any key to continue . . . _
---------------------------------------------------------
Enter two floating-point numbers:
0.5 0.5

Result = 0.318310

Press any key to continue . . . _
---------------------------------------------------------
Enter two floating-point numbers:
0.05 0.08

Result = 2.516461

Press any key to continue . . . _
---------------------------------------------------------
Enter two floating-point numbers:
0.05 0.05

Result = 3.183099

Press any key to continue . . . _
```

Chapter 12 Procedures

12.1 Write an assembly language program — not embedded in a C program — that uses a procedure to multiply two single-digit operands. Enter several operands for the multiplicand and multiplier and display the products.

```
;mul_proc.asm
;-------------------------------------------------------------
.STACK
;-------------------------------------------------------------
.DATA
PARLST  LABEL BYTE
MAXLEN  DB 5
ACTLEN  DB ?
OPFLD   DB 5 DUP(?)
PRMPT  DB 0DH, 0AH, 'Enter two single-digit integers: $'
RSLT    DB 0DH, 0AH, 'Product =        $'
;-------------------------------------------------------------
.CODE
BEGIN    PROC FAR

;set up pgm ds
      MOV   AX, @DATA    ;get addr of data seg
      MOV   DS, AX

;read prompt
      MOV   AH, 09H      ;display string
      LEA   DX, PRMPT
      INT   21H          ;dos interrupt

;kybd rtn to enter chars
      MOV   AH, 0AH      ;buffered kybd input
      LEA   DX, PARLST
      INT   21H

;get the two integers
      MOV   AL, OPFLD    ;get 1st digit (mpcnd) fm opfld
      AND   AL, 0FH      ;remove ascii bias
      MOV   AH, 00H      ;clear ah
      PUSH  AX
      MOV   BL, OPFLD+1 ;get 2nd digit (mplyr) fm opfld
      AND   BL, 0FH      ;remove ascii bias
      MOV   BH, 00H      ;clear bh
      PUSH  BX

      CALL  CALC         ;push ip.  call calc rtn
                            //continued on next page
```

```
;return to here after calc procedure
;move the sum to result
      MOV   RSLT+12, AH
      MOV   RSLT+13, AL

;display the resulting sum
      MOV   AH, 09H      ;display string
      LEA   DX, RSLT
      INT   21H

BEGIN ENDP
;------------------------------------------------------
CALC  PROC NEAR

      PUSH  BP           ;bp will be used
      MOV   BP, SP       ;bp points to tos
      MOV   AX, [BP+4]   ;get opnd a (mpcnd)
      MUL   [BP+6]       ;mul opnd b (mplyr). prod to ax
      AAM                ;ascii adjust for mul
      OR    AX, 3030H    ;convert to ascii
      POP   BP           ;restore bp
      RET   4            ;pop ip. add 4 to sp so that
                         ;sp points to initial location
CALC  ENDP
END   BEGIN
```

```
Enter two single-digit integers: 23
Product = 6
------------------------------------------------------
Enter two single-digit integers: 91
Product = 09
------------------------------------------------------
Enter two single-digit integers: 54
Product = 20
------------------------------------------------------
Enter two single-digit integers: 78
Product = 56
------------------------------------------------------
Enter two single-digit integers: 98
Product = 72
------------------------------------------------------
Enter two single-digit integers: 99
Product = 81
------------------------------------------------------
Enter two single-digit integers: 90
Product = 00
```

12.4 Write an assembly language program — not embedded in a C program — that uses a procedure to exclusive-OR six hexadecimal characters that are entered from the keyboard. The following characters are exclusive-ORed: the first and third; the second and fourth; the third and fifth; and the fourth and sixth. The keyboard data can be any hexadecimal characters; for example, 2T/}b*.

```
;xor_proc.asm
;exclusive-OR characters
;opfld  xor opfld2
;opfld2 xor opfld4
;opfld3 xor opfld5
;opfld4 xor opfld6
;------------------------------------------------------
.STACK
;------------------------------------------------------
.DATA
PARLST    LABEL BYTE
MAXLEN    DB 10
ACTLEN    DB ?
OPFLD     DB 10 DUP(?)
PRMPT     DB 0DH, 0AH, 'Enter six hexadecimal
                         characters: $'
RSLT      DB 0DH, 0AH, 'Result =         $'
;------------------------------------------------------
.CODE
BEGIN     PROC    FAR

;set up pgm ds
      MOV   AX, @DATA
      MOV   DS, AX

;read prompt
      MOV   AH, 09H
      LEA   DX, PRMPT
      INT   21H

;keyboard request rtn to enter characters
      MOV   AH, 0AH
      LEA   DX, PARLST
      INT   21H

;set up addresses
      LEA   SI, OPFLD         ;1st number
      LEA   DI, OPFLD+2       ;3rd number
      LEA   BX, RSLT+12       ;result area

      CALL  EXOR              ;push ip.  call exor rtn
                                 //continued on next page
```

```
;-----------------------------------------------------------
;return to here after exor procedure
;display result
        MOV    AH, 09H          ;display string
        LEA    DX, RSLT
        INT    21H

BEGIN ENDP

;-----------------------------------------------------------
EXOR    PROC NEAR

        MOV    CX, 4            ;loop control
LP1:    MOV    AL, [SI]
        MOV    AH, [DI]
        XOR    AL, AH
        OR     AL, 30H
        MOV    [BX], AL
        INC    SI
        INC    DI
        INC    BX
        LOOP   LP1
        RET

EXOR    ENDP

;-----------------------------------------------------------
END     BEGIN
```

```
Enter six hexadecimal characters: 123456
Result = 2662
------------------------------------------------------------
Enter six hexadecimal characters: F0E2CF
Result = 326t
------------------------------------------------------------
Enter six hexadecimal characters: B67dMt
Result = urz0
------------------------------------------------------------
Enter six hexadecimal characters: F05urM
Result = suw8
------------------------------------------------------------
Enter six hexadecimal characters: :/<?]+
Result = 60q4
```

Chapter 13 String Instructions

13.2 Determine the contents of RSLT after execution of the following program:

```
;movs_byte5.asm
;--------------------------------------------------------
.STACK

;--------------------------------------------------------
.DATA
RSLT     DB   0DH, 0AH, '123456789    $'

;--------------------------------------------------------
.CODE
BEGIN     PROC   FAR

;set up pgm ds
      MOV    AX, @DATA         ;get addr of data seg
      MOV    DS, AX            ;move addr to ds
      MOV    ES, AX            ;move addr to es

;--------------------------------------------------------
;move string  elements
      CLD
      MOV    CX, 4             ;count in cx
      LEA    SI, RSLT+2        ;addr of rslt+2 -> si as src
      LEA    DI, RSLT+4        ;addr of rslt+4 -> di as dst

REP   MOVSB                    ;move bytes to dst

;--------------------------------------------------------
;display result
      MOV    AH, 09H           ;display string
      LEA    DX, RSLT          ;put addr of rslt in dx
      INT    21H               ;dos interrupt

BEGIN ENDP
      END    BEGIN
;--------------------------------------------------------
//121212789
```

13.8 Given the program segment shown below, determine the contents of the count in register CL and the ZF flag after the program has been executed. Then write an assembly language program — not embedded in a C program — to verify the results.

```
                              . . .
.DATA
STR1     DB 'ABCDEF  $'
STR2     DB 'AB1234  $'
CL_RSLT  DB 0DH, 0AH, 'CL =        $'
FLAGS    DB 0DH, 0AH, 'ZF flag =     $'

;-------------------------------------------------------
.CODE
BEGIN    PROC FAR

;set up pgm ds and es
    MOV     AX, @DATA          ;get addr of data seg
    MOV     DS, AX             ;move addr to ds
    MOV     ES, AX             ;move addr to es

;-------------------------------------------------------
    LEA     SI, STR1           ;addr of str1 -> si
    LEA     DI, STR2           ;addr of str2 -> di
    CLD                        ;left-to-right
    MOV     CL, 6              ;count in cl

REPNE
   CMPSB                  ;compare strings while not equal
                              . . .
```

```
CL = 5
ZF flag = 1
                  //complete program on next page
```

```
;cmps_repne.asm
;use cmps with repne prefix
;-------------------------------------------------------
.STACK
;-------------------------------------------------------
.DATA
STR1     DB 'ABCDEF  $'
STR2     DB 'AB1234  $'
CL_RSLT DB 0DH, 0AH, 'CL =        $'
FLAGS    DB 0DH, 0AH, 'ZF flag =    $'
;-------------------------------------------------------
.CODE
BEGIN      PROC FAR

;set up pgm ds and es
      MOV   AX, @DATA        ;get addr of data seg
      MOV   DS, AX           ;move addr to ds
      MOV   ES, AX           ;move addr to es
;-------------------------------------------------------
      LEA   SI, STR1         ;addr of str1 -> si
      LEA   DI, STR2         ;addr of str2 -> di
      CLD                    ;left-to-right
      MOV   CL, 6            ;count in cl

REPNE
      CMPSB                  ;comp while not equal

      PUSHF                  ;push flags
      POP   AX
      AND   AL, 40H          ;isolate zf
      SHR   AL, 6            ;shift right logical 6
      OR    AL, 30H          ;add ascii bias
      MOV   FLAGS+12, AL     ;move zf to flag area
;-------------------------------------------------------
;display residual count
      OR    CL, 30H          ;add ascii bias
      MOV   CL_RSLT+7, CL    ;move to result area

      MOV   AH, 09H          ;display string
      LEA   DX, CL_RSLT      ;addr of cl_rslt -> dx
      INT   21H              ;dos interrupt
;-------------------------------------------------------
;display flags
      MOV   AH, 09H          ;display string
      LEA   DX, FLAGS        ;addr of flags -> dx
      INT   21H              ;dos interrupt

BEGIN ENDP
      END   BEGIN
```

13.9 Given the program segment shown below, obtain the count in register CL after the program executes.

```
                                    . . .
.DATA
STR      DB 'ABCDEFGHI  $'
CL_RSLT DB 0DH, 'CL =     $'
FLAGS   DB 0DH, 0AH, 'ZF flag =   $'

;------------------------------------------------------
.CODE
BEGIN   PROC   FAR

;set up pgm ds and es
      MOV    AX, @DATA         ;get addr of data seg
      MOV    DS, AX            ;move addr to ds
      MOV    ES, AX            ;move addr to es

;------------------------------------------------------
      MOV    AL, 'G'
      LEA    DI, STR           ;addr of str2 -> di
      CLD                      ;left-to-right
      MOV    CL, 9             ;count in cl

REPNE
      SCASB                    ;compare char while ≠
                                    . . .
;------------------------------------------------------
CL = 1
;------------------------------------------------------
```

Chapter 14 Arrays

14.3 Write an assembly language module embedded in a C program that adds, subtracts, and multiplies two select floating-point numbers. Three positive and three negative floating-point numbers are to be defined, then used in the calculations. Perform two calculations on each arithmetic operation and display the corresponding results.

```
//array_flp2.cpp
//add, sub, and mul two floating-point numbers
#include "stdafx.h"
int main (void)

{
//define variables
   double   flp1 = 72.5,
            flp2 = -104.8,
            flp3 = -56.7,
            flp4 = 110.5,
            flp5 = -255.3,
            flp6 = 30.6;

   double   sum1, sum2,
            diff1, diff2,
            prod1, prod2;

//switch to assembly
      _asm
      {
//addition -------------------------------------------
         FLD    flp2     //(-104.8) -> ST(0)
         FADD   flp3     //(-104.8) + (-56.7)
                         //= (-161.5) -> ST(0)
         FST    sum1     //(-161.5) -> sum1

         FLD    flp4     //(110.5) -> ST(0)
         FADD   flp5     //(110.5) +(-255.3)
                         //= (-144.8) -> ST(0)
         FST    sum2     //(-144.8) -> sum2

                            //continued on next page
```

```
//subtraction ------------------------------------------
         FLD    flp5      //(-255.3) -> ST(0)
         FSUB   flp4      //(-255.3) - (110.5)
                          //= (-365.8) - > ST(0)
         FST    diff1     //(-365.8) -> diff1

         FLD    flp6      //(30.6) -> ST(0)
         FSUB   flp2      //(30.6) - (-104.8)
                          //= (135.4) - > ST(0)
         FST    diff2     //(135.4) -> diff2

//multiplication ---------------------------------------
         FLD    flp2      //(-104.8) -> ST(0)
         FMUL   flp5      //(-104.8) x (-255.3)
                          //= 26755.44 -> ST(0)
         FST    prod1     //(26755.44) -> prod1

         FLD    flp3      //(-56.7) -> ST(0)
         FMUL   flp4      //(-56.7) x (110.5)
                          //= -6265.35 -> ST(0)
         FST    prod2     //(-6265.35) -> prod2
      }

//display results
   printf ("(-104.8) + (-56.7) = %f\n", sum1);
   printf ("(110.5) + (-255.3) = %f\n\n", sum2);
   printf ("(-255.3) - (110.5) = %f\n", diff1);
   printf ("(30.6) - (-104.8) = %f\n\n", diff2);
   printf ("(-104.8) x (-255.3) = %f\n", prod1);
   printf ("(-56.7) x (110.5) = %f\n\n", prod2);
   return 0;
}
```

```
(-104.8) + (-56.7) = -161.500000
(110.5) + (-255.3) = -144.800000

(-255.3) - (110.5) = -365.800000
(30.6) - (-104.8) = 135.400000

(-104.8) x (-255.3) = 26755.440000
(-56.7) x (110.5) = -6265.350000

Press any key to continue . . . _
```

14.5 Write a C program that calculates the cubes of the integers 1 through 10 using an array, where array [0] contains a value of 1.

```
//array_cubes.cpp
//obtain the cubes of numbers 1 through 10

#include "stdafx.h"
int main (void)
{
   int cubes[10];       //declare an array of 10 elements
   int i;               //declare an integer to count

//initialize and print the array
   for (i=1; i<11; i++)
   {
      cubes[i-1] = i*i*i;
      printf ("cubes[%d] = %d\n", i-1, cubes[i-1]);
   }

   printf ("\n");
   return 0;
}
```

```
cubes[0] = 1
cubes[1] = 8
cubes[2] = 27
cubes[3] = 64
cubes[4] = 125
cubes[5] = 216
cubes[6] = 343
cubes[7] = 512
cubes[8] = 729
cubes[9] = 1000

Press any key to continue . . . _
```

14.8 Write a C program that loads a 3×4 array with the products of the indices, then display the array in a row–column format.

```
//array_mul_indices.cpp
//load a 3 x 4 array with the products of the indices,
//then display the array in a row-column format

#include "stdafx.h"

int main (void)
{
   int array_mul[3][4];
   int i, j;

   for (i=0;i<3; i++)           //i is row index
      for (j=0; j<4; j++)       //j is column index
         array_mul[i][j] = i*j;

   for (i=0; i<3; i++)
   {
      for (j=0; j<4; j++)
         printf ("%4d", array_mul[i][j]);
         printf ("\n");
   }
      printf ("\n");
   return 0;
}
```

```
   0   0   0   0
   0   1   2   3
   0   2   4   6

Press any key to continue . . . _
```

Chapter 15 Macros

15.2 Write an assembly language program using a macro that sorts from two to ten single-digit integers in ascending numerical order that are entered from the keyboard. Enter several different sets of integers ranging from two integers to ten integers and display the results.

```
;macro_sort2.asm
;sort 2 -- 10 single-digit integers
;in ascending numerical order
;-------------------------------------------------------
;define the sort macro
SORT  MACRO
      LEA   SI, OPFLD           ;addr of first byte -> si
      MOV   DL, ACTLEN          ;# of bytes -> dx
      MOV   DH, 00H             ;... as loop ctr
      DEC   DX                  ;decrement loop ctr

LP1:  MOV   CX, DX              ;cx is loop2 ctr

LP2:  MOV   AL, [SI]            ;number -> al
      MOV   BX, CX              ;bx is base addr
      CMP   AL, [BX+SI]         ;is the number <= to al?
      JBE   NEXT                ;yes, get next number
      XCHG  AL, [BX+SI]         ;# is !<=; put # in al
      MOV   [SI], AL            ;put small # in opfld

NEXT: LOOP  LP2                 ;loop until all #s
                                ;... are compared
      INC   SI                  ;si points to next number
      DEC   DX                  ;decrement loop ctr
      JNZ   LP1                 ;compare remaining
                                ;... numbers
      ENDM

;-------------------------------------------------------
.STACK

;-------------------------------------------------------
.DATA
PARLST  LABEL   BYTE
MAXLEN  DB  15
ACTLEN  DB  ?
OPFLD   DB  15  DUP(?)
PRMPT   DB  0DH, 0AH, 'Enter numbers: $'
RSLT    DB  0DH, 0AH, 'Sorted numbers =           $'
;-------------------------------------------------------
```

//continued on next page

```
;-------------------------------------------------------
.CODE
BEGIN PROC   FAR

;set up pgm ds
      MOV   AX, @DATA          ;addr of data seg -> ax
      MOV   DS, AX             ;move addr to ds

;read prompt
      MOV   AH, 09H            ;display string
      LEA   DX, PRMPT          ;put addr of prompt in dx
      INT   21H                ;dos interrupt

;keyboard request rtn to enter characters
      MOV   AH, 0AH            ;buffered keyboard input
      LEA   DX, PARLST         ;put addr of parlst in dx
      INT   21H                ;dos interrupt

;-------------------------------------------------------
;call sort macro
      SORT

;-------------------------------------------------------
;set up src (opfld) and dst (rslt)
      LEA   SI, OPFLD
      LEA   DI, RSLT+19
      MOV   CL, ACTLEN
      MOV   CH, 00H

;move sorted numbers to result area
RSLT_AREA:
      MOV   BL, [SI]           ;opfld number to bl
      MOV   [DI], BL           ;number to result area
      INC   SI                 ;set up next source
      INC   DI                 ;set up next destination
      LOOP  RSLT_AREA          ;loop if cl != 0

;display sorted numbers
      MOV   AH, 09H            ;display string
      LEA   DX, RSLT           ;put addr of rslt in dx
      INT   21H                ;dos interrupt

BEGIN ENDP
      END   BEGIN
```

```
Enter numbers: 73
Sorted numbers = 37
-----------------------------------------------------
Enter numbers: 360
Sorted numbers = 036
-----------------------------------------------------
Enter numbers: 9845
Sorted numbers = 4589
-----------------------------------------------------
Enter numbers: 48632
Sorted numbers = 23468
-----------------------------------------------------
Enter numbers: 130474
Sorted numbers = 013447
-----------------------------------------------------
Enter numbers: 9766347
Sorted numbers = 3466779
-----------------------------------------------------
Enter numbers: 47589001
Sorted numbers = 00145789
-----------------------------------------------------
Enter numbers: 354657685
Sorted numbers = 345556678
-----------------------------------------------------
Enter numbers: 9876543210
Sorted numbers = 0123456789
```

15.5 Write a program that uses two macros. One macro generates an odd parity bit for an 8-bit code word that is entered from the keyboard. If the parity is odd, then display a 0; if the parity is even, then display a 1. The other macro removes all nonletters from data that are entered from the keyboard.

```
;macro_parity_ltrs2.asm
;use a macro to generate a parity bit for 8 bits
;... if parity is odd, display 0
;... if parity is even, display 1
;use a macro to remove all nonletters from data
;... entered from the keyboard

;-----------------------------------------------------------
;define the parity macro
PARITY   MACRO
NEXTB:MOV    AL, [SI]          ;bit -> al
      CMP    AL, 31H           ;is bit = 1?
      JNE    ZERO              ;if bit = 0, jump
      XOR    BL, 01H           ;if bit = 1, toggle parity
ZERO: INC    SI                ;si points to next bit
      LOOP   NEXTB             ;is cx != 0, get next bit
      ENDM

;-----------------------------------------------------------
;define ltrs (remove nonletters) macro
LTRS  MACRO
NEXTL:MOV    AL, [SI]          ;opfld char -> al
      CMP    AL, 41H           ;compare to uppercase A
      JL     LP1               ;if < A, jump to LP1
      CMP    AL, 5AH           ;compare to uppercase Z
      JLE    LP2               ;if >= A & <= Z, jump to LP2

      CMP    AL, 7AH           ;compare to lowercase z
      JG     LP1               ;if > z, jump to LP1
      CMP    AL, 61H           ;compare to lowercase a
      JL     LP1               ;if < a, jump to LP1

LP2:  MOV    [DI], AL          ;valid character -> dst
      INC    DI                ;di -> next dst

LP1:  INC    SI                ;get next char in opfld
      LOOP   NEXTL
      ENDM

;-----------------------------------------------------------
.STACK
;-----------------------------------------------------------
                                //continued on next page
```

```
.DATA
PARLST   LABEL BYTE
MAXLEN   DB 40
ACTLEN   DB ?
OPFLD    DB 40  DUP(?)
PRMPT    DB 0DH, 0AH, 'Enter characters: $'
RSLTP    DB 0DH, 0AH, 'Parity bit =      $'
RSLTL    DB 0DH, 0AH, 'Letters =                  $'

;-----------------------------------------------------
.CODE
BEGIN PROC  FAR

;set up pgm ds
      MOV   AX, @DATA            ;addr of data seg -> ax
      MOV   DS, AX               ;move addr to ds

;*****************************************************
;set up for the parity macro
;read prompt
      MOV   AH, 09H              ;display string
      LEA   DX, PRMPT            ;addr of prompt -> dx
      INT   21H                  ;dos interrupt

;keyboard request to enter characters
      MOV   AH, 0AH              ;buffered keyboard input
      LEA   DX, PARLST           ;addr of parlst -> dx
      INT   21H                  ;dos interrupt
;-----------------------------------------------------
;initialize parity flag to 1 and count in cx to 8
;initialize source (si) and destination (di) addr
      MOV   BL, 01H              ;set parity in bl to 1
      MOV   CX, 08H              ;number of bits is 8
      LEA   SI, OPFLD            ;addr of opfld -> si
      LEA   DI, RSLTP+11         ;addr of result -> di

;-----------------------------------------------------
;call parity macro
        PARITY
;-----------------------------------------------------
      OR    BL, 30H              ;add ascii bias to parity
      MOV   [DI], BL             ;parity -> result area

;-----------------------------------------------------
;display result
      MOV   AH, 09H              ;display string
      LEA   DX, RSLTP            ;addr of result -> dx
      INT   21H                  ;dos interrupt
                              //continued on next page
```

```
;*****************************************************
;set up for the ltrs (remove nonletters) macro
;read prompt
      MOV   AH, 09H             ;display string
      LEA   DX, PRMPT           ;addr of prompt -> dx
      INT   21H                 ;dos interrupt

;keyboard request to enter characters
      MOV   AH, 0AH             ;buffered keyboard input
      LEA   DX, PARLST          ;addr of parlst -> dx
      INT   21H                 ;dos interrupt

;get number of characters entered
      MOV   CL, ACTLEN          ;# of characters -> cl
      MOV   CH, 00H             ;cx = 00 actlen

;set up source (opfld) and destination (rslt) addr
      LEA   SI, OPFLD           ;source addr -> si
      LEA   DI, RSLTL+11        ;destination addr -> di

;-----------------------------------------------------
;call ltrs (remove nonletters) macro
        LTRS

;-----------------------------------------------------
;display result
      MOV   AH, 09H             ;display string
      LEA   DX, RSLTL           ;addr of result -> dx
      INT   21H                 ;dos interrupt

BEGIN ENDP
      END   BEGIN
```

```
Enter characters: 11001010    //parity is even
Parity bit = 1
Enter characters: ab34deFG7H
Letters = abdeFGH
------------------------------------------------------
Enter characters: 01100111    //parity is odd
Parity bit = 0
Enter characters: A678Fth9jk0Lm
Letters = AFthjkLm
------------------------------------------------------
Enter characters: 01111111    //parity is odd
Parity bit = 0
Enter characters: 65abcdef89GH
Letters = abcdefGH            //continued on next page
------------------------------------------------------
```

```
Enter characters: 00000000    //parity is even
Parity bit = 1
Enter characters: 1mno56PQr3
Letters = mnoPQr
-------------------------------------------------------
Enter characters: 11110000    //parity is even
Parity bit = 1
Enter characters: 1a23455678
Letters = a
-------------------------------------------------------
Enter characters: 00011111    //parity is odd
Parity bit = 0
Enter characters: 1234567890B
Letters = B
-------------------------------------------------------
Enter characters: 00111100    //parity is even
Parity bit = 1
Enter characters: 54689v7675t99
Letters = vt
-------------------------------------------------------
Enter characters: 10101010    //parity is even
Parity bit = 1
Enter characters: 456h8996
Letters = h
-------------------------------------------------------
Enter characters: 00110011    //parity is even
Parity bit = 1
Enter characters: 1234567890
Letters =
-------------------------------------------------------
Enter characters: 00011100    //parity is odd
Parity bit = 0
Enter characters: 6
Letters =
-------------------------------------------------------
Enter characters: 11100000    //parity is odd
Parity bit = 0
Enter characters: 23vB^&JK(t
Letters = vBJKt
```

15.8 Write a program using a macro with no parameters that multiplies two single-digit integers that are entered from the keyboard. Enter several integers and display the results.

```
;macro_mul2.asm
;obtain the product of two single-digit integers

;------------------------------------------------------
;define the mult macro
MULT  MACRO
      MOV   AL, OPFLD          ;get multiplicand
      AND   AL, 0FH            ;remove ascii bias
      MOV   AH, OPFLD+1        ;get multiplier
      AND   AH, 0FH            ;remove ascii bias

;multiply the #s, adjust using aam, add ascii bias
      MUL   AH                 ;ax = al x ah
      AAM                      ;ax = 0d 0d
      OR    AX, 3030H          ;add ascii bias
      MOV   RSLT+12, AH        ;high product -> rslt
      MOV   RSLT+13, AL        ;low product -> rslt
      ENDM

;------------------------------------------------------
.STACK

;------------------------------------------------------
.DATA
PARLST   LABEL BYTE
MAXLEN   DB 5
ACTLEN   DB ?
OPFLD    DB 5  DUP(?)
PRMPT    DB 0DH, 0AH, 'Enter two 1-digit integers: $'
RSLT     DB 0DH, 0AH, 'Product =     $'

;------------------------------------------------------
.CODE
BEGIN PROC  FAR

;set up pgm ds
      MOV   AX, @DATA          ;addr of data seg -> ax
      MOV   DS, AX             ;move addr to ds

                        //continued on next page
```

```
;read prompt
      MOV    AH, 09H              ;display string
      LEA    DX, PRMPT            ;put addr of prompt in dx
      INT    21H                  ;dos interrupt

;keyboard request rtn to enter characters
      MOV    AH, 0AH              ;buffered keyboard input
      LEA    DX, PARLST           ;put addr of parlst in dx
      INT    21H                  ;dos interrupt

;------------------------------------------------------
;call mult macro
      MULT
;------------------------------------------------------
;display product
      MOV    AH, 09H              ;display string
      LEA    DX, RSLT             ;addr of rslt -> dx
      INT    21H                  ;dos interrupt

BEGIN ENDP
      END    BEGIN
```

```
Enter two single-digit integers: 99
Product = 81
------------------------------------------------------
Enter two single-digit integers: 87
Product = 56
------------------------------------------------------
Enter two single-digit integers: 39
Product = 27
------------------------------------------------------
Enter two single-digit integers: 73
Product = 21
------------------------------------------------------
Enter two single-digit integers: 08
Product = 00
------------------------------------------------------
Enter two single-digit integers: 60
Product = 00
------------------------------------------------------
Enter two single-digit integers: 45
Product = 20
```

Chapter 16 Interrupts and Input/Output Operations

16.3 Explain why interrupt breakpoints occur only at the end of an instruction cycle.

If interrupt breakpoints occurred during an instruction cycle, for example, at the end of a CPU cycle, then control would be transferred to the ISR and the current instruction would not be completed. Therefore, interrupt breakpoints occur only at the end of an instruction cycle, that is, when the current instruction has completed execution.

16.7 Explain why DMA access to main memory has a higher priority than central processing units access to main memory.

Input/output devices equipped with DMA hardware have highest priority over central processing units when transferring data to or from main memory, because they are inherently slower and cannot have the data transfer stopped temporarily. Thus, data would be lost. Disk drives and tape drives are examples of I/O subsystems in this category.

Chapter 17 Additional Programming Examples

17.2 Write a program in assembly language — not embedded in a C program — that counts the number of times that a number occurs in an array of eight numbers. The first number entered from the keyboard is the number that is being compared to the remaining eight numbers. Display the result.

```
;count_occurrence2.asm
;count the number of times that a number
;occurs in an array of eight numbers.
;the first number entered from the keyboard
;is the number that is being compared to
;the remaining eight numbers
;---------------------------------------------------------
.STACK

;---------------------------------------------------------
.DATA
PARLST   LABEL BYTE
MAXLEN   DB 12
ACTLEN   DB ?
OPFLD    DB 12 DUP(?)
PRMPT    DB 0DH, 0AH, 'Enter 9 single-digit integers: $'
RSLT     DB 0DH, 0AH, 'Occurs =   times  $'

;---------------------------------------------------------
.CODE
BEGIN PROC FAR

;set up pgm ds
      MOV   AX, @DATA           ;addr of data seg -> ax
      MOV   DS, AX              ;put addr in ds
      LEA   DI, RSLT + 11       ;put addr of rslt in di

;read prompt
      MOV   AH, 09H             ;display string
      LEA   DX, PRMPT           ;addr of prmpt -> dx
      INT   21H                 ;dos interrupt

;keyboard rtn to enter characters
      MOV   AH, 0AH             ;buffered keyboard input
      LEA   DX, PARLST          ;load addr of parlst
      INT   21H                 ;dos interrupt

                                //continued on next page
```

```
;keyboard rtn to enter characters
      MOV   AH, 0AH           ;buffered keyboard input
      LEA   DX, PARLST        ;load addr of parlst
      INT   21H               ;dos interrupt
;------------------------------------------------------
      MOV   DL, OPFLD         ;number to compare -> dl
      LEA   SI, OPFLD + 1     ;put addr of list in si
      MOV   CX, 8             ;qty of #s to comp -> cx
      MOV   AH, 30H           ;init # of occurrences

LP1:  MOV   AL, [SI]          ;move number to al
      CMP   DL, AL            ;compare dl to al
      JE    INC_COUNT         ;if equal, jump
      INC   SI                ;point to next number
      LOOP  LP1               ;loop to compare next #
      JMP   MOVE_COUNT        ;comparison is finished

INC_COUNT:
      INC   AH                ;incr the occurrences
      INC   SI                ;point to next number
      LOOP  LP1               ;if cx != 0,
                              ;compare next number

MOVE_COUNT:
      MOV   [DI], AH          ;# of occurrences
                              ;-> result area
;------------------------------------------------------
;display result
      MOV   AH, 09H           ;display string
      LEA   DX, RSLT          ;put addr of rslt in dx
      INT   21H               ;dos interrupt

BEGIN ENDP
      END   BEGIN
```

```
Enter 9 single-digit integers; 612634656
Occurs = 3 times
------------------------------------------------------
Enter 9 single-digit integers; 123151611
Occurs = 4 times
------------------------------------------------------
Enter 9 single-digit integers; 912345679
Occurs = 1 times
------------------------------------------------------
Enter 9 single-digit integers; 712345689
Occurs = 0 times
------------------------------------------------------
                              //continued on next page
```

```
------------------------------------------------------------
Enter 9 single-digit integers; 222222222
Occurs = 8 times
------------------------------------------------------------
Enter 9 single-digit integers; 412344544
Occurs = 4 times
```

17.6 Write an assembly language module embedded in a C program to obtain the sum of cubes for the numbers 1 through 5. Display the resulting sum of cubes.

```
//sum_of_cubes7.cpp
//obtain the sum of cubes for numbers 1 through 5
#include "stdafx.h"
int main (void)
{
//define variables
   unsigned char n, sum;
   n = 1;
   sum = 1;

//switch to assembly
      _asm
      {
//------------------------------------------------------------
LP1:        INC    n
            MOV    CX, 2
            MOV    AL, n

LP2:        MUL    n
            LOOP   LP2
            ADD    sum, AL
            CMP    n, 5
            JB     LP1
      }
//print sum
   printf ("Sum of cubes 1 through 5 = %d\n\n", sum);
   return 0;
}
```

```
Sum of cubes 1 through 5 = 225

Press any key to continue . . . _
```

17.9 Write an assembly language module embedded in a C program to evaluate the expression shown below for Y using a floating-point number for the variable X. The range for X is $-3.0 \leq X \leq +3.0$. Enter several numbers for X and display the corresponding results.

$$Y = X^3 - 10X^2 + 20X + 30$$

```
//eval_expr8.cpp
//evaluate the following expression for Y using a
//floating-point number for the variable X:
//Y = X^3 -10X^2 + 20X + 30, where X has a
//range of -3.0 <= X <= +3.0

#include "stdafx.h"

int main (void)
{
//define variables
   float flp_x, flp_rslt_x3, flp_rslt_10x2,
            flp_rslt_20x, flp_rslt_y;

   int ten;
   ten = 10;

   int twenty;
   twenty = 20;

   int thirty;
   thirty = 30;

   printf ("Enter a floating-point number for x: \n");
   scanf ("%f", &flp_x);

//switch to assembly
      _asm
      {
//calculate X^3
         FLD    flp_x            //X -> ST(0)
         FMUL   flp_x            //X^2 -> ST(0)
         FMUL   flp_x            //X^3 -> ST(0)
         FST    flp_rslt_x3      //ST(0) (X^3)
                                 //-> flp_rslt_x3

                              //continued on next page
```

```
//calculate 10X^2
        FLD   flp_x             //X -> ST(0)
        FMUL  flp_x             //X^2 -> ST(0)
        FIMUL ten               //(10) x X^2 -> ST(0)
        FST   flp_rslt_10x2     //ST(0) (10X^2)
                                //-> flp_rslt_10x2

//calculate 20X
        FLD   flp_x             //X -> ST(0)
        FIMUL twenty            //(20) x X -> ST(0)
        FST   flp_rslt_20x      //ST(0) (20X)
                                //-> flp_rslt_20x

//calculate Y
        FLD   flp_rslt_x3       //(X^3) -> ST(0)
        FSUB  flp_rslt_10x2     //(X^3) - (10X^2)
                                //-> ST(0)

        FADD  flp_rslt_20x      //(X^3) - (10X^2)
                                //+ (20X) -> ST(0)

        FIADD thirty            //(X^3) - (10X^2)
                                //+ (20X) + (30)
                                //-> ST(0)

        FST   flp_rslt_y        //ST(0) (Y)
                                //-> flp_rslt_y
      }

//printf results
   printf ("\nResult X^3 = %f\n\n", flp_rslt_x3);

   printf ("\nResult 10X^2 = %f\n\n", flp_rslt_10x2);

   printf ("\nResult 20X = %f\n\n", flp_rslt_20x);

   printf ("\nResult Y = %f\n\n", flp_rslt_y);

   return 0;
}
```

```
Enter a floating-point number for x:
-1.0

Result X^3 = -1.000000

Result 10X^2 = 10.000000

Result 20X = -20.000000

Result Y = -1.000000

Press any key to continue . . . _
--------------------------------------------------
Enter a floating-point number for x:
-2.0

Result X^3 = -8.000000

Result 10X^2 = 40.000000

Result 20X = -40.000000

Result Y = -58.000000

Press any key to continue . . . _
--------------------------------------------------
Enter a floating-point number for x:
-3.0

Result X^3 = -27.000000

Result 10X^2 = 90.000000

Result 20X = -60.000000

Result Y = -147.000000

Press any key to continue . . . _
--------------------------------------------------

                           //continued on next page
```

```
Enter a floating-point number for x:
+1.0

Result X^3 = 1.000000

Result 10X^2 = 10.000000

Result 20X = 20.000000

Result Y = 41.000000

Press any key to continue . . . _
------------------------------------------------
Enter a floating-point number for x:
+2.0

Result X^3 = 8.000000

Result 10X^2 = 40.000000

Result 20X = 40.000000

Result Y = 38.000000

Press any key to continue . . . _
------------------------------------------------
Enter a floating-point number for x:
+3.0

Result X^3 = 27.000000

Result 10X^2 = 90.000000

Result 20X = 60.000000

Result Y = 27.000000

Press any key to continue . . . _
------------------------------------------------
```

//continued on next page

```
Enter a floating-point number for x:
-1.125

Result X^3 = -1.423828

Result 10X^2 = 12.656250

Result 20X = -22.500000

Result Y = -6.580078

Press any key to continue . . . _
------------------------------------------------
Enter a floating-point number for x:
+2.75

Result X^3 = 20.796875

Result 10X^2 = 75.625000

Result 20X = 55.000000

Result Y = 30.171875

Press any key to continue . . . _
```

17.12 Given the program segment shown below, obtain the results for the following floating-point numbers:

```
flp_num1 = 125.0,     flp_num2 = 245.0

flp_num1 = 650.0,     flp_num2 = 375.0

flp_num1 = 755.125,   flp_num2 = 575.150
```

```
                              . . .
FLD     flp_num1
FSQRT
FLD     flp_num2
FSQRT
FSUBP ST(1), ST(0)
FST     rslt
                              . . .

                                    //continued on next page
```

```
flp_num1 = 125.0,     flp_num2 = 245.0
Result = -4.472136

flp_num1 = 650.0,     flp_num2 = 375.0
Result = 6.130181

flp_num1 = 755.125,   flp_num2 = 575.150
Result = 3.497252
```

17.14 Write an assembly language module embedded in a C program to calculate the tangent of the following angles using the sine and cosine of the angles:

30°, 45°, and 60°

```
//tan.cpp
//calculate the tangent of the following angles:
//30 deg, 45 deg, and 60 deg
//using the sine and cosine of the angles
//tangent = sine/cosine

#include "stdafx.h"

int main (void)

{
//define angles in radians
   double angle_30, angle_45, angle_60,
            rslt_30, rslt_45, rslt_60;
   angle_30 = 0.523598775;     //30 deg must be
                               //in radians
   angle_45 = 0.785398163;     //45 deg must be
                               //in radians
   angle_60 = 1.047197551;     //60 deg must be
                               //in radians

                            //continued on next page
```

```
//switch to assembly
      _asm
      {
//calculate the tangent of 30 degrees
         FLD   angle_30         //30 deg -> ST(0)
         FSIN                   //sine 30 -> ST(0)

         FLD   angle_30         //30 deg -> ST(0)
                                //sine 30 -> ST(1)
         FCOS                   //cos 30 -> ST(0)
         FDIVP ST(1), ST(0)     //sine 30 / cos 30
                                //-> ST(1), pop
                                //sine 30 / cos 30
                                //-> ST(0)
         FST   rslt_30          //store tangent 30

//calculate the tangent of 45 degrees
         FLD   angle_45         //45 deg -> ST(0)
         FSIN                   //sine 45 -> ST(0)

         FLD   angle_45         //45 deg -> ST(0)
                                //sine 45 -> ST(1)
         FCOS                   //cos 45 -> ST(0)
         FDIVP ST(1), ST(0)     //sine 45 / cos 45
                                //-> ST(1), pop
                                //sine 45 / cos 45
                                //-> ST(0)
         FST   rslt_45          //store tangent 45

//calculate the tangent of 60 degrees
         FLD   angle_60         //60 deg -> ST(0)
         FSIN                   //sine 60 -> ST(0)

         FLD   angle_60         //60 deg -> ST(0)
                                //sine 60 -> ST(1)
         FCOS                   //cos 60 -> ST(0)
         FDIVP ST(1), ST(0)     //sine 60 / cos 60
                                //-> ST(1), pop
                                //sine 60 / cos 60
                                //-> ST(0)
         FST   rslt_60          //store tangent 60
      }
                                //continued on next page
```

```
//display results
   printf ("Tangent_30 = %f\n\n", rslt_30);

   printf ("Tangent_45 = %f\n\n", rslt_45);

   printf ("Tangent_60 = %f\n\n", rslt_60);

   return 0;
}
```

```
Tangent 30 = 0.577350

Tangent 45 = 1.000000

Tangent 60 = 1.732051

Press any key to continue . . . _
```

INDEX

Symbols

D

E

P

R

S

T

U

V

W

X

Z